Plant Therapeutics

Plant Therapeutics

Editors

Juei-Tang Cheng
I-Min Liu
Szu-Chuan Shen

MDPI • Basel • Beijing • Wuhan • Barcelona • Belgrade • Manchester • Tokyo • Cluj • Tianjin

Editors

Juei-Tang Cheng
Institute of Medical Research
Chang Jung Christian University
Tainan
Taiwan

I-Min Liu
Department of Pharmacy and Master Program
Collage of Pharmacy and Health Care
Tajen University
Pingtung
Taiwan

Szu-Chuan Shen
Graduate Program of Nutrition Science
National Taiwan Normal University
Taipei
Taiwan

Editorial Office
MDPI
St. Alban-Anlage 66
4052 Basel, Switzerland

This is a reprint of articles from the Special Issue published online in the open access journal *Plants* (ISSN 2223-7747) (available at: www.mdpi.com/journal/plants/special_issues/plant_therapeutics).

For citation purposes, cite each article independently as indicated on the article page online and as indicated below:

LastName, A.A.; LastName, B.B.; LastName, C.C. Article Title. *Journal Name* **Year**, *Volume Number*, Page Range.

ISBN 978-3-0365-5956-8 (Hbk)
ISBN 978-3-0365-5955-1 (PDF)

Contents

About the Editors

Juei-Tang Cheng

Prof. Dr. Juei-Tang Cheng specializes in drug research from bench to bedside. He received his Ph.D. from the University of Shizuoka in Japan. He then returned to his hometown to focus on teaching systemic pharmacology to students. He also wrote a pharmacology textbook in Chinese. He was appointed full professor at National Cheng Kung University's College of Medicine in 1988. He was the supervisor of 46 Masters of Science dissertations and the director of 9 Ph.D. theses at this university. Then, he was invited to Chang Jung Christian University as Chair-Professor for lectures and research development.

He has extensive experience researching herbal products. He was the first to discover that guava is effective to reverse hyperglycemia in diabetic patients and animals, and he has numerous publications on the effect of herbal products on diabetic disorders. In addition, he conducted collaborative research with clinicians to focus on cardiovascular disorders that are improved by herbal products. Thus, in 1997, he was awarded a fellowship by the American College of Clinical Pharmacology. He has previously published 468 papers in prestigious international journals such as *Diabetes* and *Diabetology*.

Prof. Cheng likes to assist with the review of scientific paper. He has served on the editorial boards of numerous international journals. He has also served as Academic Editor of *PLOS One*, Associate Editor of *Life Sciences*, and is currently the Associate Editor of *Front. Pharmacol.* Aside from that, he was named Best Reviewer by *Japan Journal of Pharmacology* and *Journal of Ethnopharmacology*.

I-Min Liu

Prof. Dr. I-Min Liu, Traditional Chinese medicine expresses its meaning through the foundation of ancient and classic Chinese scientific knowledge. Because of the different times and spaces, it is often impossible to be understood and recognized by modern science, so it is necessary to modernize traditional Chinese medicine. Professor Liu's research focuses on the prevention and treatment of diabetes complications to improve patients' quality of life, briefly described below.

In the research plan funded by the Ministry of Science and Technology of Taiwan in 2020, Professor Liu referred to the second edition of the Collection of Native Herbal Medicines of Taiwan and cooperated with the PubMed online database to select indigenous herbal medicines with potential for research and clinical application to investigate the effectiveness of improvement of diabetes-related microvascular diseases. The results of this plan may serve as a benchmark for the Government of Taiwan to promote the cultivation of Chinese medicinal plants to increase the proportion of self-production and self-use and reduce dependency on imported traditional Chinese medicine.

Professor Liu implemented a research project to explore the efficacy and mechanism of Chinese herbal medicine for clearing heat, promoting blood circulation, and removing blood stasis in relieving diabetic small vessel disease. It is found that heat-releasing herbal medicines, such as *Lycii Cortex* and *Gentiana formosana* Hayata, have antioxidant effects, reducing oxidative stress caused by a high glucose concentration. These medicinal plants have been shown to reduce the inflammatory response associated with the NF-kB transcription factor, thereby preventing damage to retinal cells caused by persistent hyperglycemia. The results confirm that some Chinese herbs are helpful for the complications of diabetes, and research and development of new plant drugs are possible.

Szu-Chuan Shen

Prof. Dr. Szu-Chuan Shen was born in 1967. In 1991 and 1993, he graduated from Department of Food Science in National Chung-Hsing University, Taiwan, for Bachelor's and Master's degree, respectively. In 2004, he completed his Ph.D. degree from the Institute of Food Science and Technology in National Taiwan University, Taiwan.

He has work experiences serve as a senior research and development scientist in the biggest food company (Uni-President Enterprises Corp., Tainan) in Taiwan for more than five years after he received his Master's degree.

At present, he is Professor and deputy Chair of the Undergraduate and Graduate Programs of Nutrition Science in National Taiwan Normal University, Taiwan. He was currently supervisor of 5 Bachelors and 4 Masters of Science Dissertations in this University.

Prof. Shen's research interest and expert field is health function and therapy of food resource/natural materials on curing or preventing human diseases, such as diabetes mellitus, non-alcoholic fatty liver disease, dementia, etc. For example, diabetes mellitus (DM) is one of the research subject in his laboratory, he investigated the hypoglycemia effect of economic fruits in Taiwan, such as guava (*Psidium guajava* Linn.) and wax apple (*Syzygium samarangense* (Blume) Merrill and Perry), propose the possible mechanisms and evaluate the feasibility for using them as a raw material in the making of anti-diabetes medicine or health food. He has published more than 50 papers in international peer-reviewed journals, such as *J. Food Sci., J. Agric. Food Chem., Phytother. Res., J. Func. Food, Nutrients, Food Res. Inter., Food Sci. & Nutr., Antioxidants, Evid.-Based Complementary Altern. Med., Front. Aging Neurosci., Food Chem., Plants, Int. J. Mol. Sci.*, etc., and more than 100 conference reports, as well as 3 book chapters of Wiley Blackwell.

Preface to "Plant Therapeutics"

This Special Issue provides recent advances in the use of plants for therapeutic purposes. This Special Issue collected the plants, including leaf, fruit, and others. This Special Issue's targets were crude plant extract and active principle purified from the plant. These studies were evaluated by the journal's regular review process, and we are pleased to publish the qualified results in this Special Issue. Additionally, we divided them into three sections: the first section contains eight reports on plant crude extracts, the second section includes seven reports on the pure compounds from plant, and the third section shows two review articles.

Plants used for therapeutic purposes are classified as preliminary in modern terms. Humans used the plant as a nutrient and/or medicine prior to the purification of active principles. Physicians in the emergency room criticized the herbal product's slow onset time. It aided in the development of a chemical agent derived from the plant. However, agents derived from the plant must adhere to the fundamental principle that the plant in question is effective in therapeutics. Unfortunately, records demonstrating the effectiveness of a plant or herb were not passed on in a positive manner. Records indicating the effectiveness of plants were kept in traditional Chinese medicine. However, the herbal mixture was the main attraction. As a result, the benefits of plants were mostly determined by local custom rather than scientific evidence.

This problem prompted researchers to provide each plant discovery. We are pleased to include in this Special Issue the screened Mexican plants and the halophytic plants growing in central Saudi Arabia. We are also pleased to have a review article from Iran identifying active principles and plants for Triple Negative Breast Cancer (TNBC). Another review article from Bangladesh will be very useful to researchers interested in renal disorders.

As a result, we have sparked scientists' interest in studying the plant for therapeutic purposes. This field necessitates network pharmacology analysis and machine-aided learning. Many disorders resistant to modern medication are looking for active principles isolated from the plants all over the world.

Juei-Tang Cheng, I-Min Liu, and Szu-Chuan Shen
Editors

plants

Editorial

Plant Therapeutics

Juei-Tang Cheng [1,*], I-Min Liu [2] and Szu-Chuan Shen [3]

1 Institute of Medical Research, Chang Jung Christian University, Tainan City 71101, Taiwan
2 Department of Pharmacy and Master Program, Collage of Pharmacy and Health Care, Tajen University, Pingtung 90741, Taiwan
3 Graduate Program of Nutrition Science, National Taiwan Normal University, Taipei 11677, Taiwan
* Correspondence: jtcheng@mail.cjcu.edu.tw

Citation: Cheng, J.-T.; Liu, I.-M.; Shen, S.-C. Plant Therapeutics. *Plants* **2022**, *11*, 2720. https://doi.org/10.3390/plants11202720

Received: 8 October 2022
Accepted: 13 October 2022
Published: 14 October 2022

Plants for therapeutics and the phytotherapy for disorders are the same thing in practice. Throughout history, humans have relied on nature material to meet their basic needs, particularly for the treatment of a wide range of diseases. Plants have thus served as the foundation of traditional medicine systems. Much plant research has been conducted in academia. Indeed, the information provided on crude extracts or active components of the plants comes from a variety of sources. Herbs are raw plant matters which are leaves, flowers, fruits, seeds, stems, and others, which may be in an entirely powdered form. Simple processes involve harvesting, drying of plant parts for preparation of phytochemicals. Increased awareness of intellectual property issues, combined with increased collaboration with the pharmaceutical industry, may result in the development of new drugs via this route. Therefore, many medicines are natural products or their derivatives [1]. This situation is likely to persist and increase for the growing prevalence of disorders that are resistant to modern medication around the world.

Ethnobotany is a subject that fills volumes of historical and biological texts but is largely ignored in contemporary texts. It is both an ancient way of life and a young and thriving scientific field. The word itself may provide the simplest definition of ethnobotany: *ethno* (people) and *botany* (science of plants). Thus, it is a study of how people from various cultures and regions use plants in their local environments. These uses can include food, medicine, fuel, shelter, and religious ceremonies in many cultures [2].

Ethnomedicine has a broad range in practices through two categories: *personalistic* systems, in which supernatural causes attributed to angry deities, ghosts, ancestors, and witches predominate, and *naturalistic* systems, in which natural causes predominate [3]. Where illness is explained in broad, impersonal in the traditional medical systems of native America, and parts of China, South Asia, Latin America. In African communities, the personalistic system appears to predominate (though not to the exclusion of naturalistic explanations). Naturalistic explanations predominate in Ayurveda, Unani, Khampo (Japan), and traditional Chinese medicine (CTM) without the exclusion of personalistic causation. Part of the belief in the later types is that the introduction of heat or cold into, or loss of, the body upsets its basic equilibrium, that is, the balance of humors, or the *dosha* of Ayurvedic medicine, or the *yin and yang* of Chinese medicine, and these must be restored if the patient is to recover. Therefore, the indigenous herbal medicine is known as a natural, herbal extract with little or no industrial processing used to treat disorders within local or regional use.

Plants play a major role in disorder management. Treatment is not limited to the sterile use of various leaves, roots, fruits, barks, grasses, and various objects such as minerals, dead insect's bones, feathers, shells, eggs, powders, and smoke from various burning objects for the cure and prevention of diseases. If a sick person is given a leaf infusion to drink, they believe not only in the plant's organic properties, but also in the magical or spiritual force infused by nature in all living things, as well as the role of their ancestors, spirits, and gods in the healing process [4]. The patient also believes in the power of the incantation and assists the handler in identifying the ingredients of the remedies given. In addition, rather than being a passive subject of therapy, the patient is an active participant in the treatment process [5]. According to the World Health Organization, despite improved access to modern medicine, a large segment of the human population still prefers traditional medicine, also known as alternative medicine. Therefore, placebo control is now critical in clinical trials of traditional medicine.

This Special Issue is going to publish the article(s) targeted on plants for therapeutics. Combination of plants and/or plant combined modern medicine are also popularly used in therapeutics. It is important to assess the renowned historical significance of products contained in the plant. We call the researchers to share the useful plant which is applied in each local market using the reliable model of disorders to indicate the effectiveness in therapeutics. Moreover, considering the high interest in plant-derived effectiveness in health as an emerging challenge for modern therapies, this Special Issue covers a wide variety of areas, aiming to contribute to the overall knowledge of medicinal plants from several aspects. These studies were evaluated by the journal's regular review process, and we are pleased to publish the qualified results in this issue. In brief, we divided them into three sections: the first contains eight reports on plant crude extracts, the second contains seven reports on plant pure compounds, and the third contains two review articles.

In the section one, water-soluble extract of Safflower, also known as thorn flower, has been identified in crude extracts to inhibit human platelet aggregation induced by ADP in vitro. The findings were consistent with traditional Chinese medicine's belief in promoting blood circulation and removing blood stasis [6]. Furthermore, using streptozotocin-nicotinamide (STZ-NA) induced type-2 diabetic rats, the aerial parts of *Eryngium longifolium* and the rhizome of *Alsophila firma* were tested for antihyperglycemic activity. For lyophilization, each plant was extracted with either water or ethanol. Then, for the first time, they confirmed the traditional use of two Mexican plants in animal model. Furthermore, aqueous (AE), methanolic (ME), and hexane (HE) extracts of *Acalypha monostachya*, another Mexican plant, were found to inhibit the growth of Human Tumor Cells in vitro. Another Mexican plant, *Inga jinicuil*, which is used in traditional medicine to treat gastrointestinal inflammation and infection, was tested in vitro. Three extracts of this plant's bark demonstrated significant antibacterial activity. Moreover, the extract of the Korean plant *Lindera obtusiloba* was shown to improve endothelial dysfunction and ameliorate the plaque development in Hyperlipidemic ApoE-Knockout Mice, primarily by reducing vascular NADPH oxidase-induced ROS generation [7]. Furthermore, a water extract of fermented rice bran has been shown to protect against liver damage and intestinal injury in elderly rats fed a high-fat diet. Finally, this section included two herbal mixtures used in clinics of traditional Chinese medicine, Yi-Gan-San and Gan-Mai-Da-Zao. Yi-Gan-San, also known as Yokukansan in Japan [8] or Shun-Ning-Yi in Taiwan, is a commercial product of Sun-Ten Pharmaceutical company that contains seven medicinal herbs per 100 g of the final product: Bupleurum 2 g, Licorice 2 g, Chuanxiong 3.2 g, Angelica 4 g, Atractylodes 4 g, Poria 4 g, and Uncaria 4 g. Yi-Gan-San has been shown to reduce Aβand Tau expression in Drosophila melanogaster using the IMR assay and Western blotting analysis. It provided an effective treatment for neurodegenerative disorders in Alzheimer's disease. Gan-Mai-Da-Zao also contained three major plants: blighted wheat (Fu Mai), licorice (Gan Cao), and jujube (Da Zao). Licorice (Gan Cao) was identified as the major herb in Gan-Mai-Da-Zao for antidepressant-like therapeutics in two rat models of depression-like disorders.

In the second section, Vescalagin isolated from Pink Wax Apple was shown to protect pancreatic β-cells from methylglyoxal-induced inflammation in rats. It suggested that Vescalagin could become a health food ingredient used to prevent diabetes. The phytochemicals and bioactivities of four major halophytic plants growing in central Saudi Arabia were then compared, including antioxidant, anticancer, and antimicrobial effects. *Lycium shawii*, a halophytic plant, has been suggested as the notable one based on phenolic and flavonoid content. Furthermore, three extracts (n-hexane, ethyl acetate, and ethanol) of *Acalypha arvensis* (Euphorbiaceae) were tested for antibacterial properties. In addition to corilagin, an active principle for the first time in this plant, chlorogenic acid, rutin, quercetin-3-O-glucoside, and caffeic acid were also discovered. Furthermore, researchers used traditional extraction methods such as the Soxhlet apparatus (SHS) and rapid solid–liquid dynamic extraction (RSLDE) to extract the principles contained in the plant *Salvia haematodes* L. AChE or BChE inhibitory activity was found in the SHN n-hexane fraction. Phytosterols, β-sitosterol and stigmasterol, have been proposed as active principles for use in Alzheimer's disease therapy. The in vitro studies were used to develop by-products of the black bean fermented soybean sauce manufacturing process as new functional foods for anti-inflammation. Furthermore, the fruit bilberry (*Vaccinium myrtillus* L.) and yeast β-glucan have been shown to help with diabetic complications. The fast-dissolving films were created to combine both. The new product provided an improved fast dissolving time and good water vapor barrier properties, suggesting its success as a novel film for packaging dry powdered pharmaceuticals.

Two review articles were arranged in the final section. One gathered information on the benefits of *Moringa oleifera* Lam., also known as the drumstick tree, in renal diseases. This review article used a meta-analysis of the database from 2011 to 15 June 2021. It was proposed that this plant reduces several pathological factors associated with kidney disease, such as inflammation and oxidative stress. Another review article suggested that natural products or active principles from plants could help treat Triple Negative Breast Cancer (TNBC) by impeding the Wnt/β-Catenin Pathway. The authors mentioned 26 different types of active principles from the plant that are useful for TNBC. This review article also introduced three herbal products including herbal mixture and the useful plants [9]. In the conclusion, it is suggested that natural bioactive compounds be evaluated in clinical trials.

This Special Issue differs from the traditional hard or soft cover book concept. This approach will make this topic more accessible to a broader range of readers. It gives us great pleasure to write a foreword for this lovely, multi-authored international publication on a unique subject. We wish this Issue great success and hope to see a regular series develop from this point forward. We also want to thank all the contributors for their outstanding efforts in making this Issue a success.

Author Contributions: All authors contributed equally to this article. All authors have read and agreed to the published version of the manuscript.

Funding: This research received no external funding.

Conflicts of Interest: The authors declare no conflict of interest.

References

1. David, B.; Wolfender, J.-L.; Dias, D.A. The pharmaceutical industry and natural products: Historical status and new trends. *Phytochem. Rev.* **2015**, *14*, 299–315. [CrossRef]
2. Gertsch, J. How scientific is the science in ethnopharmacology? Historical perspectives and epistemological problems. *J. Ethnopharmacol.* **2009**, *122*, 177–183. [CrossRef] [PubMed]
3. Sinaga, R.M.; Zuska, F.; Sitorus, P. Indegenous healer knowledge about Illness and the way to make Traditional Medicine. *Indones. J. Med. Anthropol.* **2021**, *2*, 43–47. [CrossRef]
4. Houghton, P.J. The role of plants in traditional medicine and current therapy. *J. Altern. Complement. Med.* **1995**, *1*, 131–143. [CrossRef] [PubMed]
5. Balunas, M.J.; Kinghorn, A.D. Drug discovery from medicinal plants. *Life Sci.* **2005**, *78*, 431–441. [CrossRef] [PubMed]
6. Delshad, E.; Yousefi, M.; Sasannezhad, P.; Rakhshandeh, H.; Ayati, Z. Medical uses of *Carthamus tinctorius* L. (Safflower): A comprehensive review from Traditional Medicine to Modern Medicine. *Electron. Physician* **2018**, *10*, 6672–6681. [CrossRef] [PubMed]
7. Jung, S.H.; Han, J.H.; Park, H.S.; Lee, J.J.; Yang, S.Y.; Kim, Y.H.; Heo, K.S.; Myung, C.S. Inhibition of Collagen-Induced Platelet Aggregation by the Secobutanolide Secolincomolide A from *Lindera obtusiloba* Blume. *Front. Pharmacol.* **2017**, *8*, 560. [CrossRef] [PubMed]
8. Ikarashi, Y.; Mizoguchi, K. Neuropharmacological efficacy of the traditional Japanese Kampo medicine yokukansan and its active ingredients. *Pharmacol. Ther.* **2016**, *166*, 84–95. [CrossRef] [PubMed]
9. Yang, Z.; Zhang, Q.; Yu, L.; Zhu, J.; Cao, Y.; Gao, X. The signaling pathways and targets of traditional Chinese medicine and natural medicine in triple-negative breast cancer. *J. Ethnopharmacol.* **2021**, *264*, 113249. [CrossRef] [PubMed]

plants

MDPI

Article

Safflower Extract Inhibits ADP-Induced Human Platelet Aggregation

Ping-Hsun Lu [1,2], Chan-Yen Kuo [3], Chuan-Chi Chan [4], Lu-Kai Wang [5], Mao-Liang Chen [3], I-Shiang Tzeng [3] and Fu-Ming Tsai [3,*]

1 Department of Chinese Medicine, Taipei Tzu Chi Hospital, The Buddhist Tzu Chi Medical Foundation, New Taipei City 231, Taiwan; pinghsunlu@gmail.com
2 School of Post-Baccalaureate Chinese Medicine, Tzu Chi University, Hualien 970, Taiwan
3 Department of Research, Taipei Tzu Chi Hospital, The Buddhist Tzu Chi Medical Foundation, New Taipei City 231, Taiwan; cykuo863135@gmail.com (C.-Y.K.); cater0656@hotmail.com (M.-L.C.); istzeng@gmail.com (I.-S.T.)
4 Department of Laboratory Medicine, Taipei Tzu Chi Hospital, The Buddhist Tzu Chi Medical Foundation, New Taipei City 231, Taiwan; kiki1205@tzuchi.com.tw
5 Radiation Biology Core Laboratory, Institute for Radiological Research, Chang Gung University/Chang Gung Memorial Hospital, Linkou, Taoyuan 333, Taiwan; lukai2724@cgmh.org.tw
* Correspondence: afu2215@gmail.com; Tel.: +886-2-66289779-5793

Citation: Lu, P.-H.; Kuo, C.-Y.; Chan, C.-C.; Wang, L.-K.; Chen, M.-L.; Tzeng, I-S.; Tsai, F.-M. Safflower Extract Inhibits ADP-Induced Human Platelet Aggregation. *Plants* **2021**, *10*, 1192. https://doi.org/10.3390/plants10061192

Academic Editors: Juei-Tang Cheng, I-Min Liu and Szu-Chuan Shen

Received: 12 May 2021
Accepted: 9 June 2021
Published: 11 June 2021

Abstract: Safflower extract is commonly used as a traditional Chinese medicine to promote blood circulation and remove blood stasis. The antioxidant and anticancer properties of safflower extracts have been extensively studied, but their antiaggregative effects have been less analyzed. We found that safflower extract inhibited human platelet aggregation induced by ADP. In addition, we further analyzed several safflower extract compounds, such as hydroxysafflor yellow A, safflower yellow A, and luteolin, which have the same antiaggregative effect. In addition to analyzing the active components of the safflower extract, we also analyzed their roles in the ADP signaling pathways. Safflower extract can affect the activation of downstream conductors of ADP receptors (such as the production of calcium ions and cAMP), thereby affecting the expression of activated glycoproteins on the platelet membrane and inhibiting platelet aggregation. According to the results of this study, the effect of safflower extract on promoting blood circulation and removing blood stasis may be related to its direct inhibition of platelet activation.

Keywords: safflower; hydroxysafflor yellow A; safflower yellow A; flavonoid; platelet aggregation

1. Introduction

Safflower, also known as thorn flower, is a type of chrysanthemum. The leaves are linear, thin, and long, with a special aroma and a bitter taste. Safflower can be grown in large areas and is less likely to die [1], so its survival rate is much higher than that of other plants. The safflower extract is obtained from the dried flowers of *Carthamustinctorius* L., a plant of the Asteraceae family, by removing impurities, drying in the shade, or drying under low heat. Safflower is often used as a traditional Chinese medicine to promote hormone balance, remove blemishes, regenerate new cells, circulate blood, remove bruises, and relieve pain [2–4].

The biological activity of the extract of safflower is related to the various compounds in the extract. Previous studies have indicated that the extract of safflower could be divided into lipophilic compounds and hydrophilic compounds due to different extraction methods [3]. The lipophilic compounds of safflower include fatty acids, tocopherols such as vitamin E, carotenoids, and phytosterols. Hydrophilic compounds include flavonoids such as safflower yellow A or hydroxysafflor yellow A, saponins, and other safflower nutrients such as sugars and amino acids. According to related studies of compounds in safflower,

safflower has anti-inflammatory [5–7], anticancer, and antioxidation effects [8–11] and can also be used as a potential treatment for osteoporosis and brain and liver diseases [12–16].

Platelets are formed from the cytoplasm of mature megakaryocytes in the bone marrow. Each megakaryocyte can produce 2000–7000 platelets. Platelets have a half-life of approximately 7 to 9 days and are cleared mainly by macrophages in the spleen. Under normal circumstances, platelets are biconvex, disc-shaped, can extend out a foot process, and appear irregular after stimulation [17–19]. The main physiological function of platelets is hemostasis. In general, hemostasis can be divided into three processes. First, the platelets adhere to the outer side of the truncated endothelium, a process known as "adhesion." Second, platelets change shape, turn on receptors, and secrete a chemical messenger is the process of "activation". Finally, platelets bridge each other through receptors, a process known as "aggregation". The formation of platelet emboli is related to the activation of coagulation factors and the production, deposition, and linkage of fibrin [20,21].

The processes of platelet adhesion, activation, and aggregation appear almost simultaneously after platelet activation. The adhesion of platelets refers to the adhesion of platelets and nonplatelet surfaces. The main substances involved in this process are the glycoproteins on the platelet membrane, the Von Willebrand factor in the plasma, and the collagen component in the subcutaneous tissue [22,23]. Aggregation refers to the adhesion of platelets to each other. The main substances that cause platelet aggregation are thromboxane A2 (TXA2), collagen, thrombin, and ADP [24,25]. Among them, ADP plays the most important role in inducing platelet aggregation physiologically, especially endogenous ADP released by platelets [26]. In vitro, a small amount of ADP (less than 0.9 μM) can induce platelet aggregation rapidly, but they depolymerize quickly. When a moderate amount of ADP (approximately 1 μM) was added, it could aggregate and deaggregate platelets, which could then quickly enter the second stage of the irreversible aggregation reaction. The irreversible aggregation reaction in the second stage is mainly caused by the release of endogenous ADP from platelets. If a large amount of ADP (approximately 10 μM) is added, the platelets go directly to the second stage of irreversible aggregation [27,28]. Generally, drugs that induce platelet aggregation will decrease the expression of cyclic AMP (cAMP) in platelets, which may be related to the increase in calcium ion concentration and release of endogenous ADP when the expression of cAMP in platelets decreases.

Many of the above studies have pointed to safflower's role in anti-inflammation and inhibiting cancer cell growth. However, safflower is often used in traditional Chinese medicine to promote blood circulation and remove blood stasis. The mechanism of safflower in platelet aggregation remains unclear. In this study, we confirmed that safflower has antiplatelet aggregation activity by using ADP-induced human platelet aggregation experiments. In addition, we further analyzed the active compounds of the safflower extract and the effect of the safflower extract on the signal transduction of ADP-induced platelet aggregation.

2. Results

2.1. *Safflower Extract Inhibited ADP-Induced Human Platelet Aggregation*

The results in Figure 1A show that as the dose of safflower extract increased, safflower inhibited ADP-induced human platelet aggregation in a dose-dependent manner. The inhibitory effect of safflower on platelet aggregation was 22.3–66.6%, and safflower at a dose of 100 μg/mL had the highest inhibitory effect. Many traditional Chinese medicine extracts interfere with bacterial endotoxin, so it is necessary to clarify whether endotoxin affects the platelet aggregation reaction in response to safflower extract. Polymyxin B, an inhibitor of Toll-like receptor 4, was used to observe whether endotoxin would affect the inhibition of platelet aggregation activity of the safflower extract. When platelets were treated with 50 μg/mL or 100 μg/mL safflower extract, ADP-induced platelet aggregation was significantly inhibited. When polymyxin B was added to platelets at the same time, the inhibition of ADP-induced platelet aggregation by safflower extract did not significantly change, indicating that safflower extract itself had the effect of platelet inhibition (Figure 1B). A dose of 50 μg/mL safflower extract was selected for the subsequent experiments.

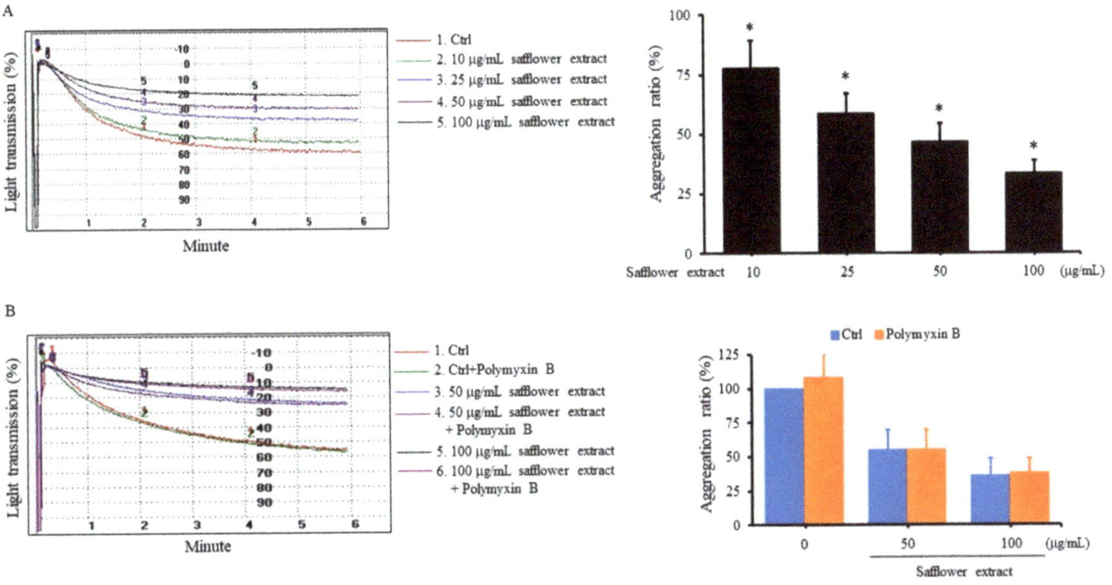

Figure 1. Safflower extract inhibited platelet aggregation. PRP was treated with different concentrations of safflower extract (**A**) or treated with 50–100 μg/mL safflower extract and polymyxin B (**B**). ADP was added and analyzed by an aggregometer. * $p < 0.05$ compared with the control group.

2.2. All Compounds Found in Safflower Inhibited ADP-Induced Human Platelet Aggregation

Since safflower extract inhibits platelet aggregation, we next wanted to explore which safflower extract compounds play an important role, and we selected three compounds available for further analysis. When human platelets were treated with 25 μM hydroxysafflor yellow A, safflower yellow A, or luteolin, ADP-induced platelet aggregation was significantly inhibited by 70.9%, 42.3%, and 74.9%, respectively. When platelets were treated with each high concentration (100 μM) of compound found in the safflower extract, almost no platelet aggregation reaction was found (Figure 2). These above results showed that these compounds found in safflower extract all have a strong antiplatelet aggregation effect. A dose of 25 μM of each compound found in the safflower extract was chosen for follow-up research.

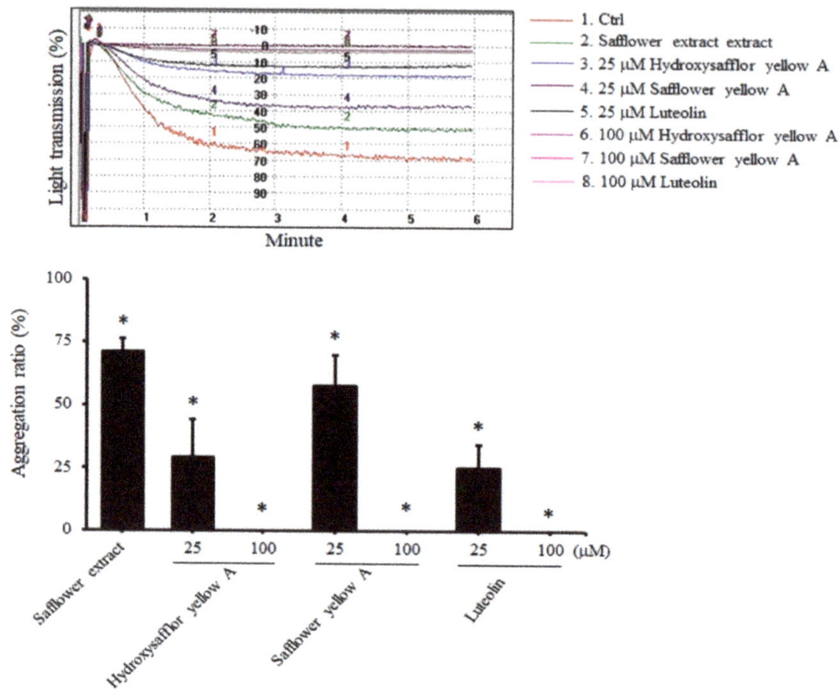

Figure 2. The compounds found in the safflower extract inhibited platelet aggregation. PRP was treated with 50 μg/mL safflower extract or 25–100 μg/mL compounds found in the safflower extract, and ADP was added and analyzed by an aggregometer. * $p < 0.05$ compared with the control group.

2.3. *Safflower Extract and Its Compounds Inhibit Platelet Aggregation through P2Y1 and P2Y12 Receptors*

ADP induces platelet aggregation mainly through P2Y1 and P2Y12 receptors on the platelet membrane [26]. Next, we wanted to understand which receptor on platelets the safflower extract mainly targeted to inhibit platelet aggregation. When platelets were treated with the P2Y1 antagonist A2P5P, A2P5P significantly inhibited platelet aggregation by 42.3%, and the safflower extract and its compounds combined with A2P5P further inhibited ADP-induced platelet aggregation by 79.5–98.3% (Figure 3A), which means that the safflower extract and its compounds inhibited platelet aggregation through the P2Y12 pathway. Similarly, when platelets were treated with the P2Y12 antagonist clopidogrel, clopidogrel slightly inhibited platelet aggregation by 16%, and safflower extract and its compounds combined with clopidogrel further inhibited ADP-induced platelet aggregation by 57.1–90.3% (Figure 3B), which indicated that safflower extract and its compounds can also inhibit platelet aggregation through the P2Y1 pathway. When platelets were treated with A2P5P and clopidogrel at the same time, no aggregation reaction occurred (Figure 3C), confirming the importance of P2Y1 and P2Y12 receptors in the induction of platelets by ADP.

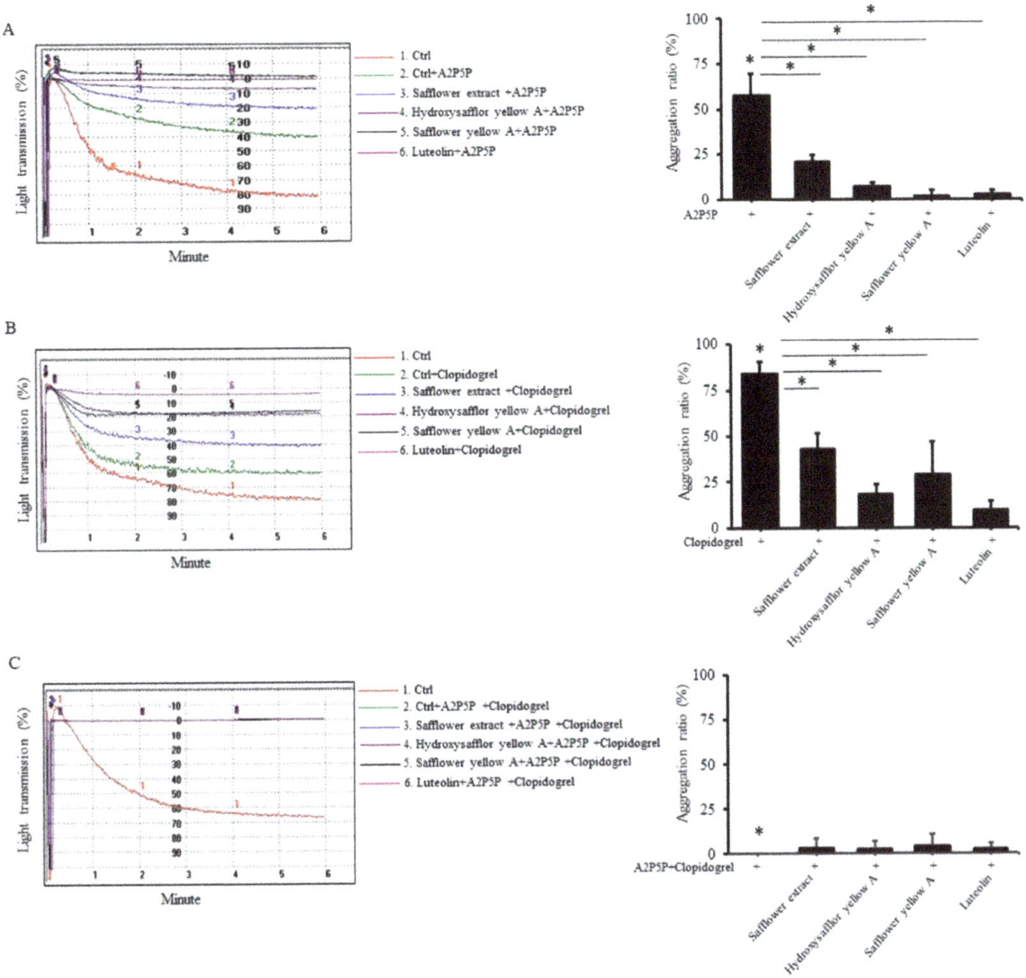

Figure 3. The effect of the ADP receptor antagonist on the inhibition of platelet aggregation by safflower extract. PRP was treated with 50 μg/mL safflower extract or 25 μg/mL of compounds found in safflower extract, 10 μM ADP and 1 mM A2P5P (**A**), 1 mM clopidogrel (**B**), or A2P5P+clopidogrel (**C**) added and analyzed by an aggregation analyzer. * $p < 0.05$ between two groups.

2.4. Safflower Extract and Its Compounds Inhibit the Activation of Calcium Ion Influx and the Production of cAMP Regulated by ADP

The results of experiments using the abovementioned antagonists found that safflower extract and its compounds inhibit the activation of P2Y1 and P2Y12 receptors on the platelet membrane. Next, we wanted to explore whether safflower extract and its compounds truly affect the downstream activation pathways involved in P2Y1 and P2Y12 receptors. We directly analyzed the safflower extract, and its compounds affected the activation of calcium influx induced by the P2Y1 receptor and cAMP production regulated by the P2Y12 receptor. ADP induced an increase in intracellular calcium ions in platelets by 1.84-fold, while safflower extract and its compounds inhibited the increase in calcium ion concentration induced by ADP by 31.3–46.4% (Figure 4A). When platelets were treated with forskolin, the cAMP concentration in platelets was significantly increased by 6.12-fold, and when platelets were added to ADP, the cAMP concentration in platelets was significantly decreased by

61.4%. When the platelets were exposed to safflower extract and its compounds at the same time, the decrease in cAMP production caused by ADP increased significantly by 84.5–237.9% (Figure 4B). The above results show that the safflower extract and its units can inhibit the participation of P2Y1 and P2Y12 in internal signal transduction in platelets.

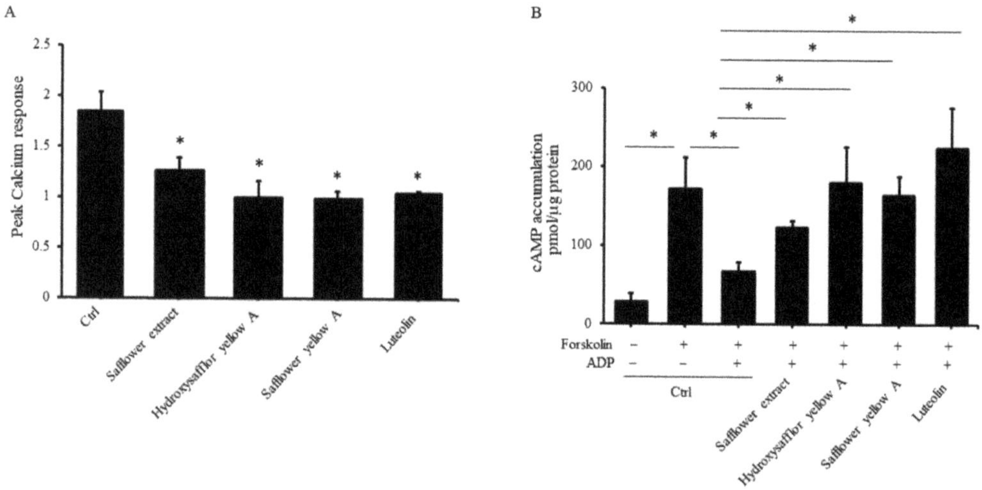

Figure 4. Effects of safflower extract on the production of calcium ions and cAMP in platelets. (**A**) PRP was treated with safflower extract or compounds found in safflower extract for 1 h and then reacted with Fura-2AM at 37 °C. Isolated platelets were collected, and calcium responses induced by ADP were measured. (**B**) PRP was treated with safflower extract or compounds found in safflower extract for 1 h. The purified platelets were collected and reacted with 10 μM forskolin and 10 μM ADP at 37 °C. Cell lysates were prepared, and the cAMP levels were measured by enzyme immunoassay. * $p < 0.05$ between two groups.

2.5. Safflower Extract and Its Compounds Inhibit the Production of Thromboxane A2 (TXA2) and Arachidonic Acid (AA)

After participating in calcium ion activation and regulation of cAMP production on the platelet membrane through the ADP receptor, the phospholipid on the membrane is then degraded, and AA is released, which then generates TXA2 and induces the formation of glycoprotein on the membrane to participate in the aggregation of platelets [29]. Therefore, in addition to calcium ion activation and cAMP production, we also wanted to explore whether safflower extract and its compounds could also affect the production of TXA2 and AA in platelets. Because of the poor stability of TXA2, TXB2 production was used to reflect the expression level of TXA2 in platelets. As shown in Figure 5A, fibrinogen increased the expression of TXB2 in platelets, and the addition of ADP had an additive effect. Compared with untreated platelets, TXB2 was increased 13.4-fold when platelets were treated with fibrinogen and ADP. However, when safflower extract or its compounds were treated with platelets at the same time, the increased yields of TXB2 induced by fibrinogen and ADP were significantly inhibited by 39.4–73%. In terms of AA production, fibrinogen and ADP can induce platelets to produce AA by 97.2%. Similarly, when platelets were simultaneously treated with safflower extract or its compounds, the production of AA was inhibited by 21.1–47.3% (Figure 5B).

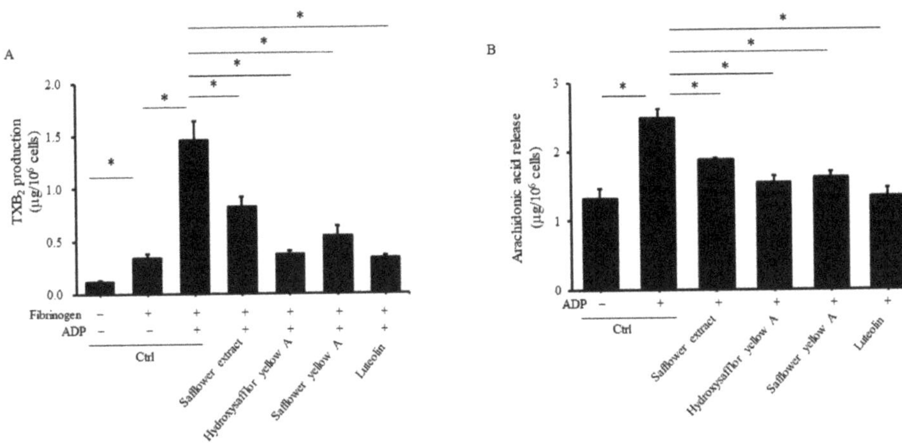

Figure 5. Effects of safflower extract on the production of TXB_2 and AA in human platelets. PRP was treated with safflower extract or compounds found in safflower extract for 1 h. Isolated platelets were collected and then stimulated with 3 μM fibrinogen and 10 μM ADP at 37 °C. The content of TXB2 (**A**) or AA (**B**) was determined by enzyme immunoassay. * $p < 0.05$ between two groups.

2.6. Safflower Extract and Its Compounds Inhibit the Formation of PAC-1 in Platelets

Glycoprotein IIb/IIIa (also called CD41/CD61, PAC-1 epitope or integrin αIIbβ3) on the platelet membrane is mainly produced after stimulation of platelets by ADP and can combine with fibrinogen to cause platelet aggregation. In addition to the above intraplatelet signal transduction tests, we finally wanted to investigate whether safflower extract and its compounds caused changes in glycoprotein expression on the platelet membrane. In order to simulate the living conditions, whole blood was used for drug treatment and staining analysis, and platelets were selected on a flow cytometer for membrane protein analysis. Although the platelet cell population accounts for less than 1% of blood samples, these glycoproteins on the platelet membrane can still be found to be altered by ADP stimulation (Supplementary Figure S1). In whole blood treatment with ADP, CD61 and PAC-1 production on the platelet membrane was significantly increased. However, when whole blood was treated with safflower extract or its compounds, CD61 on the platelet membrane decreased slightly only in the blood treated with safflower yellow A, but no significant changes were found in the other groups (Figure 6). In contrast, safflower extract or all its compounds inhibited the expression of PAC-1 on the ADP-induced platelet membrane (Figure 6). Combined with the above research results, safflower extract or its compounds can inhibit a continuous signal transduction pathway induced by ADP in platelets, thus affecting the formation of related glycoproteins on the membrane and thereby inhibiting platelet aggregation.

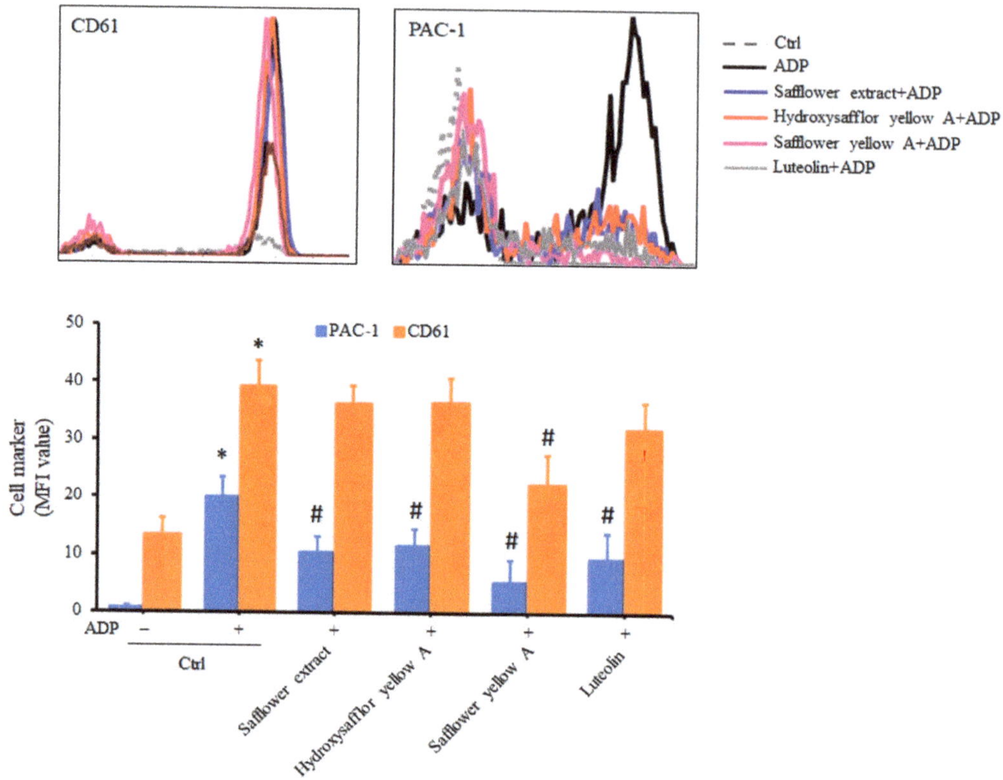

Figure 6. Effects of safflower extract on the surface activation markers of human platelets. Diluted whole blood was treated with safflower extract or compounds found in safflower extract for 1 h, stained with fluorescent antibody, and stimulated with ADP. After platelet fixation, the content of activation markers on the platelet membrane was analyzed by flow cytometry. * $p < 0.05$ when compared to the control group. # $p < 0.05$ when compared to the ADP treatment group.

3. Discussion

Our results indicated that safflower extract could inhibit platelet aggregation mainly in the following aspects: ADP receptor transduction and expression of PAC-1 glycoprotein on the platelet membrane, calcium ion activation, and the cAMP, AA, and TXA2 contents regulated by ADP in intracellular platelets were all significantly inhibited by safflower extract (Figure 7). These results suggest that safflower extract extensively inhibits platelet aggregation, though not specifically, in the process of platelet activation. Its curative effect as a traditional Chinese medicine for promoting blood circulation and removing blood stasis may also be related to its inhibition of platelet activation, thereby inhibiting the formation of thrombi.

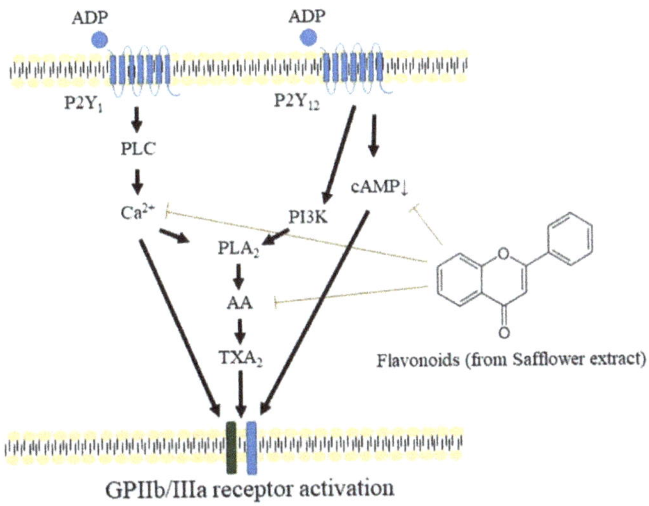

Figure 7. Schematic diagram of the mechanism of safflower extract inhibiting platelet aggregation. The compounds found in safflower extract inhibit the activation of calcium ions in platelets and the transduction of cAMP pathways caused by ADP, thereby inhibiting the expression of TXA_2 in platelets, the activation of membrane protein GPIIb/IIIa, and the aggregation of platelets.

Although safflower has the effect of promoting blood circulation and removing blood stasis, in vivo experiments were mostly conducted with various compounds of safflower extract as materials. Oral feeding carthamins yellow at 100 mg/Kg and 200 mg/kg in living rats can significantly reduce blood fluidity [30]. In rats, sublingual intravenous injection of 1.5, 3.0, and 6.0 mg/kg of hydroxysafflor yellow A significantly reduced thrombosis, while the treatment of 3.0 and 6.0 mg/kg of hydroxysafflor yellow A significantly reduced the infarct size [31]. In addition, 200 mg/kg/d safflower yellow or hydroxysafflor yellow A in diet-induced obese mice can effectively inhibit fat mass by 57.8–61.6% [32]. Although it has been documented that safflower extract can be converted into substances in liver metabolism [3,4], the concentrations of either safflower extract or these compounds found in safflower extract in the blood and their half-life in the blood after administration are still unknown. In any case, the concentration of the compounds used in the in vitro experiment was similar to the dose used in this study [31].

The compounds found in safflower extract used in this study, whether hydroxysafflor yellow A, safflower yellow A, or luteolin, are all converted into flavonoids during metabolism. Flavonoids can be found in many green vegetables or fruits, and their effects have been extensively studied. Because of their unique benzene ring structure, flavonoids are believed to have a strong antioxidant effect. Studies have shown that they inhibit the metabolic pathway of AA (including PLA2 and COX2) and modulate the activation of transcription factors involved in inflammation, such as NF-κB, GATA-3, and STAT-6 [33,34]. In terms of anticancer research, flavonoids have been found to regulate several important factors related to the growth of cancer cells, such as epidermal growth factor receptors, platelet-derived growth factor receptors, vascular endothelial growth factor receptors, and cyclin-dependent kinases [35,36]. However, flavonoids are not well absorbed by the human body, so even though they have many effects in in vitro studies, except for some evidence of anticancer effects [37,38], none of them have significant clinical effects in the treatment of anti-inflammatory or cardiovascular diseases [39].

In addition to their anti-inflammatory and anticancer effects, flavonoids have also been shown to directly bind to TXA2 receptors to directly inhibit platelet activation [40]. Based on the above research results, when the AA metabolic pathway is inhibited by flavonoids, the production of precursors is inhibited and thus indirectly affects the production of TXA2 in platelets. Concordant with this study, almost all of the various signaling transmitters regulated by ADP, whether calcium ions or cAMP, were affected by safflower extract, suggesting that safflower extract has a wide range of targets and thus has anti-inflammatory and anticancer effects and efficacy against cardiovascular diseases. However, the various compounds of safflower extract are mainly metabolized in liver tissue to produce flavonoids. Therefore, the importance of other metabolites or degradants produced by each compound of safflower extract in inhibiting platelet agglutination cannot be excluded, in addition to the unique structure of flavonoids in the various compounds of safflower extract [41].

In summary, we found that safflower extract and several safflower extract compounds can inhibit ADP-regulated human platelet aggregation. The production of calcium ions, cAMP, AA, and TXA2 regulated by ADP receptors, whether P2Y1 or P2Y12, would be inhibited by safflower extract or compounds found in safflower extract. We further found that the formation of the PAC-1 complex, a glycoprotein involved in the aggregation of platelet membranes, was also inhibited by safflower extract or several safflower extract compounds. Therefore, the use of safflower extract in traditional Chinese medicine as a way to promote blood circulation and remove blood stasis may be related to its inhibition of platelet aggregation.

4. Materials and Methods

4.1. Preparation of the Aqueous Extract of Safflower

Safflower powder was purchased from Shun Ten Pharmaceutical Co., Ltd. (New Taipei City, Taiwan). A total of 10 g of safflower powder was dissolved in 100 mL of distilled water, boiled, and filtered. The water-soluble safflower extract was sterilized by an autoclave and then stored at 4 °C.

4.2. Preparation of Platelet-Rich Plasma (PRP), Platelet-Poor Plasma (PPP), and Purified Platelets

Human blood was obtained from healthy adults through venipuncture in a vacuum tube containing sodium citrate. All participants who participated in this study provided informed consent, and this study was approved by the Taipei Tzu Chi Hospital Institutional Review Board. The obtained blood was centrifuged at 180× *g* at room temperature for 20 min to obtain PRP. The PRP was centrifuged at 1500× *g* for 15 min, and the PPP and isolated platelets were obtained in the supernatant and sediment, respectively. The isolated platelets were resuspended in modified calcium-free Tyrode buffer for subsequent tests.

4.3. Measurement of Platelet Aggregation

The platelet processing of safflower extract and the measurement of platelet aggregation were performed using the light transmittance aggregometer platform (PAP-8E, Platelet Aggregation Profiler, Bio/Data Corporation, Horsham, PA, USA) with adjustable temperature and continuous stirring. PRP was treated with safflower extract, hydroxysafflor yellow A (ChemFaces Biochemical Co., Ltd. Hubei, China), safflower yellow A (ChemFaces), or luteolin (ChemFaces) at 37 °C for 1 h. After setting the reference value of each sample with PPP, PRP was treated with 100 μM A2P5P (Santa Cruz Biotechnology, Santa Cruz, CA, USA) or clopidogrel (Santa Cruz Biotechnology) for 3 min. Finally, 10 μM ADP (Sigma-Aldrich, St. Louis, MO, USA) was added, and the aggregation reaction was monitored for 6 min.

4.4. Detection of Calcium Ions and cAMP in Platelets

PRP was treated with safflower extract or compounds found in safflower extract at 37 °C for 1 h, after which 3 μM Fura-2AM (Molecular Probes, Invitrogen, Carlsbad, CA, USA) was added for 45 min, and the isolated platelets were obtained by centrifugation. The method of measuring calcium ions has been described previously [42]. For the measurement of cAMP, isolated platelets were obtained from control or safflower extract-treated PRP by centrifugation, and then 10 μM forskolin and 10 μM ADP were added and incubated at 37 °C for 5 min. After washing with PBS, the cells were destroyed with 0.1 N HCl, and the cell extracts were obtained by centrifugation. cAMP analysis was performed using EIA technology and in accordance with the manufacturer's instructions (cyclic AMP EIA Kit, Cayman Chemical, Ann Arbor, MI).

4.5. Determination of Thromboxane B2 (TXB2) and Arachidonic acid (AA) in Platelets

PRP was treated with safflower extract or compounds found in safflower extract for 1 h, and isolated platelets were obtained by centrifugation. Then, 3 μM fibrinogen and 10 μM ADP were added and placed at 37 °C for 3 min for stimulation. The sample was then placed in dry-ice-cooled ethanol to quickly freeze the sample to stop the reaction. The content of TXB2 in each sample was measured in accordance with the manufacturer's instructions (Cayman Chemical). Alternatively, isolated platelets obtained by the treatment of safflower extract or compounds found in safflower extract with 10 μM ADP and fibrinogen stimulation for 30 min were centrifuged, and the supernatant was obtained to measure the content of AA (Cusabio Biotech, Hubei, China).

4.6. Measurement of Platelet Surface Activation Markers

The obtained blood was diluted with PBS at a ratio of 1:2. The diluted blood was treated with safflower extract or compounds found in safflower extract for 1 h and then treated with fluorescent antibodies (PE-conjugated CD61 and FITC-conjugated PAC-1, BD Pharmingen, San Jose, CA, USA) with 20 μM ADP at room temperature for 20 min. The diluted blood samples were immediately fixed with 400 μL of 1% formaldehyde/PBS, and the expression of surface antigen was analyzed by flow cytometry (FACScan, Becton Dickinson, Flanklin Lakes, NJ, USA).

4.7. Statistical Analysis

Each experimental data point was obtained from at least three samples. The experiment was performed with at least three replicates and presented as ±SD of the mean. Statistical analysis was performed using one-way ANOVA and Dunnett's posttest. A p-value of less than 0.5 between groups was considered significant.

Supplementary Materials: The following are available online at https://www.mdpi.com/article/10.3390/plants10061192/s1, Figure S1: Effects of safflower extract on the expression of CD61 and PAC-1 of human platelets.

Author Contributions: P.-H.L., C.-Y.K., and C.-C.C. designed and conducted the experiments, L.-K.W. and M.-L.C. designed experiments and analyzed data, I.-S.T. performed statistical data analysis, and F.-M.T. conceived and designed experiments and completed this manuscript. All authors have read and agreed to the published version of the manuscript.

Funding: This study was supported by grants (TCRD-TPE-110-13 and TCRD-TPE-110-49) from the Taipei Tzu Chi Hospital through the Buddhist Tzu Chi Medical Foundation, Taipei, Taiwan.

Institutional Review Board Statement: This research project was reviewed and approved by Taipei Tzu Chi Hospital Institutional Review Board. The applicant was Ping-Hsun Lu, the case number was 09-XD-075, and the title of the subject was "Study of the effect of Chinese herbs for eliminating blood statis on human platelets activity".

Informed Consent Statement: All subjects participating in this study have understood the content of the study and agreed to participate in this study.

Data Availability Statement: All data used to support the results of this study are included in the article.

Acknowledgments: The authors thank the Core Laboratory of the Buddhist Tzu Chi General Hospital for support.

Conflicts of Interest: The authors declare no conflict of interest.

Abbreviations

TXA2	thromboxane A2
cAMP	cyclic AMP
PRP	platelet-rich plasma
PPP	platelet-poor plasma
TXB2	thromboxane B2
AA	arachidonic acid

References

1. Hussain, M.I.; Al-Dakheel, A.J. Effect of salinity stress on phenotypic plasticity, yield stability, and signature of stable isotopes of carbon and nitrogen in safflower. *Environ. Sci. Pollut. Res. Int.* **2018**, *25*, 23685–23694. [CrossRef]
2. Delshad, E.; Yousefi, M.; Sasannezhad, P.; Rakhshandeh, H.; Ayati, Z. Medical uses of *Carthamus tinctorius* L. (Safflower): A comprehensive review from Traditional Medicine to Modern Medicine. *Electron. Physician* **2018**, *10*, 6672–6681. [CrossRef] [PubMed]
3. Mani, V.; Lee, S.K.; Yeo, Y.; Hahn, B.S. A Metabolic Perspective and Opportunities in Pharmacologically Important Safflower. *Metabolites* **2020**, *10*, 253. [CrossRef]
4. Asgarpanah, J.; Kazemivash, N. Phytochemistry, pharmacology and medicinal properties of *Carthamus tinctorius* L. *Chin. J. Integr. Med.* **2013**, *19*, 153–159. [CrossRef] [PubMed]
5. Lu, Q.Y.; Ma, J.Q.; Duan, Y.Y.; Sun, Y.; Yu, S.; Li, B.; Zhang, G.M. Carthamin yellow protects the heart against Ischemia/Reperfusion Injury with reduced reactive oxygen Species Release and Inflammatory Response. *J. Cardiovasc. Pharmacol.* **2019**, *74*, 228–234. [CrossRef] [PubMed]
6. Guo, X.; Zheng, M.; Pan, R.; Zang, B.; Gao, J.; Ma, H.; Jin, M. Hydroxysafflor yellow A (HSYA) targets the platelet-activating factor (PAF) receptor and inhibits human bronchial smooth muscle activation induced by PAF. *Food Funct.* **2019**, *10*, 4661–4673. [CrossRef]
7. Du, S.; Deng, Y.; Yuan, H.; Sun, Y. Safflower yellow B protects brain against cerebral ischemia reperfusion injury through AMPK/NF-kB pathway. *Evid. Based Complement Alternat. Med.* **2019**, *2019*, 7219740. [CrossRef]
8. Zhang, J.; Li, J.; Song, H.; Xiong, Y.; Liu, D.; Bai, X. Hydroxysafflor yellow A suppresses angiogenesis of hepatocellular carcinoma through inhibition of p38 MAPK phosphorylation. *Biomed. Pharmacother.* **2019**, *109*, 806–814. [CrossRef] [PubMed]
9. Ma, Y.; Feng, C.; Wang, J.; Chen, Z.; Wei, P.; Fan, A.; Wang, X.; Yu, X.; Ge, D.; Xie, H.; et al. Hydroxyl safflower yellow A regulates the tumor immune microenvironment to produce an anticancer effect in a mouse model of hepatocellular carcinoma. *Oncol. Lett.* **2019**, *17*, 3503–3510. [CrossRef]
10. Yu, S.Y.; Lee, Y.J.; Kim, J.D.; Kang, S.N.; Lee, S.K.; Jang, J.Y.; Lee, H.K.; Lim, J.H.; Lee, O.H. Phenolic composition, antioxidant activity and anti-adipogenic effect of hot water extract from safflower (*Carthamus tinctorius* L.) seed. *Nutrients* **2013**, *5*, 4894–4907. [CrossRef] [PubMed]
11. Cho, S.H.; Jang, J.H.; Yoon, J.Y.; Han, C.D.; Choi, Y.; Choi, S.W. Effects of a safflower tea supplement on antioxidative status and bone markers in postmenopausal women. *Nutr. Res. Pract.* **2011**, *5*, 20–27. [CrossRef] [PubMed]
12. Jang, H.O.; Park, Y.S.; Lee, J.H.; Seo, J.B.; Koo, K.I.; Jeong, S.C.; Jin, S.D.; Lee, Y.H.; Eom, H.S.; Yun, I. Effect of extracts from safflower seeds on osteoblast differentiation and intracellular calcium ion concentration in MC3T3-E1 cells. *Nat. Prod. Res.* **2007**, *21*, 787–797. [CrossRef]
13. Alam, M.R.; Kim, S.M.; Lee, J.I.; Chon, S.K.; Choi, S.J.; Choi, I.H.; Kim, N.S. Effects of Safflower seed oil in osteoporosis induced-ovariectomized rats. *Am. J. Chin. Med.* **2006**, *34*, 601–612. [CrossRef]
14. Kim, J.H.; He, M.T.; Kim, M.J.; Yang, C.Y.; Shin, Y.S.; Yokozawa, T.; Park, C.H.; Cho, E.J. Safflower (*Carthamus tinctorius* L.) seed attenuates memory impairment induced by scopolamine in mice via regulation of cholinergic dysfunction and oxidative stress. *Food Funct.* **2019**, *10*, 3650–3659. [CrossRef]
15. Zhang, L.; Zhou, Z.; Zhai, W.; Pang, J.; Mo, Y.; Yang, G.; Qu, Z.; Hu, Y. Safflower yellow attenuates learning and memory deficits in amyloid beta-induced Alzheimer's disease rats by inhibiting neuroglia cell activation and inflammatory signaling pathways. *Metab. Brain Dis.* **2019**, *34*, 927–939. [CrossRef] [PubMed]
16. Zhang, Y.; Guo, J.; Dong, H.; Zhao, X.; Zhou, L.; Li, X.; Liu, J.; Niu, Y. Hydroxysafflor yellow A protects against chronic carbon tetrachloride-induced liver fibrosis. *Eur. J. Pharmacol.* **2011**, *660*, 438–444. [CrossRef]
17. van der Meijden, P.E.J.; Heemskerk, J.W.M. Platelet biology and functions: New concepts and clinical perspectives. *Nat. Rev. Cardiol.* **2019**, *16*, 166–179. [CrossRef]
18. Ghoshal, K.; Bhattacharyya, M. Overview of platelet physiology: Its hemostatic and nonhemostatic role in disease pathogenesis. *Sci. World J.* **2014**, *2014*, 781857. [CrossRef]
19. Golebiewska, E.M.; Poole, A.W. Platelet secretion: From haemostasis to wound healing and beyond. *Blood Rev.* **2015**, *29*, 153–162. [CrossRef]
20. Xu, X.R.; Carrim, N.; Neves, M.A.; McKeown, T.; Stratton, T.W.; Coelho, R.M.; Lei, X.; Chen, P.; Xu, J.; Dai, X.; et al. Platelets and platelet adhesion molecules: Novel mechanisms of thrombosis and anti-thrombotic therapies. *Thromb. J.* **2016**, *14*, 29. [CrossRef] [PubMed]
21. Ruggeri, Z.M.; Mendolicchio, G.L. Adhesion mechanisms in platelet function. *Circ. Res.* **2007**, *100*, 1673–1685. [CrossRef] [PubMed]

22. Reininger, A.J.; Heijnen, H.F.; Schumann, H.; Specht, H.M.; Schramm, W.; Ruggeri, Z.M. Mechanism of platelet adhesion to von Willebrand factor and microparticle formation under high shear stress. *Blood* **2006**, *107*, 3537–3545. [CrossRef]
23. Tomokiyo, K.; Kamikubo, Y.; Hanada, T.; Araki, T.; Nakatomi, Y.; Ogata, Y.; Jung, S.M.; Nakagaki, T.; Moroi, M. Von Willebrand factor accelerates platelet adhesion and thrombus formation on a collagen surface in platelet-reduced blood under flow conditions. *Blood* **2005**, *105*, 1078–1084. [CrossRef] [PubMed]
24. Gerhards, C.; Uhlig, S.; Etemad, M.; Christodoulou, F.; Bieback, K.; Kluter, H.; Bugert, P. Expression of ADP receptor P2Y12, thromboxane A2 receptor and C-type lectin-like receptor 2 in cord blood-derived megakaryopoiesis. *Platelets* **2021**, *32*, 618–625. [CrossRef] [PubMed]
25. Zhou, L.; Schmaier, A.H. Platelet aggregation testing in platelet-rich plasma: Description of procedures with the aim to develop standards in the field. *Am. J. Clin. Pathol.* **2005**, *123*, 172–183. [CrossRef]
26. Woulfe, D.; Yang, J.; Brass, L. ADP and platelets: The end of the beginning. *J. Clin. Investig.* **2001**, *107*, 1503–1505. [CrossRef]
27. Jarvis, G.E.; Humphries, R.G.; Robertson, M.J.; Leff, P. ADP can induce aggregation of human platelets via both P2Y(1) and P(2T) receptors. *Br. J. Pharmacol.* **2000**, *129*, 275–282. [CrossRef] [PubMed]
28. Bandyopadhyay, S.K.; Azharuddin, M.; Dasgupta, A.K.; Ganguli, B.; SenRoy, S.; Patra, H.K.; Deb, S. Probing ADP Induced Aggregation Kinetics During Platelet-Nanoparticle Interactions: Functional Dynamics Analysis to Rationalize Safety and Benefits. *Front. Bioeng. Biotechnol.* **2019**, *7*, 163. [CrossRef]
29. Shah, B.H.; Rasheed, H.; Rahman, I.H.; Shariff, A.H.; Khan, F.L.; Rahman, H.B.; Hanif, S.; Saeed, S.A. Molecular mechanisms involved in human platelet aggregation by synergistic interaction of platelet-activating factor and 5-hydroxytryptamine. *Exp. Mol. Med.* **2001**, *33*, 226–233. [CrossRef]
30. Li, H.X.; Han, S.Y.; Wang, X.W.; Ma, X.; Zhang, K.; Wang, L.; Ma, Z.Z.; Tu, P.F. Effect of the carthamins yellow from *Carthamus tinctorius* L. on hemorheological disorders of blood stasis in rats. *Food Chem. Toxicol.* **2009**, *47*, 1797–1802. [CrossRef]
31. Zhu, H.B.; Zhang, L.; Wang, Z.H.; Tian, J.W.; Fu, F.H.; Liu, K.; Li, C.L. Therapeutic effects of hydroxysafflor yellow A on focal cerebral ischemic injury in rats and its primary mechanisms. *J. Asian Nat. Prod. Res.* **2005**, *7*, 607–613. [CrossRef] [PubMed]
32. Yan, K.; Wang, X.; Pan, H.; Wang, L.; Yang, H.; Liu, M.; Zhu, H.; Gong, F. Safflower Yellow and Its Main Component HSYA Alleviate Diet-Induced Obesity in Mice: Possible Involvement of the Increased Antioxidant Enzymes in Liver and Adipose Tissue. *Front. Pharmacol.* **2020**, *11*, 482. [CrossRef] [PubMed]
33. Maleki, S.J.; Crespo, J.F.; Cabanillas, B. Anti-inflammatory effects of flavonoids. *Food Chem.* **2019**, *299*, 125124. [CrossRef]
34. Choy, K.W.; Murugan, D.; Leong, X.F.; Abas, R.; Alias, A.; Mustafa, M.R. Flavonoids as Natural Anti-Inflammatory Agents Targeting Nuclear Factor-Kappa B (NFkappaB) Signaling in Cardiovascular Diseases: A Mini Review. *Front. Pharmacol.* **2019**, *10*, 1295. [CrossRef]
35. Wang, Z.F.; Liu, J.; Yang, Y.A.; Zhu, H.L. A Review: The Anti-inflammatory, Anticancer and antibacterial properties of four kinds of Licorice flavonoids isolated from Licorice. *Curr. Med. Chem.* **2020**, *27*, 1997–2011. [CrossRef]
36. Singh, R.P.; Agarwal, R. Natural flavonoids targeting deregulated cell cycle progression in cancer cells. *Curr. Drug Targets* **2006**, *7*, 345–354. [CrossRef] [PubMed]
37. Romagnolo, D.F.; Selmin, O.I. Flavonoids and cancer prevention: A review of the evidence. *J. Nutr. Gerontol. Geriatr.* **2012**, *31*, 206–238. [CrossRef] [PubMed]
38. Rodriguez-Garcia, C.; Sanchez-Quesada, C.; Gaforio, J.J. Dietary flavonoids as cancer chemopreventive agents: An updated review of human studies. *Antioxidants* **2019**, *8*, 137. [CrossRef] [PubMed]
39. Wang, X.; Ouyang, Y.Y.; Liu, J.; Zhao, G. Flavonoid intake and risk of CVD: A systematic review and meta-analysis of prospective cohort studies. *Br. J. Nutr.* **2014**, *111*, 1–11. [CrossRef]
40. Guerrero, J.A.; Lozano, M.L.; Castillo, J.; Benavente-Garcia, O.; Vicente, V.; Rivera, J. Flavonoids inhibit platelet function through binding to the thromboxane A2 receptor. *J. Thromb. Haemost.* **2005**, *3*, 369–376. [CrossRef] [PubMed]
41. Zhao, F.; Wang, P.; Jiao, Y.; Zhang, X.; Chen, D.; Xu, H. Hydroxysafflor Yellow A: A Systematical Review on Botanical Resources, Physicochemical Properties, Drug Delivery System, Pharmacokinetics, and Pharmacological Effects. *Front. Pharmacol.* **2020**, *11*, 579332. [CrossRef] [PubMed]
42. Wu, C.C.; Tsai, F.M.; Chen, M.L.; Wu, S.; Lee, M.C.; Tsai, T.C.; Wang, L.K.; Wang, C.H. Antipsychotic Drugs Inhibit Platelet Aggregation via P2Y 1 and P2Y 12 Receptors. *Biomed Res. Int.* **2016**, *2016*, 2532371. [PubMed]

plants

Article

Hypoglycemic Effect of Two Mexican Medicinal Plants

Adolfo Andrade-Cetto *, Fernanda Espinoza-Hernández, Gerardo Mata-Torres and Sonia Escandón-Rivera

Laboratory of Ethnopharmacology, School of Sciences, National Autonomous University of Mexico, Mexico City 04510, Mexico; f.artemisa.ehdz@ciencias.unam.mx (F.E.-H.); gerardom.torres@ciencias.unam.mx (G.M.-T.); soniaer@ciencias.unam.mx (S.E.-R.)

* Correspondence: aac@ciencias.unam.mx

Abstract: Type 2 diabetes is a worldwide prevalent disease that is due to a progressive loss of adequate β-cell insulin secretion, frequently against a background of insulin resistance. In Mexican traditional medicine, the therapeutic use of hypoglycemic plants to control the disease is a common practice among type 2 diabetic patients. In the present work, we examined the traditional use of the aerial parts of *Eryngium longifolium* and the rhizome of *Alsophila firma*, consumed by people use over the day (in fasting state) to control their blood glucose levels, therefore, we aimed to assess the acute hypoglycemic effect of both plants. First, basic phytochemical profiles of both plants were determined and, subsequently, acute toxicity tests were carried out. Then, in vivo hypoglycemic tests were performed in streptozotocin-nicotinamide (STZ-NA) induced hyperglycemic Wistar rats and finally the effect of the plants on three enzymes involved in glucose metabolism was assayed in vitro. Through HPLC-DAD chromatography, caffeic acid, chlorogenic acid, rosmarinic acid, isoflavones, and glycosylated flavonoids were identified in *E. longifolium*, while the possible presence of flavanones or dihydroflavonols was reported in *A. firma*. Both plants exhibited a statistically significant hypoglycemic effect, without a dose-dependent effect. Furthermore, they inhibited glucose 6-phosphatase and fructose 1,6-bisphosphatase in in vitro assays, which could be associated with the hypoglycemic effect in vivo. Thus, this study confirmed for the first time the traditional use of the aerial part of *E. longifolium* and the rhizome of *A. firma* as hypoglycemic agents in a hyperglycemic animal model. In addition, it was concluded that their ability to regulate hyperglycemia could involve the inhibition of hepatic glucose output, which mainly controls glucose levels in the fasting state.

Keywords: traditional medicine; hypoglycemic plants; *Eryngium longifolium*; *Alsophila firma*; glucose 6-phosphatase; fructose 1,6-bisphosphatase

Citation: Andrade-Cetto, A.; Espinoza-Hernández, F.; Mata-Torres, G.; Escandón-Rivera, S. Hypoglycemic Effect of Two Mexican Medicinal Plants. *Plants* **2021**, *10*, 2060. https://doi.org/10.3390/plants10102060

Academic Editors: Juei-Tang Cheng, I-Min Liu and Szu-Chuan Shen

Received: 9 September 2021
Accepted: 27 September 2021
Published: 29 September 2021

1. Introduction

Traditional medicine (TM), as defined by the Word Health Organization (WHO), includes medication therapies that involve the use of herbal medicines, animal parts and/or minerals; the term is used to refer to various forms of indigenous medicine [1]. In the case of Mexico, TM has not been incorporated into the national health care system and conversely the official system is mainly based on allopathic medicine. However, many people still rely in the use of medicinal plants to treat health problems, thus the use of plants in diseases like type 2 diabetes (T2D) is a common practice [2,3]. The American Diabetes Association (ADA) states that T2D is due to a progressive loss of adequate β-cell insulin secretion, frequently on the background of insulin resistance, which manifests clinically as hyperglycemia. Once hyperglycemia occurs, patients are at risk of developing chronic complications, such as microvascular and macrovascular diseases, and acute coronary syndrome [4]. T2D accounts for 90–95% of all diabetes, the last report of the International Diabetes Federation (IDF) ranks Mexico as sixth in the world, with 12.8 million T2D patients [5]. In T2D patients, two sources that contribute to raise blood glucose levels are food ingestion, which increases glucose levels in postprandial state, and the liver glucose output, which promotes high glucose levels in the fasting state. In these metabolic processes, the intestinal α-glucosidases, and the hepatic glucose 6-phosphatase (G6Pase) and fructose 1,6-bisphosphatase (FBPase) enzymes play an important role [6,7].

Mexico stands out among the mega-diverse countries, being the fourth nation in terms of species richness with the presence of all the climates of the planet [8]. It is calculated that more than 6000 plants are used for medicinal purposes. Moreover, around 30 million people live in rural areas, about 12 million belong to one of the 86 ethnolinguistic indigenous groups that make up about 10% of the population [9], and 80% of the population are poor or vulnerable (due to social deprivation or low income) [10]. These aspects, combined with the richness in medicinal plants and the large number of ethnic groups, makes Mexico an important source of traditionally used medicinal plants. Considering the high amount of T2D patients, the perfect combination is presented for the exploration and study of traditionally used hypoglycemic plants. Hence, two plants used in the municipalities of Tlanchinol and Huejutla de Reyes (Hidalgo), located in central Mexico, were selected for this study. Both plants were already reported by our group as hypoglycemic species based on previous ethnopharmacological field work [3].

Eryngium longifolium Cav. (Apiaceae) (Figure 1a), which is native and endemic to Mexico, is a herb of up to 3 m in height with simple, lobed, or spiny-toothed to linear leaves, varied venation, flower in bracted heads, and globose or ovoid fruit [11]. In addition to being used in the treatment of diabetes, it has also been reported as a diuretic, emmenagogue, and alexiteric [12]. More information of the plant can be found at https://enciclovida.mx/especies/171163-eryngium-longifolium (access on 25 September 2021).

(**a**) (**b**)

Figure 1. Photographs of the study plants ready for consumption: (**a**) Dried aerial part of *E. longifolium*; (**b**) Rhizome of *A. firma*.

On the other hand, *Alsophila firma* (Baker) D.S.Conant (Cyatheaceae) (Figure 1b) is an arborescent fern with spiny stems up to 10 m high, yellowish-brown in color, with a blade up to 3.3 × 1.5 m, 2-pinnate-pinnatifid. It can be found in temperate and humid forests. It sometimes presents adventitious buds, veins without trichomes or with few irregular dark squamules at their base abaxially; globose to subglobose indusian, with or without an umbo, very delicate, fleeting at maturity, usually glabrous, diaphanous to yellowish-brown [11]. Scales from the fronds are used to stop hemorrhages [13]. More information about the plant can be found at https://enciclovida.mx/especies/151254-alsophila-firma (access on 25 September 2021).

Until now, no pharmacological studies about these species are reported in the international literature. Therefore, the present work aims to contribute to the overall knowledge of *E. longifolium* and *A. firma*, two plants therapeutically used by T2D patients as hypoglycemics. The acute hypoglycemic effect of the traditionally used aqueous extract and the ethanol-water (EtOH) extract as well as their possible dose-dependent effect were tested in hyperglycemic rats. Additionally, the effect of the extracts on two of the main enzymes involved in the gluconeogenesis pathway (G6Pase and FBPase) as well as the main enzyme involved in the intestinal carbohydrate breakdown (α-glucosidase) was tested. Also, acute toxicity tests were carried out and basic phytochemical profiles of the extracts that had a higher biological activity in the in vivo tests were provided.

2. Results

2.1. Ethnobotany

According to the information obtained from the specialists, sellers, and diabetic patients, the traditional use of *E. longifolium* by the patients from the town of Tlanchinol and patients who visit the Huejutla de Reyes market to control their blood glucose levels was confirmed. Both locations are in Hidalgo, Mexico. Particularly, the traditional healer Isabel Escalante recommended the use of the plant *E. longifolium*, known by its Spanish name "piñuela". It is prepared as an infusion, in which 20 g of the dry plant (aerial part) are placed in 500 mL of boiling water. When the infusion reaches room temperature, it is consumed over the day as so-called "agua de uso".

On the other hand, the specialist Guadalupe Vite specifically recommends *A. firma* for controlling blood glucose levels of diabetics in Tlanchinol, where the plant is known by its Nahuatl name "peshma". The patients also prepare an infusion with 20 g of the dried rhizome (powder) in 500 mL of boiling water. After filtration, it is also consumed as "agua de uso".

It is important to note that since both plants are consumed as "agua de uso", they are mainly consumed in the fasting state, namely they are drunk throughout the day instead of normal water and not with the meals.

2.2. Chromatographic Profiles

According to the in vivo outcomes, the *E. longifolium* EtOH extract and the *A. firma* aqueous extract presented the best hypoglycemic activity. The HPLC fingerprint profile of the *E. longifolium* EtOH extract shows the most important signals at 254 and 320 nm (Figure 2). Seven major peaks (at 6.3, 6.9, 11.9, 13.0, 16.2, 19.2 and 23.6 min from 1 to 7, respectively) and several minor peaks were observed. The UV spectra of 1, 2 and 4 revealed characteristic signs of known phenolic acids which were identified by comparing the retention time and the UV spectrum with commercial standards (caffeic acid, **1**; chlorogenic acid, **2**; rosmarinic acid, **3**). Peaks 3 and 4 have the same spectral characteristics, while peaks 5 and 6 shows UV spectral typical signs of isoflavones, with an intense Band II absorption (λ_{max} 254 nm in both cases) and Band I with a very low intensity (λ_{max} 316 and 312 nm, respectively) [14] and, peak 7 shows signs of glycosylated flavonoids as previously reported in other *Eryngium* species [15,16].

The HPLC fingerprint profile of the *A. firma* aqueous extract shows the most important signals at 254 and 280 nm (Figure 3), i.e., peaks with characteristic phenolic UV spectra probably due to flavanones or dihydroflavonols (peaks 8 to 14). Peak 10 in particular has a UV spectrum with this property [14].

2.3. Acute Oral Toxicity Tests

To assess a possible acute toxic effect of the tested extracts, acute toxicity tests were carried out according to the guidelines for the Testing of Chemicals of the Organisation for Economic Co-operation and Development (OECD) [17]. The results showed no physical or behavioral abnormalities after oral administration of the maximum dose of 2000 mg/kg body weight (b.w.) of each extract. Also, no deaths were reported. Therefore, traditional doses, and elevated-traditional doses used for this study are within the safe range, indicating that the acute exposure of these plants does not generate toxic effects and the LD_{50} is greater than the maximum dose used.

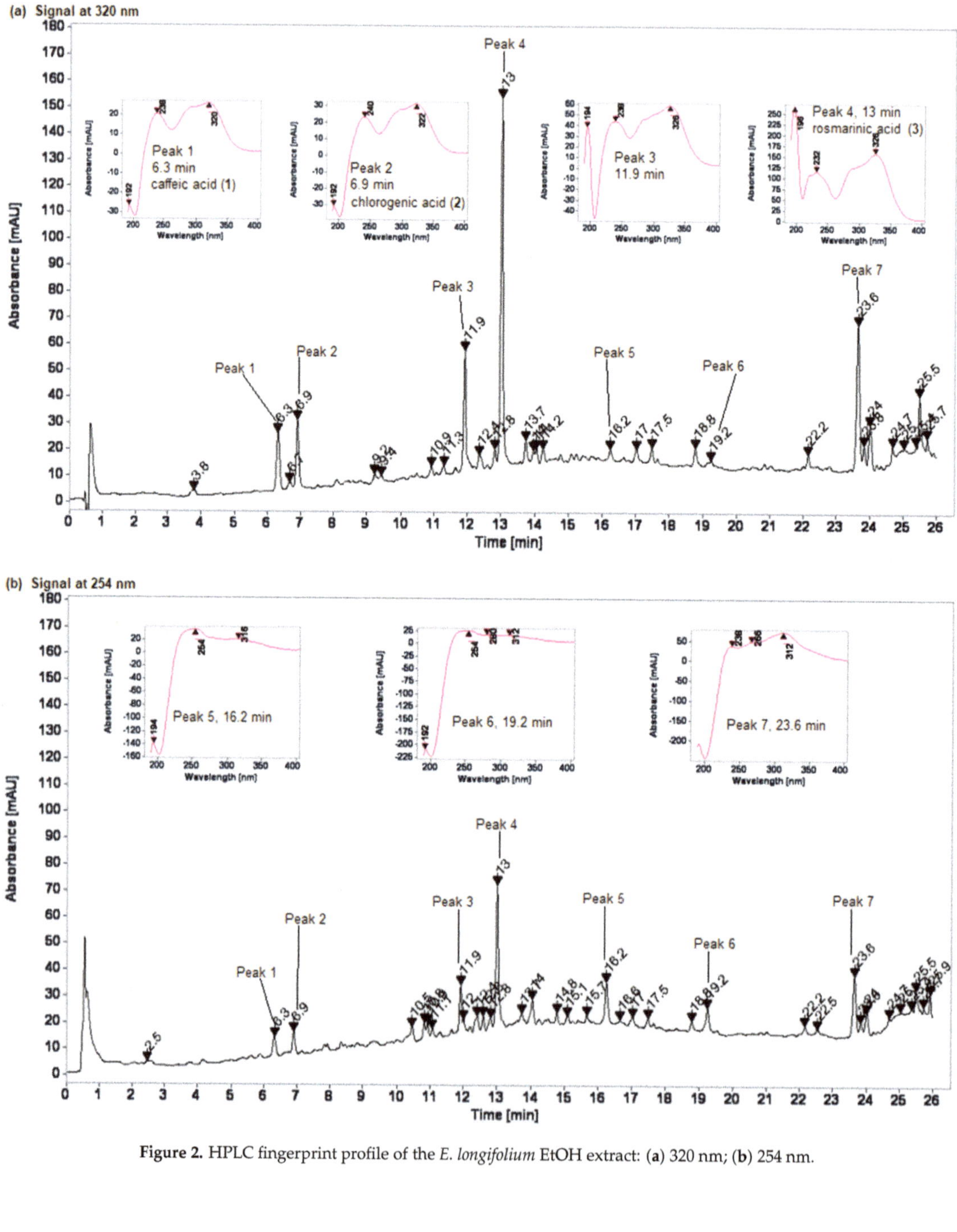

Figure 2. HPLC fingerprint profile of the *E. longifolium* EtOH extract: (**a**) 320 nm; (**b**) 254 nm.

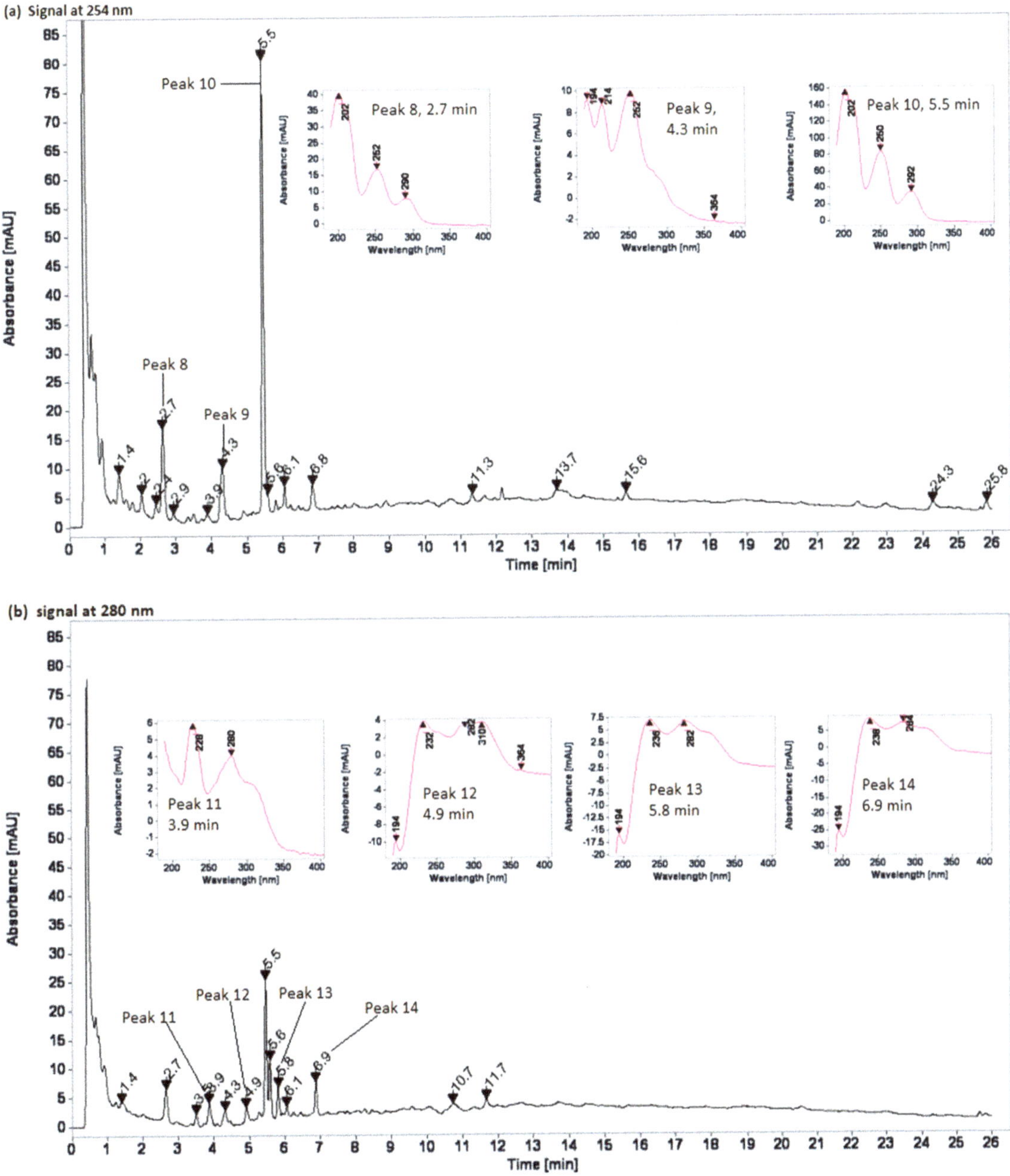

Figure 3. HPLC fingerprint profile of the *A. firma* aqueous extract: (**a**) 254 nm; (**b**) 280 nm.

2.4. Hypoglycemic Effect of Plant Extracts

As shown in Table 1, both normoglycemic control and negative hyperglycemic control exhibited stable and unchanged blood glucose values throughout the acute test, indicating that physiological solution (vehicle) administration did not alter this parameter over time despite the physiological condition of the animals.

Table 1. Blood glucose values from 3-h acute tests expressed as mg/dL (mean ± SEM, $n = 9$).

Group		Dose	0 min	60 min	120 min	180 min
Normoglycemic Control		n/a	120 ± 3 ^{a}	118 ± 2 ^{a}	116 ± 2 ^{a}	119 ± 3 ^{a}
Hyperglycemic Control		n/a	192 ± 4	190 ± 5 ^{b}	188 ± 6 ^{b}	193 ± 5 ^{b}
Hyperglycemic + Glibenclamide Control		5 mg/kg	187 ± 5	152 ± 6 ^{a,*}	128 ± 3 ^{a,*}	124 ± 3 ^{a,*}
Hyperglycemic + *A. firma* Rhizome	Aqueous Extract	16 mg/kg	195 ± 4	157 ± 5 ^{a,*}	136 ± 4 ^{a,*}	132 ± 3 ^{a,*}
		160 mg/kg	190 ± 3	159 ± 5 *	150 ± 3 ^{a,b,*}	140 ± 2 ^{a,*}
	EtOH Extract	37 mg/kg	199 ± 5	166 ± 4 *	149 ± 4 ^{a,b,*}	147 ± 5 ^{a,b,*}
		374 mg/kg	205 ± 4	184 ± 5 ^{b}	158 ± ^{a,b,*}	153 ± 4 ^{a,b,*}
Hyperglycemic + *E. longifolium* Aerial Part	Aqueous Extract	30 mg/kg	197 ± 6	169 ± 7 *	148 ± 4 ^{a,b,*}	149 ± 4 ^{a,,b*}
		310 mg/kg	204 ± 7	164 ± 7 ^{a,*}	135 ± 4 ^{a,*}	136 ± 5 ^{a,*}
	EtOH Extract	32 mg/kg	186 ± 6	156 ± 5 ^{a,*}	141 ± 3 ^{a,*}	121 ± 3 ^{a,*}
		318 mg/kg	193 ± 5	152 ± 4 ^{a,*}	141 ± 3 ^{a,*}	132 ± 3 ^{a,*}

Rows: *—statistically significant difference versus its initial time at $p < 0.05$. Columns: ^{a}—indicates statistically significant difference versus hyperglycemic group in that time at $p < 0.05$; ^{b}—indicates statistically significant difference versus glibenclamide control in that time at $p < 0.05$.

Comparing both groups, the hyperglycemic control shown statistical differences against normoglycemic control in all times through the 3 h. The pancreatic damage caused by STZ administration was such that blood glucose levels rose significantly around 190 mg/dL in the hyperglycemic control compared with those observed in the normoglycemic control. However, the oral administration of the hypoglycemic agent glibenclamide turned out in a significant decrease in these levels from the first hour, reaching normoglycemic levels at the second hour of the test. In addition, glibenclamide shown statistical differences against hyperglycemic control from 60 to 180 min. These outcomes support the idea that this hyperglycemic model is adequate to assess the hypoglycemic effect of mixtures or isolated compounds since it can respond to hypoglycemic drugs, such as glibenclamide, despite its reduced insulin levels.

Regarding plant extracts, all of them could effectively decrease blood glucose levels though none displayed a dose-dependent effect, namely no statistical differences between the doses of each type of extract were observed. The aqueous extract of *A. firma* at its traditional dose shown a better glycemic control than its EtOH one at both doses, while *E. longifolium* reached normoglycemic levels at the end of the experiment with its two types of extracts, mainly with the EtOH one at its traditional dose. Particularly, although the EtOH extract of *A. firma* high dose could improve hyperglycemia by significantly lowering around 50 mg/dL of blood glucose, it controlled overall glycemia in a modest way as compared with the other extracts after 3 h of treatment, as shown in Figure 4. Taken together, these results support the traditional use of *A. firma* and *E. longifolium* for the treatment of T2D. It is suggested that future experiments on their mechanism of action be investigated using the traditional doses of the aqueous extract of *A. firma*, and the EtOH extract of *E. longifolium* since both were as effective as glibenclamide in lowering glucose levels.

2.5. Inhibition of Key Hyperglycemia-Related Enzymes

Impaired hepatic glucose production, specifically gluconeogenesis, is the main source of fasting hyperglycemia and is one of the contributors of postprandial hyperglycemia in diabetic patients [6]. Therefore, the *A. firma* aqueous extract and the *E. longifolium* EtOH extract were evaluated on two of the principal enzymes involved in this pathway: G6Pase and FBPase. Regarding G6Pase, each plant extract effectively inhibited its activity. As shown in Figure 5, even though the control chlorogenic acid had the strongest effect on this enzyme, both extracts could decrease the enzyme activity in the same order of magnitude. The *A. firma* aqueous extract diminished G6Pase activity by 87% at the highest concentration (5000 µg/mL), while *E. longifolium* EtOH extract by 75%.

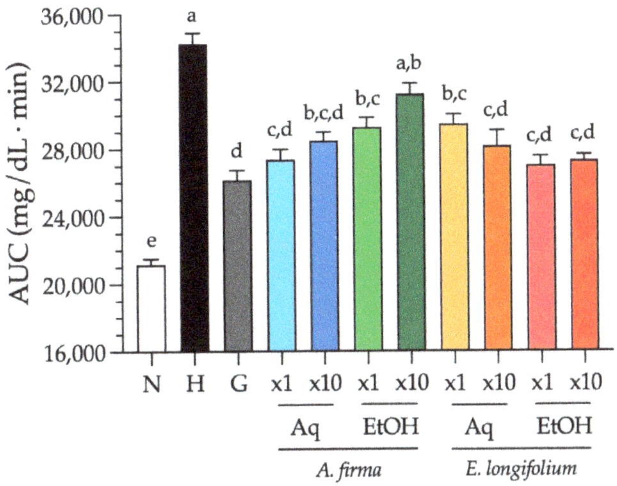

Figure 4. Blood glucose AUC from 3-h acute tests (mean ± SEM, n = 9). Different letters over bars indicate statistically significant difference among groups at $p < 0.05$ (a > b > c > d > e). N–normoglycemic control; H–hyperglycemic control; G–hyperglycemic + glibenclamide control; x1–traditional dose; x10–traditional high dose; Aq–aqueous extract; EtOH–ethanol-water extract.

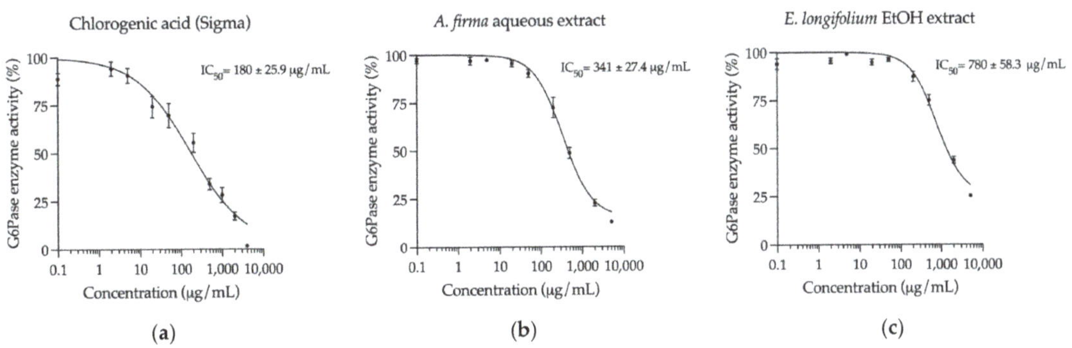

Figure 5. Concentration-response inhibition curves on G6Pase (mean ± SEM, n = 6): (**a**) Curve of chlorogenic acid (control); (**b**) Curve of *A. firma* aqueous extract; (**c**) Curve of *E. longifolium* EtOH extract.

In contrast to what was observed for G6Pase activity, both extracts exhibited a stronger inhibitory effect on FBPase activity than the control adenosine 5′-monophosphate (AMP), as shown in Figure 6. Interestingly, the *E. longifolium* EtOH extract inhibited the enzyme activity by 100% at the highest concentration (5000 μg/mL), being more potent than AMP (92%), while the *A. firma* aqueous extract reduced the FBPase activity by 86%.

Finally, the effect of the *A. firma* aqueous extract and the *E. longifolium* EtOH extract on intestinal α-glucosidase enzymes was assessed. Among other mechanisms, these enzymes are responsible for the postprandial hyperglycemia in diabetic patients since they hydrolyze diet's oligosaccharides for their subsequent absorption into simple monomers after their consumption [7]. As shown in Figure 7, neither the *A. firma* aqueous extract nor the *E. longifolium* EtOH extract exerted a substantial inhibition on intestinal α-glucosidase enzymes, suggesting that these plants are not capable of producing a significant reduction of postprandial hyperglycemia through the inhibition of carbohydrate breakdown.

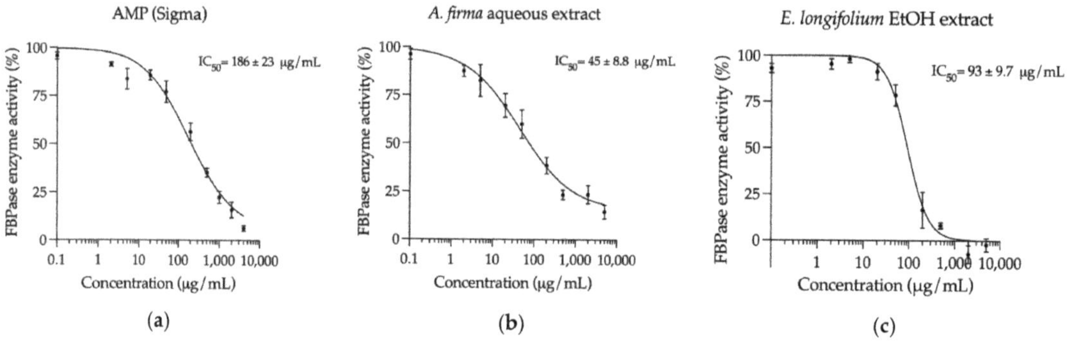

Figure 6. Concentration-response inhibition curves on FBPase (mean ± SEM, n = 6): (**a**) Curve of AMP (control); (**b**) Curve of *A. firma* aqueous extract; (**c**) Curve of *E. longifolium* EtOH extract.

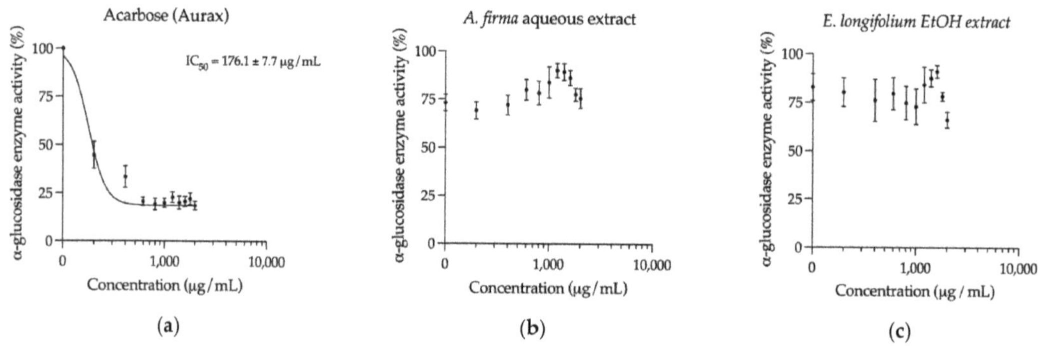

Figure 7. Concentration-response inhibition curves on intestinal α-glucosidase enzymes (mean ± SEM, n = 6): (**a**) Curve of acarbose (control); (**b**) Curve of *A. firma* aqueous extract; (**c**) Curve of *E. longifolium* EtOH extract.

3. Discussion

T2D is an emerging disease in Mexico. In the 19th century, only one death per year was attributable to diabetes. Since 1950, there has been an increase in the mortality rates of this disease, which went from 0.2 to 31.7 per 100,000 inhabitants in 1990. In 2019, the prevalence was 13.5, placing the country in sixth place in the world [5,18]. Moreover, the use of medicinal plants to treat any disease is part of the Mexican idiosyncrasy and a disease such as T2D, with high prevalence, is no exception. As a result of ethnopharmacological research, we found that the Mexican population (herbal specialists, plant sellers, and T2D patients) are looking for new medicinal plants to counteract the high glucose levels produced by the disease.

This phenomenon is observed in other Latin American countries. For instance, it is documented that 91% of the Cakchiquels, a Guatemalan ethnicity, with T2D use medicinal plants in addition to their medical prescription [19]. The local usage of medicinal plants is related to plant diversity and traditional knowledge [20], therefore, the richness of plant species in America gives the opportunity to use a great variety of plants for medicinal purposes.

In 2005, Barbosa-Filho et al. reported 224 hypoglycemic plants supported by scientific studies from North, Central, and South American countries [21]. However, only in that year, 306 Mexican species popularly used as hypoglycemic agents were documented [22], and by 2020, it was estimated the use of about 800 plants in Mexico for treating T2D [23]. These growing numbers indicate that although traditionally used plants continue to be identified through ethnopharmacological approaches, works that support their hypoglycemic effect, their phytochemical composition, and their associated mechanisms of action are still lacking.

In Mexico, T2D affects nearly 13 million people [5], who usually treat it with hypoglycemic plants. Particularly, we detected the use of the aerial part of *E. longifolium* and the rhizome of *A. firma* to control the disease among healers and T2D patients in the municipalities of Tlanchinol and Huejutla, in the state of Hidalgo. For this purpose, an infusion is prepared and consumed over the day, mainly in the postabsorptive state, where glucose production by the liver plays a crucial role in maintaining blood glucose levels [24]. Our findings suggest that the hypoglycemic effect exerted in vivo by *E. longifolium* and *A. firma* could be correlated with the regulation of hepatic glucose output since both plants were able to inhibit two of the key enzymes that participate in gluconeogenesis: G6Pase and FBPase.

The sustained hypoglycemic effect of the EtOH extract of *E. longifolium* and the aqueous extract of *A. firma* is comparable to that observed for glibenclamide. Likewise, other species used in Central and South America have proved to decrease blood glucose levels in the same way as this hypoglycemic drug, such as the ethanolic leaf extract of *Phyllanthus acidus* (L.) Skeels (Phyllanthaceae) [25], the methanolic extract of *Eclipta prostrata* (L.) L. (Compositae) [26], the ethanolic bark extract of *Croton guatemalensis* Lotsy (Euphorbiaceae), and the ethanolic leaf extract of *Solanum americanum* Mill. (Solanaceae) [27]. Furthermore, other South American species like *Clusia latipes* Planch. & Triana (Clusiaceae) [28] and *Terminalia phaeocarpa* Eichler (Combretaceae) [29] shown to possess inhibitory activity on α-glucosidases, which was reported as not significant in the Mexican species in this study. However, even though both plants presented a modest inhibition of α-glucosidases (around 25%), this mechanism could contribute to the hypoglycemic effect in the postprandial state in a synergistic way, namely the possible reduction of postprandial hyperglycemia by these plants could be mainly due to other mechanisms, such as insulin secretion, improvement of insulin function, and blocking of intestinal absorption, rather than inhibition of carbohydrate breakdown. In this regard, it is necessary to perform carbohydrate load curves in further works with the aim to prove the anti-hyperglycemic effect of these medicinal plants.

Moreover, we confirmed that the STZ-NA hyperglycemic model is suitable for testing plant extracts. Although this model does not present all the characteristics of T2D, it exhibits a stable hyperglycemia that can be controlled with a hypoglycemic agent, such as a sulfonylurea, due to the responsiveness to insulin secretagogues [30]. Nevertheless, different type of models could be considered to assess other therapeutic aspects of these plants, such as the improvement of insulin resistance.

As can be observed in the phytochemical profile of *E. longifolium* EtOH extract (Figure 1), phenolic acids and flavonoids are the predominant components, which agrees with the chemical profiles previously reported in other *Eryngium* species [31]. Rosmarinic acid (**4**) was identified by comparing it with a commercial standard and it was noted as the most abundant in the extract. This important antioxidant has also been identified as the most abundant compound in a hydroalcoholic extract of *E. viviparum* [32] and the second one in an aqueous extract of *E. cymosum* [33]. On the other hand, flavonol glycosides have been isolated and identified in several species of *Eryngium* [31], while isoflavones are less common. Ayuso et al. reported, in 2020, the presence of three tectorigenin glycosides for the first time in the genus [32]. However, more chemical analysis is needed to identify this kind of flavonoids in *E. longifolium*.

On the other hand, Figure 2 shown the phytochemical profile of *A. firma* aqueous extract. It can be noted that the compounds have very high polarity and present characteristics of flavonoids that exhibit no conjugation between the A- and B-rings [14]. Moreover, it should be noted that the intensity of the signals obtained in this chromatographic profile are low, which may indicate the presence of one or more abundant compounds that do not absorb UV light. Further phytochemical analyzes are required to identify the major compounds.

According to the phytochemical profiles provided for both plants, most compounds that can be noted are phenolic compounds. In addition to having strong antioxidant properties, these types of phytochemicals are associated to favorable effects on glucose and lipid metabolism [34,35]. In fact, they can be found not only in medicinal plants, but also in functional foods consumed worldwide, having a good impact in human health and disease [36]. The chronic administration of the major compound identified in the *E. longifolium* EtOH extract, rosmarinic acid (**4**), has shown to reduce the activity of G6Pase and FBPase in STZ-hyperglycemic rats fed with a high-fat diet [37], while chlorogenic acid (**2**) is a well-known inhibitor of G6Pase and a weak inhibitor of α-glucosidases [38].

The principal objective of the current study was to evaluate the acute hypoglycemic effect of the traditionally used *E. longifolium* and *A. firma*, but it would be necessary to assess their sustained effect over time and possible chronic toxicological effects to evaluate their safety in further experiments. We did not detect any conservation issues with these plants, and in the future they could be cultivated in their natural habitats to produce standardized hypoglycemic agents. The present work stablishes the bases of the hypoglycemic activity of these two plants used in Mexican traditional medicine.

4. Materials and Methods

4.1. Ethnobotany

In a previous ethnobotanical work [3], we detected the use of *E. longifolium* and *A. firma* by T2D patients and two herbal specialists in the town of Tlanchinol, Hidalgo. While *E. longifolium* was reported in the original work, *A. firma* was not included in the cited study because at that time only its common name was known, without a correct botanical identification. To corroborate the previous ethnobotanical information in this work, direct interviews were performed to the two herbal specialists from the town of Tlanchinol, five sellers from the Huejutla market, and 10 diabetic patients who visited the specialists or the market. All of them confirmed the use of the plants for the treatment of T2D. For this work, interviews were carried out to corroborate the part used, dose, preparation, and administration of the plants. The interviews were in free format, asking specific questions about the recollection, preparation, used doses, and administration form of the plants. With the help of the specialist Isabel Escalante, *E. longifolium* was purchased in the Huejutla market, and a voucher specimen (IMSS16170) was deposited in the Mexican Institute of Social Security (IMSS) herbarium, Mexico City. The rhizome of *A. firma* was collected near Tlanchinol, in "La Tangente", with the help of the specialist Guadalupe Vite, and a voucher specimen (Etnof228) was deposited in the UNAM School of Sciences herbarium, (Mexico City, Mexico).

4.2. Elaboration of Traditional Extracts and Dose Calculation

Aqueous and EtOH extracts of the plants were prepared. For the aqueous extracts, 20 g of the dry and ground plant material stirred in 500 mL of boiling water for 15 min and then the mixture was filtered and lyophilized. For the EtOH extracts, 20 g were added to 500 mL of an ethanol-water mixture (1:1) and then heated at 40 °C for 4 h; subsequently, it was filtered and subjected to evaporation in a rotary evaporator (Büchi, Flawil, Switzerland); finally, the whole process was repeated once more and at the end both extracts were mixed and lyophilized.

The traditional dose of each extract was calculated based on the 20-g consumption of a plant by a 70-kg person, namely from the initial 20 g, the yield of each extract was obtained and expressed as mg/kg. To assess a possible dose-dependent response, a high dose was calculated by multiplying the traditional dose by 10 for each type of extract (elevated-traditional dose).

4.3. Chemicals and Reagents

Nicotinamide (N0636), streptozotocin (S0130), ethylenediaminetetraacetic acid (EDS), 4-(2-hydroxyethyl)piperazine-1-ethanesulfonic acid (H3375), imidazole (I0250), sodium dodecyl sulfate (L3771), ascorbic acid (A7506), malachite green (M6880), Tween® 20 (P1379), Tris-HCl (T3253), $MgCl_2$ (M9272), glucose 6-phosphate (G7879), fructose 1,6-bisphosphate (F6803), adenosine 5′-monophosphate (A2252), 4-nitrophenyl α-D-glucopyranoside (N1377), and intestinal acetone powders from rat (I1630) were purchased from Sigma-Aldrich (Steinheim, Germany). Ammonium molybdate (AT0330-5) was bought from Tecsiquim (Mexico City, Mexico).

4.4. HPLC Analysis

The HPLC profiles were obtained using a 1260 HPLC instrument (Agilent, San Jose, CA, USA) equipped with a G1311B Quaternary Pump, a G1367E Autosampler and an Agilent G1315C UV diode array detector (DAD). System control, data collection, and data processing were accomplished using OpenLAB LC 1260 chromatography software. Elution was carried out at a flow rate of 0.35 mL/min with water as solvent A containing 0.1% formic acid and acetonitrile (MeCN) as solvent B, the elution gradient was carried out by starting with a mixture of 99:1 (A:B), increasing the amount of solvent B as follows: 75:25 (A:B) at 14 min, 70:30 (A:B) at 14–18 min, 65:35 (A:B) at 18–22 min, 5:95 (A:B) at 22–27 holding this mixture for a min and 99:1 (A:B) at 28–30. The separation was carried out using a Luna Omega Polar C18, 50 × 2.1 mm of internal diameter., 1.6 µm) reverse phase column (Phenomenex, San Jose, CA, USA). The column temperature was kept at 35 °C. Working solutions of samples (extracts and standards) were prepared by dissolving 10 mg and 15 mg of *E. cymosum* and *A. firma* extracts, respectively, with the appropriate solvent for each extract (1mL of H2O for *A. firma* and 2 mL of a mixture of H2O:MeCN:MeOH; 50:25:25 for *E. longifolium*) and 2 mg of standard in 5 mL of methanol (MeOH) and diluted to 100 µg/mL. All samples were filtered on membrane filters (PTFE, 0.20 µm) and injected (3 µL). For UV detection, the wavelength program was set at an acquisition of λ 240, 254, 280, 320 and 365 nm; the UV spectra were recorded from 230 at 400 for *E. longifolium* profile and from 180 at 390 nm for *A. firma*. The identification of caffeic acid (**1**), chlorogenic acid (**2**) and rosmarinic acid (**3**) in *E. longifolium* was carried out as previously described [33].

4.5. Experimental Animals

For the in vivo hypoglycemic tests, ninety-nine 8-week-old Wistar rats were used, while twenty 8-week-old BALB/c mice were acquired for the acute toxicity tests. All animals were obtained from the bioterium of the School of Sciences, UNAM, Mexico City, Mexico and maintained with free access to food and water under standard conditions (25 °C, 55% humidity and 12 h light:12 h dark periods). All methods applied to the animals in this study were approved by the Academic Ethics and Scientific Responsibility Commission (CEARC) of the School of Sciences, UNAM, Mexico City, Mexico (P_2021_05_01) and carried out according to the Guide for the Care and Use of Laboratory Animals, Washington DC, USA [39].

4.6. Acute Oral Toxicity Tests

To test the acute safety of the plant extracts, acute toxicity tests were performed by orally giving a maximum single dose of 2000 mg/kg b.w. to five BALB/c mice for each extract, according to the OECD guidelines 425 [17]. First, one mouse was administered with the tested extract and closely observed for 30 min to detect any behavioral or physical abnormality. Next, it was observed every four hours for 24 h and then frequently for 14 days. Based on the outcomes, the maximum single dose was given to the remaining four mice to repeat the procedure previously described.

4.7. Induction of Hyperglycemia

The STZ-NA hyperglycemic model [40] was selected to test the hypoglycemic effect of the plant extracts. In brief, Wistar rats were fasted for 12 h and then administered intraperitoneally with 150 mg/kg b.w. of a fresh NA solution prepared with physiological solution. Fifteen min later, they were injected intravenously with 65 mg/kg b.w. of a STZ solution prepared one day before in 0.1 M acetate buffer, pH 4.5. Animals with non-fasting blood glucose above 180 mg/dL were selected one week after the induction to perform the experiments.

4.8. Assessment of Hypoglycemic Effect

The animals were divided into 11 groups with nine individuals each: the normoglycemic control (1) and the negative hyperglycemic control (2) were administered with physiological solution; the positive hyperglycemic control (3) was given the drug glibenclamide (Euglucon®, 5 mg/kg b.w.); two hyperglycemic groups were administered with *A. firma* rhizome aqueous extract, one with the traditional dose (4) and another with the traditional high dose (5) (16 mg/kg b.w. and 160 mg/kg b.w., respectively); two hyperglycemic groups were given *A. firma* rhizome EtOH extract, one with the traditional dose (6) and another with the traditional high dose (7) (37 mg/kg b.w. and 374 mg/kg b.w., respectively); two hyperglycemic groups were administered with *E. longifolium* aerial part aqueous extract, one with the traditional dose (8) and another with the traditional high dose (9) (30 mg/kg b.w. and 310 mg/kg b.w., respectively); and two hyperglycemic groups were given *E. longifolium* aerial part EtOH extract, one with the traditional dose (10) and another with the traditional high dose (11) (32 mg/kg b.w. and 318 mg/kg b.w., respectively). Both glibenclamide and extracts were dissolved in physiological solution.

Blood glucose levels were measured at baseline and, after the treatment administration by gavage, monitored every hour for 3 h [41]. Samples were obtained from the tail vein and quantified in duplicate in using glucometers (Accutrend® Plus, Roche Diagnostics International AG, Rotkreuz, Switzerland).

4.9. Glucose 6-phosphatase Inhibition Assay

The activity of G6Pase enzyme was assessed by a colorimetric test which detects the blue phosphomolybdate formation [42,43]. Microsomal fractions obtained from livers of Wistar rats were resuspended in buffer (40 mM imidazole, 250 mM sucrose, pH 7) and mixed with extracts or chlorogenic acid (control) at concentrations ranging from 2 µg/mL to 5000 µg/mL. The reaction was started with the addition of 20 mM glucose 6-phosphate and incubated at 20 °C for 20 min. At the end of incubation, 900 µL of stop solution (0.42% ammonium molybdate in 1 N H_2SO_4, 10% SDS, and 10% ascorbic acid) was added to the reaction mixture and incubated at 45 °C for 20 min. The absorbances of reaction mixtures were obtained at 830 nm. Two independent experiments in triplicate were carried out.

4.10. Fructose 1,6-bisphosphatase Inhibition Assay

To measure the inhibition of FBPase activity, a previous procedure was followed [33,44]. Before starting the assay, the color reagent was prepared by adding 0.12% malachite green in 5 volumes of water and 1 volume of H_2SO_4. On the day of the assay, 1 volume of 7.5% molybdate the ammonium was mixed with 4 volumes of the color reagent. Then, 0.17% of TWEEN 20 was added. The reaction assay was prepared by adding buffer (5 µM EDTA, 5 mM $MgCl_2$, 50 mM Tris-HCl, pH 7.2) enriched with 0.1 mM fructose 1,6-bisphosphate and inhibitory samples (AMP or extract) at concentrations ranging from 2 µg/mL to 5000 µg/mL. Cytosolic supernatant (diluted 1:10) obtained from livers of Wistar rats was used to start the reaction, and then it was incubated for 15 min at 20 °C. After that, color reagent was added, and a second incubation was made for 10 min at 20 °C. The absorbances were obtained at 630 nm. Two independent experiments were performed in triplicate.

4.11. α-glucosidase Inhibition Assay

A modified method of a previously performed α-glucosidase assay was used [33,45]. First, in 100-µL reaction volumes, 0.1 M phosphate buffer at pH 6.8, inhibitor samples (acarbose (Aurax®, Mexico City, Mexico) or extracts) at concentrations ranging from 200 µg/mL to 2000 µg/mL, and an α-glucosidase solution prepared from rat intestinal acetone powder were placed and incubated at 35 °C for 3 min. Afterwards, 2 mM 4-nitrophenyl α-D-glucopyranoside (p-NPG) was added to start the enzymatic activity and, subsequently, the reaction was incubated at 35 °C for 30 min. Finally, the absorbances were determined at 405 nm. Two independent experiments in triplicate were performed.

4.12. Statistical Analysis

Data were represented as mean ± SEM. For the 3-h acute tests, data were assessed for normal distribution and log transformed as necessary. Then, ordinary one-way ANOVA and Tukey's post-hoc tests were carried out to compare the means among groups in each time, while repeated measures ANOVA and Dunnet's post-hoc tests were performed to compare the means with their baseline. The respective non-parametric tests were applied if normality was not obtained, even after log transformation. *P*-values less than 0.05 were considered significant. Additionally, the areas under the curve (AUC) of blood glucose were calculated and compared using ordinary one-way ANOVA and Tukey's post-hoc tests.

To obtain the IC_{50} values, absorbances were transformed to activity percentage as follows:

$$\text{Enzyme activity (\%)} = (A_S - A_{SB})/(A_C - A_{CB}) \times 100 \quad (1)$$

where, A_S is the absorbance of the inhibitor sample at a specific concentration, A_{SB} is the blank of the inhibitor sample, A_C is the highest absorbance (without inhibitor), and A_{CB} is the blank of the highest absorbance. Then, percentage values were plotted on concentration-response curves to find the best fitting non-lineal regression model (three or four parameters).

5. Conclusions

The traditional therapeutic use of the aerial part of *E. longifolium* and the rhizome of *A. firma* as hypoglycemic agents was confirmed in a hyperglycemic animal model. Furthermore, both plants inhibited two key enzymes involved in the liver glucose output, which mainly controls glucose levels in the fasting state. For *E. longifolium*, the presence of rosmarinic and chlorogenic acids could explain the bioactivity. Further studies are needed to discard other action mechanisms that can act in a synergistic way to produce the observed hypoglycemic effect.

Author Contributions: Conceptualization, A.A.-C.; methodology, A.A.-C.; formal analysis, F.E.-H., G.M.-T. and S.E.-R.; investigation, A.A.-C.; resources, A.A.-C.; data curation, F.E.-H., G.M.-T. and S.E.-R.; writing—original draft preparation, A.A.-C.; writing—review and editing, A.A.-C., F.E.-H., G.M.-T. and S.E.-R.; visualization, A.A.-C.; supervision, A.A.-C.; project administration, A.A.-C.; funding acquisition, A.A.-C. All authors have read and agreed to the published version of the manuscript.

Funding: This project was sponsored by DGAPA, PAPIIT IN226719.

Institutional Review Board Statement: The study was conducted according to the guidelines of the Declaration of Helsinki and approved by the Ethics Committee of School of Sciences, UNAM, Mexico City, Mexico (approval number P_2021_05_01, approval date 17 June 2021).

Informed Consent Statement: Informed consent was obtained from all subjects who participate in the interviews.

Data Availability Statement: Data are available upon request.

Acknowledgments: We acknowledge Rocio Trejo and Grisel Jeanette Pérez Martignon for their help with the animal experiments, Grisel Pérez Martignon for the *Eryngium* picture, Christian Alan Cabello-Hernández for maintaining the animals at the Bioterium, and Ramiro Cruz and Ernesto Velázquez Montes for the plant identification.

Conflicts of Interest: The authors declare no conflict of interest.

References

1. World Health Organization (WHO). *WHO Traditional Medicine Strategy 2014–2023*; World Health Organization (WHO): Geneva, Switzerland, 2013; pp. 1–76.
2. Andrade-Cetto, A.; Becerra-Jiménez, J.; Martínez-Zurita, E.; Ortega-Larrocea, P.; Heinrich, M. Disease-Consensus Index as a tool of selecting potential hypoglycemic plants in Chikindzonot, Yucatán, México. *J. Ethnopharmacol.* **2006**, *107*, 199–204. [CrossRef]
3. Andrade-Cetto, A. Ethnobotanical study of the medicinal plants from Tlanchinol, Hidalgo, México. *J. Ethnopharmacol.* **2009**, *122*, 163–171. [CrossRef]
4. American Diabetes Association 2. Classification and diagnosis of diabetes: Standards of medical care in diabetes-2021. *Diabetes Care* **2021**, *44*, S15–S33. [CrossRef]
5. International Diabetes Federation (IDF). *IDF Diabetes Atlas*; International Diabetes Federation (IDF): Brussels, Belgium, 2019; pp. 1–168.
6. Rizza, R.A. Pathogenesis of fasting and postprandial hyperglycemia in type 2 diabetes: Implications for therapy. *Diabetes* **2010**, *59*, 2697–2707. [CrossRef]
7. Ghani, U. Re-exploring promising α-glucosidase inhibitors for potential development into oral anti-diabetic drugs: Finding needle in the haystack. *Eur. J. Med. Chem.* **2015**, *103*, 133–162. [CrossRef]
8. CONABIO. *Capital Natural de México*; Sarukhán, J., Koleff, P., Carabias, J., Soberón, J., Dirzo, R., Llorente-Bousquets, J., Halffter, G., González, R., March, I., Mohar, A., et al., Eds.; Comisión Nacional Para El Conocimiento Y Uso de la Biodiversidad: Mexico City, Mexico, 2017; ISBN 9786077607021.
9. INEGI Lengua Indígena. Available online: https://www.inegi.org.mx/temas/lengua/ (accessed on 26 August 2021).
10. CONEVAL Pobreza en México. Available online: https://www.coneval.org.mx/socialdeprivationorincomeMedicion/Paginas/PobrezaInicio.aspx (accessed on 29 August 2021).
11. CONABIO Enciclo Vida. Available online: http://enciclovida.mx/especies/185075 (accessed on 3 September 2021).
12. Palá, J. *Contribución Al Conocimiento de Los Aceites Esenciales Del Género "Eryngium" L, En La Península Ibérica*; Universidad Complutense De Madrid: Madrid, Spain, 2002.
13. Uphof, J. *Dictionary of Economic Plants*; Weinheim, H. R. Engelmann (J. Cramer): New York, NY, USA, 1959.
14. Mabry, T.J.; Markham, K.R.; Thomas, M.B. *The Systematic Identification of Flavonoids*; Springer: Berlin/Heidelberg, Germany, 1970; pp. 165–167. ISBN 978-3-642-88460-3.
15. Cádiz-Gurrea, M.d.l.L.; Fernández-Arroyo, S.; Joven, J.; Segura-Carretero, A. Comprehensive characterization by UHPLC-ESI-Q-TOF-MS from an *Eryngium bourgatii* extract and their antioxidant and anti-inflammatory activities. *Food Res. Int.* **2013**, *50*, 197–204. [CrossRef]
16. Zhang, Z.; Li, S.; Ownby, S.; Wang, P.; Yuan, W.; Zhang, W.; Scott Beasley, R. Phenolic compounds and rare polyhydroxylated triterpenoid saponins from *Eryngium yuccifolium*. *Phytochemistry* **2008**, *69*, 2070–2080. [CrossRef] [PubMed]
17. OECD/OCDE (Testing Guidelines 425). *OECD Guidelines for the Testing of Chemicals: Acute Oral Toxicity—Up and Down Procedure (UDP)*; OECD Publishing: Paris, France, 2008; Volume 27.
18. Rojas Martínez, R.; Aguilar Salinas, C. Epidemiología de la diabetes mellitus en México. *Rev. Fac. Med. UNAM* **1994**, *37*, 15–28.
19. Cruz, E.C.; Andrade-Cetto, A. Ethnopharmacological field study of the plants used to treat type 2 diabetes among the Cakchiquels in Guatemala. *J. Ethnopharmacol.* **2015**, *159*, 238–244. [CrossRef] [PubMed]
20. Caballero-Serrano, V.; McLaren, B.; Carrasco, J.C.; Alday, J.G.; Fiallos, L.; Amigo, J.; Onaindia, M. Traditional ecological knowledge and medicinal plant diversity in Ecuadorian Amazon home gardens. *Glob. Ecol. Conserv.* **2019**, *17*, e00524. [CrossRef]
21. Barbosa-Filho, J.M.; Vasconcelos, T.H.C.; Alencar, A.A.; Batista, L.M.; Oliveira, R.A.G.; Guedes, D.N.; Falcão, H.d.S.; Moura, M.D.; Diniz, M.F.F.M.; Modesto-Filho, J. Plants and their active constituents from South, Central, and North America with hypoglycemic activity. *Rev. Bras. Farmacogn.* **2005**, *15*, 392–413. [CrossRef]
22. Andrade-Cetto, A.; Heinrich, M. Mexican plants with hypoglycaemic effect used in the treatment of diabetes. *J. Ethnopharmacol.* **2005**, *99*, 325–348. [CrossRef] [PubMed]
23. Escandón-Rivera, S.M.; Mata, R.; Andrade-Cetto, A. Molecules isolated from Mexican hypoglycemic plants: A review. *Molecules* **2020**, *25*, 4145. [CrossRef] [PubMed]
24. Dimitriadis, G.D.; Maratou, E.; Kountouri, A.; Board, M.; Lambadiari, V. Regulation of Postabsorptive and Postprandial Glucose Metabolism by Insulin-Dependent and Insulin-Independent Mechanisms: An Integrative Approach. *Nutrients* **2021**, *13*, 159. [CrossRef] [PubMed]
25. Tan, S.P.; Tan, E.N.Y.; Lim, Q.Y.; Nafiah, M.A. *Phyllanthus acidus* (L.) Skeels: A review of its traditional uses, phytochemistry, and pharmacological properties. *J. Ethnopharmacol.* **2020**, *253*, 112610. [CrossRef] [PubMed]
26. Feng, L.; Zhai, Y.Y.; Xu, J.; Yao, W.F.; Cao, Y.D.; Cheng, F.F.; Bao, B.H.; Zhang, L. A review on traditional uses, phytochemistry and pharmacology of *Eclipta prostrata* (L.). *J. Ethnopharmacol.* **2019**, *245*, 112109. [CrossRef]
27. Andrade-Cetto, A.; Cruz, E.C.; Cabello-Hernández, C.A.; Cárdenas-Vázquez, R. Hypoglycemic Activity of Medicinal Plants Used among the Cakchiquels in Guatemala for the Treatment of Type 2 Diabetes. *Evid.-Based Complement. Altern. Med.* **2019**, *2019*, 2168603. [CrossRef]
28. Silva-Rivas, R.; Bailon-Moscoso, N.; Cartuche, L.; Romero-Benavides, J.C. The antioxidant and hypoglycemic properties and phytochemical profile of *Clusia latipes* extracts. *Pharmacogn. J.* **2020**, *12*, 144–149. [CrossRef]

29. Gomes, J.H.d.S.; Mbiakop, U.C.; Oliveira, R.L.; Stehmann, J.R.; de Pádua, R.M.; Cortes, S.F.; Braga, F.C. Polyphenol-rich extract and fractions of Terminalia phaeocarpa Eichler possess hypoglycemic effect, reduce the release of cytokines, and inhibit lipase, α-glucosidase, and α-amilase enzymes. *J. Ethnopharmacol.* **2021**, *271*, 113847. [CrossRef]
30. Szkudelski, T. Streptozotocin-nicotinamide-induced diabetes in the rat. Characteristics of the experimental model. *Exp. Biol. Med.* **2012**, *237*, 481–490. [CrossRef]
31. Wang, P. Phytochemical Constituents and Pharmacological Activities of *Eryngium* L. (Apiaceae). *Pharm. Crop.* **2012**, *3*, 99–120. [CrossRef]
32. Ayuso, M.; Pinela, J.; Dias, M.I.; Barros, L.; Ivanov, M.; Calhelha, R.C.; Soković, M.; Ramil-Rego, P.; Barreal, M.E.; Gallego, P.P.; et al. Phenolic composition and biological activities of the in vitro cultured endangered *Eryngium viviparum* J. Gay. *Ind. Crop. Prod.* **2020**, *148*, 112325. [CrossRef]
33. Espinoza-Hernández, F.; Andrade-Cetto, A.; Escandón-Rivera, S.; Mata-Torres, G.; Mata, R. Contribution of fasting and postprandial glucose-lowering mechanisms to the acute hypoglycemic effect of traditionally used *Eryngium cymosum* F. Delaroche. *J. Ethnopharmacol.* **2021**, 114339. [CrossRef]
34. Saadeldeen, F.S.A.; Niu, Y.; Wang, H.; Zhou, L.; Meng, L.; Chen, S.; Sun-Waterhouse, D.; Waterhouse, G.I.N.; Liu, Z.; Kang, W. Natural products: Regulating glucose metabolism and improving insulin resistance. *Food Sci. Hum. Wellness* **2020**, *9*, 214–228. [CrossRef]
35. Xu, L.; Li, Y.; Dai, Y.; Peng, J. Natural products for the treatment of type 2 diabetes mellitus: Pharmacology and mechanisms. *Pharmacol. Res.* **2018**, *130*, 451–465. [CrossRef]
36. Lin, D.; Xiao, M.; Zhao, J.; Li, Z.; Xing, B.; Li, X.; Kong, M.; Li, L.; Zhang, Q.; Liu, Y.; et al. An overview of plant phenolic compounds and their importance in human nutrition and management of type 2 diabetes. *Molecules* **2016**, *21*, 1374. [CrossRef]
37. Jayanthy, G.; Subramanian, S. Rosmarinic acid, a polyphenol, ameliorates hyperglycemia by regulating the key enzymes of carbohydrate metabolism in high fat diet—STZ induced experimental diabetes mellitus. *Biomed. Prev. Nutr.* **2014**, *4*, 431–437. [CrossRef]
38. Yan, Y.; Zhou, X.; Guo, K.; Zhou, F.; Yang, H. Use of Chlorogenic Acid against Diabetes Mellitus and Its Complications. *J. Immunol. Res.* **2020**, *2020*, 1–6. [CrossRef]
39. National Research Council (US); Committee for the Update of the Guide for the Care and Use of Laboratory Animals. *Guide for the Care and Use of Laboratory Animals*, 8th ed.; The National Academies Press: Washington, DC, USA, 2011.
40. Masiello, P.; Broca, C.; Gross, R.; Roye, M.; Manteghetti, M.; Hillarire-Buys, D.; Novelli, M.; Ribes, G.; Hillaire-buys, D.; Novelli, M.; et al. Development of a new model in adult rats administered streptozotocin and nicotinamide. *Diabetes* **1998**, *47*, 224–229. [CrossRef]
41. Andrade-Cetto, A.; Medina-Hernández, A.E. Hypoglycemic effect of *Bromelia plumieri* (E. Morren) L.B. Sm., leaves in STZ-NA-induced diabetic rats. *Front. Pharmacol.* **2013**, *4*, 36. [CrossRef]
42. Arion, W.J. [7] Measurement of intactness of rat liver endoplasmic reticulum. In *Methods in Enzymology*; Fleischer, S., Fleischer, B., Abelson, J., Simon, M., Eds.; Academic Press: Cambridge, MA, USA, 1989; Volume 174, pp. 58–67.
43. Andrade-Cetto, A.; Cárdenas-Vázquez, R. Gluconeogenesis inhibition and phytochemical composition of two *Cecropia* species. *J. Ethnopharmacol.* **2010**, *130*, 93–97. [CrossRef] [PubMed]
44. Tashima, Y.; Yoshimura, N. Control of rabbit liver fructose-1, 6-diphosphatase activity by magnesium ions. *J. Biochem.* **1975**, *78*, 1161–1169. [CrossRef] [PubMed]
45. Escandón-Rivera, S.; González-Andrade, M.; Bye, R.; Linares, E.; Navarrete, A.; Mata, R. α-Glucosidase inhibitors from *Brickellia cavanillesii*. *J. Nat. Prod.* **2012**, *75*, 968–974. [CrossRef] [PubMed]

plants

MDPI

Article

Cytotoxic Effect In Vitro of *Acalypha monostachya* Extracts over Human Tumor Cell Lines

Gloria A. Guillén-Meléndez [1,†], Sheila A. Villa-Cedillo [1,†], Raymundo A. Pérez-Hernández [2], Uziel Castillo-Velázquez [3], Daniel Salas-Treviño [1], Odila Saucedo-Cárdenas [1,4], Roberto Montes-de-Oca-Luna [1], Christian A. Gómez-Tristán [1], Aimé Jazmín Garza-Arredondo [5], Diana Elisa Zamora-Ávila [6], María de Jesús Loera-Arias [1,*] and Adolfo Soto-Domínguez [1,*]

1 Departamento de Histología, Facultad de Medicina, Universidad Autónoma de Nuevo León, Monterrey C.P. 64460, NL, Mexico; gloria.guillenmln@uanl.edu.mx (G.A.G.-M.); svilla.me0121@uanl.edu.mx (S.A.V.-C.); daniel.salastr@uanl.edu.mx (D.S.-T.); odila.saucedocr@uanl.edu.mx (O.S.-C.); robertomontesdeocaln@uanl.edu.mx (R.M.-d.-O.-L.); christian.gomeztrs@uanl.edu.mx (C.A.G.-T.)

2 Departamento de Química, Facultad de Ciencias Biológicas, Universidad Autónoma de Nuevo León, San Nicolás de los Garza C.P. 64455, NL, Mexico; raymundo.perezhrz@uanl.edu.mx

3 Departamento de Inmunología, Facultad de Medicina Veterinaria, Universidad Autónoma de Nuevo León, Escobedo C.P. 66050, NL, Mexico; uziel.castillovl@uanl.edu.mx

4 Departamento de Genética Molecular, Centro de Investigación Biomédica del Noreste (CIBIN) del IMSS, Monterrey C.P. 66720, NL, Mexico

5 Departamento de Reproducción, Facultad de Medicina Veterinaria, Universidad Autónoma de Nuevo León, Escobedo C.P. 66050, NL, Mexico; aime.garzaarr@uanl.edu.mx

6 Departamento de Genética, Facultad de Medicina Veterinaria, Universidad Autónoma de Nuevo León, Escobedo C.P. 66050, NL, Mexico; diana.zamoravl@uanl.edu.mx

* Correspondence: mdjesus.loeraars@uanl.edu.mx (M.d.J.L.-A.); adolfo.sotodmn@uanl.edu.mx (A.S.-D.); Tel.: +52-(81)-8329-4195 (M.d.J.L.-A. & A.S.-D.)

† Equal contribution.

Citation: Guillén-Meléndez, G.A.; Villa-Cedillo, S.A.; Pérez-Hernández, R.A.; Castillo-Velázquez, U.; Salas-Treviño, D.; Saucedo-Cárdenas, O.; Montes-de-Oca-Luna, R.; Gómez-Tristán, C.A.; Garza-Arredondo, A.J.; Zamora-Ávila, D.E.; et al. Cytotoxic Effect In Vitro of *Acalypha monostachya* Extracts over Human Tumor Cell Lines. *Plants* **2021**, *10*, 2326. https://doi.org/10.3390/plants10112326

Academic Editors: Juei-Tang Cheng, I-Min Liu and Szu-Chuan Shen

Received: 30 September 2021
Accepted: 11 October 2021
Published: 28 October 2021

Abstract: *Acalypha monostachya (A. monostachya)* is a plant that is used in traditional medicine as a cancer treatment; however, its effect has not been validated. In this study, the potential cytotoxic effects and morphological changes of *A. monostachya* were evaluated in human tumor cell lines. The aqueous (AE), methanolic (ME), and hexane (HE) extracts were obtained, and flavonoid-type phenolic compounds were detected, which indicates an antineoplastic effect. We observed a time-dependent and concentration-selective toxicity in human tumor cells. Additionally, the ME and HE showed the greatest cytotoxic effect at minimum concentrations compared to the AE, which showed this effect at the highest concentrations. All extracts induced significant morphological changes in tumor cells. The HeLa (cervix carcinoma) cells were more sensitive compared to the MDA-MB-231 (triple-negative breast cancer) cells. In conclusion, we demonstrated a cytotoxic in vitro effect of *A. monostachya* extracts in tumoral human cell lines. These results show the potential antineoplastic effects of *A. monostachya* in vitro. Hereafter, our lab team will continue working to usefully isolate and obtain the specific compounds of *A. monostachya* extracts with cytotoxic effects on tumor cells to find more alternatives for cancer treatment.

Keywords: human tumor cell lines; cytotoxic effect; plant extracts; *Acalypha monostachya*

1. Introduction

According to the World Health Organization, cancer is defined as a process of uncontrolled cell growth and dissemination that can appear practically anywhere in the body [1]. In 2020, there were 19,292,789 cases of cancer presented around the world, of which 9,958,133 resulted deaths from this pathology [2]. In 2020 in Latin America and the Caribbean, the incidence of cancer in both sexes was 1,470,274 cases, and there was a mortality of 713, 414 [2]. In Mexico, there were 195,499 new cases and 90,222 deaths in

2020 [3]. Currently, there are multiple treatments for cancer depending on the type and stage of cancer. The most common methods are surgery, radiation therapy, and chemotherapy; other methods include immunotherapy, targeted therapy, hormone therapy, stem cell transplantation, and precision medicine [4].

For centuries, the knowledge surrounding the medicinal usage of plants has been passed down from generation to generation, and this knowledge has evolved based on observations, experience, and trial and error experiments. One of the advantages of the use of medicinal plants is their availability and the fact that they are culturally acceptable, as these plants are commonly used by indigenous populations because of these qualities [5].

Acalypha is the fourth largest genus in the *Euphorbiaceae* family, with approximately 450 to 570 species. Many of these species are used as medicinal plants, with their use occurring mainly in Africa and on the Mascarene Islands. The entire plant, including the leaves, stems, and roots, is commonly used in traditional remedies [6]. The leaves of the *Acalypha* species are succulent with sap stems, which tend to fall off with age. They are alternate, petiolate, or subsessile and have an entire sheet, crenate, or are toothed. The staminate flowers have four to eight stamens and vermiform anthers. The pistillate flowers are often prominently bracts with three sepals, three carpels, and one ovule per carpel and each divided style. Several species of *Acalypha* share the characteristic of allomorphic pistillate flowers and fruits [7].

Most *Acalypha* species are used as medicinal plants in West and East Africa, especially in Nigeria. Each part of the plant, including the leaf, stem, and roots, is used to make mixtures and decoctions to treat various ailments. *Acalypha* species such as *A. wilkesiana Müll. Arg.*, *A. communis Müll. Arg.*, and *A. indica* L. are used in folk medicine as diuretics, anthelmintics, and for respiratory problems such as bronchitis, asthma, and pneumonia [8]. *A. wilkesiana*, *A. indica*, and *A. hispida Burm.f.* are common species found in Mauritius [9]. The Mauritian population uses *A. indica* leaves as well as whole plant as a treatment for skin infections such as scabies and dermatitis; *A. wilkesiana* is used to control diabetes, dysentery, and asthma. *A. integrifolia* is used as an astringent, purgative, and to eliminate intestinal worms as well as to cure various skin infections [9,10].

Acalypha monostachya (*A. monostachya*) is a perennial herb found in the southwestern United States and Mexico. It is used as a medicinal plant by the inhabitants of San Rafael and Zapotitlán Salinas, and Puebla, Mexico against skin eruptions, wounds, and diarrhea. In northern Mexico, it can be found in Bustamante, Nuevo León and, is commonly called the "Cancer Herb" and is used in the form of an infusion by boiling the leaves and inflorescences; it is also used in conjunction with *Bougainvillea glabra* for the treatment of colic and external tumors [11]. A study with a methanol extract (ME) of *A. monostachya* showed antimicrobial and antioxidant activities as well as toxicity against *A. salina* [12]. Another study evaluated the presence of micro and macronutrients in *A. monostachya* and concluded that it contains large amounts of Mg^{+2}, $Fe^{+3,}$ and Zn^{+2}, so an analysis of its potential antioxidant effect is recommended [13,14]. This may also be confirmed by the fact that strong zinc accumulation has been detected in a closely related species (*A. alopecuroidea*) [15].

It has been observed by a phytochemical screening and through antimicrobial effect tests that the aqueous extract (AE) of *A. monostachya* contains carbohydrates and flavonoids as well as antimicrobial activity against *Pseudomonas aeruginosa* and *Staphylococcus aureus*. In the ethanolic extract (EE), positivity for flavonoids and terpenes and antimicrobial activity against the aforementioned bacteria and *Escherichia coli* were obtained [16].

Other studies have demonstrated the anticancer activity of the *Acalypha* genus, methanolic (ME), hexane (HE), and chloroform extracts of *A. indica*. These extracts were not cytotoxic to Vero (non-tumor kidney) cells and exhibited anticancer activity against NCIH187 (lung carcinoma) cells. Likewise, L-quebrachitol was purified from the extract, which was characterized by NMR [17].

A moderate cytotoxic effect of *A. fruticose* AE was demonstrated, which inhibited the proliferation of MDA-MB-435S (melanoma cell line) and Hep3B (hepatocellular carcinoma) cells in addition to the protection of DNA against oxidative damage induced with H_2O_2 [18]. *A. wilkesiana* exhibits an antiproliferative effect on U87MG (likely glioblastoma cells), A549 (lung carcinoma cells), and MCR5 (non-tumor lung) cells with their ethyl acetate extract, and a morphological study confirmed apoptosis and DNA damage [19]. The growth inhibition effect on MDA-MB-468 and MCR5 cells has been evaluated using the EE of *A. wilkesiana* [20].

Ethyl acetate extract (EE) combined with β, γ, and δ tocotrienols treatments of *A. wilkesiana* has potent antiproliferative effects on A549 and U87MG cells [21]. Studies indicate cytotoxic activity in HepG2 (hepatocellular carcinoma) and MCF7 (breast carcinoma) cells using EE and *A. wilkesiana* fractions showing 75.8% and 87.1% inhibition, respectively. Antioxidant activity was also observed by performing a DPPH test [22]. On the other hand, the anti-proliferative activity of the ME and fractions of *A. californica* were analyzed. Particularly, the hexanoic fraction inhibited RAW 264.7, HeLa, and L929 cells. This fraction was analyzed by molecular exclusion chromatography, obtaining terpenes and steroids, and its residual fraction contains tannins. Through HPLC, the presence of the compounds β-sitosterol and sigmasterol was demonstrated [23].

At this time, the use of medicinal plants as therapeutic agents is widely extended; *A. monostachya* is a plant that is used in traditional medicine as a cancer treatment; however, its effect has not been validated. For this reason, in the present study, we analyzed whether this plant has an in vitro antineoplastic effect on human tumor cell lines, specifically on two of the most frequent types of cancer that affect women in the world: breast cancer and cervical cancer.

2. Results

2.1. Taxonomic Identification of the Plant

The plant *A. monostachya* was identified as a perennial herb that is approximately 10 to 40 cm tall with numerous branching stems with blades that are the same length and width as the petioles (0.5 cm to 2.5 cm long). It has red male and female inflorescences on the same plant: the terminal male ones, and the axillary or terminal female ones.

2.2. Extract Yields and Phytochemical Screening

The yields obtained from the extracts AE, ME, and HE by the maceration method were 9.8%, 10.7%, and 2.2% *w*/*w*, respectively. All three extracts showed a green-brown color. Phytochemical analysis showed the presence of unsaturations, phenols, coumarins, lactones, flavonoids, saponins, aromatic compounds, carbohydrates, and carbonyl groups in the three extracts. Only ME and HE showed the presence of steroids and terpenoids. AE and ME were positive for sesquiterpene lactones and only AE showed alkaloids (Table 1).

Table 1. Partial phytochemical screening of the crude extracts of *A. monostachya*.

Phytoconstituents	Test	Observations	AE	ME	HE
Unsaturations	Potassium permanganate test	Brown precipitate	+++	++	+
Phenols	Ferric chloride test	Green color	++	+++	+
Terpenoids and steroids	Salkowski test	Reddish-brown ring formation	−	++	++
Coumarins and lactones	Sodium hydroxide test	Yellow color	++	++	+
Sesquiterpene lactones	Baljet's test	Orange color	+	+	−
Flavonoids	Sulphuric acid test	Reddish color	+	+	++
Alkaloids	Dragendorff test	Orange precipitate	+	−	−
Saponins	Foam test	Presence of stable foam	+++	+	+
Carbohydrates	Molisch´s test	Purple ring formation	+++	++	+
Aromatic compounds	Formaline test	Red color	+	+	+
Carbonyl group	2-4 dinitrophenylhy-drazine test	Orange color	++	+	+

− Not detected; + slightly positive reaction; ++ positive reaction; and +++ strong positive reaction. The first columns indicate the name of the secondary metabolite and the test to perform the detection. The next column shows the observations for each reaction followed by the results obtained in the extracts. Aqueous extract (AE), methanolic extract (EM), hexanoic extract (HE).

The results obtained in the phytochemical analysis for each one of the extracts were compared with negative control, which was made up of the solvent used in each test, and subsequently, the corresponding reagents were added without the presence of any crude extract. Then, through a semi-quantitative analysis based on colorimetric and precipitation reactions, the presence of different groups of secondary metabolites was determined. The results were represented semi-quantitatively with crosses according to the intensity observed in each reaction (Table 1).

2.3. *A. Monostachya Extracts Induce Morphological Changes in Cultured Human Tumor Cells*

To determine the cytotoxic effect of the extracts on tumor and non-tumor cells, light microscopy observations were performed after treatments at concentrations of 0, 10, 50, 50, 100, 300, and 500 μg/mL for 12, 24, 48, and 72 h. Figures 1–3 show bright-field micrographs of Vero, HeLa, and MDA-MB-231 cell lines, respectively, after exposure to the concentrations of the extracts for 24 h.

Concentration-dependent changes were observed upon exposure to the three extracts, mainly in tumor cells; HeLa cells were the most affected after 24 h (Figures 1–3) compared to MDA-MB-231. These changes were accentuated in HE-treated cells followed by those treated with ME and AE. On the other hand, reduced confluence was observed when tumor cells were treated for 12 h with the highest concentration (500 μg/mL), mainly with the HE, suggesting a cytotoxic and antiproliferative effect. Morphological changes were also observed as the cellular contraction, acquiring a rounded shape caused by the loss of adhesion. At 48 and 72 h, confluence decreased considerably, and morphological changes increased, suggesting a time-dependent effect.

These results were compared with those of the control cells, which were treated with vehicle; no such changes occurred in the control cells. In non-tumorigenic Vero cells, a slight decrease in confluency was observed at a concentration of 500 μg/mL after 24 h.

2.4. *A. Monostachya Extracts Induce Changes in the Nuclear Morphology of Human Tumor Cells*

To evaluate confluence variations caused by the effect of the extracts, the cell adhesion area to the plate was quantified by a nuclei labeling assay with DAPI, proving the relationship between adherence and cell viability. At 12 h, the results were consistent with those observed by light microscopy. Tumor cells showed changes in nuclear morphology corresponding to cell death, described as chromatin condensation (karyolysis), rounding, decrease in size (pyknosis), and an intense fluorescence signal (Figure 4).

Likewise, after calculating the percentage of the area, the results were represented in graphs showing the significant differences vs. Vero cells (Figure 5) and the behavior of the cells against the three extracts at the different times: 12, 24, 48, and 72 h (Figure 6). It was observed that HE significantly decreased the percentage of area at 24 h when using the concentrations of 300 and 500 µg/mL. In the case of HE and ME, this effect was observed at 24 h, starting at the 10 µg/mL concentration, and was accentuated in HE. Significant differences were observed vs. Vero cells, suggesting selective toxicity for tumoral cells.

2.5. A. Monostachya Extracts Decrease the Viability in Human Tumor Cells

In addition to quantifying viable cells by establishing a relationship with cell adhesion, we decided to perform an MTT assay to correlate mitochondrial activity with viability and thus observe cell energy metabolism behavior against the extracts. In this case, viable cells were able to metabolize the tetrazolium salt and reduce it to obtain the formazan, a purple-blue compound, whose absorbance readings allowed us obtain the following results: At 24 h, the results were similar to those observed at 12 h: AE showed significant differences against the control cell line when treated with the concentration of 300 µg/mL, and interestingly, ME and HE showed a cytotoxic effect from the concentration of 50 µg/mL onwards. At this exposure time, a concentration-dependent response began to be observed. In contrast, MDA-MB-231 cells showed resistance and, in some cases, higher relative viability than the Vero cell line (Figure 7). The longer the exposure time (48 and 72 h) to the three extracts, the more decreased the relative viability was (Figure 8).

The behavior of the cells with the extracts was compared as a function of time, and it should be noted that the HeLa cells, after 24 h, presented a lower percentage of relative viability after exposure to AE, ME, and HE. The MDA-MB-231 cells showed irregular behavior at 12 and 24 h. These results indicate a selective time- and concentration-dependent cytotoxic effect (Figure 6).

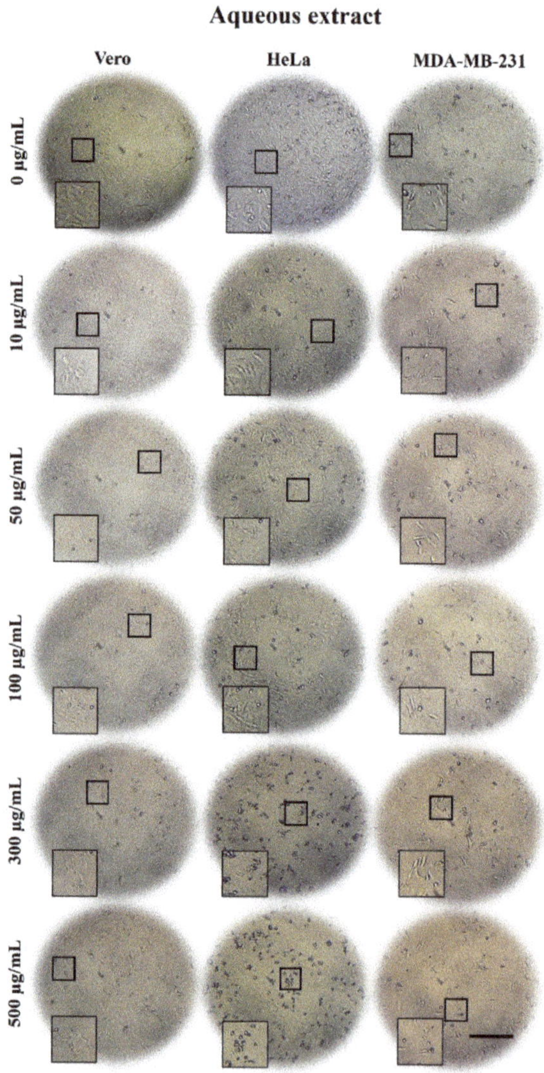

Figure 1. Micrographs of Vero, HeLa, and MDA-MB-231 cells exposed for 24 h to 0, 10, 50, 100, 300, and 500 µg/mL of *A. monostachya* AE. The tumor cells showed decrease confluence and morphological alterations such as rounding and loss of adhesion upon exposure to the treatments. These changes were accentuated at concentrations of 300 µg/mL and 500 µg/mL. HeLa cells showed greater changes compared to MDA-MB-231 and Vero cells, which showed minimal changes at this time point. These results were time- and concentration-dependent. Scale bar = 200 µm.

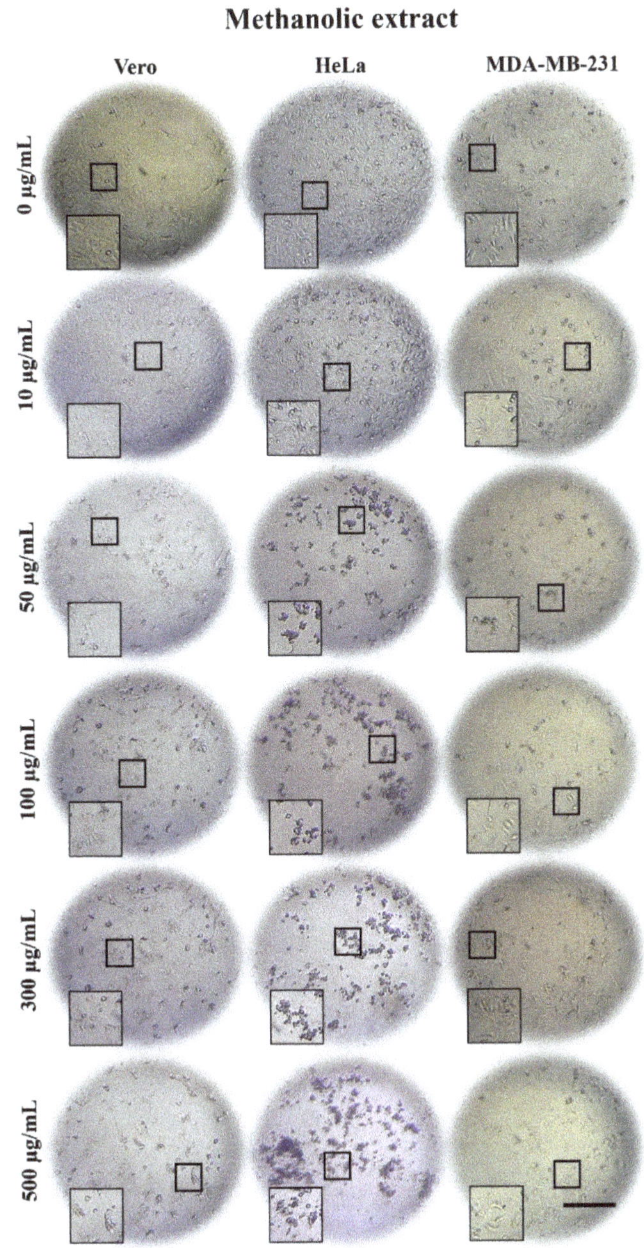

Figure 2. Micrographs of Vero, HeLa, and MDA-MB-231 cells exposed for 24 h to 0, 10, 50, 100, 300, and 500 µg/mL of *A. monostachya* ME. The tumor cells showed a marked decrease in confluence and morphological alterations as well as loss of adhesion after exposure to the extract. These changes were present from the concentrations of 50 µg/mL and higher. The HeLa cells showed more evident changes compared to Vero cells, which showed minimal changes, followed by the MDA-MB-231 cells. These alterations were time- and concentration-dependent. Scale bar = 200 µm.

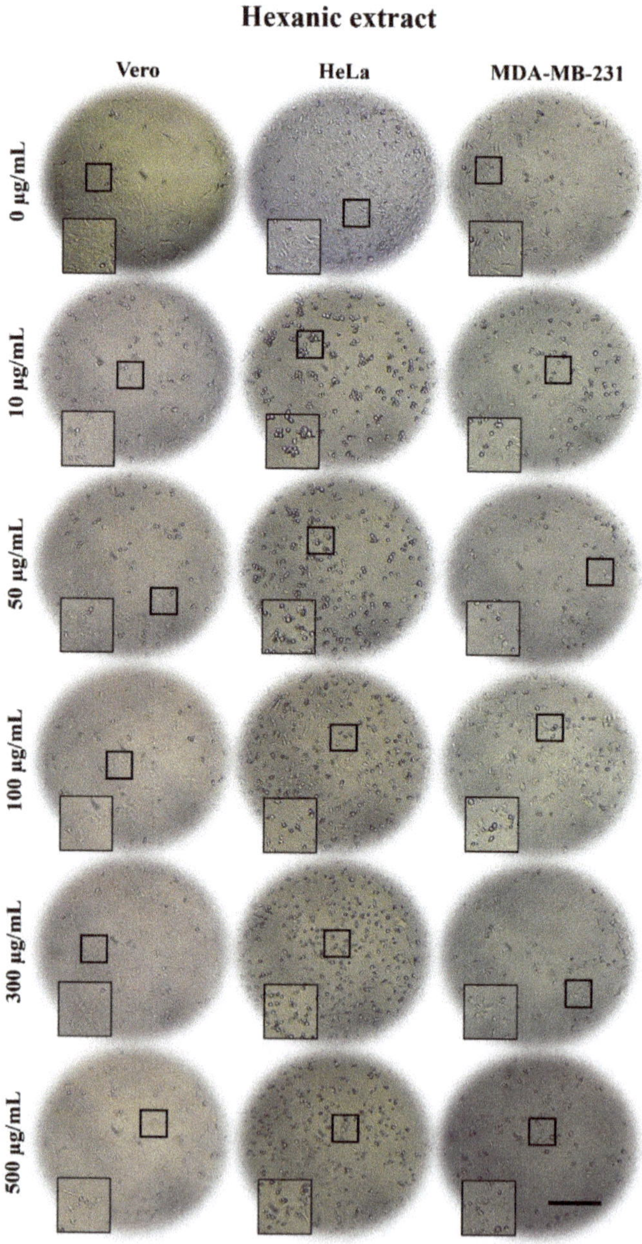

Figure 3. Micrographs of Vero, HeLa, and MDA-MB-231 cells exposed for 24 h to 0, 10, 50, 100, 300, and 500 µg/mL of *A. monostachya* HE. The tumor cells showed evident morphological alterations such as rounding and loss of adhesion as well as a marked decrease in confluence after exposure to the lowest concentration (10 µg/mL) of extract. The HeLa cells showed more evident changes vs. Vero cells, followed by the MDA-MB-231 cells, which showed these changes at higher concentrations. These alterations were time- and concentration-dependent. Scale bar = 200 µm.

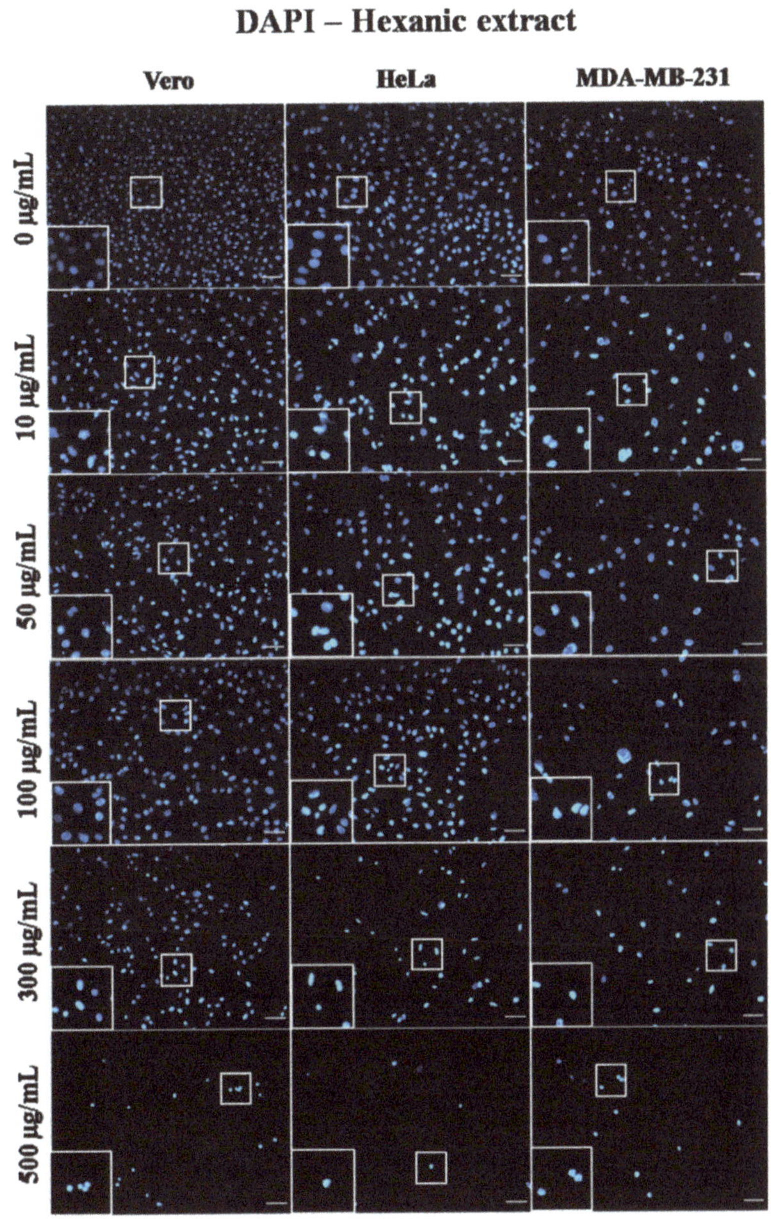

Figure 4. Micrographs of Vero, HeLa, and MDA-MB-231 cells exposed for 24 h to 0, 10, 50, 100, 300, and 500 µg/mL of *A. monostachya* HE. Nuclei labeling with DAPI. The nuclei of tumor cells showed obvious morphological alterations, such as pyknosis, rounding, and intense fluorescence as well as a marked decrease in confluence after exposure to the lowest concentration (10 µg/mL) of the extract. The HeLa cells showed more evident changes compared to Vero cells followed by the MDA-MB-231 cells, which showed these changes at higher concentrations. These alterations were also time- and concentration-dependent. Scale bar = 50 µm.

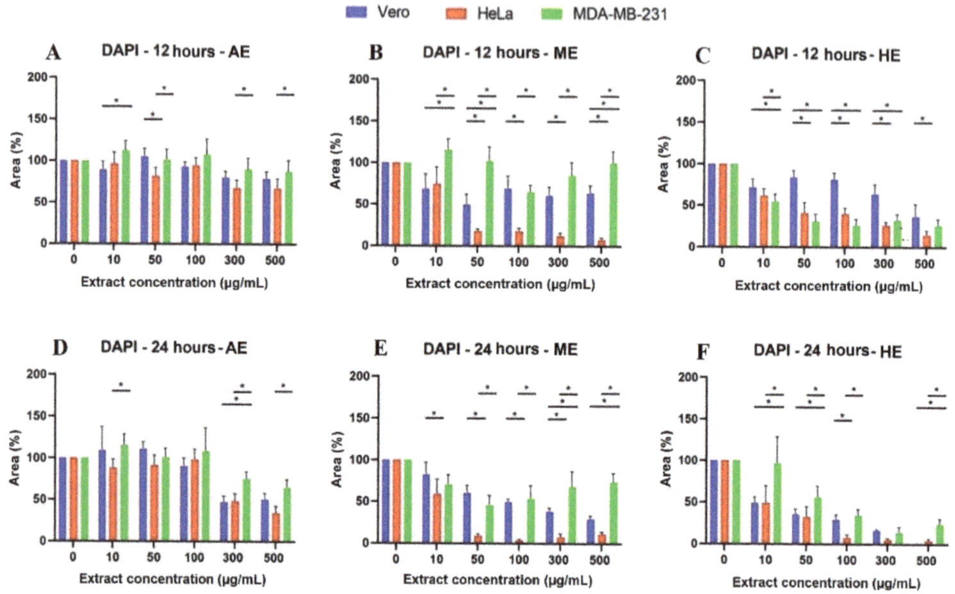

Figure 5. Quantitative analysis of nuclei labeling with DAPI in cells treated for 12 and 24 h with *A. monostachya* extracts. Cell area percentages are shown for Vero, HeLa, and MDA-MB-231 lines. HeLa and MDA-MB-231 lines exposed to concentrations of 0, 10, 50, 50, 100, 300, and 500 µg/mL of AE (*A. monostachya*) are shown for µg/mL of AE (**A**,**D**), ME (**B**,**E**), and HE (**C**,**F**). $n = 8$. * $p < 0.05$, Tukey's multiple comparison test vs. Vero cells.

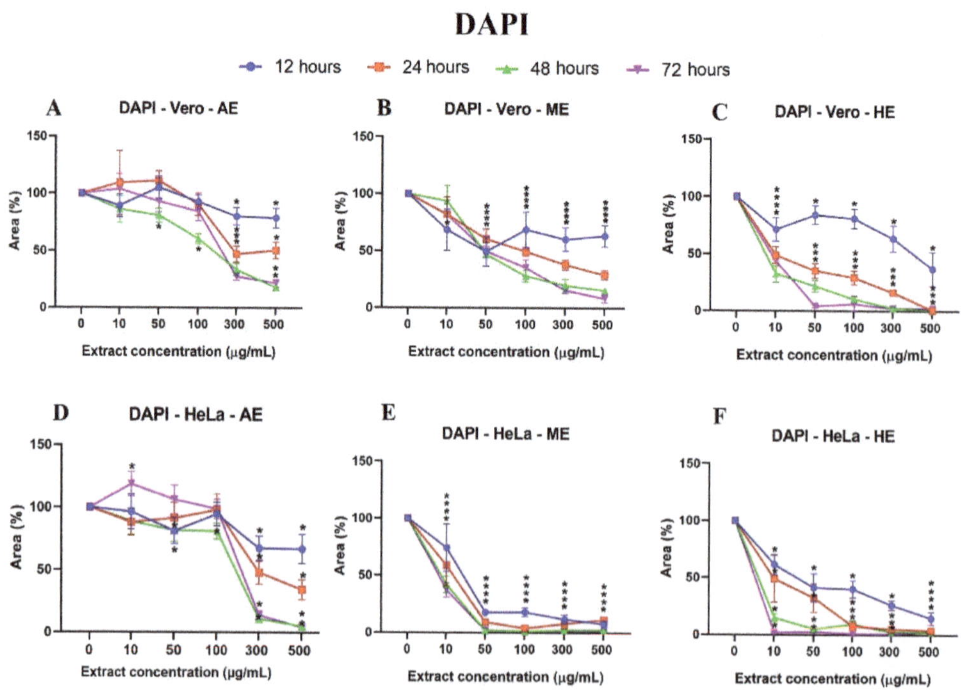

Figure 6. *Cont.*

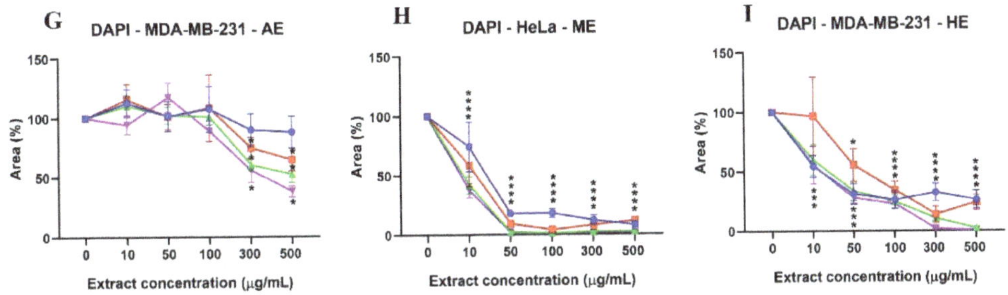

Figure 6. Cell behavior against different concentrations of *A. monostachya* extracts by nuclei labeling with DAPI. Comparisons of the effects of AE, ME, and HE on the cell lines: Vero (**A–C**), HeLa (**D–F**), and MDA-MB-231 (**G–I**) after exposure to the extracts at 12 h (blue line), 24 h (red line), 48 h (green line), and 72 h (purple line). The graphs show the % area as a function of the concentration of the extracts. It can be seen that HE has a greater cytotoxic effect on the tumor cell lines at 24 h and later. The HeLa cells showed a lower percentage of the area after exposure to the extracts. * $p < 0.05$, Tukey's multiple comparison test vs. vehicle-treated cells.

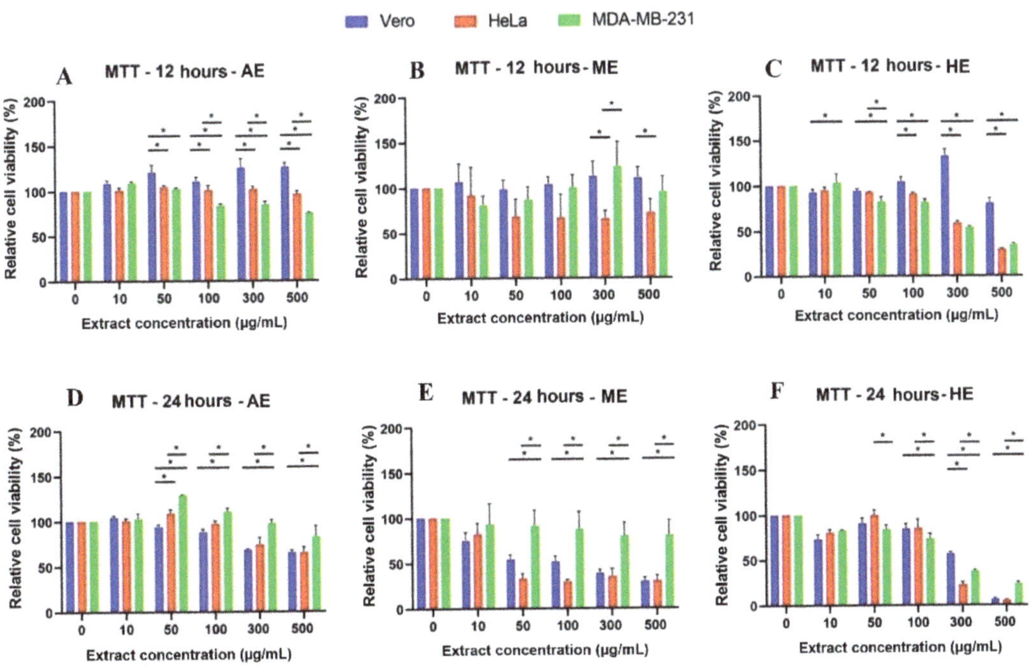

Figure 7. Cytotoxicity analysis by MTT assay in cells treated for 12 and 24 h with different concentrations of *A. monostachya* extracts. Relative cell viability percentages are shown for Vero, HeLa, and MDA-MB-231 lines. HeLa and MDA-MB-231 lines exposed to concentrations of 0, 10, 50, 50, 100, 300, and 500 µg/mL of AE (*A. monostachya*) are shown µg/mL of AE (**A**,**D**), ME (**B**,**E**), and HE (**C**,**F**). $n = 8$. * $p < 0.05$, Tukey's multiple comparison test vs. Vero cells.

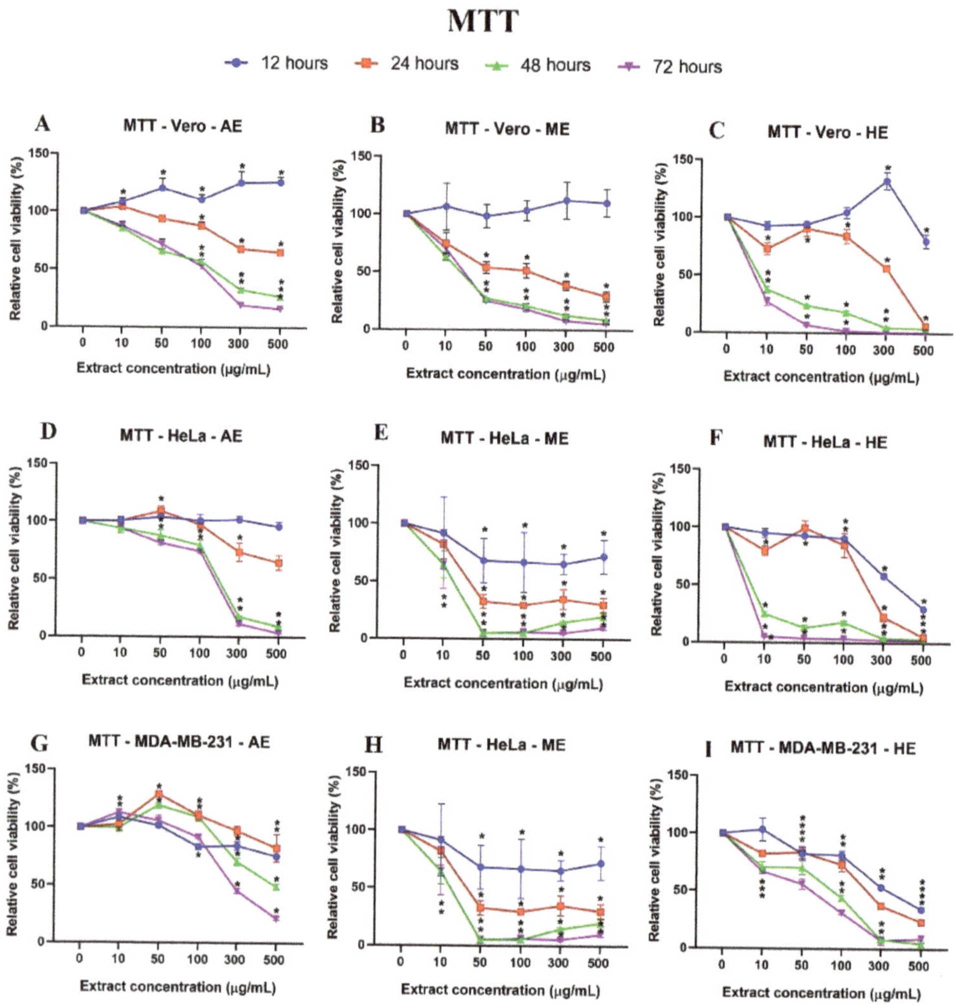

Figure 8. Cell behavior against different concentrations of *A. monostachya* extracts by MTT assay. Comparisons of the effects of AE, ME, and HE on the cell lines: Vero (**A–C**), HeLa (**D–F**), and MDA-MB-231 (**G–I**) after exposure to the extracts at 12 (blue line), 24 (red line), 48 (green line), and 72 h (purple line) are presented. The graphs show the % area as a function of the concentration of the extracts. It can be seen that HE has a greater cytotoxic effect on the tumor cell lines at 24 h and later. The HeLa cells showed greater sensitivity to the extracts. * $p < 0.05$, Tukey's multiple comparison test vs. vehicle-treated cells.

2.6. Human Tumor Cells Show More Morphological and Nuclear Alterations upon Exposure to A. Monostachya Extracts

After evaluating the morphological changes by light microscopy, fluorescence microscopy, and cytotoxicity by MTT assay, a morphological analysis was performed on semi-thin sections. For this purpose, the effect was evaluated at the concentrations at which the major morphological and viability changes were observed in the previous assays: for AE 300 and 500 µg/mL, and for ME and HE, the 10 and 50 µg/mL concentrations were used, respectively.

First, the morphological characteristics of the untreated cells were evaluated, the Vero cells showed heterochromatic nuclei, a prominent nucleolus, and homogeneous cytoplasm. The tumor cells presented similar characteristics, but multiple mitotic events were observed, such as larger cellular and nuclear size compared to control, some presented two nucleoli. Moreover, the treated cells, especially tumor cells, showed vacuolizations in the cytoplasm when exposed to the lowest concentration that was tested (50 μg/mL and 300 μg/mL) of the three extracts (Figure 9).

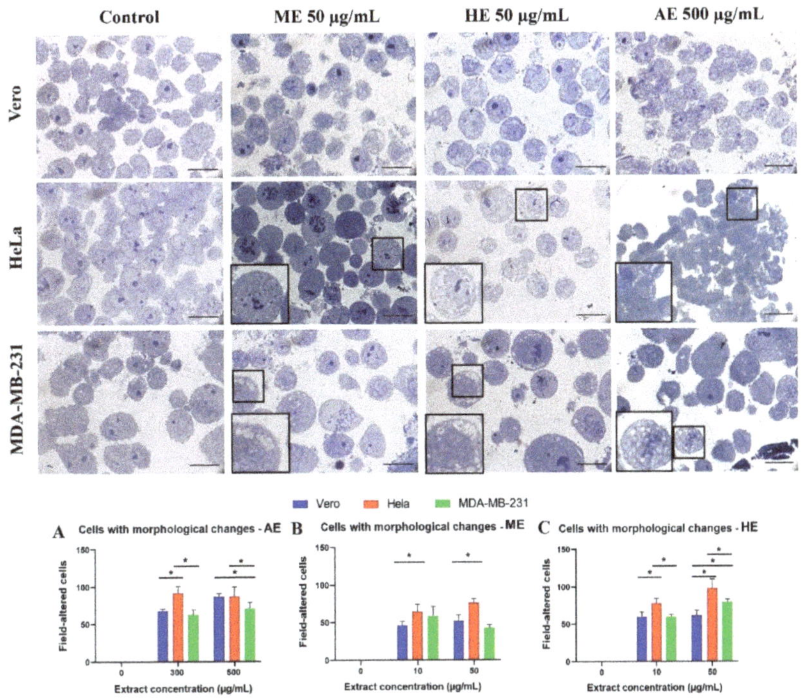

Figure 9. Evaluation of morphological changes in semi-thin sections stained with toluidine blue obtained from cells treated for 24 h with *A. monostachya* extracts. Note that upon exposure to the extracts, changes in nuclear morphology (condensation and displacement of chromatin towards the periphery), and cytoplasmic morphology (accumulation of vacuoles) were observed. Alterations in cell size were also observed, especially in tumor cells treated with HE. The HeLa cells showed more evident changes vs. the Vero cells followed by the MDA-MB-231 cells. The amplifications show changes in nuclear morphology and cytoplasm. Scale bar = 20 μm. Quantization of cells with morphological alterations related to cell death after exposure to *A. monostachya* extracts. Vero (blue bar), HeLa (red bar), and MDA-MB-231 (green bar) both after exposure to concentrations of 0, 10, 50, 300, and 500 μg/mL of AE (**A**), ME (**B**), and HE (**C**). $n = 8$. * $p < 0.05$, Tukey's multiple comparison test vs. Vero cells.

Effects such as chromatin condensation were accentuated at the highest concentration were also observed and tended to be concentrated at the periphery of the nucleus, indicating that these results also occurred in a concentration-dependent manner. Interestingly, tumor cells treated with 50 and 500 μg/mL of the three extracts showed greater morphological alterations. Subsequently, cells showing morphological changes indicative of cell death were quantified. The results were in agreement with previous assays where it was observed that the HeLa cells presented significant differences vs. the control as well as a higher number of cells with morphological changes related to cell death compared to MDA-MB-231 and Vero cells (Figure 9).

3. Discussion

Plant secondary metabolites are chemical compounds produced by the plant cell through metabolic pathways derived from primary metabolism. They have been shown to have biological effects that provide the basis for traditional medicine in ancient communities [24]. In the results obtained in this study, the presence of flavonoid-type phenolic compounds stands out, which coincides with previous studies that also detected phenolic compounds in the ME of *A. monostachya*, from which benzoic acids, flavones, and flavonols were isolated. In addition, unsaturated fatty acids (linolenic and linoleic acid) were isolated from the HE [12]. These compounds are related to antimicrobial and antioxidant activity, and they are also possibly related to the cytotoxic activity demonstrated in our study.

Concerning the other *Acalypha* species, other types of compounds were also detected in the AE and ME, which are obtained from the leaves, stems, and roots of *A manniana*, *A. hispida*, *A. racemosa*, *A. marginata*, *A. indica*, and *A. alnifolia* that correspond to alkaloids, phenols, flavonoids, hydroxyanthraquinones, saponins, terpenes, tannins, and steroids, highlighting the presence of flavonoids in the whole plant [25–28]. These results coincide with our study by detecting groups of similar compounds in *A. monostachya*. The hexane fraction obtained from a crude ME of *A. hispida* showed the presence of flavonoids, carbohydrates, anthraquinones, cardiac glycosides, proteins, alkaloids, and the absence of tannins, sterols, and saponins [29]. In contrast, in our study, saponins were detected in low quanties in *A. monostachya*. In this regard, variations of secondary metabolites that occur between each species or even between the same species were found and are due to various factors, including the season, age, phenological state of the plant, water status, amount of nutrients, geographical location, and even the conditions in which the plant was collected [30]. These factors could explain the differences in the metabolites detected in our study.

Moreover, it has been shown that *A. indica* contains non-polar compounds with anti-inflammatory and anticancer activity in the HE and ethyl acetate extract showing the powerful inhibition of lipoxygenases and cyclooxygenases [31]; these compounds may be related to the high cytotoxic activity detected in the HE from *A. monostachya*. Likewise, the ME of *A. indica* contains L-Quebrachitol, a compound used for the synthesis of anticancer drugs [15], which may also be related to the activity of the ME in our study.

The morphological changes observed with light microscopy after the treatments that were previously described in response to the exposure of plant extracts [32]. These alterations can be described as membrane contraction, rounding, loss of contact with neighboring cells, the formation of membrane blisters, and apoptotic bodies, which are related to cell death due to apoptosis [32]. This indicates that, in this study, the observed cellular response is mainly attributed to the phenomenon of cellular death induced by the extracts of *A. monostachya*; however, more studies are required to confirm our results. In this aspect, the MTT assay is one of the most widely used methods to analyze cell proliferation and viability; it is absorbed by endocytosis and reduced by mitochondrial enzymes and endosomal/lysosomal compartments in order to be transported to cell surfaces, forming needle-shaped crystals of formazan. It has been previously reported that MTT does not cause injury or induce cell death [33]. The results of our study correlate with previous studies describing that the crude HE of *A. indica* has an IC50 of 50 μg/mL, and cytotoxic effects were observed from the concentration of 10 μg/mL to 100 μg/mL, where a slight increase in viability was observed [34].

Previous reports show that the *Acalypha* extracts obtained from non-polar solvents exhibited the highest antiproliferative activity and selective toxicity toward tumor cell lines [35] as well as the extracts of *A. monostachya*. This also suggests that the bioactive compound(s) of *Acalypha* are chemically non-polar compounds. One of the typical characteristics of cell death due to apoptosis is chromatin shrinkage and condensation, which was appreciated in our analysis with DAPI staining. Furthermore, it is reported that apoptotic cells present nuclear condensation and chromatin cleavage upon exposure to plant extracts [36]. These results were also observed in our investigation, with the DAPI nuclear staining of the tumor cells being exposed to the three extracts of *A. monostachya*. These results also suggest that the type of cell death that is induced is oriented towards apoptosis, according to our observations.

As mentioned above, among the morphological changes found by these extracts are those related to apoptosis, such as DNA fragmentation. Among the possible mechanisms of action that are proposed, it has been found that *A. wilkesiana* must simultaneously activate different mechanisms that cause single-stranded and double-stranded DNA breaks that lead to apoptosis. The constituents of the extracts can trigger different apoptotic pathways in different cancer phenotypes and can be specific to the cancer cell [19]. In accordance with these reports, we considered that the components of our extracts could also be inducing apoptosis by inducing DNA damage, as observed in the DAPI nuclei contrast assay; nevertheless, we need further molecular analysis to verify these findings.

On the other hand, the ME, HE, and chloroformic extracts of *A. indica* were not cytotoxic for Vero cells, but they were toxic for the tumor NCIH-187 cell line [17]; this agrees with our results, where we observed that the ME and HE of *A. monostachya* did not show a cytotoxic effect on Vero cells either. Interestingly, an antiproliferative effect of the AE of *A. fruticosa* on MDA-MB-435s cells has been reported [18]. In our study, this effect was not observed for *A. monostachya* AE. This could be due to the differences between species of the same genus, *Acalypha*.

Furthermore, cell morphology changes when the homeostasis conditions of the cells are affected by different processes, including cell death. These morphological alterations involve both the nucleus and the cytoplasm and are similar in all types of cells and species [37]. The main characteristics of apoptosis are the condensation of chromatin and nuclear fragmentation; the condensation forms a structure in the shape of a crescent or ring [38], which was observed in the semi-fine sections and with the DAPI stain. A characteristic of cell death due to apoptosis occurs when cells lose contact with neighboring cells and begin to form bulges on the plasma membrane known as bullae; the cells shrink, and finally, the bullae become well-known apoptotic bodies [39]. These characteristics were appreciated using light microscopy in a concentration-dependent manner.

Furthermore, necrosis differs in that it presents various morphological characteristics, such as dilation of the organelles, and in some cases, chromatin condensation and inflammation occur. In the case of autophagic and non-lysosomal cell death, the first is characterized by numerous vacuoles in the cytoplasm being filled with cellular debris, and the second shows the dilatation of organelles and empty spaces. Unlike apoptosis, the cell membrane becomes permeable very early on [40–42]. Interestingly, the results observed in the semi-fine sections indicate that the tumor cells treated with the extracts of *A. monostachya* have characteristics of both types of cell death, apoptosis, and necrosis.

There are also mitochondrial alterations that are described as markers of early apoptosis. When the permeability of the inner mitochondrial membrane is lost, fluid enters causing dilation, which has been observed to be similar to vacuolization under light microscopy [43]. The foregoing tests indicate that based on the observations of the alterations in nuclear morphology in greater detail and to know the composition of the vacuoles observed in our study, the evaluatation of these changes using transmission electron microscopy is suggested. There are other types of non-apoptotic cell death, which have been recently described that also involve the formation of cytoplasmic vacuoles, which are related to exposure to ischemic injury, cytotoxic compounds, or pathogens; these are paraptosis, oncosis, and methuosis [44]. Its characteristics include inflammation and cytoplasmic vacuolization. Taken together, the evidence described above suggests that it is necessary to perform tests that indicate the type of cell death that takes place in response to exposure to the extracts.

The MDA-MB-231 cell line showed sensitivity to phenanthrene derivatives isolated from *Juncus gerardii* [45]. The cytotoxicity observed and the resistance of the MDA-MB-231 cells to extracts of *A. monostachya* in our study have been observed previously when exposing the same cell line to extracts of *Origanum majorana* L.; there is an expression of survivin, which is a therapeutic target against breast cancer [46], which could explain the behavior that was observed by the cell line against extracts of *A. monostachya*. Together, these results provide information that is needed to continue the study of *A. monostachya* extracts. Our results demonstrated that *A. monostachya* extracts have a cytotoxic and morphological effect that is evidenced by nuclear and cytoplasmic morphological changes.

4. Materials and Methods

4.1. Plant Material

The aerial part (stems, leaves, and inflorescences) of *A. monostachya* were used, which werecollected from the Loma Larga Oriente in the municipality of San Pedro Garza García, Nuevo León (25°39′21.9″ N, −100°20′07.3.14″ W) in November 2020. The plant was identified and authenticated by Dr. Marco A. Guzmán-Lucio in the Facultad de Ciencias Biológicas, UANL. A specimen was deposited at the herbarium of this Faculty with accession number 030641.

4.2. A. Monostachya Extracts

The plant material was dried in the shade at room temperature (RT) for 3 days and was then ground using a manual grain mill (Victoria). The extraction of the plant material was performed by maceration at RT (23 ± 2 °C) with constant stirring at 200 RPM for 24 h using 20 g of plant and 400 mL of solvents (distilled water, absolute methanol, and n-hexane). The ME and HE were filtered and concentrated under reduced pressure (150 mbar) in a rotary evaporator (Yamato Scientific CO. LTD. RE-200) at 45 °C for 35 min.

To obtain the AE, an infusion was made with 20 g of the ground plant and 400 mL of distilled water at 95 °C and was stirred at 200 RPM for 1 h. After this time, the infusion was filtered and sterilized with a Corning® syringe filter with a pore size of 0.22 μm and was subsequently subjected to evaporation in an incubator for 3 days at 40 °C. Once the solvent had been completely removed, the extract yield was obtained using the following formula:

$$\text{Yield } (\%p/p) = \frac{\text{WE}}{\text{IW}} \times 100 \tag{1}$$

where:

WE: Weight of the extract obtained.

IW: Initial weight of the plant material.

Finally, the extracts were stored in glass flasks and were protected from light at 4 °C.

4.3. Partial Characterization (Partial Phytochemical Screening)

One of the first stages in the investigation of medicinal plants is the realization of a partial characterization, also called phytochemical screening, which consists of mixing the plant material with suitable solvents for the extraction and qualitative determination of the compounds present in the plant, which are mainly from the group of secondary metabolites. This is to provide the information that guides the research of the fractionation and/or isolation of bioactive compounds [47].

For this, colorimetric and precipitation reactions were performed. The results were semi-quantitative and were represented by crosses relating the intensity of the observed reaction with the presence of the plant compound: (−) not detected; (+) slightly positive reaction; (++) positive reaction; and (+++) strong positive reaction.

4.3.1. Instaurations ($KMnO_4$ Test)

An amount of 2 mg of the extracts was dissolved in 2 mL of water, acetone, or methanol, and three drops of 2% $KMnO_4$ in water were added. The test was considered positive when discoloration or formation of a brown precipitate was observed, a result of the formation of MnO_2.

4.3.2. Carbonyl Group (2-4 Dinitrophenylhydrazine Test)

A 5 mg sample of the extracts was dissolved in 2 mL of ethanol, and 1 mL of a saturated solution of 2-4 dinitrophenylhydrazine in 6N HCl was added. The formation of a yellow or orange precipitate indicated the presence of the carbonyl group.

4.3.3. Phenolic Compounds (Vegetable Tannins) ($FeCl_3$ Test)

An amount of 2 mg of the extracts was dissolved in 2 mL of water or ethanol, and then three drops of 2.5% $FeCl_3$ in water were added. The appearance of a red, blue-violet, or green precipitate was considered positive.

4.3.4. Steroids and Terpenes (Salkowski Test)

An amount of 2 mg of each extract was dissolved in 2 mL of chloroform, and subsequently, 2 mL of H_2SO_4 was added. A positive result was considered for sterols and methyl sterols when a red-brown ring was formed at the interface.

4.3.5. Carbohydrates

Molish Test

Molish's reagent (1% alpha-naphthol in ethanol) was added dropwise to 2 mg of the extracts and then 2 mL of H_2SO_4 was added through the walls of the tube. The test was considered positive when a purple-colored ring formed at the interface.

Coumarin Test

An amount of 2 mg of the extracts was dissolved in 2 mL of ethanol and 10% NaOH was added dropwise. The test was considered positive if a yellow coloration was present and if it disappeared when the solution is acidified.

Lactone Test

An amount of 2 mg of the extracts was dissolved and 2 mL of an alcoholic solution of 10% NaOH was added. The test was considered positive when there was a turn to yellow or orange color that was lost or disappeared when a few drops of HCl were added, indicating the presence of a lactonic ring.

4.3.6. Sesquiterpene Lactones (Baljet Test)

A amount of 2 mg of the extracts was dissolved in 2 mL of ethanol, and three drops of the mixed solution were added, with a positive result being indicated if it turned from orange to dark red. The 1:1 mixed solution consisted of Solution A, which contained 1% $C_6H_3N_3O_7$ in ethanol and Solution B, which contained 10% NaOH.

4.3.7. Flavonoids (H_2SO_4 Test)

An amount of 2 mg of the extract was dissolved in 2 mL of H_2SO_4, and a positive result was indicated by yellow coloration for flavonoids, orange-cherry for flavones, red-bluish for chalcones, and red-purple for quinones.

4.3.8. Alkaloids (Dragendorff Test)

The Munier and Macheboeuf modification [48] was used to determine the presence of alkaloids. An amount of 2 mf of the extract were dissolved in 2 mL of ethanol, and three drops of Dragendorff reagent were added. To prepare the reagent, two solutions were used: Solution A, which contained 0.85 g of $Bi(NO_3)_3$, which were mixed with 10 mL of CH_3COOH and 40 mL of water, and Solution B, which contained 8 g of KI dissolved in 20 mL of water. The reagent was prepared by mixing 5 mL of A, 4 mL of B, and 100 mL of water; the test was considered positive if a persistent red-orange coloration was present for 24 h.

4.3.9. Saponins

$NaHCO_3$ Test

The aqueous solution of 10% $NaHCO_3$ was prepared, and then 2 mg of the extracts were dissolved in 2 mL of water or ethanol, and three drops of concentrated H_2SO_4 were added. It was stirred slightly, and three drops of the $NaHCO_3$ solution were added. The appearance of bubbles and their permanence for more than 1 min indicated the presence of saponins.

Salkowski Test for Saponins

An amount of mg of the extract were dissolved in 2 mL of chloroform and 2 mL of H_2SO_4 was added thereto. The test was considered positive by the appearance of red color.

Aromaticity (H_2SO_4- CH_2O Test)

A mixture of 1 mL of concentrated H_2SO_4 was prepared with a drop of CH_2O (formaldehyde). An amount of 5 mg of the dissolved extract was dissolved in 1 mL of non-aromatic solvent (ethanol), and three drops of the above mixture were added, and when a red-violet color appeared, the test was considered positive.

4.4. Cell Lines

4.4.1. Vero Cell Line

All cell lines were purchased from the American Type Culture Collection (ATCC) (Manassas VA, USA). Non-tumor Vero cells (ATCC: CCL-81) werederived from the kidney of an adult green monkey (*Cercopithecus aethiops*); they are adherent cells and with epithelial morphology. The cells were incubated in Advanced DMEM medium, 1× supplemented with 4% v/v of inactivated fetal bovine serum (FBS), 1% v/v of penicillin/streptomycin, and 1% of L-glutamine.

4.4.2. HeLa Cell Line

Cells derived from adenocarcinoma of the human cervix (ATCC: CCL-2) from a 31-year-old Black female patient. They are adherent cells, with epithelial morphology and are positive for cytokeratin. They have been reported to contain human papillomavirus 18 (HPV-18) sequences [49]; additionally, these cells express low levels of p53 and normal levels of pRB [50]. The cells were incubated in 1X Advanced DMEM medium supplemented with 4% v/v of inactivated FBS, 1% penicillin/streptomycin, and 1% L-glutamine.

4.4.3. MDA-MB-231 Cell Line

These cells were derived from human mammary adenocarcinoma (ATCC: CRM-HTB-26) from a 51-year-old Caucasian female patient. They are adherent cells of epithelial morphology and express the WNT7B oncogene. They express epidermal growth factor (EGF) and transforming growth factor-alpha (TGF α) [51]. Cells were incubated in 1× DMEM medium supplemented with 10% v/v of inactivated FBS, pyruvate 1×, 1% v/v penicillin/streptomycin, and 1% L-glutamine.

All cell lines were incubated at 37 °C in a 5% CO_2 atmosphere.

4.5. Preparation of Working Solutions

From the crude extracts, stock solutions of 10 mg/mL were prepared to dissolve this concentration of each extract 50 µL of 100% dimethyl-sulfoxide (DMSO) and 950 µL of culture medium to obtain a DMSO concentration of 5%. Subsequently, solutions of 1 mg/mL were prepared to make a 1:10 dilution of the stock solution, with a DMSO concentration of 0.5%, this is the minimum non-toxic concentration [52]. Finally, from these solutions, the dilutions were prepared at concentrations of 0, 10, 50, 100, 300, and 500 µg/mL for the cell treatments.

4.6. Cytotoxicity Test with MTT(3-(4,5-Dimethylthiazol-2-yl)-2,5-Diphenyltetrazolium Bromide)

Mitochondrial function can be evaluated based on the activity of the reductases found in the organelle and are directly related to cell viability, and the MTT assay can be used to measure such activity. This compound is a yellow tetrazolium salt, which is reduced to formazan by the action of mitochondrial enzymes, especially succinate dehydrogenase, resulting in a purple-blue product that can be evaluated by spectrophotometry [53]. For this assay, 5×10^4 cells per well ($n = 7$) were incubated for 24 h in a 96-well plate to allow adhesion. Subsequently, the treatments of each extract were applied at concentrations of 10, 50, 100, 300, and 500 μg/mL diluted in 100 μL of culture medium. Cell culture medium was used as a negative control (vehicle). The results were analyzed at 12, 24, 40, and 72 h.

Two hours before the end of the corresponding incubation times, representative photomicrographs of each concentration were taken at 40× with a Southern Precision Instrument inverted microscope to observe the effect of the extracts on the cell morphology. Subsequently, 15 μL of MTT (3 mg/mL) were added to each well, and the cells were incubated for 2 h at 37 °C; then, the medium was removed, and the MTT developer reagent (4 mM HCl, 0.040% NP40 in isopropanol) was added. The plate was shaken at 125 RPM for 10 min, and an absorbance analysis was performed at 590 nm with a 620 nm reference filter in an iMark™ microplate reader. The absorbance value is directly proportional to the number of metabolically active cells, which is an indirect measure of cell viability.

4.7. DAPI Nucleus Labeling Assay

To relate the results obtained in the MTT assay, the use of the fluorescent dye DAPI (4′,6-diamidino-2-phenylindole) that binds to the nucleic acids of adhered cells, allowed us to estimate the number of cells viable related to cell adhesion. For this, 96-well plate cultures were performed under the same conditions as MTT assay treatments ($n = 8$). At the end of the incubation times, the medium was removed, two washes of 200 μL were performed with 1 × PBS, and subsequently, the cells were fixed with methanol/acetone at a ratio of 1: 1 for 20 min at 4 °C. After the fixer was removed, two washes were performed again with 1X PBS, and 100 μL of DAPI (100 ng/mL) were added for 15 min at RT in the darkness. At the end of this period, the excess DAPI was removed, and the plate was dried and observed under fluorescence microscopy. Eight micrographs were taken per concentration at a total magnification of 100×, which were evaluated with ImageJ® software version 1.51, (NIH) which calculated the percentage of the area covered by the adhered cells in each of the wells

4.8. Morphological Analysis

To evaluate the presence of morphological changes in the cells due to exposure to extracts of *A. monostachya*, 1×10^6 cells were seeded in 60 mm dishes and were incubated for 24 h to allow adherence. Subsequently, the treatments of 0, 10, and 50 μg/mL of EM and EH and 0, 300, and 500 μg/mL of EA were applied and diluted in culture medium at a total of 3 mL per dish and were incubated for 24 and 48 h. At the end of the incubation time, the cells were harvested using 0.25% trypsin, two washes were performed with 1× PBS, and later, they were fixed with 2.5% glutaraldehyde buffered in 0.1 M cacodylate buffer (pH 7.4) for 24 h. Next, the cells were washed two times with cacodylate buffer for 5 min with centrifugation at 14,000 RMP; they were decanted, and a post-fixation was performed with OsO_4 at 2% for 12 h; they were washed again and dehydrated with a gradient of ketones (30%, 50%, 70%, 90%, and 100%) for 5 min and embedded in middle epoxy resin for 72 h at 60 °C. From the epoxy blocks, 350 nm thick semi-fine sections were obtained by ultramicrotomy and subsequent staining with 1% toluidine blue for 1 min. One hundred cells were counted per section when analyzing the presence of changes in the nuclear morphology, chromatin condensation, and vacuolization in the cytoplasm to obtain the percentage of positive cells in each treatment.

4.9. Statistical Analysis

For the descriptive statistics, the quantitative variables were summarized by calculating the mean, and as a measure of dispersion, the standard deviation (SD). In the inferential statistics, the Kolmogorov–Smirnov normality test was applied to evaluate the distribution of the data, with all variables having a normal distribution. Subsequently, an ANOVA and a Tukey test were performed to evaluate the presence of the differences between the groups. All graphs, calculations, and statistical analyses were made using GraphPad Prism version 8.0 software for Windows (GraphPad Software, Inc., San Diego, CA, USA).

5. Conclusions

This research provides information on the effect of crude extracts of different polarities obtained from *A. monostachya* on human tumor lines, demonstrating that it possesses time- and concentration-dependent cytotoxic activity. This was evidenced by evaluating the relative viability and the presence of morphological alterations when the extracts were exposed. Likewise, this cytotoxicity is selective for tumor cell lines, highlighting that the HeLa cells presented more cytoplasmic and nuclear morphological changes as well as a lower percentage of relative viability when exposed to extracts compared to MDA-MB-231 cells, which responded in the same way to higher concentrations. This activity can be attributed to the phenolic compounds detected by the phytochemical screening being carried out.

The perspectives of this work are oriented to perform a fractionation guided by bioassay to obtain a fraction with the bioactive compound and thus isolate and purify it. With the data obtained, the IC50 of the crude extracts and subsequently their different fractions can be calculated. Likewise, it is convenient to evaluate the effect of the extracts in other tumor cell lines to compare the induced changes. Finally, it is possible to further evaluate the type of cell death to establish a mechanism by which the bioactive compounds present in *A. monostachya* act.

Author Contributions: Conceptualization: G.A.G.-M., S.A.V.-C., R.A.P.-H., U.C.-V., M.d.J.L.-A. and A.S.-D.; methodology: G.A.G.-M., S.A.V.-C., R.A.P.-H., U.C.-V.; D.S.-T., M.d.J.L.-A. and A.S.-D.; software, G.A.G.-M., S.A.V.-C., R.A.P.-H., U.C.-V., D.S.-T., M.d.J.L.-A. and A.S.-D.; validation, G.A.G.-M., S.A.V.-C., R.A.P.-H., U.C.-V., D.S.-T., M.d.J.L.-A. and A.S.-D.; formal analysis, G.A.G.-M., S.A.V.-C., R.A.P.-H., U.C.-V., D.S.-T., O.S.-C., R.M.-d.-O.-L., C.A.G.-T., M.d.J.L.-A. and A.S.-D.; investigation, G.A.G.-M., S.A.V.-C., R.A.P.-H., U.C.-V., D.S.-T., O.S.-C., R.M.-d.-O.-L., M.d.J.L.-A. and A.S.-D.; resources, G.A.G.-M., S.A.V.-C., R.A.P.-H., U.C.-V., O.S.-C., R.M.-d.-O.-L., M.d.J.L.-A. and A.S.-D.; data curation, G.A.G.-M., S.A.V.-C., R.A.P.-H., U.C.-V., C.A.G.-T., M.d.J.L.-A. and A.S.-D.; writing—original draft preparation, G.A.G.-M., S.A.V.-C., R.A.P.-H., M.d.J.L.-A., A.S.-D.; writing—review and editing, G.A.G.-M., S.A.V.-C., R.A.P.-H., U.C.-V., D.S.-T., O.S.-C., R.M.-d.-O.-L., C.A.G.-T., A.J.G.-A., D.E.Z.-Á., M.d.J.L.-A. and A.S.-D.; visualization, G.A.G.-M., S.A.V.-C., R.A.P.-H., U.C.-V., D.S.-T., C.A.G.-T., M.d.J.L.-A. and A.S.-D.; supervision, G.A.G.-M., S.A.V.-C., R.A.P.-H., U.C.-V., D.S.-T., O.S.-C., R.M.-d.-O.-L., M.d.J.L.-A. and A.S.-D.; project administration, G.A.G.-M., S.A.V.-C., R.A.P.-H., U.C.-V., O.S.-C., R.M.-d.-O.-L., M.d.J.L.-A. and A.S.-D.; funding acquisition, G.A.G.-M., S.A.V.-C., R.A.P.-H., U.C.-V., O.S.-C., R.M.-d.-O.-L., A.J.G.-A., D.E.Z.-Á., M.d.J.L.-A. and A.S.-D. All authors have read and agreed to the published version of the manuscript.

Funding: This study was partially supported by PAICYT 2021. UANL. México, grant SA1864-21, and GAGM is a recipient of the CONACyT (Consejo Nacional de Ciencia y Tecnología) México scholarship 1006252.

Institutional Review Board Statement: The research protocol was approved by the Ethics Committee in the Faculty of Medicine, UANL with approval number HT21–00002. We did not perform studies involving humans or animals.

Informed Consent Statement: Not applicable.

Data Availability Statement: The data and materials supporting the conclusions of this article are included within the article.

Acknowledgments: We acknowledge the work of Eduardo Estrada-Castillón and Marco A. Guzmán-Lucio for their assistance in plant identification and to R. Smith for reviewing the English manuscript.

Conflicts of Interest: The authors declare no conflict of interest. The funders had no role in the design of the study; in the collection, analyses, or interpretation of data; in the writing of the manuscript; or in the decision to publish the results.

References

1. WHO. Health Topics: Cancer. Available online: https://www.who.int/topics/cancer/es/ (accessed on 15 March 2021).
2. Globocan: All Cancers. Available online: https://gco.iarc.fr/today/data/factsheets/cancers/39-All-cancers-fact-sheet.pdf (accessed on 15 March 2021).
3. Globocan: Source Mexico. Available online: https://gco.iarc.fr/today/data/factsheets/populations/484-mexico-fact-sheets.pdf (accessed on 15 March 2021).
4. National Cancer Institute: Types of Cancer Treatment. Available online: https://www.cancer.gov/about-cancer/treatment/types (accessed on 10 May 2020).
5. Karunamoorthi, K.; Jegajeevanram, K.; Vijayalakshmi, J.; Mengistie, E. Traditional Medicinal Plants: A Source of Phytotherapeutic Modality in Resource-Constrained Health Care Settings. *J. Evid.-Based Complement. Altern. Med.* **2013**, *18*, 67–74. [CrossRef]
6. Seebaluck, R.; Gurib-Fakim, A.; Mahomoodally, F. Medicinal plants from the genus *Acalypha* (Euphorbiaceae) A review of their ethnopharmacology and phytochemistry. *J. Ethnopharmacol.* **2015**, *159*, 137–157. [CrossRef] [PubMed]
7. Soladoye, M.O.; Sonibare, M.A.; Rosanwo, T.O. Phytochemical and morphometric analysis of the genus *Acalypha* Linn. (Euphorbiaceae). *J. Appl. Sci.* **2008**, *8*, 3044–3049. [CrossRef]
8. Emeka, P.M.; Badger-Emeka, L.I.; Fateru, F. In vitro antimicrobial activities of *Acalypha* ornate leaf extracts on bacterial and fungal clinical isolates. *J. Herb. Med.* **2012**, *2*, 136–142. [CrossRef]
9. Gurib-Fakim, A.; Gueho, J. *Plantes Médicinal de Maurice*, 2nd ed.; Edition deL'Ocean Indien: Beau Bassin-Rose Hill, Mauritius, 1996; pp. 71–77.
10. Gurib-Fakim, A.; Brendler, J. *Medicinal and aromatic plants of Indian Ocean Islands*, 1st ed.; MedPharm: Stuttgart, Germany, 2004; pp. 87–89.
11. Estrada-Castillón, E.; Villarreal-Quintanilla, J.Á.; Rodríguez-Salinas, M.M.; Encinas-Domínguez, J.A.; González-Rodríguez, H.; Figueroa, G.R.; Arévalo, J.R. Ethnobotanical Survey of Useful Species in Bustamante, Nuevo León, Mexico. *Hum. Ecol.* **2018**, *46*, 117–132. [CrossRef]
12. Canales, M.; Hernández, T.; Rodríguez, M.; Flores, O.; Jiménez, E.; Hernández, L.; Hernández, M.; Trejo, N.; Ramírez, J.E.; Martínez, K. Evaluation of the antimicrobial activity of *Acalypha monostachya* Cav. (Euphorbiales: Euphorbiaceae). *Afr. J. Pharm. Pharmacol.* **2011**, *5*, 640–647. [CrossRef]
13. Maiti, R.; González, R.H.; Kumari, A. Nutrient Profile of Native Woody Species and Medicinal Plants in Northeastern Mexico: A Synthesis. *J. Bioprocess. Biotech.* **2016**, *6*, 283.
14. Maiti, R.; Rodriguez, H.G.; Degu, H.D.; Kumari, C.A.; Sarkar, N.C. Macro and Micronutrients of 44 Medicinal Plant Species Used Traditionally in Nuevo Leon, Mexico. *Int. J. Bio-Resour. Stress Manag.* **2016**, *7*, 1054–1062. [CrossRef]
15. Ramírez, A.; García, G.; Werner, O.; Navarro-Pedreño, J.; Ros, R.M. Implications of the Phytoremediation of Heavy Metal Contamination of Soils and Wild Plants in the Industrial Area of Haina, Dominican Republic. *Sustainability* **2021**, *13*, 1403. [CrossRef]
16. Macías, K.L.; Juárez, B.I.; Cárdenas, N.C.; Aguirre, J.R.; Jasso, Y. Evaluación de plantas tradicionalmente utilizadas en la desinfección de heridas. *Rev. Mex. Cienc. Farm.* **2009**, *40*, 5–10.
17. Sanseera, D.; Niwatananun, W.; Liawruangrath, B.; Liawruangrath, S.; Baramee, A.; Trisuwan, K.; Pyne, S.G. Antioxidant and anticancer activities from aerial parts of *Acalypha indica* Linn. *CMU J. Nat. Sci.* **2012**, *11*, 157–168.
18. Rajkumar, V.; Guha, G.; Ashok Kumar, R. Therapeutic potential of *Acalypha fruticosa*. *Food. Chem. Toxicol.* **2010**, *48*, 1709–1713. [CrossRef]
19. Lim, S.W.; Ting, K.N.; Bradshaw, T.D.; Zeenathul, N.A.; Wiart, C.; Khoo, T.J.; Loh, H.S. *Acalypha wilkesiana* extracts induce apoptosis by causing single strand and double strand DNA breaks. *J. Ethnopharmacol.* **2011**, *138*, 616–623. [CrossRef] [PubMed]
20. Chu, J.H.-Y. Investigating In Vitro Anticancer Properties of Malaysian Rainforest Plants: *Acalypha wilkesiana Mull*, Arg. *Archidendron ellipticum* (Blume) Hassk. *Duabanga grandiflora* Walp. *Pseuduvaria macrophylla* (Oliv.) Merr. Ph.D. Thesis, University of Nottingham, Nottingham, UK, 2014.
21. Abubakar, I.B.; Lim, S.; Loh, H. Synergistic Apoptotic Effects of Tocotrienol Isomers and *Acalypha wilkesiana* on A549 and U87MG Cancer Cells Cancer is a group of disease caused by internal and external factors characterized by uncontrolled growth and spread of abnormal cells, resultant. *Trop. Life Sci. Res.* **2018**, *29*, 229–238. [CrossRef]
22. El-Raey, M.A.; Mohamed, T.K.; El-Kashak, W.A.; Fayad, W.O. Phenolic constituents and biological activities of *Acalypha wilkesiana* forma tricolor muell arg seeds. *Int. J. Pharmacogn. Phytochem. Res.* **2016**, *8*, 386–392.
23. Rascón-Valenzuela, L.; Jiménez-Estrada, M.; Velázquez-Contreras, C.; Garibay-Escobar, A.; Vilegas, W.; Campaner, L.; Robles-Zepeda, R.E. Chemical composition and antiproliferative activity of *Acalypha californica*. *Ind. Crops Prod.* **2015**, *69*, 48–54. [CrossRef]

24. Rehab, A.; Hussein; El-Anssary, A.A. Plants Secondary Metabolites: The Key Drivers of the Pharmacological Actions of Medicinal Plants, Herbal Medicine, Philip F. Builders. Available online: https://www.intechopen.com/chapters/61866 (accessed on 16 May 2020).
25. Noumedem, J.A.; Tamokou, J.; Teke, G.N.; Momo, R.C.; Kuete, V.; Kuiate, J.R. *Phytochemical Analysis, Antimicrobial and Radical-scavenging Properties of Acalypha manniana Leaves*; SpringerPlus: Berlin/Heidelberg, Germany, 2013; Volume 2, p. 503.
26. Iniaghe, O.M.; Malomo, S.O.; Adebayo, J.O. Proximate Composition and Phytochemical Constituents of Leaves of Some *Acalypha* Species. *Pak. J. Nutr.* **2009**, *8*, 256–258. [CrossRef]
27. Nazri, N.M.; Hazali, N.; Ibrahim, M.; Masri, M.; Ayob, M.K. Preliminary studies on *Acalypha indica*: Proximate analysis and phytochemical screening. *Int. J. Pharm. Pharm. Sci.* **2016**, *8*, 406–408.
28. Evanjelene, V.K.; Natarajan, D. In Vitro Antioxidant and Phytochemical Analysis of *Acalypha alnifolia* Klein Ex Willd. *IOSR J. Pharm. Biol. Sci.* **2012**, *1*, 43–47. [CrossRef]
29. Onocha, P.A.; Oloyede, G.K.; Afolabi, Q.O. Phytochemical investigation, cytotoxicity and free radical scavenging activities of non-polar fractions of *Acalypha hispida* (leaves and twigs). *EXCLI J.* **2011**, *10*, 1–8.
30. Sepúlveda-Vázquez, J.; Torres-Acosta, F.; Sandoval-Castro, C.; Martínez-Puc, J.; Chan-Pérez, J.I. La importancia de los metabolitos secundarios en el control de nematodos gastrointestinales en ovinos con énfasis en Yucatán, México. *J. Selva Andina Anim. Sci.* **2018**, *5*, 79–95. [CrossRef]
31. Reddy, T. Exploring the Anti-inflammatory and Anti-cancer compounds from the leaves of *Acalypha indica*. *IOSR J. Pharm. Biol. Sci.* **2012**, *4*, 1–7. [CrossRef]
32. Moongkarndi, P.; Kosem, N.; Kaslungka, S.; Luanratana, O.; Pongpan, N.; Neungton, N. Antiproliferation, antioxidation and induction of apoptosis by *Garcinia mangostana* (mangosteen) on SKBR3 human breast cancer cell line. *J. Ethnopharmacol.* **2004**, *90*, 161–166. [CrossRef] [PubMed]
33. Kuete, V.; Karaosmanoğlu, O.; Sivas, H. *Anticancer Activities of African Medicinal Spices and Vegetables in Medicinal Spices and Vegetables from África, 1st. ed.*; Academic Press: Dschang, Cameroon, 2017; pp. 271–297.
34. Chekuri, S.; Panjala, S.; Anupalli, R.R. Cytotoxic activity of *Acalypha indica* L. hexane extract on breast cancer cell lines (MCF-7). *J. Phytopharmacol.* **2017**, *6*, 264–268. [CrossRef]
35. Motadi, L.R.; Choene, M.S.; Mthembu, N.N. Anticancer properties of *Tulbaghia violacea* regulate the expression of p53-dependent mechanisms in cancer cell lines. *Sci. Rep.* **2020**, *10*, 12924. [CrossRef] [PubMed]
36. Rahman, S.N.S.A.; Wahab, N.a.; AbdMalek, S.N. In Vitro Morphological Assessment of Apoptosis Induced by Antiproliferative Constituents from the Rhizomes of *Curcuma zedoaria*. *Evid.-Based Complementary Altern. Med.* **2013**, *2013*, 257108.
37. Häcker, G. The morphology of apoptosis. *Cell Tissue Res.* **2000**, *301*, 5–17. [CrossRef]
38. Majno, G.; Joris, I. Apoptosis, oncosis, and necrosis. An overview of cell death. *Am. J. Pathol.* **1995**, *146*, 3–15. [PubMed]
39. Hengartner, M.O. Apoptosis: Corralling the corpses. *Cell* **2001**, *104*, 325–328. [CrossRef]
40. Clarke, P.G. Developmental cell death: Morphological diversity and multiple mechanisms. *Anat. Embryol.* **1990**, *181*, 195–213. [CrossRef]
41. Ziegler, U.; Groscurth, P. Morphological Features of Cell Death. *Physiology* **2004**, *19*, 124–128. [CrossRef] [PubMed]
42. Gonzalez-Polo, R.A. The apoptosis/autophagy paradox: Autophagic vacuolization before apoptotic death. *J. Cell Sci.* **2005**, *118*, 3091–3102. [CrossRef]
43. Soto-Domínguez, A.; Ballesteros-Elizondo, R.G.; Santoyo-Pérez, M.E.; Rodríguez-Rocha, H.; García-Garza, R.; Nava-Hernández, M.P.; Saucedo-Cárdenas, O. Peroxisomicine A1 (toxin T-514) induces cell death of hepatocytes in vivo by triggering the intrinsic apoptotic pathway. *Toxicon* **2018**, *154*, 79–89. [CrossRef] [PubMed]
44. Shubin, A.V.; Demidyuk, I.V.; Komissarov, A.A.; Rafieva, L.M.; Kostrov, S.V. Cytoplasmic vacuolization in cell death and survival. *Oncotarget* **2016**, *7*, 55863–55889. [CrossRef] [PubMed]
45. Stefkó, D.; Kúsz, N.; Barta, A.; Kele, Z.; Bakacsy, L.; Szepesi, Á.; Fazakas, C.; Wilhelm, I.; Krizbai, I.A.; Hohmann, J.; et al. Gerardiins A-L and Structurally Related Phenanthrenes from the Halophyte Plant *Juncus gerardii* and Their Cytotoxicity against Triple-Negative Breast Cancer Cells. *J. Nat. Prod.* **2020**, *83*, 3058–3068. [CrossRef]
46. Hanane, M.; Mouhcine, E.; Ahmed, M.; Mohammed, E.; Laila, B.; Mohammed, A. Cytotoxicity of the Aqueous Extract and Organic Fractions from *Origanum majorana* on Human Breast Cell Line MDA-MB-231 and Human Colon Cell Line HT-29. *Adv. Pharmacol. Pharm. Sci.* **2018**, *2018*, 9.
47. Mendoza, N.; Escamilla, E. Introduction to Phytochemicals: Secondary Metabolites from Plants with Active Principles for Pharmacological Importance, Phytochemicals—Source of Antioxidants and Role in Disease Prevention. 2018. Available online: https://www.intechopen.com/books/phytochemicals-source-of-antioxidants-and-role-in-disease-prevention/introduction-to-phytochemicals-secondary-metabolites-from-plants-with-active-principles-for-pharmaco (accessed on 8 June 2021).
48. Kikugawa, K.; Hayatsu, H.; Ukita, T. Modifications of nucleosides and nucleotides. *Biochim. Biophys. Acta* **1967**, *134*, 221–231. [CrossRef]
49. Popescu, N.C.; DiPaolo, J.A.; Amsbaugh, S.C. Integration sites of human papillomavirus 18 DNA sequences on HeLa cell chromosomes. *Cytogenet Cell Genet* **1987**, *44*, 58–62. [CrossRef]
50. Hoppe, F.; Butz, K. Repression of endogenous p53 transactivation function in HeLa cervical carcinoma cells by human papillomavirus type 16 E6, human mdm-2, and mutant p53. *J. Virol* **1993**, *67*, 3111–3117. [CrossRef]

51. Mur, C.; Martínez-Carpio, P.A.; Fernández-Montolí, M.E.; Ramon, J.M.; Rosel, P.; Navarro, M.A. Growth of MDA-MB-231 cell line: Different effects of TGF-beta (1), EGF and estradiol depending on the length of exposure. *Cell Boil. Int.* **1998**, *22*, 679–684. [CrossRef]
52. Chen, X.; Thibeault, S. Effect of DMSO concentration, cell density and needle gauge on the viability of cryopreserved cells in three dimensional hyaluronan hydrogel. In Proceedings of the 35th Annual International Conference of the IEEE Engineering in Medicine and Biology Society (EMBC), Osaka, Japan, 3–7 July 2013.
53. Van Tonder, A.; Joubert, A.M.; Cromarty, A.D. Limitations of the 3-(4,5-dimethylthiazol-2-yl)-2,5-diphenyl-2H-tetrazolium bromide (MTT) assay when compared to three commonly used cell enumeration assays. *BMC Res. Notes* **2015**, *8*, 47. [CrossRef] [PubMed]

plants

MDPI

Article

Preliminary Phytochemical Profile and Bioactivity of *Inga jinicuil* Schltdl & Cham. ex G. Don

Ammy Joana Gallegos-García [1,2], Carlos Ernesto Lobato-García [1,*], Manasés González-Cortazar [2,*], Maribel Herrera-Ruiz [2], Alejandro Zamilpa [2], Patricia Álvarez-Fitz [3], Ma Dolores Pérez-García [2], Ricardo López-Rodríguez [1], Ever A. Ble-González [1], Eric Jaziel Medrano-Sánchez [1], Max R. Feldman [4], Alejandro Bugarin [4] and Abraham Gómez-Rivera [1,*]

1 División Académica de Ciencias Básicas, Universidad Juárez Autónoma de Tabasco, Carretera Cunduacán-Jalpa Km. 0.5, Cunduacán 86690, Tabasco, Mexico; joana90102010@gmail.com (A.J.G.-G.); ricardo.lopezr@ujat.mx (R.L.-R.); ble_49@hotmail.com (E.A.B.-G.); ericsanz123@gmail.com (E.J.M.-S.)

2 Centro de Investigación Biomédica del Sur, Instituto Mexicano del Seguro Social, Argentina No. 1, Col. Centro, Xochitepec 62790, Morelos, Mexico; cibis_herj@yahoo.com.mx (M.H.-R.); azamilpa_2000@yahoo.com.mx (A.Z.); lola_as@yahoo.com.mx (M.D.P.-G.)

3 Laboratorio de Toxicología, Cátedra CONACyT-Universidad Autónoma de Guerrero, Av. Lázaro Cárdenas s/n. Col. La Haciendita, Chilpancingo 39070, Guerrero, Mexico; paty_fitz@hotmail.com

4 Department of Chemistry and Physics, Florida Gulf Coast University, Fort Myers, FL 33965, USA; mrfeldman2018@eagle.fgcu.edu (M.R.F.); abugarin@fgcu.edu (A.B.)

* Correspondence: carlos.lobato@ujat.mx (C.E.L.-G.); gmanases2000@gmail.com (M.G.-C.); abgori@gmail.com (A.G.-R.); Tel.: +52-(777)-361-2155 (M.G.-C.); +52-(993)-358-1500 (ext. 4711) (A.G.-R.)

Citation: Gallegos-García, A.J.; Lobato-García, C.E.; González-Cortazar, M.; Herrera-Ruiz, M.; Zamilpa, A.; Álvarez-Fitz, P.; Pérez-García, M.D.; López-Rodríguez, R.; Ble-González, E.A.; Medrano-Sánchez, E.J.; et al. Preliminary Phytochemical Profile and Bioactivity of *Inga jinicuil* Schltdl & Cham. ex G. Don. *Plants* **2022**, *11*, 794. https://doi.org/10.3390/plants11060794

Academic Editors: Juei-Tang Cheng, I-Min Liu and Szu-Chuan Shen

Received: 17 February 2022
Accepted: 11 March 2022
Published: 17 March 2022

Publisher's Note: MDPI stays neutral with regard to jurisdictional claims in published maps and institutional affiliations.

Abstract: Several Mesoamerican cultures have used *Inga jinicuil* as traditional medicine for the treatment of gastrointestinal, inflammatory, and infectious issues. The aims of this contribution were to elucidate the phytochemical profile of the organic extracts from the bark and leaves of *I. jinicuil* and to assess the anti-inflammatory and antibacterial properties of these extracts. The preliminary chemical profile was determined by HPLC-PDA and GC-MS; the anti-inflammatory activity was evaluated with a mouse ear edema model, whereas the antibacterial activity was screened against several bacteria. The phytochemical profile of both organs (bark and leaves) of *I. jinicuil* led to the identification of 42 compounds, such as polyphenolic, flavonoids, triterpenes, prenol-type lipids, and aliphatic and non-aliphatic esters. This molecular diversity gave moderate anti-inflammatory activity (67.3 ± 2.0%, dichloromethane bark extract) and excellent antibacterial activity against *Pseudomona aeruginosa* and methicillin-resistant *Sthaphylococcus aureus* (MIC values of <3.12 and 50 μg/mL, respectively). These results contribute to the chemotaxonomic characterization and the rational use in traditional medicine of *Inga jinicuil* Schltdl & Cham. ex G. Don.

Keywords: *Inga jinicuil*; phytochemical profile; HPLC-PDA; GC-MS; anti-inflammatory; antimicrobial

1. Introduction

Anti-inflammatory and antimicrobial properties have been attributed to a great diversity of plants used in traditional medicine, from which many commercial drugs have been developed [1]. These properties have been related to the presence of secondary metabolites such as tannins, terpenes, and flavonoids, among many others [2]. Currently, medicinal plants are a valuable alternative and, in agreement with the WHO strategies on complementary and traditional medicine, it is necessary to perform studies aimed at identifying their bioactive compounds and confirming their pharmacological activity in order to guarantee their safe, effective, and rational use [3].

Even though there are several alternatives for the treatment of inflammation, some anti-inflammatory drugs, such as aspirin, ketorolac, naproxen, or piroxicam, have adverse effects (e.g., the risk of developing intestinal bleeding) [2]; therefore, a constant search for new anti-inflammatory treatments is critical in order to achieve an increased pharmacological response with the lowest degree of unwanted side effects [4]. The rise of microbial strains resistant to current antibiotics similarly presses the medical field to find new, effective compounds. These issues have led to the search for alternatives derived from natural sources to help in either the prevention or treatment of infectious problems [5].

Related to the above statements, a promissory prospect is the tropical species *Inga jinicuil* Schltdl & Cham. ex G. Don, known in Mexico as "cuijinicuil", "cuajicuil", or "jinicuil", and named in the Maya-Chontal language as "bujte". This plant belongs to the *Leguminosae* family and is classified as a multi-purpose tree in Mesoamerican indigenous communities, where it is mainly used as a shade tree in agroecosystems for cocoa and coffee plantations [6,7]. It is also an ornamental tree present in many gardens, parks, and roads, and it is highly recommended for repopulating watersheds [6]. The cotton pulp that covers the seeds can be consumed fresh or used in jellies and drinks [8]. Some indigenous communities of the Maya-Chontal region in Mexico and in the Amazon boil the seeds in salt water and use it as an appetizer or complement in traditional stews [6,7]. The aerial parts are used for healing purposes for the treatment of parasitic and infectious problems [6,8]. A mixture of seeds and leaves is also used as both an antidiarrheal and antirheumatic remedy [6]; in rural communities of Veracruz and Tabasco, Mexico, this plant is also utilized for gastrointestinal diseases by taking an infusion made from the pod and bark [7,8].

There are few reports on the phytochemical and biological activity of the *Inga* genus. For instance, a recent report highlights the antibacterial activity of the organic extracts from the leaves of *I. semialata*, which had an inhibitory effect on the growth of *Staphylococcus aureus*, *Staphylococcus epidermidis*, *Micrococcus luteus*, and *Klebsiella pneumoniae* associated with recurrent infections; the analysis of the extracts revealed the presence of gallic acid, epicatechin, and rutin [9]. There is also a series of reports aimed at the phytochemical and pharmacological analysis of *I. edulis* and *I. laurina*, where antimicrobial and antioxidant activities have been associated mainly with the presence of phenolic compounds [10–13]. In the case of *I. edulis*, the dichloromethane extract from leaves exerts a moderate antibacterial activity (MIC 7.0 mg/mL) against two strains of *S. aureus*; whereas for *I. laurina*, its effect against some strains of aerobic and anaerobic micro-organisms has been reported. The chemical composition of the active extract was determined by GC-MS, finding terpenoids, fatty acids, and esters [10–13]. Despite the extensive use of *I. jinicuil* in traditional medicine in southeastern Mexico, only the antimicrobial activity of hexanic and chloroform extracts from the seeds has been reported. These extracts proved to be active against *S. aureus* 25,923 and *Listeria monocytogenes* 244, with an MIC of 100 μg/mL for each micro-organism [14]; however, to date, no studies have been found about the phytochemical composition of the bark and leaves of this plant, nor on the evaluation of its anti-inflammatory activity. Therefore, the objectives of this work were to analyze the phytochemical profile of organic extracts from the bark and leaves of *I. jinicuil* via chromatographic methods, to evaluate their anti-inflammatory activity in the phorbol ester (TPA)-induced mouse ear edema test, and to expose its antibacterial activity against strains of clinical importance.

2. Results and Discussion

The yield of the extracts from *Inga jinicuil* are shown in Table 1. Three types of extracts (in order of increasing polarity) were acquired from bark (Hexane **Ij-BH**, Dichloromethane **Ij-BD**, & Hydroalcoholic **Ij-BHac**) and three more from leaves (Hexane **Ij-LH**, Dichloromethane **Ij-LD**, & Hydroalcoholic **Ij-LHac**). It was found that, in general, the extracts from leaves provided higher yields when compared to bark extracts, with the hydroalcoholic extract from leaves being the one with the highest percentage.

Table 1. Percentages obtained from *Inga jinicuil* extracts.

Extract	*n*-Hexane (% Yield)	Dichloromethane (% Yield)	Hydroalcoholic (% Yield)
Bark extract	0.095	0.82	0.25
Leaf extract	0.95	1.02	4.65

2.1. HPLC and UV-Vis Spectra Analysis of Polar Extracts from *Inga jinicuil*

HPLC analysis helped to determine the presence of terpenic and flavonoid-type compounds in both the dichloromethane and hydroalcoholic extracts from *I. jinicuil*. The chromatograms of the four polar extracts (**Ij-LD**, **Ij-LHac**, **Ij-BD**, and **Ij-BHac**) are presented in Figure 1. A total of 21 peaks related to terpenic and flavonoid-type compounds were observed. Table 2 presents a summary of the following information: retention times, main absorption bands of the UV-Vis spectra, and the presence of each peak in the four extracts analyzed.

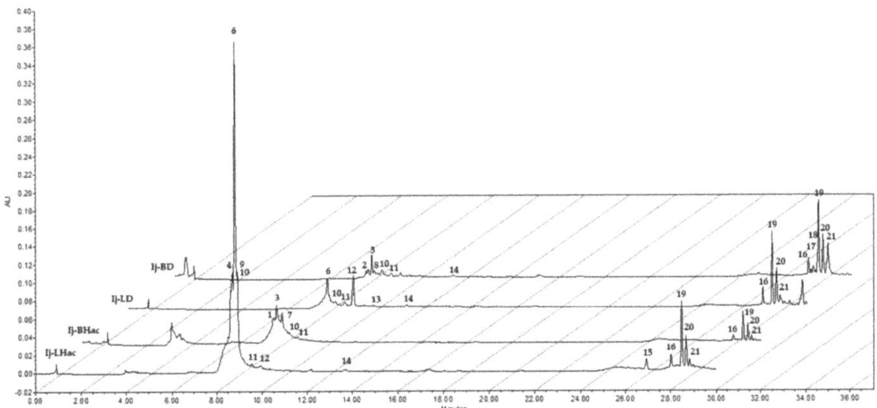

Figure 1. HPLC chromatograms of bark (**Ij-BD**, **Ij-BHac**) and leaf (**Ij-LD**, **Ij-LHac**) extracts. The peaks are numbered in ascending order according to their retention times (λ = 270 nm).

As seen in Figure 1, the peaks in the chromatograms can be differentiated into two main groups based on their retention time (t_R): the first group consists of 14 peaks with t_R between 8 and 12 min, while the second group has 7 peaks with t_R from 26 to 29 min.

Based on the retention times and the absorption bands in the UV-Vis spectra (Figures S1–S4) of the peaks shown in Table 2, it was possible to perform a preliminary analysis of each of the metabolites present in the extracts by comparing those parameters with known standards and data from the literature. Accordingly, for Peak **1**, the observed absorption bands at 220.4, 261.6, and 294.7 nm were equal with those of the protocatechuic acid standard, which when analyzed in identical chromatographic conditions presented the same t_R (8.46 min). Since Peak **2** showed chromatographic behavior similar to **1** along with the analysis of the UV-Vis bands and the literature, it is inferred that it may be a derivative of protocatechuic acid [15,16].

Regarding Peaks **3** and **7**, when their t_R and UV-Vis absorption bands were compared with the gallic acid standard (t_R = 7.46 min, λ_{max} = 220.4, 272.2 nm), a good match was found in the UV-Vis spectrum; however, the differences in retention times suggested the presence of gallic acid derivatives [17,18].

The analysis of the UV-Vis bands for Peaks **4**, **6**, and **9** indicated that these compounds may be of the flavonoid type; this inference was strengthened when they were compared with an apigenin standard (t_R = 16.76 min, λ_{max} = 211, 267.5, 338.6 nm) and an identical match was found in their UV-Vis spectra. The differences in retention times led us to potentially consider these peaks as glycosylated analogs of this flavone [19,20].

Table 2. Preliminary phytochemical profile by HPLC-UV-Vis analysis of polar extracts from bark and leaves of *I. jinicuil*.

Peak	Retention Time (min)	Absorption Bands (nm)	Extract(s) *	Compound Affinity **	Ref.
1	8.46	220.4, 261.6, 294.7	■	Protocatechuic acid	Standard [15,16]
2	8.58	249.8, 273.6	●	Protocatechuic acid derivative	Standard [15,16]
3	8.58	218.1, 276.9	■	Gallic acid derivative	Standard [17,18]
4	8.66	212.2, 251.5, 352.9	□	Glycosylated Flavone. Apigenin derivative	Standard [19,20]
5	8.75	219.2, 249.8, 273.4	●	Lignane	Standard [21,22]
6	8.81	215.7, 269.8, 337.4	○ □	Glycosylated Flavone. Apigenin derivative	Standard [19,20]
7	8.85	219.2, 279.3	■	Gallic acid derivative	Standard [17,18]
8	8.86	215.7, 308.9	●	Coumaric acid derivative	Standard [23]
9	8.91	207.5, 269.8, 335.1	□	Glycosylated Flavone. Apigenin derivative	[19,20]
10	9.18	249.8	○ □ ● ■	Terpene	[24]
11	9.58	245.1	○ □ ● ■	Terpene	[24]
12	9.96	209.9, 294.7, 338.6	○ □	Coumarin derivative	[25–27]
13	10.03	276.9	○	Epigallocatechin Gallate derivative	Standard [28]
14	12.30	235.7, 266.3	○ □ ●	Terpene	[24]
15	26.91	219.2, 273.4, 293.5	□	Vanillic acid derivative	[29,30]
16	28.01	204, 248.6	○ □ ● ■	Terpene	[24]
17	28.11	278.1	●	Epigallocatechin Gallate derivative	[28]
18	28.21	245.1, 278.1, 327.9	●	Coumarin derivative	[25–27]
19	28.43	201.7, 261.6	○ □ ● ■	Salicylate derivative	Standard [31]
20	28.65	200.5, 263.9	○ □ ● ■	Salicylate derivative	Standard [31]
21	28.81	263.9	○ □ ● ■	Salicylate derivative	Standard [31]

* Extracts: Bark extracts, ● (**Ij-BD**), ■ (**Ij-BHac**); Leaf extracts, ○ (**Ij-LD**), □ (**Ij-LHac**). ** Compounds were suggested by a preliminary comparison of retention time (t_R) and UV-Vis bands with standards and literature data.

For the case of Peak **5**, its UV-Vis pattern was comparable to those reported for lignane-type compounds. Similarly, Peak **8** presented characteristic bands associated with derivatives of coumaric acid [23]. On the other hand, for Peaks **12** and **18**, their UV-Vis spectra were characteristic of those reported for coumarin-type compounds [25–27].

Regarding the analysis of the UV-Vis spectra of Peaks **10**, **11**, **14**, and **16**, it was possible to associate them with previous reports for terpene derivatives [24]. Likewise, Peaks **13** and **17** were consistent with bibliographical data of epigallocatechin gallate derivatives [28], and Peak **15** may be associated with vanillic acid derivatives [29,30]. Finally, the absorption bands of Peaks **19**, **20**, and **21** were associated with salicylate derivatives [31].

Considering the above information, Peaks **10**, **11**, **16**, **19**, **20**, and **21**, attributed to terpenes and salicylates, were detected in the four extracts analyzed, whereas Peak **14**, which was also recognized as a terpene, was found in three extracts (absent in **Ij-BHac**). The two leaf extracts shared the presence of Peaks **6** and **12**, which were consistent with apigenin and coumarin derivatives, respectively. Despite this, as can be appreciated in Table 2, around 60% of the metabolites identified in the polar extracts of the bark and leaves of *Inga jinicuil* were only found in one extract.

It should be noted that, to date, no reports have been found on secondary metabolites present in bark or leaves from *I. jinicuil*, so this work represents a first approach for the phytochemical study of these organs of the plant. However, there are reports about the phytochemical composition for other species of the *Inga* genus, where the presence of a high content of polyphenols with an important antioxidant capacity has been demonstrated [31]. For *I. semialata* and *I. edulis*, the analysis of leaf extracts allowed the identification of compounds such as: epicatechin, apigenin C-di-hexoside, myricetin-*O*-hexose-deoxyhexose, myricetin-O-deoxyhexose, and vicenin-2 [9,32]. Likewise, other studies on leaf extracts from *I. edulis*, reported the presence of four triterpenes (lupeol, α-amirin, olean-18-ene acid, and frideline), three flavonoids, eight phenolic acids, an anthocyanin derived from delphinidin-3-glycoside, and a mixture of five acylated anthocyanins. It is important to highlight the fact that gallic acid, methyl gallate, protocatechuic acid, and quercetin were also identified [33]. For *I. laurina*, there is a presence of flavonoids 3-O-(2″-O-galloyl)-α-rhamnopyranoside and myricetin-3-rhamnoside in leaf extracts [19], as well as gallic acid, myricetin derivatives, quercetin glycoside, and glycoside myricetin-3-O-rhamnosid from ethanolic extracts of leaves from this plant [20].

In view of the above-mentioned studies, our preliminary assessment of the phytochemical profile of *Inga jinicuil* allowed the identification of a chemotaxonomic resemblance with other species of the same genus, since a shared presence of phenolic and terpenic compounds, such as gallates, protocatechuic acid, and its derivatives, as well as flavonoids such as apigenin, can be recognized. It should be emphasized that in published work, the phytochemical research reports on *I semialata*, *I. eludis*, and *I. laurina* refer mainly to polar leaf extracts, whereas the phytochemical analysis of bark has been oriented to non-polar extracts (as discussed below). Therefore, this report also contributes to the identification of secondary metabolites in polar extracts from this organ for a species of the *Inga* genus.

2.2. Chemical Profile of Hexane Extracts from Bark and Leaves of I. jinicuil by GC-MS

The analysis of the GC-MS chromatograms of **Ij-BH** and **Ij-LH** [Figure 2A,B] allowed the identification of 21 compounds, where 7 of them were only found in the bark extracts, 11 compounds only appeared in the analysis of the leaf extracts, and 3 werecommon to the extracts of both organs. Table 3 presents a list of the compounds detected arranged according to their elution order. The most abundant compounds detected for **Ij-BH** were prenol, α-tocopherol (relative abundance: 40.49%), and triterpene 24-methylenecycloartan-3-one (38.61%); these compounds represented approximately 80% of the content of this extract. For the **Ij-LH** extract, the triterpenes included lup-20 (29)-en-3-one (26.74%) and lupeol (16.44%), as well as the aliphatic compound hentriacontane (16.66%), all of which constituted nearly 60% of its metabolic content. The compounds identified in both extracts were hexadecanoic acid methyl ester, hexadecanoic acid ethyl ester, and octadecanoic acid methyl ester. It is worth mentioning that these three compounds were found in greater abundance in **Ij-BH**.

It is worth noting that this is the first report of a CG-MS analysis of hexanic extracts from *Inga jinicuil*. However, similar studies have been documented for other species of the *Inga* genus; such is the case for *I. edulis*, where triterpene compounds including lupeol and stigmasterol, as well as aliphatic compounds, have been identified from extracts of the bark and leaves [33,34]. Likewise, extracts from the bark and leaves of *I. laurina* have shown the presence of terpenes such as phytol, the aliphatic nonacosane, and esterified aliphatic acids [35], whereas in a hexanic fraction obtained from the leaves of *I. semialata*, the main compounds isolated were triterpenes, such as lupeol, α-amyrin, oleanolic acid, and friedelin [30]. In this report, the presence of esterified aliphatic acids was identified and, as in other species of the *Inga* genus, the presence of lupeol has been established. However, the following compounds: hentriacontane, α-tocopherol, lup-20 (29)-en-3-one and 24-methylenecycloartan-3-one, are reported for the first time for this genus.

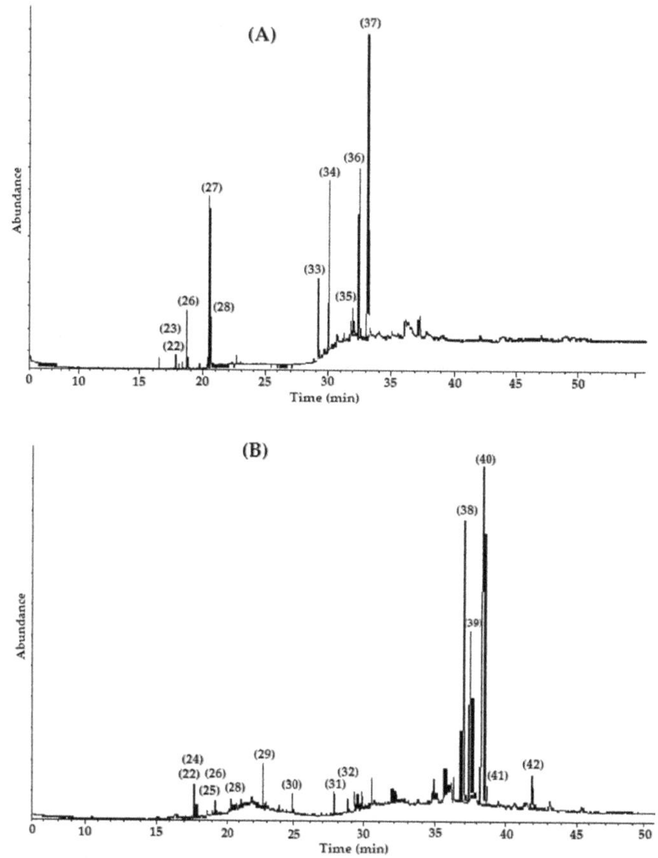

Figure 2. GC chromatograms of the hexanic extracts: (**A**) **Ij-BH** and (**B**) **Ij-LH**. The peaks are numbered in ascending order according to their retention times.

2.3. Anti-Inflammatory Activity of Organic Extracts from *Inga jinicuil*

The results corresponding to the anti-inflammatory study of the organic extracts are presented in Figure 3. At the same dose of 1.0 mg/ear, all of the extracts showed anti-inflammatory activity. For the bark extracts, the percentages of inhibition were: **Ij-BH** 34.6 ± 3.0%, **Ij-BD** 67.3 ± 2.0%, and **Ij-BHac** 24.4 ±1.0%, and for leaf extracts, the corresponding percentages were: **Ij-LH** 34.9 ± 1.3%, **Ij-LD** 23.0 ± 1.0%, and **Ij-LHac** 49.6 ± 1.0%. For indomethacin (**Indo**), which was employed as the reference drug, the inhibition percentage was 75.5 ± 2.2%. As can be seen, the two extracts with the greatest anti-inflammatory activity were **Ij-BD** followed by **Ij-LHac**, and the statistical comparison between the anti-inflammatory activities of the extracts and the reference drug revealed significant differences ($p < 0.05$). No extract reached an effect equal to or greater than that of **Indo** (Indomethacin). However, the comparison using the Tukey test of the effect of the extracts and the reference drug showed that there were no significant differences ($p < 0.05$) between some of the extracts, such as **Ij-BH** compared to **Ij-LH** and **Ij-BHac** compared to **Ij-LD**.

Table 3. Phytochemicals identified in hexanic extracts from the bark (**Ij-BH**) and leaves (**Ij-LH**) of *Inga jinicuil* by GC-MS.

Peak	Retention Time (min)	Molecular Weight (amu)	Extract(s) (% in the Sample) *	Compound **
22	17.80 17.75	268.5 268.5	▲ (1.07), △ (1.18)	2-pentadecanone,6,10,14-trimethyl
23	17.80	296.5	▲ (1.07)	3,7,11,15-Tetramethyl-2-hexadecen-1-ol
24	18.55	270.5	△ (1.04)	Hexadecanoic acid, methyl ester
25	18.61	276.3	△ (0.88)	7,9-Di-tert-butyl-1-oxaspiro(4,5)deca-6,9-diene-2,8-dione
26	18.68, 18.61	270.5	▲ (3.83), △ (0.88)	Hexadecanoic acid, ethyl ester
27	20.51	296.5	▲ (11.74)	Phytol
28	20.60, 20.47	298.5	▲ (3.14), △ (0.44)	Octadecanoic acid, methyl ester
29	22.53	324.5	△ (1.35)	4,8,12,16-tetramethylheptadecan-4-olide
30	24.63	390.6	△ (0.80)	1,2-benzenedicarboxylic acid diisooctyl ester
31	27.76	380.6	△ (1.62)	15-Tetracosenoic acid, methyl ester
32	29.19	518.7	△ (1.02)	Tetradecanoic acid, 3,3a,4,6a,7,8,9,10,10a, 10b-decahydro-3a, 10a, dihydroxy-5-(hydroxymethyl)-2, 10-dimethyl-3-oxobenz [e] azulen-8-yl ester
33	29.21	410.7	▲ (5.98)	Squalene
34	30.05	408.8	▲ (12.55)	Nonacosane
35	31.95	416.7	▲ (3.74)	*β*-Tocopherol
36	32.44	436.8	▲ (16.66)	Hentriacontane
37	33.07	430.7	▲ (40.49)	α-Tocopherol
38	36.91	424.7	△ (26.74)	Lup-20 (29)-en-3-one
39	37.39	426.7	△ (16.43)	Lupeol
40	38.35	438.7	△ (38.61)	24-Methylenecycloartan-3-one
41	38.64	412.7	△ (2.27)	Stigmast-4-en-3-one
42	41.85	440.7	△ (4.99)	9,19-Cyclolanostan-3-ol,24-methylene-, (3β)-

* Extracts: Bark extract (▲ (**Ij-BH**) and leaf extract (△ (**Ij-LH**). ** Compared with the National Institute of Standards and Technology (NIST) 1.7 Library.

Even when species such as *I. laurina*, *I. edulis*, *I. marginata*, and *I. jinicuil* are employed to treat stomach and inflammatory disorders in traditional medicine, few studies have been conducted to confirm their attributed pharmacological properties. However, recent reports have shown the presence of flavonoids and other phenolic compounds in several of these species that may be associated with pharmacological effects [36]. The present study represents a preliminary approach in the assessment of the anti-inflammatory activity of *I. jinicuil*, with the bark extracts exerting a more consistent effect and **Ij-BD** showing the highest activity. It is noteworthy to mention that the chemical profile of this extract showed the presence of salicylates, terpenoids, and derivatives of epigallocatechin gallate, as well as derivatives of protocatechuic and coumaric acids, which may be associated with its biological effect [24,28,31]. In the case of the extracts from leaves, **Ij-LHac** showed the best inhibitory effect, and the analysis of its metabolic content revealed the presence of polyphenolic compounds, terpenoids, coumarins, vanillic acid derivatives, and flavonoid-type compounds such as apigenin derivatives, all of which have reported anti-inflammatory effects [19,34,37]. Furthermore, previous reports regarding several of the metabolites present in both extracts have postulated an anti-inflammatory activity that proceeds via the inhibition of cyclooxygenase enzymes (COX) [38–40], which happens to be the known mechanism of the reference drug (**Indo**) [40]. Finally, it is important to mention that the two extracts with the highest activity have the presence of terpenes and salicylates in common; these compounds are recognized for their analgesic and anti-inflammatory effects [24,31].

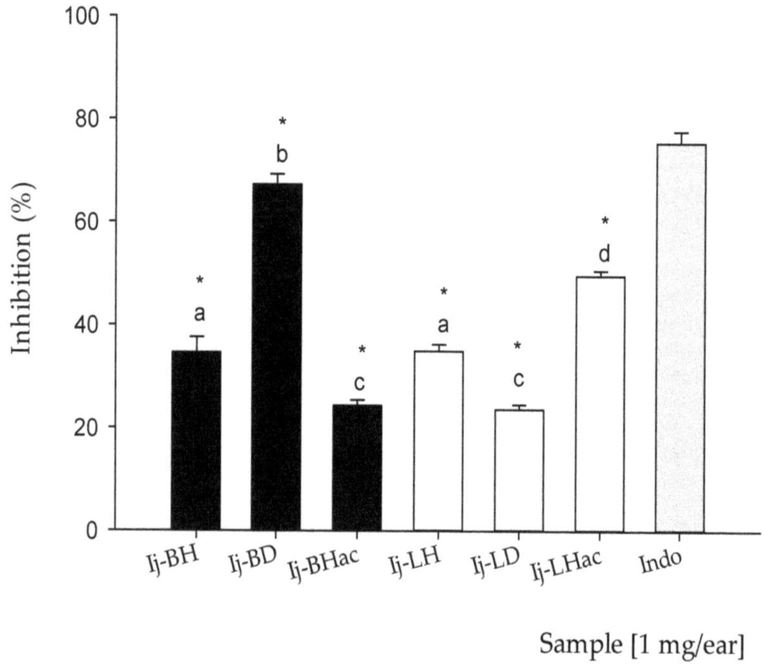

Figure 3. Percentage inhibition of inflammation (%) of **Ij-BH**, **Ij-BD**, **Ij-BHac**, **Ij-LH**, **Ij-LD**, and **Ij-LHac** extracts from *Inga jinicuil* and Indo (Indomethacin) in edema induced by TPA in mouse ear at 1.0 mg/ear. Values are presented as means ± standard error of the means (SEM). $n = 5$. ANOVA, with post-test Dunnet with * $p \leq 0.05$ in comparison with Indo and Tukey test, where different letters indicate significant differences among them.

2.4. Antibacterial Activity of *Inga jinicuil* Organic Extracts

The antibacterial activity of bark and leaf *I. jinicuil* extracts were evaluated on clinically important micro-organisms. As showed in Table 4, the three extracts of bark (**Ij-BH**, **Ij-BD**, and **Ij-BHac**) exhibited excellent activity against *Pseudomona aeruginosa* (**Pa**) and methicillin-resistant *Staphylococcus aureus* (**Sa1**), with MIC values of <3.12 and 50 µg/mL, respectively. Only **Ij-BH** showed activity against *Staphylococcus epidermidis* (**Se1**). Similarly, **Ij-LD** and **Ij-LHac** had good activity against *Pseudomona aeruginosa* (**Pa;** MIC < 3.12 µg/mL), methicillin-resistant *Staphylococcus aureus* (**Sa1**; MIC = 50 µg/mL) and *Staphylococcus epidermidis* (**Se1**; MIC = 200 µg/mL).

The results obtained are interesting considering that in 2017 the WHO published a list of "priority pathogens" resistant to antibiotics, which include *Pseudomonas aeruginosa* (resistant to carbapenems) and *Staphylococcus aureus* (resistant to methicillin), emphasizing the urgent need for the search for new agents against these micro-organisms [5].

The antibacterial activity of bark and leaf extracts against *Pseudomona aeruginosa* (**Pa**) and methicillin-resistant *Staphylococcus aureus* (**Sa1**) can be attributed to the presence of several secondary metabolites: hentriacontane and α-tocopherol in **Ij-BH,** and polyphenols, flavonoids, and terpenoids in both **Ij-BD** and **Ij-BHac** [41]. Special attention may be paid to the presence of gallate and coumarin derivatives, since their antibacterial mechanism has been described at the cell membrane level by repressing the transport system of proteins and inhibiting the biofilm formation in clinical strains of **Sa1** [17,42,43]. Furthermore, coumarin derivatives are considered as potential antibacterial agents that act as inhibitors to several binding proteins of **Sa1** and potential competitive inhibitors of the DNA-gyrase [44,45]. It is worth noting that the chemical moiety responsible for the antibacterial activity of coumarins is the basic structure of benzopyrone, which resembles the structure of benzopyridone present in antibacterial drugs derived from quinolone [44,45]. Therefore, the wide range of chemical structures found in these extracts may represent a potential source of molecular templates for new antibacterial drugs.

Table 4. Antibacterial activity (MIC µg/mL) of extracts from *Inga jinicuil.*

	Bacterial Strains									
	Gram- Positive					Gram- Negative				
Extract	**Sa1**	**Sa2**	**Se1**	**Se2**	**Sh**	**Ec1**	**Ec2**	**Ef**	**Kp**	**Pa**
Ij-LH	>200	>200	>200	>200	>200	>200	>200	>200	>200	>200
Ij-LD	**50**	>200	**200**	>200	>200	>200	>200	>200	>200	**<3.12**
IjLHac	**50**	>200	>200	>200	>200	>200	>200	>200	>200	**<3.12**
Ij-BH	**50**	>200	**50**	>200	>200	>200	>200	>200	>200	**<3.12**
Ij-BD	**50**	>200	>200	>200	>200	>200	>200	>200	>200	**<3.12**
IjBHac	**50**	>200	>200	>200	>200	>200	>200	>200	>200	**<3.12**
C1	*	*	*	*	*	*	*	*	*	*
C2	*	*	*	*	*	*	*	*	*	*
C+	–	–	–	–	–	–	–	–	–	–

Sa1: methicillin-resistant *S. aureus*; **Sa2**: *S. aureus*; **Se1**: *S. epidermis*; **Se2**: *S. epidermis*; **Sh**: clinically isolated *S. haemolyticus*; **Ec1**: *E. coli* **Ec2**: *E. cloacae*; **Ef**: *E. fecalis*; **Kp**: *K. pneumoniae*; **Pa**: *P. aeruginosa*; **C1** and **C2**: controls of viability (*: bacterial growth); **C+**: positive control (Gentamicine 100 µg/mL; –: not bacterial growth).

3. Materials and Methods

3.1. Plant Material and Extraction of Inga jinicuil

Aerial parts of *Inga jinicuil* Schltdl & Cham. ex G. Don were collected in July 2019, in Libertad, Cunduacán, Tabasco, Mexico (10 m.a.s.l., latitude 18°10′53.06 N, longitude 93°22′28.13 W). A specimen was deposited at the Herbarium of the Academic Division of Biological Sciences of the Universidad Juárez Autónoma de Tabasco for its taxonomic identification (voucher number: 36576).

Plant material was dried at room temperature in the dark for 72 h, with drying and spraying in Pulvex MP300 milled (4–6 mm). The extracts were obtained by maceration with *n*-hexane, dichloromethane, and a 60:40 ethanol:water mixture 1:4; the maceration procedure was performed three times for each solvent in order to ensure an exhaustive extraction. These extracts were filtered, concentrated in a rotary evaporator (Heidolph G3, Schwabach, Germany), and then lyophilized (Heto Drywinner DW3) to give the bark (*n*-Hexane **Ij-BH**, Dichloromethane **Ij-BD**, Hydroalcoholic **Ij-BHac**) and leaf extracts (*n*-Hexane **Ij-LH**, Dichloromethane **Ij-LD**, Hydroalcoholic **Ij-LHac**).

3.2. HPLC Analysis

Chromatographic analysis was carried out in a Waters 2695 separation module system with a Waters 2695 photodiode matrix detector and Empower Pro software (Waters Corporation, Milford, MA, USA). Chemical separation was performed using a Supelcosil LC-F column (4.6 mm × 250 mm i.d., particle size 5 µm) (Sigma-Aldrich, Bellefonte, PA, USA). The mobile acid phase was performed using 0.5% triflouroacetic, aqueous solution (solvent A), and acetonitrile (solvent B) gradient: 0–1 min, 0% of B; 2–3 min, 5% of B; 4–20 min, 30% of B; 21–23 min, 50% of B; 24–25 min, 80% of B; 26–27 min, 100% of B; 28–30 min, 0% of B. The flow rate was 0.9 mL/min with a volume of 10 µL sample. Absorbance was measured at 270 nm [46]. A preliminary identification of the peaks resolved was performed by comparison with t_R and UV-Vis characteristic bands of known standards and literature data.

3.3. GC-MS Analysis of Hexane Extracts

The chemical composition of **Ij-BH** and **Ij -LH** was analyzed on Gas Chromatography-Mass Spectrometry (GC-MS) equipment, consisting of an Agilent 6890 plus gas chromatograph coupled to a simple quadrupole mass spectrometry detector, model 5972N (Agilent Technology, Santa Clara, CA, USA).

Volatile compounds were separated on an HP 5MS capillary column (25 m long, 0.2 mm i.d., with 0.3-µm film thickness). Oven temperature was set at 40 °C for 2 min, then programmed at 40–260 °C for 10 °C/min, and maintained for 20 min at 260 °C. Mass detector conditions were as follows: interphase temperature, 200 °C, and mass acquisition range, 20–550. Injector and detector temperatures were set at 250 and 280 °C, respectively. Splitless injection mode was carried out with 1 µL of each fraction (3 mg/mL solution). The carrier gas was helium at a flow rate of 1 mL/min. The identification of volatiles was performed, comparing their mass spectra with those of the National Institute of Standards and Technology (NIST) 1.7 Library and comparing these with data from the literature [47].

3.4. *Pharmacological Activity*

3.4.1. Anti-Inflammatory Activity

Male ICR mice with a weight range of 25–30 g, from Envigo RMS, S.A. de C.V., were used throughout the experiments. These animals were maintained in the Bioterium of Centro de Investigación Biomédica Del Sur (CIBIS-IMSS) under a 12 h light-dark cycle and constant temperature (23–25 °C) with free access to food and water. The animals were treated under the Mexican federal regulations for care and use of laboratory animals, NOM-062-ZOO-1999 Guidelines [48], and international ethical guidelines for the care and use of experimental animals [49]; the number of animals (n = 5) and the intensity of the noxious stimuli utilized were the minimum necessary to demonstrate the consistent effects of the pharmacological treatments. The animal studies were approved by the Ethics Committee of the Mexican Social Security Institute (R-2020-1702-008).

Auricular inflammation was induced following the method previously described [50]. The dose evaluated for the extracts was 1.0 mg/ear. A control group received acetone as vehicle, and Indomethacin (Indo, Sigma-Aldrich, Toluca, Mexico) 1.0 mg/ear was utilized as an anti-inflammatory positive control. All treatments were dissolved in acetone and applied topically on both ears immediately after the solution of 12-*O*-tetradecanoylphorbol-13-acetate (TPA, Sigma-Aldrich, Toluca, México) as an inflammatory agent. Six hours after the administration of TPA, the animals were euthanized by cervical dislocation.

Circular sections 6 mm in diameter were taken from both the treated (t) and non-treated (nt) ears, which were weighed to determine the inflammation. The percentage of inhibition was obtained employing the expression below:

$$\% \text{ Inhibition} = [\text{Dw control} - \text{Dw treated}/\text{Dw control}] \times [100]$$

where Dw = wt − wnt; wt is the weight of the section of the treated ear; and wnt the weight of untreated ear section.

3.4.2. Antibacterial Activity

The extracts were evaluated against bacterial strains ATCC: methicillin-resistant *Staphylococcus aureus* (**Sa1**; ATCC 43330, *Staphylococcus aureus* (**Sa2**; ATCC 29213), *Staphylococcus epidermis* (**Se1**; ATCC 12228), *Staphylococcus epidermis* (**Se2**; ATCC 35984), *Enterococcus fecalis* (**Ef**; ATCC 29212), *Escherichia coli* (**Ec1**; ATCC 25922), *Enterobacter cloacae* (**Ec2**; ATCC 700323), *Klebsiella pneumoniae* (**Kp**; ATCC 700603), *Pseudomona aeruginosa* (**Pa**; ATCC 27853), and the clinically isolated *Staphylococcus haemolyticus* (**Sh**; 1038). The strains were reseeded in antibiotic agar No. 1 (Bioxon, Mexico) for 24 h at 37 °C. The strain of clinical isolate was provided from the General Hospital of Acapulco, State of Guerrero, Mexico, to the Bacteria Bank of the Autonomous University of Guerrero (UAGro).

For the trials, cultures with 24 h of incubation (37 °C) were used and about 3–4 colonies were taken of each strain and diluted in Müeller–Hinton broth (MHb; Bioxon, Toluca, Mexico). The inoculums were adjusted using the 0.5 MacFarland scale (1.5×10^8 UFC/mL). Subsequently, dilution with distilled water was performed to obtain 1×10^4 UFC/mL.

The MIC of extracts was determined by the microtiter broth dilution method [51]. Briefly, the samples (50 mg/mL) were dissolved in a DMSO–water mixture (20:80), and the tested concentrations were 3.37, 6.75, 12.5, 25, 50, 100, and 200 μg/mL. The samples were added to sterile microplates of 96 wells, along with 200 μL of MHb and 2 μL of inoculum (1×10^4 UFC/mL). The viability controls used were: MHb + DMSO + inoculum and MHb+ inoculum; Gentamicin (**C**$^+$, 100 μg/mL; Sigma Aldrich, Mexico) was employed as the reference antibiotic. The plates were incubated at 37 °C for 24 h, and after incubation the MIC was determined by adding 30 μL of a solution (0.05%) of 3-(4,5-dimethylthiazol-2-yl)-2,5, diphenyl tetrazolium bromide (MTT, Sigma-Aldrich, Hong Kong, China) in every well, of which purple development was observed if there was viability of bacteria and colorless if there was no feasibility. All assays were performed in triplicate.

3.5. Statistical Analysis

For the analysis of the anti-inflammatory activity, the data were expressed as the mean ± standard error of the mean (SEM), and statistical significance was determined using an analysis of variance (ANOVA) with a confidence level of 95% (* $p \leq 0.05$), followed by the one-tailed Dunnet test compared to Indo and the Tukey test. All analyses was performed using IBM SPSS statistics ver. 23.0 statistical program (GraphPad Software, IBM, San Diego, CA, USA).

4. Conclusions

This report presents the biological activity of organic extracts obtained from the bark and leaves of *I. jinicuil*. The anti-inflammatory activity tests showed moderate to good effects, with the dichloromethane extract from bark showing the highest activity, followed by the hexanic extract from leaves. Based on the findings of anti-inflammatory activity, it is possible to propose the exploration of the potential antinociceptive effect of the tested extracts, using an appropriate pharmacological model. Likewise, it was found that the three extracts from the bark of this plant have excellent antibacterial activity (primarily against methicillin-resistant *Staphylococcus aureus* and *Pseudomonas aeruginosa* strains) and this leads us to consider the possibility of extending antibacterial activity tests to a greater number of microbiological strains of clinical interest. On the other hand, it should be mentioned that, to our knowledge, this is the first approach to the phytochemical profiling of bark and leaves of *I. jinicuil*, which is consistent with the chemotaxonomic profiles reported for other species of *Inga* and suggest the presence of polyphenolic compounds, flavonoids, triterpenes, and lipid prenols, as well as aliphatic and esterified aliphatic lipids; these natural products may be responsible for both bioactivities assessed in this work. These results allow predicting a wide potential for future studies aimed at the isolation and structural characterization of compounds that might serve as molecular templates with specific biological activities. Finally, it is important to highlight that these results systematically contribute to the use in traditional Mexican medicine of a highly important sociocultural and nutritional species such as *Inga jinicuil* Schltdl & Cham. ex G. Don.

Supplementary Materials: The following are available online at https://www.mdpi.com/article/10.3390/plants11060794/s1, Figure S1: UV-spectra of the main compounds of (Ij-BD) *I. jinicuil*; Figure S2: UV-spectra of the main compounds of (Ij-BHac) *I. jinicuil*; Figure S3: UV-spectra of the main compounds of (Ij-LD) *I. jinicuil*; Figure S4: UV-spectra of the main compounds of (Ij-LHac) *I. jinicuil*.

Author Contributions: Conceptualization, M.G.-C., C.E.L.-G. and A.G.-R.; methodology, A.J.G.-G., R.L.-R. and P.Á.-F.; validation, M.H.-R. and M.D.P.-G.; formal analysis, A.Z.; investigation, A.G.-R. and A.B.; resources, A.J.G.-G., R.L.-R. and M.R.F.; data curation, E.J.M.-S. and M.R.F.; writing—original draft preparation, A.J.G.-G., R.L.-R. and E.A.B.-G.; writing—review and editing, A.G.-R., C.E.L.-G. and A.B.; visualization, M.H.-R. and P.Á.-F.; supervision, M.D.P.-G.; project administration, M.G.-C., A.G.-R. and C.E.L.-G.; funding acquisition, A.G.-R., M.G.-C. and A.Z. All authors have read and agreed to the published version of the manuscript.

Funding: This research was funded by PRODEP, grant number UJAT-EXB-242.

Institutional Review Board Statement: The animal studies were approved by the Ethics Committee of the Mexican Social Security Institute (R-2020-1702-008).

Informed Consent Statement: Not applicable.

Data Availability Statement: Data is contained within the article and Supplementary Material.

Acknowledgments: The authors wish to thank Arturo Pérez, Jonathan Orduño, and Ixchel Palacios (CIBIS-IMSS), and Dora Elena Aguilar Dominguez and Wilbert Pérez Fuentes (DACB-UJAT) for technical and analytical assistance; al Consejo Nacional de Ciencia y Tecnologia (CONACYT) por la beca otorgada con No. de CVU: 703106; R.L.-R. (478597) was supported by CONACYT postdoctoral fellowship (866998).

Conflicts of Interest: The authors declare no conflict of interest.

References

1. Maione, F.; Russo, R.; Khan, H.; Mascolo, N. Medicinal plants with anti-inflammatory activities. *Nat. Prod. Res.* **2015**, *30*, 1343–1352. [CrossRef] [PubMed]
2. Heydari, H.; Saltan, I.G.; Eryilmaz, M.; Bahadir, A.Ö.; Yilmaz, S.S.; Tekin, M.; Çoban, T. Antimicrobial and Anti-Inflammatory Activity of Some *Lathyrus* L. (Fabaceae) Species Growing in Turkey. *Turk. J. Pharm. Sci.* **2019**, *16*, 240–245. [CrossRef] [PubMed]
3. WHO. *WHO Traditional Medicine Strategy: 2014–2023*; WHO Press, World Health Organization: Hong Kong, China, 2013; pp. 15–19.
4. Ghasemian, M.; Owlia, S.; Owlia, M.B. Review of Anti-Inflammatory Herbal Medicines. *Adv. Pharmacol. Sci.* **2016**, *2016*, 9130979. [CrossRef] [PubMed]
5. WHO. Global Priority List of Antibiotic-Resistant Bacteria to Guide Research, Discovery, and Development of New Antibiotics. Available online: https://www.who.int/medicines/publications/WHO-PPL-Short_Summary_25Feb-ET_NM_WHO.pdf (accessed on 11 January 2022).
6. Vargas, S.G.; Pire, R. *Inga jinicuil Schtdl*; Multiuso, Á., Ed.; Universidad Juárez Autónoma de Tabasco: Villahermosa, Mexico, 2017; pp. 2–4., ISBN 978-607-606-393-4.
7. Alejandro, M.A.M.; Campillo, L.M.G.; Méndez, R.M. El Uso de las Plantas Medicinales en las Comunidades Maya-Chontales de Nacajuca, Tabasco, México. *Polibotánica* **2010**, *29*, 213–2062. Available online: http://www.scielo.org.mx/scielo.php?script=sci_arttext&pid=S1405-27682010000100011&lng=es&nrm=iso (accessed on 13 January 2022).
8. Maldonado, M.F. *Flora Medicinal del Estado de Tabasco: Uso, Manejo y Conservación*, 2nd ed.; Instituto para el Desarrollo de Sistemas de Producción del Trópico Húmedo de Tabasco: Villahermosa, Tabasco, Mexico, 2005; p. 50, ISBN 968-7991-24-0.
9. Maziero, M.; Lovato, M.O.; Lorenzoni, V.V.; Moraes, G.G.; Dornelles, R.C.; Sagrillo, M.R.; Horner, R.; Manfron, M.P. Phytochemical Study, an Evaluation of the Antioxidant Potential and the Antimicrobial Activity of *Inga semialata* (Vell.) C. Mart. Hydroalcohol Extract. *Nat. Prod. Res.* **2019**, *34*, 192–196. [CrossRef]
10. Pompeu, D.R.; Rogez, H.; Monteiro, K.M.; Tinti, S.V.; Carvalho, J.E. Capacidad antioxidante e triagem farmacológica de extratos brutos de folhas de *Byrsonima crassifolia* e de *Inga edulis*. *Acta Amaz.* **2012**, *42*, 165–172. [CrossRef]
11. Dib, H.X.; de Oliveira, D.G.L.; de Oliveira, C.F.R.; Taveira, G.B.; de Oliveira Mello, E.; Verbisk, N.V.; Chang, M.R.; Corrêa, D., Jr.; Gomes, V.M.; Macedo, M.L.R. Biochemical Characterization of a Kunitz Inhibitor from *Inga edulis* Seeds with Antifungal Activity against *Candida* spp. *Arch. Microbiol.* **2019**, *201*, 223–233. [CrossRef]
12. Alves do Nascimento, V.H.; Guimarães Sobrinho, A.C.; De Oliveira Souza, C.; Silva de Souza, J.N.; Sousa, C.L. Determination of Phenolic Compounds with Antimicrobial Activity of *Byrsonima crassifolia* and *Inga edulis* Leaves Extracts. *Ens. Ciênc.* **2021**, *25*, 21–28. [CrossRef]
13. Lima, N.; Santos, V.; La Porta, F. Chemodiversity, Bioactivity and Chemosystematics of the Genus *Inga* (FABACEAE): A Brief Review. *Rev. Virtual Quim.* **2018**, *10*, 459–473. [CrossRef]
14. Rugerio, M.O.; Hernández, M.H.H.; Suárez, R.A.; Vargas, D.M.E.; Daniel, A.B. Estudio químico y antimicrobiano de *Inga jinicuil* del estado de Tlaxcala. In Proceedings of the 7th Reunión Internacional de Investigación de Productos Naturales "Dr. Pedro Joseph-Nathan", Revista Latinoamérica de Química, Morelia, Mexico, 18–20 May 2011; Navarete, A., Ed.; Mixim: Naucalpan de Juárez, Mexico, 2011; C-95, p. 149. Available online: https://www.yumpu.com/es/document/view/27637927/latinoamericana-de-quimica-revista-latinoamericana-de- (accessed on 13 January 2022).
15. Sun, J.; Liang, F.; Bin, Y.; Li, P.; Duan, C. Screening Non-Colored Phenolics in Red Wines Using Liquid Chromatography/Ultraviolet and Mass Spectrometry/Mass Spectrometry Libraries. *Molecules* **2007**, *12*, 679–693. [CrossRef]
16. Robbins, R.J. Phenolic Acids in Foods: An Overview of Analytical Methodology. *J. Agric. Food Chem.* **2003**, *51*, 2866–2887. [CrossRef] [PubMed]
17. Gabe, V.; Kacergius, T.; Abu-Lafi, S.; Kalesinskas, P.; Masalha, M.; Falah, M.; Abu-Farich, B.; Melninkaitis, A.; Zeidan, M.; Rayan, A. Inhibitory Effects of Ethyl Gallate on *Streptococcus mutans* Biofilm Formation by Optical Profilometry and Gene Expression Analysis. *Molecules* **2019**, *24*, 529. [CrossRef] [PubMed]

18. Martins, C.M.; de Morais, S.A.L.; Martins, M.M.; Cunha, L.C.S.; da Silva, C.V.; Martins, C.H.G.; Leandro, L.F.; de Oliveira, A.; de Aquino, F.J.T.; do Nascimento, E.A.; et al. Chemical Composition, Antifungal, and Cytotoxicity Activities of *Inga laurina* (Sw.) Willd Leaves. *Sci. World J.* **2019**, *2019*, 9423658. [CrossRef] [PubMed]
19. Falcoski, T.O.R.; Lima, N.M.; Navegante, G.; Serafim, R.B.; Sorbo, J.M.; Valente, V.; Santos, V.N.C.; Santos, R.A.; Silva, D.H.S.; Soares, C.P. Genotoxicity, Cytotoxicity and Chemical Profile from *Inga laurina* (Fabaceae). *Nat. Prod. Res.* **2021**, *35*, 676–680. [CrossRef] [PubMed]
20. Kumar, S.; Pandey, A.K. Chemistry and Biological Activities of Flavonoids: An Overview. *Sci. World J.* **2013**, *2013*, 162750. [CrossRef]
21. Sun, Y.; Zeng, Q.-H.; Lu, H.-Q.; Meng, F.-C.; Shen, Y.; Zeng, W.-Y.; Chi, H.; Zhou, Y.-Q.; Chen, M. Two New Lignans from *Zanthoxylum armatum*. *Nat. Prod. Res.* **2020**, *2020*, 1–6. [CrossRef] [PubMed]
22. Guo, K.; Tong, C.; Fu, Q.; Xu, J.; Shi, S.; Xiao, Y. Identification of Minor Lignans, Alkaloids, and Phenylpropanoid Glycosides in *Magnolia officinalis* by HPLC-DAD-QTOF-MS/MS. *J. Pharm. Biomed. Anal.* **2019**, *170*, 153–160. [CrossRef]
23. Holser, R.A. Principal Component Analysis of Phenolic Acid Spectra. *ISRN Spectrosc.* **2012**, *2012*, 493203. [CrossRef]
24. Renuka, B.; Sanjeev, B.; Ranganathan, D. Evaluation of Phytoconstituents of *Caralluma nilagiriana* by FTIR and UV-VIS Spectroscopic Analysis. *J. Pharmacogn. Phytochem.* **2016**, *5*, 105–108. Available online: https://www.phytojournal.com/archives?year=2016&vol=5&issue=2&ArticleId=813 (accessed on 18 January 2022).
25. Fatima, S.; Mansha, A.; Asim, S.; Shahzad, A. Absorption Spectra of Coumarin and Its Derivatives. *Chem. Pap.* **2022**, *76*, 627–638. [CrossRef]
26. Hroboňová, K.; Sádecká, J. Coumarins Content in Wine: Application of HPLC, Fluorescence Spectrometry, and Chemometric Approach. *J. Food Sci. Technol.* **2020**, *57*, 200–209. [CrossRef] [PubMed]
27. Li, G.J.; Wu, H.J.; Wang, Y.; Hung, W.L.; Rouseff, R.L. Determination of Citrus Juice Coumarins, Furanocoumarins and Methoxylated Flavones Using Solid Phase Extraction and HPLC with Photodiode Array and Fluorescence Detection. *Food Chem.* **2019**, *271*, 29–38. [CrossRef] [PubMed]
28. Sun, S.; He, G.; Yu, H.; Yang, J.; Borthakur, D.; Zhang, L.; Shen, S.; Das, U.N. Free Zn2+ Enhances Inhibitory Effects of EGCG on the Growth of PC-3 Cells. *Mol. Nutr. Food Res.* **2008**, *52*, 465–471. [CrossRef] [PubMed]
29. Wodnicka, A.; Huzar, E. Synthesis and Photoprotective Properties of New Salicylic and Vanillic Acid Derivatives. *Curr. Chem. Lett.* **2017**, *6*, 125–134. [CrossRef]
30. Lima, N.M.; de Marqui, S.R.; Silva, D.H.S. Phytochemical, Metabolic Profiling and Antiparasitic Potential from *Inga semialata* Leaves (Fabaceae). *Nat. Prod. Res.* **2020**, *2020*, 1–6. [CrossRef]
31. Hussain, M.A.; Badshah, M.; Iqbal, M.S.; Tahir, M.N.; Tremel, W.; Bhosale, S.V.; Sher, M.; Haseeb, M.T. HPMC-Salicylate Conjugates as Macromolecular Prodrugs: Design, Characterization, and Nano-Rods Formation: Rapid Communication. *J. Polym. Sci. A Polym. Chem.* **2009**, *47*, 4202–4208. [CrossRef]
32. Alves, G.A.D.; Fernandes, S.D.; Venteu, T.T.; de Souza, R.O.; Rogez, H.; Fonseca, M.J.V. Obtainment of an Enriched Fraction of *Inga edulis*: Identification Using UPLC-DAD-MS/MS and Photochemopreventive Screening. *Prep. Biochem. Biotechnol.* **2020**, *50*, 28–36. [CrossRef]
33. Tchuenmogne, A.M.T.; Donfack, E.V.; Kongue, M.D.T.; Lenta, B.N.; Ngouela, S.; Tsamo, E.; Sidhu, N.; Dittrich, B.; Laatsch, H. Ingacamerounol, A New Flavonol and Other Chemical Constituents from Leaves and Stem Bark of *Inga edulis* Mart. *Bull. Korean Chem. Soc.* **2013**, *34*, 3859–3862. [CrossRef]
34. Lima, N.M.; Andrade, T.J.A.S.; Silva, D.H.S. Dereplication of Terpenes and Phenolic Compounds from *Inga edulis* Extracts Using HPLC-SPE-TT, RP-HPLC-PDA and NMR Spectroscopy. *Nat. Prod. Res.* **2022**, *36*, 488–492. [CrossRef]
35. Furtado, F.B.; de Aquino, F.J.T.; Nascimento, E.A.; de Martins, C.M.; de Morais, S.A.L.; Chang, R.; Cunha, L.C.S.; Leandro, L.F.; Martins, C.H.G.; Martins, M.M.; et al. Seasonal Variation of the Chemical Composition and Antimicrobial and Cytotoxic Activities of the Essential Oils from *Inga laurina* (Sw.) Willd. *Molecules* **2014**, *19*, 4560–4577. [CrossRef]
36. Marqui, S.; Santos, L.; Young, M.; Torres, L.; Bolzani, V.; Moraes, M.; Soares, C.; Silva, D. Chemopreventive Potential of Rhamnosyl Depsides from *Inga laurina*. *Planta Med.* **2010**, *76*, 47. [CrossRef]
37. Nogueira, K.M.; de Souza, L.K.M.; Medeiros, J.V.R. Technological Prospection of Anti-Inflammatory Vanilic Acid Activity, with Emphasis on Its Semisynthetic Derivative Isopropyl Vanilate. *Res. Soc. Dev.* **2021**, *10*, e35910313451. [CrossRef]
38. Enciso, E.; Arroyo, J. Efecto antiinflamatorio y antioxidante de los flavonoides de las hojas de Jungia rugosa Less (matico de puna) en un modelo experimental en ratas. *An. Fac. Med.* **2013**, *72*, 231. [CrossRef]
39. Howes, M.J.R. Phytochemicals as Anti-Inflammatory Nutraceuticals and Phytopharmaceuticals. In *Immunity and Inflammation in Health and Disease*; Chatterjee, S., Jungraithmayr, W., Bagchi, W., Eds.; Academic Press: Cambridge, UK, 2018; pp. 363–388. [CrossRef]
40. Molehin, O.R.; Adeyanju, A.A.; Adefegha, S.A.; Oyeyemi, A.O.; Idowu, K.A. Protective Mechanisms of Protocatechuic Acid against Doxorubicin-Induced Nephrotoxicity in Rat Model. *J. Basic Clin. Physiol. Pharmacol.* **2019**, *30*, 20180191. [CrossRef] [PubMed]
41. Pistelli, L.; Bertoli, A.; Noccioli, C.; Mendez, J.; Musmanno, R.A.; Di Maggio, T.; Coratza, G. Antimicrobial Activity of *Inga fendleriana* Extracts and Isolated Flavonoids. *Nat. Prod. Commun.* **2009**, *4*, 1679–1683. [CrossRef]

42. Favela-Hernández, J.M.J.; Clemente-Soto, A.F.; Balderas-Rentería, I.; Garza-González, E.; del Camacho-Corona, M.R. Potential Mechanism of Action of 3'-Demethoxy-6-O-Demethyl-Isoguaiacin on Methicillin Resistant *Staphylococcus aureus*. *Molecules* **2015**, *20*, 12450–12458. [CrossRef]
43. Wang, D.; Jin, Q.; Xiang, H.; Wang, W.; Guo, N.; Zhang, K.; Tang, X.; Meng, R.; Feng, H.; Liu, L.; et al. Transcriptional and Functional Analysis of the Effects of Magnolol: Inhibition of Autolysis and Biofilms in *Staphylococcus aureus*. *PLoS ONE* **2011**, *6*, e26833. [CrossRef]
44. Pisano, M.B.; Kumar, A.; Medda, R.; Gatto, G.; Pal, R.; Fais, A.; Era, B.; Cosentino, S.; Uriarte, E.; Santana, L.; et al. Antibacterial Activity and Molecular Docking Studies of a Selected Series of Hydroxy-3-Arylcoumarins. *Molecules* **2019**, *24*, 2815. [CrossRef]
45. Kampranis, S.C.; Gormley, N.A.; Tranter, R.; Orphanides, G.; Maxwell, A. Probing the Binding of Coumarins and Cyclothialidines to DNA Gyrase. *Biochemistry* **1999**, *38*, 1967–1976. [CrossRef]
46. De Araújo, R.S.A.; Barbosa, F.J.M.; Scotti, M.T.; Scotti, L.; da Cruz, R.M.D.; Falcão, S.V.D.S.; de Siqueira, J.J.P.; Mendonça, J.F.J.B. Modulation of Drug Resistance in *Staphylococcus aureus* with Coumarin Derivatives. *Scientifica* **2016**, *2016*, 6894758. [CrossRef]
47. Adam, R.P. *Identification of Essential Oils Components by Gas Chromatography/Mass Spectrometry*, 4th ed.; Allured Business Media: Carol Stream, IL, USA, 2007; pp. 1–804. ISBN 978-193-263-32-14.
48. SAGARPA. Norma Oficial Mexicana NOM-062-ZOO-1999. Especificaciones Técnicas para la Producción, Cuidado y Uso de los Animales de Laboratorio (Technical Specifications for the Production, Care, and Use of Laboratory Animals). *México Diario Oficial de la Federación*. 22 August 2001. Available online: https://www.gob.mx/cms/uploads/attachment/file/203498/NOM-062-ZOO-1999_220801.pdf (accessed on 2 February 2022).
49. Zimmermann, M. Ethical guidelines for investigations of experimental pain in conscious animals. *Pain* **1983**, *16*, 109–110. [CrossRef]
50. Payá, M.; Ferrándiz, M.L.; Sanz, M.J.; Bustos, G.; Blasco, R.; Rios, J.L.; Alcaraz, M.J. Study of the Antioedema Activity of Some Seaweed and Sponge Extracts from the Mediterranean Coast in Mice. *Phytother. Res.* **1993**, *7*, 159–162. [CrossRef]
51. Salazar, P.D.T.; Castro, A.N.; Moreno, G.M.E.; Nicasio, T.M.P.; Perez, H.J.; Alvarez, F.P. Antibacterial and Anti-Inflammatory Activity of Extracts and Fractions from *Agave cupreata*. *Int. J. Pharmacol.* **2017**, *13*, 1063–1070. [CrossRef]

plants

MDPI

Article

A Standardized *Lindera obtusiloba* Extract Improves Endothelial Dysfunction and Attenuates Plaque Development in Hyperlipidemic ApoE-Knockout Mice

Sang-Hyun Ihm [1,2,†], Sin-Hee Park [1,3,†], Jung-Ok Lee [4], Ok-Ran Kim [1,3], Eun-Hye Park [1,3], Kyoung-Rak Kim [4], Jong-Hoon Kim [5], Byung-Hee Hwang [1,3], Ho-Joong Youn [1,3], Min-Ho Oak [6,*] and Kiyuk Chang [1,3,*]

1 College of Medicine, Cardiovascular Research Institute for Intractable Disease, The Catholic University of Korea, Seoul 06591, Korea; limsh@catholic.ac.kr (S.-H.I.); akqmf02@gmail.com (S.-H.P.); okrane@hanmail.net (O.-R.K.); park1119@catholic.ac.kr (E.-H.P.); hbhmac@gmail.com (B.-H.H.); 93015195@cmcnu.or.kr (H.-J.Y.)
2 Division of Cardiology, Department of Internal Medicine, Bucheon St. Mary's Hospital, The Catholic University of Korea, Seoul 06591, Korea
3 Division of Cardiology, Department of Internal Medicine, Seoul St. Mary's Hospital, The Catholic University of Korea, Seoul 06591, Korea
4 Research and Development Center, Han Wha Pharma, Co., Ltd., Chuncheon 24468, Korea; leejeongok@gmail.com (J.-O.L.); krkim@hwpharm.com (K.-R.K.)
5 Research Center, YangJi Chemicals, Suwon 16229, Korea; jonghoonk@hwpharm.com
6 College of Pharmacy, Mokpo National University, Muan-gun 58554, Korea
* Correspondence: mhoak@mokpo.ac.kr (M.-H.O.); kiyuk@catholic.ac.kr (K.C.)
† Equal contributors.

Citation: Ihm, S.-H.; Park, S.-H.; Lee, J.-O.; Kim, O.-R.; Park, E.-H.; Kim, K.-R.; Kim, J.-H.; Hwang, B.-H.; Youn, H.-J.; Oak, M.-H.; et al. A Standardized *Lindera obtusiloba* Extract Improves Endothelial Dysfunction and Attenuates Plaque Development in Hyperlipidemic ApoE-Knockout Mice. *Plants* **2021**, *10*, 2493. https://doi.org/10.3390/plants10112493

Academic Editors: Juei-Tang Cheng, I-Min Liu and Szu-Chuan Shen

Received: 21 October 2021
Accepted: 11 November 2021
Published: 18 November 2021

Publisher's Note: MDPI stays neutral with regard to jurisdictional claims in published maps and institutional affiliations.

Abstract: *Lindera obtusiloba* extract (LOE), a traditional herbal medicine used to enhance blood circulation and to reduce inflammation, induced NO-mediated endothelium-dependent relaxation, and reduced the formation of reactive oxygen species (ROS). The study investigated whether LOE improves endothelial dysfunction and reduces plaque inflammation and progression by inhibiting ROS generation in a mouse model of atherosclerosis. Eight-week-old apolipoprotein E-deficient (apoE$^{-/-}$) mice fed with a western diet (WD) were randomized into different groups by administering vehicle (0.5% carboxymethylcellulose (CMC)), LOE (100 mg/kg/day), or losartan (30 mg/kg/day) by gavage until the age of 28 weeks. Fourteen male C57BL/6 mice that were fed normal chow and treated with CMC were used as negative controls. Similar to losartan treatment, LOE treatment induced the concentration-dependent relaxation of aorta rings in WD-fed apoE$^{-/-}$ mice. LOE treatment significantly reduced the vascular ROS formation and expression of NADPH oxidase subunits, including p22phox and p47phox. Compared with WD-fed apoE$^{-/-}$ mice, mice exposed to chronic LOE treatment exhibited reductions in plaque inflammation-related fluorescence signals and atherosclerotic lesions. These effects were greater than those of losartan treatment. In conclusion, LOE treatment improves endothelial dysfunction and reduces plaque inflammation as well as lesion areas by reducing vascular NADPH oxidase-induced ROS generation in a mouse model of atherosclerosis.

Keywords: *Lindera obtusiloba*; apolipoprotein E-deficient mice; atherosclerosis; endothelium

1. Introduction

Atherosclerosis is a narrowing of the arteries caused by the buildup of plaque. It is characterized by endothelial dysfunction, oxidative stress and chronic low-grade inflammation [1–3]. Endothelial dysfunction is the initial event leading to atherogenesis and is associated with increased generation of NADPH oxidase-derived reactive oxygen species (ROS) in vessel walls [4,5]. Increased vascular ROS generation, including hydrogen peroxide and superoxide anions, contributes to the formation of atherosclerotic plaque by inducing oxidative stress, reducing nitric oxide (NO) bioavailability, and promoting proinflammatory responses [3,6,7]. Vascular NADPH oxidase represents an important source of ROS in numerous cardiovascular diseases involving the pathogenesis of atherosclerosis [8–10]. Indeed, Nox2 or p47phox protein deficiency reduced the formation of atherosclerotic plaque in high-fat diet-fed apolipoprotein E-deficient (apoE$^{-/-}$) mice [11,12], indicating that NADPH oxidase was crucial for progression of atherosclerotic lesions. Furthermore, inflammation is related to all aspects of atherosclerosis, including the formation of foam cells, the progression and disruption of plaque, and the formation of thrombus [13]. Upregulation of adhesion molecules, cytokines, and chemokines expressed in endothelial cells and their complex interaction promotes leukocyte infiltration into the vascular wall, followed by transendothelial migration, which triggers atherogenesis [14,15].

Lindera obtusiloba, which is widely distributed in East Asian countries, has long been used as a traditional herbal medicine to augment circulation in the body, possibly via vasodilation; to suppress inflammation; and to prevent hepatic injury [16]. The extracts were derived from different parts of *L. obtusiloba* and their biologically active compounds such as butenolides, polyphenols, flavonoids, lignans, and neolignans exhibit antioxidant effects. The cytotoxic, anti-allergic, neuroprotective, antithrombotic, and anti-inflammatory effects of these bioactive compounds have been reported [17–20]. Thus, *L. obtusiloba* possesses an abundance of bioactive compounds, especially antioxidants, which have been studied in several diseases associated with oxidative stress. In addition, the *L. obtusiloba* extract (LOE) inhibits adipogenesis via persistent Wnt signaling and diminishes the tumor necrosis factor α- and lipopolysaccharide-induced IL-6 secretion by preadipocytes, suggesting its therapeutic potential in metabolic syndrome and obesity [21]. Our previous study demonstrated that LOE induces NO-mediated endothelium-dependent relaxation, reduces ROS generation in isolated aortic rings, and prevents hypertension and endothelial dysfunction induced by angiotensin II in rats [22]. In addition, we reported that LOE improves vascular oxidative stress and endothelial dysfunction most likely via normalization of the angiotensin system in diabetic mice [23]. Overall, these findings suggest that LOE has the potential to inhibit ROS generation, to improve endothelial dysfunction in vessel walls, and to reduce inflammatory cytokine production.

Therefore, the aim of the present study was to assess whether LOE improves endothelial dysfunction and prevents the development of atherosclerosis by reducing vascular ROS generation in an experimental model of atherosclerosis, the apoE$^{-/-}$ mice. In particular, the effect of the LOE intake was determined on (1) the endothelium-dependent vascular relaxation of mice aortic rings, (2) the vascular generation of ROS and NADPH oxidase subunits in aortic sections, and (3) the plaque inflammation of aortas and atherosclerotic plaque burden in the aortic sinus.

2. Results

2.1. LOE Improves Endothelial Dysfunction in WD-Fed apoE$^{-/-}$ Mice

The effect of LOE on endothelial dysfunction was investigated by monitoring endothelium-dependent vascular relaxation. Compared with aortic rings isolated from chow diet-fed C57BL/6 mice, those obtained from WD-fed apoE$^{-/-}$ mice showed decreased relaxation to acetylcholine (Ach), suggesting impaired endothelium-dependent vascular relaxation (Figure 1). Both LOE and losartan treatment significantly improved the endothelium-dependent vascular relaxation of aorta rings derived from WD-fed apoE$^{-/-}$ mice (Figure 1).

2.2. LOE Decreases Excessive Vascular ROS Formation by Inhibiting NADPH Oxidase Subunits in WD-Fed apoE$^{-/-}$ Mice

To determine the in vivo vascular pro-oxidant effects of LOE therapy, the vascular ROS in the aortic walls of apoE$^{-/-}$ mice and C57BL/6 mice were evaluated using the redox-sensitive fluorescent probe dihydroethidine (DHE). Almost no fluorescence signal was observed in the aortic walls of chow diet-fed C57BL/6 mice (Figure 2). In contrast, there was a significant increase in the DHE fluorescence signal in aortic plaques of WD-fed apoE$^{-/-}$ mice compared with chow diet-fed C57BL/6 mice (Figure 2). Similar to losartan treatment (18.1 ± 0.5 vs. 30.8 ± 1.8, 41.3 ± 1% reduction), LOE administration markedly reduced DHE fluorescence in WD-fed apoE$^{-/-}$ mouse aortas (14.6 ± 1.6 vs. 30.8 ± 1.8 and 52.6 ± 5% reduction) (Figure 2).

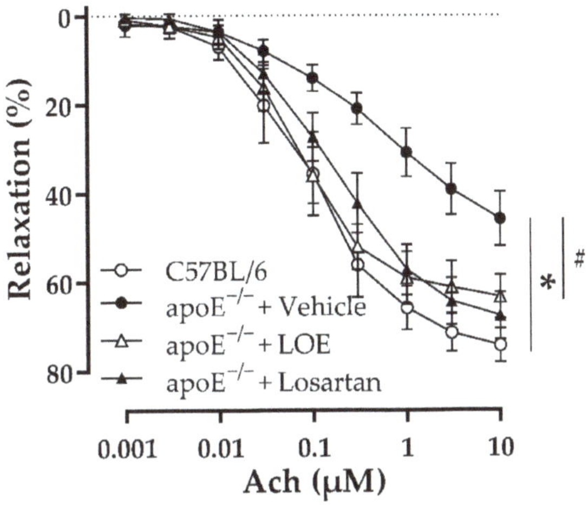

Figure 1. LOE and losartan treatments improve endothelium-dependent relaxation in response to Ach in the aortas of apoE$^{-/-}$ mice. Aortic rings (3–4 mm in length) with endothelium derived from the indicated groups of mice were suspended in organ baths containing oxygenated Krebs solution and precontracted with phenylephrine (100 nM) before the construction of concentration–relaxation curves to Ach (1 nM–10 μM). The results are shown as mean ± SEM (n = 5–7). * $p < 0.05$ indicates a significant difference between the apoE$^{-/-}$ vehicle group versus the C57BL/6 group and # $p < 0.05$ the apoE$^{-/-}$ LOE group or apoE$^{-/-}$ losartan group versus the apoE$^{-/-}$ vehicle group.

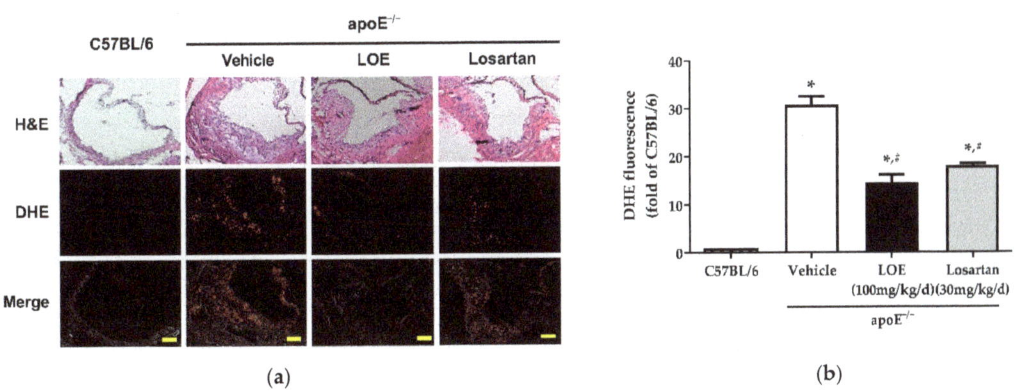

Figure 2. LOE and losartan treatments prevent vascular generation of reactive oxygen species in aortic sections derived from apoE$^{-/-}$ mice. The aortic sections were exposed to dihydroethidine (DHE, 2.5 μM), a redox-sensitive fluorescent dye, for 30 min. Subsequently, ethidium fluorescence was evaluated via confocal microscopy. (**a**) Representative images show H&E staining (**top**), DHE staining in red (**middle**), and merged images (**bottom**). (**b**) Corresponding cumulative data. The scale bar represents 100 μm. The results are shown as mean ± SEM (n = 4–6). * $p < 0.05$ versus the C57BL/6 group and # $p < 0.05$ versus the apoE$^{-/-}$ vehicle group.

In addition, since ROS generation in apoE$^{-/-}$ mice has been correlated with the upregulation of NADPH oxidase activity, the activity of NADPH oxidase was assessed based on the expression of its subunits, including p22phox and p47phox [11,24]. The fluorescence signals of p22phox and p47phox were markedly increased in WD-fed apoE$^{-/-}$ mice; however, LOE intake significantly reduced the levels of both NADPH oxidase subunits (p22phox: 9.5 ± 1.1 vs. 39.1 ± 1.9; p47phox: 8.6 ± 1.2 vs. 37 ± 0.3) (Figure 3).

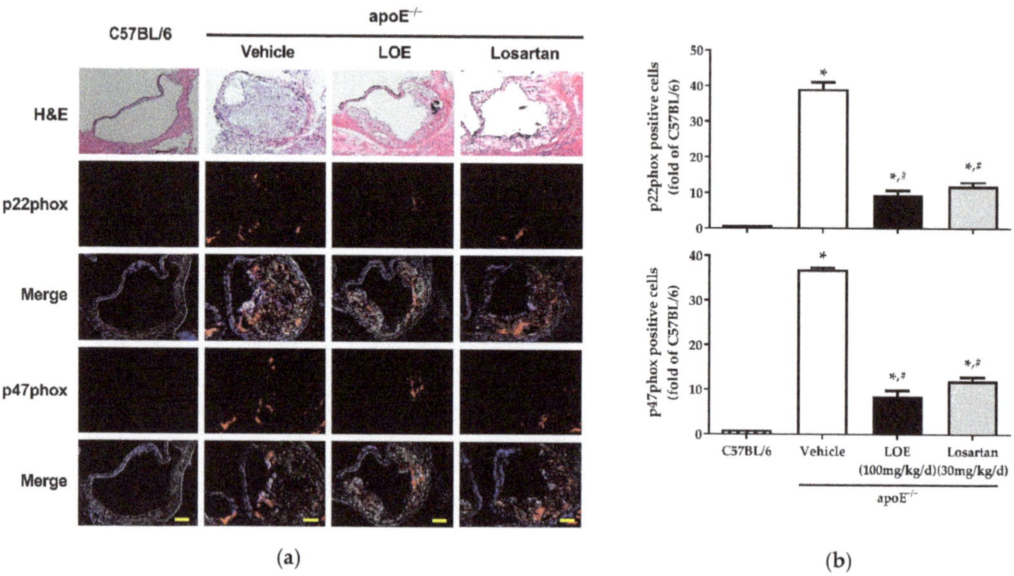

Figure 3. LOE and losartan treatments inhibit NADPH oxidase subunits p22phox and p47phox in aortic sections obtained from apoE$^{-/-}$ mice. The expression of the NADPH oxidase subunits p22phox and p47phox was determined by confocal microscopy using a purified polyclonal antibody and a fluorescence-tagged secondary antibody. Nuclei were stained with DAPI (blue). (**a**) Representative images show H&E staining (**top**), p22phox and p47phox staining in red (**middle**), and merged fluorescence staining and DAPI (**bottom**). (**b**) Corresponding cumulative data. The scale bar represents 100 μm. The results are shown as mean ± SEM (n = 4–6). * $p < 0.05$ versus the C57BL/6 group and # $p < 0.05$ versus the apoE$^{-/-}$ vehicle group.

2.3. LOE Suppresses Inflammation in Murine Aortic Atherosclerosis

WD-fed apoE$^{-/-}$ mice exhibited abundant atherosclerotic plaques, especially in the aortic root, arch, and abdominal aorta along the renal artery branches, while atherosclerotic plaques were rarely observed in the aortas of C57BL/6 mice. The plaque inflammation-modulating effects of LOE were assessed via an ex vivo fluorescence reflectance imaging (FRI) study performed to measure inflammation in the aortic atherosclerotic plaques. After 20 weeks of treatment, the aortas of WD-fed apoE$^{-/-}$ mice showed markedly increased inflammation compared with those of C57BL/6 mice (Figure 4). LOE treatment significantly reduced the degree of plaque inflammation in WD-fed apoE$^{-/-}$ mice (594.7 ± 429.8 vs. 1971.0 ± 99.9 AU), whereas losartan treatment did not (Figure 4).

2.4. LOE Reduces Atherosclerotic Plaque Burden in apoE$^{-/-}$ Mice in WD-Fed apoE$^{-/-}$ Mice

Hematoxylin and eosin (H&E) staining showed atherosclerotic plaques predominantly in the aortic sinus. The plaque area in the aortas of WD-fed apoE$^{-/-}$ mice was increased compared with that of aortas derived from C57BL/6 mice (Figure 5). While losartan failed to reduce the atherosclerotic lesion area in WD-fed apoE$^{-/-}$ mice (0.47 ± 0.07 vs. 0.55 ± 0.03 mm^2), LOE treatment significantly reduced the aortic plaque area (0.33 ± 0.03 vs. 0.55 ± 0.03 mm^2) (Figure 5).

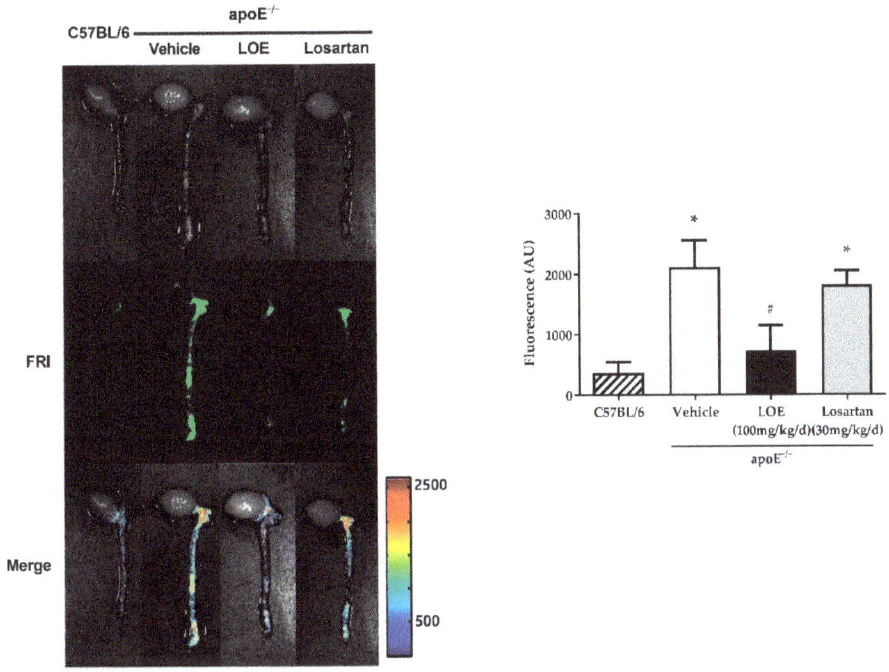

(a) (b)

Figure 4. Effect of LOE and losartan treatments on plaque inflammation in the aortas of apoE$^{-/-}$ mice. LOE treatment significantly reduced the degree of plaque inflammation in western diet-fed apoE$^{-/-}$ mice, whereas losartan treatment did not. In vivo imaging of atherosclerotic plaque inflammation was assessed by fluorescence reflectance imaging (FRI) using AP-HGC-Cy5.5 nanoparticles. The extent of plaque inflammation was measured with an IVIS-200 FRI system. (**a**) Middle: FRI; bottom: merged images. (**b**) Corresponding cumulative data. The results are expressed as mean ± SEM (n = 4–6). $^{*}p < 0.05$ versus the C57BL/6 group and $^{\#}p < 0.05$ versus the apoE$^{-/-}$ vehicle group.

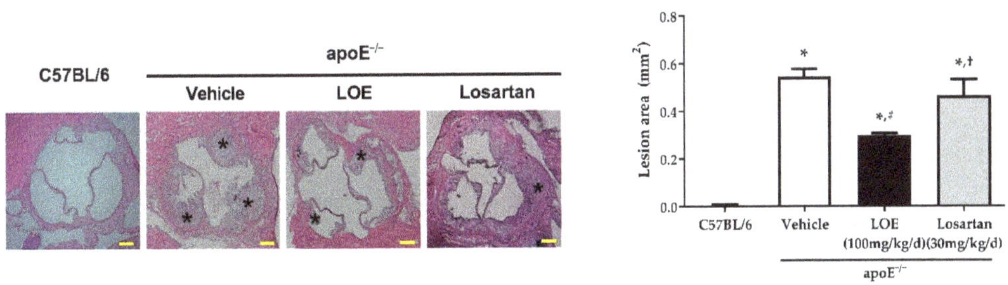

(a) (b)

Figure 5. Effect of LOE and losartan treatments on the atherosclerotic plaque areas of aortic sections obtained from apoE$^{-/-}$ mice. LOE treatment significantly reduced the extent of aortic plaque, whereas losartan failed to reduce the area of atherosclerotic lesion in western diet-fed apoE$^{-/-}$ mice. The atherosclerotic plaque area (mm^2) was measured as the area between the internal elastic lamina and the lumen and quantified by ImageJ after H&E staining. (**a**) H&E staining. Asterisk (*) indicates atherosclerotic plaque. (**b**) Corresponding cumulative data. The scale bar represents 200 μm. The results are expressed as mean ± SEM (n = 4–6). $^{*}p < 0.05$ versus the C57BL/6 group, $^{\#}p < 0.05$ versus the apoE$^{-/-}$ vehicle group and $^{\dagger}p < 0.05$ versus the apoE$^{-/-}$ LOE group.

3. Discussion

The remnants of apolipoprotein B-containing lipoprotein within the arterial walls [25] and subsequent oxidative modification trigger an inflammatory response and endothelial dysfunction [26,27], leading to the formation of atherosclerotic plaques. Current therapeutics against atherosclerotic cardiovascular diseases cannot completely prevent vascular ROS-induced atherogenesis. The major findings of the present study indicate that LOE decreases vascular oxidative stress by suppressing the expression of NADPH oxidase, resulting in improved endothelial dysfunction and in the prevention of atherosclerotic inflammation and lesion progression in a mouse model of atherosclerosis.

A damaged endothelium disturbs the balance between vasoconstrictors and vasodilators and leads to multiple events that promote or exacerbate atherosclerosis [28]. Several studies have suggested that a number of polyphenol-rich natural products are capable of improving endothelium-dependent relaxation by enhancing the endothelial production of vasoprotective factors including NO and EDH and ultimately prevent endothelial dysfunction in cardiovascular diseases including hypertension [29–31]. Our previous studies revealed that LOE was a potential vasorelaxant acting via NO, and that LOE-induced relaxation was attenuated by PI3-kinase/Akt pathway inhibitors in isolated aortic rings. LOE induced a time-dependent phosphorylation of Akt at Ser473 and eNOS at Ser1177 in endothelial cells [22]. The findings suggest that LOE was an activator of the PI3-kinase/Akt-dependent eNOS phosphorylation at Ser 1177. In addition, LOE treatment attenuated endothelial dysfunction and hypertension induced by angiotensin II in rats [22]. Long-term administration of LOE to diabetic mice restored the abolished endothelium-dependent relaxation to Ach in aortic rings and ameliorated hyperglycemia partially [23]. The current findings based on aortic rings isolated from apoE$^{-/-}$ mice, an experimental model of atherosclerosis, clearly indicate that chronic LOE treatment improved endothelium-dependent vascular relaxation to Ach in WD-fed apoE$^{-/-}$ mice to a degree comparable to losartan. Losartan is a well-known angiotensin II receptor antagonist with anti-hypertensive activity and is used to treat hypertension and heart failure. Previous studies have shown that losartan exhibits anti-atherosclerotic effects in apoE$^{-/-}$ mice by reducing lipid accumulation and macrophage infiltration as well as by inhibiting LDL lipid peroxidation [32,33].

Endothelial dysfunction is triggered by high levels of NADPH oxidase-derived superoxide anions, which react with NO, thereby decreasing its bioavailability in the arterial wall. NADPH oxidase, which was detected in neutrophils, is found in vascular endothelial cells and smooth muscle cells, and essential source of superoxide formation in several animal models of vascular disease [34]. Furthermore, p22phox, one of the NADPH oxidase components, has been identified in atherosclerotic coronary arteries of humans [35]. In this study, we demonstrated that LOE inhibited the increased levels of vascular oxidative stress in apoE$^{-/-}$ mice as indicated by the pronounced DHE staining in all of the aortic plaques and the arterial wall as well as the excessive vascular expression of NADPH oxidase subunits, including p22phox and p47phox. The findings indicate that the potential effect of LOE intake on endothelium-dependent vascular relaxation is most likely associated with its ability to reduce vascular oxidative stress, partially via decreasing the NADPH oxidase expression in apoE$^{-/-}$ mice. Indeed, previous studies suggested that tea and grape-derived polyphenols downregulate the expression of NADPH oxidase subunits such as p22phox and nox1, and tea polyphenols induce upregulation of catalase expression in vascular cells [29,36]. Since endothelial dysfunction is identified ahead of structural alterations in the arterial wall, it most likely acts as an early signaling event in the pathological initiation and progression of atherosclerosis [28].

Inflammation, which is mediated by various factors such as cytokines, adhesion molecules, and NO, is considered a vital component of atherogenesis [1,3]. Several studies have shown that dietary polyphenols, especially theaflavin and quercetin, ameliorate atherosclerosis by improving inflammation and bioavailability of NO in apoE$^{-/-}$ mice [37,38]. We employed molecular FRI as a sensing platform to measure plaque inflammation after injecting AP-HGC-Cy5.5 nanoparticles as molecular imaging agents. These nanoparticles selectively distinguish atherosclerotic plaques by binding to the IL-4 receptor on macrophages, endothelial cells, and smooth muscle cells. Therefore, this molecular imaging tool facilitates the visualization of early atherosclerotic plaque lesions [39]. We detected that increased plaque inflammation in WD-fed apoE$^{-/-}$ mice and LOE treatment for 20 weeks resulted in marked regression of plaque inflammation, even more than losartan treatment. A reduction in plaque inflammation in LOE-treated apoE$^{-/-}$ mice decreased the aortic plaque area, suggesting anti-atherogenic and anti-inflammatory effects of LOE in apoE$^{-/-}$ mice.

4. Materials and Methods

4.1. Plant Extraction and Standardization

Plant extraction and standardization were performed as described previously [22]. Briefly, the *L. obtusiloba* stems were collected in the vicinity of Hongcheon, Korea, and the voucher specimen (no. YJP-14) was stored at the Herbarium of KIST Gangneung Institute (Gangneung, Korea). *L. obtusiloba* was identified by Dr. Sang Hoon Jung, KIST Gangneung Institute, Gangneung, Korea. Short and dried branches were segmented into small slices and pulverized with a commercially available food mixer. Using 50% ethanol, 15.8 kg of *L. obtusiloba* powder was extracted four times for 4 h at 70 °C. The ethanolic solution was vaporized to dryness under vacuum to obtain 1.7 kg of the total extract (yield 7.0%). The polyphenol content in the LOE was evaluated via the Folin–Ciocalteau assay and was approximately 23.7% described as (−)-epicatechin equivalents. Previous study has reported that the major compounds in LOE were hyperin, isoquercitrin, guaijaverin, avicularin, and quercitrin [22,23]. Among them, hyperin and isoquercitrin were determined as standard substances for standardization of LOE. An HPLC analysis of LOE yielded a linear calibration curve for the standard compounds, which include hyperin and isoquercitrin. The main compounds including hyperin and isoquercitrin were identified in *L. obtusiloba* stems and designated as standard compounds by the Korea Food and Drug Administration (KFDA). LOE (10 µL) in 50% aqueous methanol (10 mg/mL), hyperin (16 µg/mL), and isoquercitrin (7 µg/mL) was loaded onto an Agilent 1200 series HPLC system. Separation was performed using a C18 reverse-phase column (4.6 × 250 mm, 10 µm, Shiseido, Tokyo, Japan) at 35 °C and eluted at a flow rate of 1.0 mL/min using a mobile phase of 17% aqueous acetonitrile acidified with 0.1% trifluoroacetic acid. Chromatographic profiles were recorded at 254 nm. The correlation coefficient was 0.998 for the respective standard curve. Hyperin and isoquercitrin in the LOE were identified and quantified at 15.33 min and 16.48 min, respectively, as shown in Supplementary Figure S1. The results revealed that 100 mg of LOE contains 0.138 mg of hyperin and 0.055 mg of isoquercitrin. These concentrations are within the appropriate ranges approved by the KFDA. All other quality control test results including heavy metals and pesticide residues were within acceptable limits.

4.2. Animals

This study complied with the Guide for the Care and Use of Laboratory Animals published by the US National Institutes of Health (NIH Publication No. 85-23, revised 1996) and was approved by the IACUC of Gyeonggi Biocenter (Approval No. 2009-09-09). For this study, 14 male C57BL/6 mice (6 weeks old, 21–24 g) and 42 male apoE$^{-/-}$ mice (6 weeks old, 20–25 g) were purchased from Central Lab (Seoul, Korea) and acclimated with normal chow diet for 2 weeks. At the age of 8 weeks, apoE$^{-/-}$ mice were randomized into 3 groups; fed a western diet (WD: Research Diets, Inc. (New Brunswick, NJ, USA) composed of (wt/wt) 20% protein, 50% carbohydrate, 21% fat, and 0.15% cholesterol); and administered the vehicle (0.5% carboxymethylcellulose (CMC); apoE$^{-/-}$ vehicle group, n = 14), LOE (100 mg/kg per day; apoE$^{-/-}$ LOE group, n = 14), or losartan (30 mg/kg per day; apoE$^{-/-}$ losartan group, n = 14) by gavage until the age of 28 weeks. C57BL/6 mice exposed to a normal chow diet and treated with 0.5% CMC by gavage were used as negative controls. The dose of losartan was selected based on previous studies indicating that a maximal functional effect without hemodynamic compromise can be obtained with this dose (data not shown). After 20 weeks of treatment, blood was collected under pentobarbital (50 mg/kg, i.p.) anesthesia, and the aorta and the heart were removed. The blood was centrifuged (3000 rpm, 10 min), and serum was obtained and stored at −70 °C until use.

4.3. Vascular Reactivity

Aortic rings (3–4 mm in length) were obtained and mounted in myographs and placed in organ baths containing Krebs bicarbonate solution (in mM: 119 NaCl, 4.7 KCl, 1.18 KH_2PO_4, 1.18 $MgSO_4$, 1.25 $CaCl_2$, 25 $NaHCO_3$, and 11 D-glucose, pH 7.4), which was oxygenated (95% O_2; 5% CO_2) and warmed to 37 °C to measure isometric tension. Following equilibration for 60 min under a resting tension of 1.0 g, the maximal contraction was measured by monitoring vasoconstriction evoked using a potassium-rich Krebs solution (80 mM) for 10 min. Subsequently, the rings were washed for 60 min and contracted with phenylephrine (100 nM). After a washout and a 30-min equilibration period, the rings were contracted again with increasing concentrations of phenylephrine to approximately 80% of the maximal contraction. The relaxation induced by cumulative treatment with Ach (1 nM–10 μM) on a half-logarithmic scale was measured to yield a concentration-relaxation curve to Ach. Concentration response is expressed as a percentage of contraction by phenylephrine.

4.4. Determination of Vascular ROS Formation

The in situ ROS synthesis was measured using the oxidative fluorescent dye DHE (Sigma–Aldrich, Milwaukee, WI, USA) as previously described [4]. Thoracic aortas from all groups were embedded into an optimal cutting temperature (OCT) compound (O.C.T. Tissue-Tek, Sakura Finetek, Torrance, CA, USA) and then frozen in liquid nitrogen for cryostat sectioning. The frozen aortas were sliced into 5 μm thick sections, followed by incubation with DHE (2.5 μM) in a humidified light-protected chamber for 30 min at 37 °C. Images were examined using a confocal microscope (LSM 510 META, Carl Zeiss, Inc., Overkochen, Germany) with a 20× epifluorescence objective. Mean intensities are expressed as arbitrary densitometric units.

4.5. Immunohistochemical Analysis of NADPH Oxidase Subunit Expression

Thoracic aorta sections (5 μm) were sliced from paraffin blocks fixed with 4% paraformaldehyde. Antigen retrieval was performed by heating the sections in a 60 °C oven with a 10 mM citrate buffer overnight. After cooling, the sections were incubated with 3% hydrogen peroxide for 10 min to inhibit endogenous peroxidases. A rabbit polyclonal anti-p22phox antibody (Santa Cruz Biotechnology, Santa Cruz, CA, USA) was used at a 1:50 dilution for p22phox immunostaining. An Alexa Fluor 488-conjugated secondary antibody (Invitrogen Corp., Carlsbad, CA, USA) was used at a 1:2000 dilution. For p47phox immunostaining, the sections were incubated with a rabbit polyclonal anti-p47phox antibody (Santa Cruz Biotechnology, Santa Cruz, CA, USA) diluted at 1:50 in 0.5 M TBST at 4 °C overnight. An Alexa Fluor 555-conjugated secondary antibody (Invitrogen Corp., Carlsbad, CA, USA) was used at a 1:2000 dilution. Nuclear counterstaining was accomplished with DAPI at a 1:10,000 dilution. The sections were stored in the dark until they were analyzed using a confocal microscope (LSM 510 META, Carl Zeiss, Inc., Overkochen, Germany).

4.6. Macroscopic Fluorescence Reflectance Imaging of Plaque Inflammation

To investigate the ability of LOE to reduce plaque inflammation in apo$E^{-/-}$ mice, we employed ex vivo FRI to evaluate aortic plaque inflammation. Twenty-four hours before imaging, molecular imaging agents targeted against inflammatory endothelial cells (atherosclerotic plaque-homing peptide (AP)-hydrophobically modified glycol chitosan (HGC)-Cy5.5 nanoparticles; kindly provided by Dr. KM Kim from KIST (Korea)) were intravenously injected via the tail vein (10 mg/kg) [39]. The extent of plaque inflammation in apoE$^{-/-}$ mice compared with control C57BL/6 mice and the plaque inflammation-modulating effects of LOE compared with losartan were determined via the ex vivo FRI of the extracted aortas. Immediately before imaging, the mice were sacrificed via cervical dislocation and perfused with 20 mL saline. The aortas, which were connected to the hearts, were then excised and imaged in the Cy5.5 channel on an IVIS-200 FRI system (Xenogen Corp., Alameda, CA, USA).

4.7. Measurement of H&E Staining and the Aortic Atherosclerotic Plaque Area

After sacrifice, the mice were perfused with PBS through the left ventricle. For histopathological analysis, isolated aortic roots and right carotid arteries were embedded in an OCT compound. The frozen sections of the embedded aortic roots and carotid arteries were obtained. Adjacent sections were stained with H&E for general morphological analysis. The aortic lesion area (mm^2) was quantified using ImageJ software (NIH). The images of aortic plaque were calibrated using a hemocytometer, with 1 mm considered equal to 1600 pixels. The aortic plaque area was estimated as the area between the internal elastic lamina and the lumen on H&E sections.

4.8. Statistical Analysis

Statistical analysis was performed using SPSS software (version 11, SPSS, Inc., Chicago, IL, USA). All values are reported as the mean $\pm$ SEM. Differences in the measured values among multiple groups were analyzed via analysis of variance, followed by Bonferroni's multiple comparison. For all statistical analyses, a p-value less than 0.05 was considered statistically significant.

5. Conclusions

Treating WD-fed apoE$^{-/-}$ mice with LOE improves endothelial dysfunction by reducing NADPH oxidase expression and the formation of ROS. Consequently, LOE treatment is associated with the prevention of atherosclerotic inflammation and plaque development in apoE$^{-/-}$ mice. Altogether, the present findings indicate that LOE might be an attractive herbal candidate for the development of atherosclerosis associated with endothelial dysfunction and vascular oxidative stress.

Supplementary Materials: The following are available online at https://www.mdpi.com/article/10.3390/plants10112493/s1, Figure S1: Representative HPLC chromatograms of hyperin and isoquercitrin (a), and LOE (b) under conditions described in the Materials and Methods section. Peaks for hyperin and isoquercitrin are indicated.

Author Contributions: Conceptualization, methodology, validation, investigation, and writing—review and editing, S.-H.I.; methodology, software, validation, formal analysis, data curation, writing—original draft preparation, and visualization, S.-H.P.; methodology, investigation, and formal analysis, O.-R.K. and E.-H.P.; funding acquisition, project administration, and software, J.-O.L., K.-R.K., and J.-H.K.; validation, and writing—review and editing, B.-H.H. and H.-J.Y.; conceptualization, writing—review and editing, supervision, and funding acquisition, M.-H.O. and K.C. All authors have read and agreed to the published version of the manuscript.

Funding: This study was funded in part by the Ministry of Knowledge Economy (MKE), Korea, under the Leading Industry Development for Gangwon Economic Region (LIDGER) program (No. 70007579) and Basic Science Research Program through the National Research Foundation of Korea (NRF) funded by the Ministry of Education (NRF-2018R1D1A1B07049375).

Institutional Review Board Statement: The study was conducted according to the guidelines of the Care and Use of Laboratory Animals published by the US National Institutes of Health (NIH Publication No. 85-23, revised 1996) and was approved by the IACUC of Gyeonggi Biocenter (Approval No. 2009-09-09).

Informed Consent Statement: Not applicable.

Data Availability Statement: All the data generated and analyzed during this study are included in this article.

Conflicts of Interest: The authors declare no conflict of interest. The funders had no role in the design of the study; in the collection, analyses, or interpretation of data; in the writing of the manuscript; or in the decision to publish the results.

References

1. Libby, P. Inflammation in atherosclerosis. *Nature* **2002**, *420*, 868–874. [CrossRef]
2. Quyyumi, A.A. Prognostic value of endothelial function. *Am. J. Cardiol.* **2003**, *91*, 19–24. [CrossRef]
3. Stocker, R.; Keaney, J.F., Jr. Role of oxidative modifications in atherosclerosis. *Physiol. Rev.* **2004**, *84*, 1381–1478. [CrossRef]
4. Miller, F.J.; Gutterman, D.D.; Rios, C.D.; Heistad, D.D.; Davidson, B.L. Superoxide Production in Vascular Smooth Muscle Contributes to Oxidative Stress and Impaired Relaxation in Atherosclerosis. *Circ. Res.* **1998**, *82*, 1298–1305. [CrossRef]
5. Taddei, S.; Virdis, A.; Ghiadoni, L.; Salvetti, G.; Bernini, G.; Magagna, A.; Salvetti, A. Age-Related Reduction of NO Availability and Oxidative Stress in Humans. *Hypertension* **2001**, *38*, 274–279. [CrossRef]
6. Förstermann, U.; Xia, N.; Li, H. Roles of Vascular Oxidative Stress and Nitric Oxide in the Pathogenesis of Atherosclerosis. *Circ. Res.* **2017**, *120*, 713–735. [CrossRef]
7. Tian, R.; Ding, Y.; Peng, Y.Y.; Lu, N. Inhibition of Myeloperoxidase- and Neutrophil-Mediated Hypochlorous Acid Formation in Vitro and Endothelial Cell Injury by (−)-Epigallocatechin Gallate. *J. Agric. Food Chem.* **2017**, *65*, 3198–3203. [CrossRef]
8. Ding, Y.; Tian, R.; Yang, Z.; Chen, J.; Lu, N. NADPH oxidase-dependent degradation of single-walled carbon nanotubes in macrophages. *J. Mater. Sci. Mater. Med.* **2017**, *28*, 7. [CrossRef]
9. Drummond, G.R.; Selemidis, S.; Griendling, K.K.; Sobey, C.G. Combating oxidative stress in vascular disease: NADPH oxidases as therapeutic targets. *Nat. Rev. Drug Discov.* **2011**, *10*, 453–471. [CrossRef]
10. Lassègue, B.; San Martín, A.; Griendling, K.K. Biochemistry, physiology, and pathophysiology of NADPH oxidases in the cardiovascular system. *Circ. Res.* **2012**, *110*, 1364–1390. [CrossRef] [PubMed]
11. Barry-Lane, P.A.; Patterson, C.; van der Merwe, M.; Hu, Z.; Holland, S.M.; Yeh, E.T.; Runge, M.S. p47phox is required for atherosclerotic lesion progression in ApoE$^{(-/-)}$ mice. *J. Clin. Investig.* **2001**, *108*, 1513–1522. [CrossRef]
12. Judkins, C.P.; Diep, H.; Broughton, B.R.; Mast, A.E.; Hooker, E.U.; Miller, A.A.; Selemidis, S.; Dusting, G.J.; Sobey, C.G.; Drummond, G.R. Direct evidence of a role for Nox2 in superoxide production, reduced nitric oxide bioavailability, and early atherosclerotic plaque formation in ApoE$^{-/-}$ mice. *Am. J. Physiol.-Heart Circ. Physiol.* **2010**, *298*, H24–H32. [CrossRef]
13. Raggi, P.; Genest, J.; Giles, J.T.; Rayner, K.J.; Dwivedi, G.; Beanlands, R.S.; Gupta, M. Role of inflammation in the pathogenesis of atherosclerosis and therapeutic interventions. *Atherosclerosis* **2018**, *276*, 98–108. [CrossRef] [PubMed]
14. Lahera, V.; Goicoechea, M.; de Vinuesa, S.G.; Miana, M.; de las Heras, N.; Cachofeiro, V.; Luño, J. Endothelial dysfunction, oxidative stress and inflammation in atherosclerosis: Beneficial effects of statins. *Curr. Med. Chem.* **2007**, *14*, 243–248. [CrossRef] [PubMed]
15. Lee, Y.W.; Kim, P.H.; Lee, W.H.; Hirani, A.A. Interleukin-4, Oxidative Stress, Vascular Inflammation and Atherosclerosis. *Biomol. Ther.* **2010**, *18*, 135–144. [CrossRef]
16. Yook, C. *Medical Plants of Korea.*; Jinmyeong Publishing Co.: Seoul, Korea, 1989; p. 184.
17. Freise, C.; Querfeld, U. The lignan (+)-episesamin interferes with TNF-α-induced activation of VSMC via diminished activation of NF-kB, ERK1/2 and AKT and decreased activity of gelatinases. *Acta Physiol.* **2015**, *213*, 642–652. [CrossRef]
18. Hong, C.O.; Lee, H.A.; Rhee, C.H.; Choung, S.Y.; Lee, K.W. Separation of the antioxidant compound quercitrin from Lindera obtusiloba Blume and its antimelanogenic effect on B16F10 melanoma cells. *Biosci. Biotechnol. Biochem.* **2013**, *77*, 58–64. [CrossRef]
19. Jung, S.H.; Han, J.H.; Park, H.S.; Lee, J.J.; Yang, S.Y.; Kim, Y.H.; Heo, K.S.; Myung, C.S. Inhibition of Collagen-Induced Platelet Aggregation by the Secobutanolide Secolincomolide A from Lindera obtusiloba Blume. *Front. Pharmacol.* **2017**, *8*, 560. [CrossRef] [PubMed]
20. Lee, B.W.; Ha, J.H.; Shin, H.G.; Jeong, S.H.; Kim, J.H.; Lee, J.; Park, J.Y.; Kwon, H.J.; Jung, K.; Lee, W.S.; et al. Lindera obtusiloba Attenuates Oxidative Stress and Airway Inflammation in a Murine Model of Ovalbumin-Challenged Asthma. *Antioxidants* **2020**, *9*, 563. [CrossRef]
21. Freise, C.; Erben, U.; Neuman, U.; Kim, K.; Zeitz, M.; Somasundaram, R.; Ruehl, M. An active extract of Lindera obtusiloba inhibits adipogenesis via sustained Wnt signaling and exerts anti-inflammatory effects in the 3T3-L1 preadipocytes. *J. Nutr. Biochem.* **2010**, *21*, 1170–1177. [CrossRef] [PubMed]
22. Lee, J.O.; Oak, M.H.; Jung, S.H.; Park, D.H.; Auger, C.; Kim, K.R.; Lee, S.W.; Schini-Kerth, V.B. An ethanolic extract of Lindera obtusiloba stems causes NO-mediated endothelium-dependent relaxations in rat aortic rings and prevents angiotensin II-induced hypertension and endothelial dysfunction in rats. *Naunyn Schmiedebergs Arch. Pharmacol.* **2011**, *383*, 635–645. [CrossRef] [PubMed]
23. Lee, J.O.; Auger, C.; Park, D.H.; Kang, M.; Oak, M.H.; Kim, K.R.; Schini-Kerth, V.B. An ethanolic extract of Lindera obtusiloba stems, YJP-14, improves endothelial dysfunction, metabolic parameters and physical performance in diabetic db/db mice. *PLoS ONE* **2013**, *8*, e65227. [CrossRef]

24. Buday, A.; Őrsy, P.; Godó, M.; Mózes, M.; Kökény, G.; Lacza, Z.; Koller, Á.; Ungvári, Z.; Gross, M.-L.; Benyó, Z.; et al. Elevated systemic TGF-β impairs aortic vasomotor function through activation of NADPH oxidase-driven superoxide production and leads to hypertension, myocardial remodeling, and increased plaque formation in apoE$^{-/-}$ mice. *Am. J. Physiol.-Heart Circ. Physiol.* **2010**, *299*, H386–H395. [CrossRef] [PubMed]
25. Tabas, I.; Williams, K.J.; Borén, J. Subendothelial Lipoprotein Retention as the Initiating Process in Atherosclerosis. *Circulation* **2007**, *116*, 1832–1844. [CrossRef]
26. Antoniades, C.; Cunnington, C.; Antonopoulos, A.; Neville, M.; Margaritis, M.; Demosthenous, M.; Bendall, J.; Hale, A.; Cerrato, R.; Tousoulis, D.; et al. Induction of vascular GTP-cyclohydrolase I and endogenous tetrahydrobiopterin synthesis protect against inflammation-induced endothelial dysfunction in human atherosclerosis. *Circulation* **2011**, *124*, 1860–1870. [CrossRef] [PubMed]
27. Tabas, I. Macrophage death and defective inflammation resolution in atherosclerosis. *Nat. Rev. Immunol.* **2010**, *10*, 36–46. [CrossRef] [PubMed]
28. Davignon, J.; Ganz, P. Role of Endothelial Dysfunction in Atherosclerosis. *Circulation* **2004**, *109*, III-27–III-32. [CrossRef]
29. Sarr, M.; Chataigneau, M.; Martins, S.; Schott, C.; El Bedoui, J.; Oak, M.H.; Muller, B.; Chataigneau, T.; Schini-Kerth, V.B. Red wine polyphenols prevent angiotensin II-induced hypertension and endothelial dysfunction in rats: Role of NADPH oxidase. *Cardiovasc. Res.* **2006**, *71*, 794–802. [CrossRef]
30. Zanetti, M.; Gortan Cappellari, G.; Burekovic, I.; Barazzoni, R.; Stebel, M.; Guarnieri, G. Caloric restriction improves endothelial dysfunction during vascular aging: Effects on nitric oxide synthase isoforms and oxidative stress in rat aorta. *Exp. Gerontol.* **2010**, *45*, 848–855. [CrossRef]
31. Zhang, H.; Morgan, B.; Potter, B.J.; Ma, L.; Dellsperger, K.C.; Ungvari, Z.; Zhang, C. Resveratrol improves left ventricular diastolic relaxation in type 2 diabetes by inhibiting oxidative/nitrative stress: In vivo demonstration with magnetic resonance imaging. *Am. J. Physiol.-Heart Circ. Physiol.* **2010**, *299*, H985–H994. [CrossRef]
32. Keidar, S.; Attias, J.; Smith, J.; Breslow, J.L.; Hayek, T. The angiotensin-II receptor antagonist, losartan, inhibits LDL lipid peroxidation and atherosclerosis in apolipoprotein E-deficient mice. *Biochem. Biophys. Res. Commun.* **1997**, *236*, 622–625. [CrossRef] [PubMed]
33. Lee, B.S.; Choi, J.Y.; Kim, J.Y.; Han, S.H.; Park, J.E. Simvastatin and losartan differentially and synergistically inhibit atherosclerosis in apolipoprotein e(−/−) mice. *Korean Circ. J.* **2012**, *42*, 543–550. [CrossRef] [PubMed]
34. Guzik, T.J.; West, N.E.; Black, E.; McDonald, D.; Ratnatunga, C.; Pillai, R.; Channon, K.M. Functional effect of the C242T polymorphism in the NAD(P)H oxidase p22phox gene on vascular superoxide production in atherosclerosis. *Circulation* **2000**, *102*, 1744–1747. [CrossRef]
35. Azumi, H.; Inoue, N.; Takeshita, S.; Rikitake, Y.; Kawashima, S.; Hayashi, Y.; Itoh, H.; Yokoyama, M. Expression of NADH/NADPH oxidase p22phox in human coronary arteries. *Circulation* **1999**, *100*, 1494–1498. [CrossRef]
36. Ying, C.J.; Xu, J.W.; Ikeda, K.; Takahashi, K.; Nara, Y.; Yamori, Y. Tea polyphenols regulate nicotinamide adenine dinucleotide phosphate oxidase subunit expression and ameliorate angiotensin II-induced hyperpermeability in endothelial cells. *Hypertens. Res.* **2003**, *26*, 823–828. [CrossRef] [PubMed]
37. Loke, W.M.; Proudfoot, J.M.; Hodgson, J.M.; McKinley, A.J.; Hime, N.; Magat, M.; Stocker, R.; Croft, K.D. Specific dietary polyphenols attenuate atherosclerosis in apolipoprotein E-knockout mice by alleviating inflammation and endothelial dysfunction. *Arterioscler. Thromb. Vasc. Biol.* **2010**, *30*, 749–757. [CrossRef]
38. Loke, W.M.; Proudfoot, J.M.; Stewart, S.; McKinley, A.J.; Needs, P.W.; Kroon, P.A.; Hodgson, J.M.; Croft, K.D. Metabolic transformation has a profound effect on anti-inflammatory activity of flavonoids such as quercetin: Lack of association between antioxidant and lipoxygenase inhibitory activity. *Biochem. Pharmacol.* **2008**, *75*, 1045–1053. [CrossRef] [PubMed]
39. Park, K.; Hong, H.-Y.; Moon, H.J.; Lee, B.-H.; Kim, I.-S.; Kwon, I.C.; Rhee, K. A new atherosclerotic lesion probe based on hydrophobically modified chitosan nanoparticles functionalized by the atherosclerotic plaque targeted peptides. *J. Control Release* **2008**, *128*, 217–223. [CrossRef]

plants

Article

Effects of the Water Extract of Fermented Rice Bran on Liver Damage and Intestinal Injury in Aged Rats with High-Fat Diet Feeding

Ting-Yu Chen [1], Ya-Ling Chen [2], Wan-Chun Chiu [2], Chiu-Li Yeh [2], Yu-Tang Tung [3], Hitoshi Shirakawa [4], Wei-Tzu Liao [5] and Suh-Ching Yang [1,2,6,7,*]

1 Graduate Institute of Metabolism and Obesity Sciences, Taipei Medical University, Taipei 11031, Taiwan; ma48109006@tmu.edu.tw
2 School of Nutrition and Health Sciences, Taipei Medical University, Taipei 11031, Taiwan; ylchen01@tmu.edu.tw (Y.-L.C.); wanchun@tmu.edu.tw (W.-C.C.); clyeh@tmu.edu.tw (C.-L.Y.)
3 Graduate Institute of Biotechnology, National Chung Hsing University, Taichung 40227, Taiwan; peggytung@nchu.edu.tw
4 Laboratory of Nutrition, Graduate School of Agricultural Science, Tohoku University, Sendai 980-8857, Japan; shirakah@m.tohoku.ac.jp
5 Chian-E Biomedical Technology Corporation, Taipei 11031, Taiwan; cindyxox.tw@gmail.com
6 Research Center of Geriatric Nutrition, College of Nutrition, Taipei Medical University, Taipei 11031, Taiwan
7 Nutrition Research Center, Taipei Medical University Hospital, Taipei 11031, Taiwan
* Correspondence: sokei@tmu.edu.tw; Tel.: +886-2-2736-1661 (ext. 6553); Fax: +886-2-2737-3112

Citation: Chen, T.-Y.; Chen, Y.-L.; Chiu, W.-C.; Yeh, C.-L.; Tung, Y.-T.; Shirakawa, H.; Liao, W.-T.; Yang, S.-C. Effects of the Water Extract of Fermented Rice Bran on Liver Damage and Intestinal Injury in Aged Rats with High-Fat Diet Feeding. *Plants* **2022**, *11*, 607. https://doi.org/10.3390/plants11050607

Academic Editors: Juei-Tang Cheng, I-Min Liu and Szu-Chuan Shen

Received: 28 January 2022
Accepted: 18 February 2022
Published: 24 February 2022

Publisher's Note: MDPI stays neutral with regard to jurisdictional claims in published maps and institutional affiliations.

Abstract: The purpose of this study was to investigate the protective effects of the water extract of fermented rice bran (FRB) on liver damage and intestinal injury in old rats fed a high-fat (HF) diet. Rice bran (RB) was fermented with *Aspergillus kawachii*, and FRB was produced based on a previous study. Male Sprague Dawley rats at 36 weeks of age were allowed free access to a standard rodent diet and water for 8 weeks of acclimation then randomly divided into four groups (six rats/group), including a normal control (NC) group (normal diet), HF group (HF diet; 60% of total calories from fat), HF + 1% FRB group (HF diet + 1% FRB *w/w*), and HF + 5% FRB group (HF diet + 5% FRB *w/w*). It was found that the antioxidant ability of FRB was significantly increased when compared to RB. After 8 weeks of feeding, the HF group exhibited liver damage including an increased non-alcoholic fatty liver disease score (hepatic steatosis and inflammation) and higher interleukin (IL)-1β levels, while these were attenuated in the FRB-treated groups. Elevated plasma leptin levels were also found in the HF group, but the level was down-regulated by FRB treatment. An altered gut microbiotic composition was observed in the HF group, while beneficial bacteria including of the Lactobacillaceae and Lachnospiraceae had increased after FRB supplementation. In conclusion, it was found that FRB had higher anti-oxidative ability and showed the potential for preventing liver damage induced by a HF diet, which might be achieved through regulating imbalanced adipokines and maintaining a healthier microbiotic composition.

Keywords: water extract of fermented rice bran; non-alcoholic fatty liver disease; high-fat diet; aged rat

1. Introduction

The World Health Organization (WHO) reported that more than 1 billion people globally were 60 years and older in 2020 [1], and over 1.9 billion adults were overweight and obese (with a body-mass index (BMI) $\geq$25 kg/m^2) in 2016 [2]. According to data from the 2013–2016 Nutrition and Health Survey in Taiwan (NAHSIT), the prevalence of being overweight and obese (defined as a BMI of $\geq$24 kg/m^2) and having metabolic syndrome (METS) were more elevated in the elderly than in adults [3]. There were also a number of risk factors and chronic disease signs caused by or related to obesity which were ranked in the top ten leading causes of death in Taiwan [4]. These conditions revealed that the incidence of chronic diseases such as obesity and METS had greatly increased with aging.

Non-alcoholic fatty liver disease (NAFLD) is considered the most common chronic liver disease in the world, which affects about one-quarter of the world's population [5] and 11.4~41% of the population in Taiwan [6]. The two-hit hypothesis is commonly used to describe the pathophysiology of NAFLD. The first hit is mainly correlated with insulin resistance (IR), which stimulates the release of free fatty acids (FAs; FFAs) from adipocytes, thereby leading to excess accumulation of triglycerides (TGs) in the liver [7]. Oxidative stress triggers the second hit of NAFLD, accompanied by the generation of lipid peroxidation and proinflammatory cytokines such as tumor necrosis factor (TNF)-α, interleukin (IL)-1β, and IL-6, and endoplasmic reticulum (ER) stress, which accelerate the progression of NAFLD [8]. The mechanisms of dysregulated lipid metabolism play key roles in NAFLD progression. In NAFLD, decreased adiponectin levels and increased leptin levels result in an adipokine imbalance [9]. Adiponectin is an adipocyte-derived anti-inflammatory adipokine, which is expressed on hepatic membranes by adiponectin receptor 2 (AdipoR2) and increases activation of adenosine monophosphate-activated protein kinase α (AMPKα) and the nicotinamide adenine dinucleotide-dependent deacetylase, sirtuin-1 (SIRT1) [10]. It was reported that decreased serum adiponectin levels and liver AdipoR2 protein levels were observed after high-fat (HF) diet intake, which inactivates AMPKα and SIRT1, thereby inhibiting FA oxidation [11]. In addition, a lower expression of peroxisome proliferator activated receptor (PPAR)-α was found in NAFLD, causing the reduced expression of carnitine palmitoyltransferase (CPT)-1, which in turn inhibits mitochondrial β-oxidation [12]. Sterol regulatory element-binding protein (SREBP)-1c, mainly a lipogenic transcription factor which is usually increased in NAFLD, can up-regulate expressions of lipogenesis-associated enzymes such as stearoyl coenzyme A desaturase 1 (SCD1), FA synthase (FAS), and acetyl-CoA carboxylase (ACC) [13,14]. On the other hand, gut dysbiosis is also recognized as a crucial risk factor in NAFLD, which causes dysfunction of the gut endothelial barrier, elevation of intestinal permeability, and enabled translocation of gut-derived endotoxins, thereby mediating liver inflammation [15]. In addition, bacterial diversity, beneficial microorganisms, and the availability of total short-chain FAs (SCFAs) are reduced in the gut microbiota of the elderly [16]. Moreover, dietary pattern is an important factor in determining the composition of the intestinal microbiome [17]. A previous study showed that the intake of dietary fiber increased Prevotella abundances and the production of gut metabolites (e.g., SCFAs) [18], while a HF diet resulted in an increase in the Bacteroides enterotype [19] and the Firmicutes/Bacteroidetes (F/B) ratio [20]. Consequently, nutrients appear to be involved in modulating the gut microbiotic structure in order to balance changes in the intestinal microbiota in the elderly.

Rice bran (RB), which contributes about 10% to the whole grain weight, is a by-product of the rice milling process [21]. In Taiwan, 110,000~140,000 tons of RB are produced annually in the milling process. However, most RB is used as animal feed or is discarded as an agricultural waste. Utilizing agricultural byproducts or wastes including RB is a recent trend of reducing the environmental burden and saving resources [22]. During the past few years, previous studies reported that RB is rich in bioactive components, including vitamin E, vitamin B complex, γ-oryzanol, ferulic acid, plant sterols, phytic acid, among others, which possess antioxidant, anti-inflammatory, antihyperlipidemic, and hepatoprotective activities [23,24]. Studies also indicated that fermented RB (FRB) can effectively enhance nutritional values, such as protein, dietary fiber, and total phenolic contents compared to non-FRB [25]. Hence, RB has gradually been developed into a low-cost and functional food ingredient. The potential protective effects of FRB against metabolic disorders were investigated in several studies, including METS, hypertension [25], diabetes [26], inflammatory bowel disease (IBD) [27,28], and tumor progression [29], which validated the functional properties of FRB in improving lipid metabolism impairment, regulating glucose levels [25], suppressing oxidative stress [28], attenuating inflammatory responses [26,28], and affecting the gut microbiota [27]. These effects are also involved in regulating NAFLD development.

However, the ameliorating effects of FRB on NAFLD are still controversial. Possible reasons include differences in fermentation processes, bacterial species, and animal models.

Based on the above background, this study established a HF diet-fed aged rat model to simulate the conditions of obesity and metabolic disorders in senior citizens. However, the effect of RB from Taiwan fermented by *Aspergillus kawachii* has not been discussed in aged rats with HF diet consumption. Therefore, as shown in Figure 1, the purpose of this study was to investigate whether FRB can attenuate HF diet-induced NAFLD development in aged rats, as well as explore the underlying mechanisms of the protective effects of FRB against NAFLD.

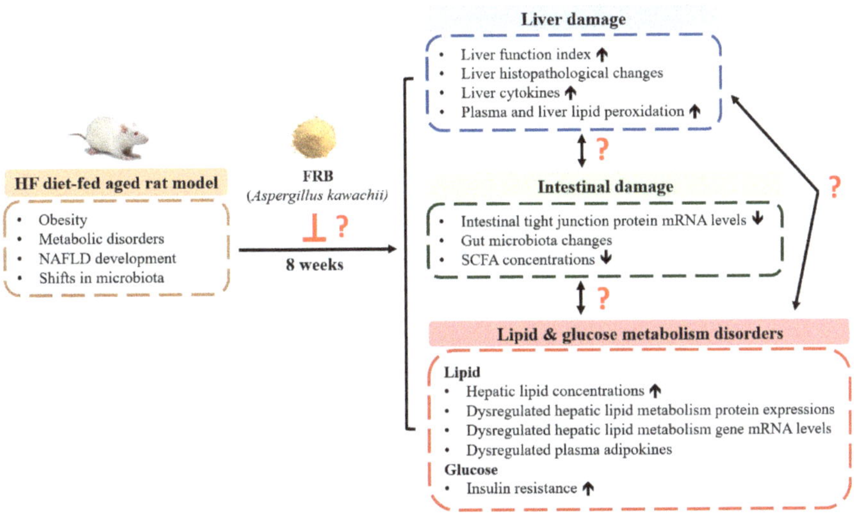

Figure 1. The research motivation and specific aims. NAFLD, non-alcoholic fatty liver disease; HF, high-fat; FRB, water extract of fermented rice bran; SCFA, short-chain fatty acid.

2. Results

2.1. Antioxidant Ability of FRB

As shown in Table 1, FRB had higher antioxidant capacity than RB. Additionally, the radical-scavenging activity in FRB was higher than that of RB.

Table 1. Antioxidant ability of the water extract of fermented rice bran (RB; FRB).

	RB	FRB
Total antioxidant capacity (mM)	65.38 ± 2.49	77.38 ± 0.27
Inhibition ratio (% per 100 mg)	9.05	64.10
Trolox equivalent (per 100 mg)	1578.13	10,096.88

All values are presented as the mean ± standard deviation. RB, water extract of rice bran; FRB, water extract of fermented rice bran.

2.2. Food Intake and Final Body Weights (BWs)

Food intake showed no differences among the groups (Figure 2A). Additionally, the HF group had significantly higher caloric intake and food efficiency ratio than the NC group, while there were no differences between the HF group and FRB treatment groups (Figure 2B,C). BW gain and final BWs of the HF group were significantly higher than those of the NC group. However, BW gain and final BWs of the FRB treatment groups only showed a lower trend than the HF group (Figure 2D,E).

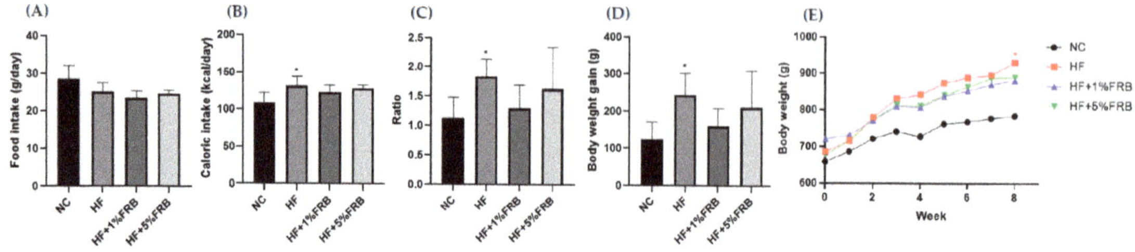

Figure 2. Effects of the water extract of fermented rice bran (FRB) on food intake and body weight in aged rats with high-fat (HF) diet feeding. (**A**) Food intake, (**B**) caloric intake, (**C**) food efficiency ratio (FER), (**D**) body weight gain and (**E**) body weight changes over time. The FER was calculated by applying the equation: FER = (body weight gain (g)/food intake (g)). Values are presented as the mean ± standard deviation (n = 6). * $p < 0.05$ vs. the normal control (NC) group; significance between two groups was determined using Student's t-test. In the HF diet-fed groups, significant differences between groups were determined by a one-way ANOVA with Fisher's post hoc test.

2.3. Liver Damage Indicators

2.3.1. Liver Function Index

Compared to the NC group, the HF group only showed a trend of higher plasma aspartate aminotransferase (AST) and alanine aminotransferase (ALT) activities. Meanwhile, only a decreasing trend was observed in the FRB treatment groups when compared to the HF group (Figure 3).

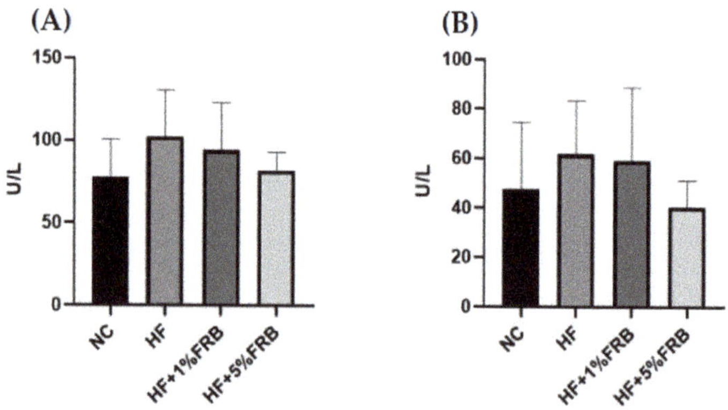

Figure 3. Effects of water extract of fermented rice bran (FRB) on liver function index in aged rats with high-fat (HF) diet feeding. Plasma level of (**A**) aspartate aminotransferase (AST) and (**B**) alanine aminotransferase (ALT). Values are presented as the mean ± standard deviation (n = 6). * $p < 0.05$ vs. the normal control (NC) group; significance between two groups was determined using Student's t-test. In the HF diet-fed groups, differences between groups were determined by a one-way ANOVA with Fisher's post hoc test.

2.3.2. Liver Histopathological Examinations

Liver steatosis and inflammatory cell infiltration in the HF group were elevated relative to the NC group, whereas the extents of steatosis and inflammation were ameliorated by the FRB intervention (Figure 4A). Moreover, the HF group showed a significantly higher NAFLD score compared to the NC group. In addition, the NAFLD score was significantly reduced with FRB supplementation, especially in the HF+5% FRB group, compared to the HF group (Figure 4B, Table 2).

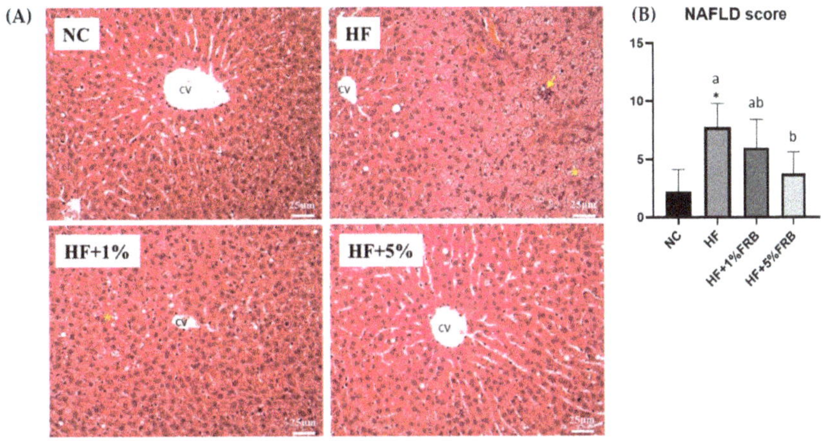

Figure 4. Effects of the water extract of fermented rice bran (FRB) on liver histopathological changes in aged rats with high-fat (HF) diet feeding. (**A**) H&E staining. Magnification: ×200. Scale bar, 25 μm. (Yellow arrow: inflammatory cell infiltration; yellow star: lipid droplets.) (**B**) Histopathological analysis scores. Values are presented as the mean ± standard deviation (n = 4). * $p < 0.05$ vs. the normal control (NC) group; significance between two groups was determined using Student's t-test. In the HF diet-fed groups, different letters indicate significant differences between groups at $p < 0.05$ by a one-way ANOVA with Fisher's post hoc test. NAFLD, nonalcoholic fatty liver disease.

Table 2. Histopathological analysis scores.

Item	Definition	Score	Groups			
			NC	HF	HF + 1% FRB	HF + 5% FRB
Macrovesicular steatosis	<5% 5~33% 33~66% >66%	0 1 2 3	0.75 ± 0.96	0.25 ± 0.50	0.25 ± 0.50	0.00 ± 0.00
Microvesicular steatosis	<5% 5~33% 33~66% >66%	0 1 2 3	0.75 ± 0.50	1.75 ± 0.50 *	1.25 ± 0.50	1.00 ± 0.82
Hepatocellular hypertrophy	<5% 5~33% 33~66% >66%	0 1 2 3	0.50 ± 0.58	2.50 ± 0.58 *	1.75 ± 0.96	2.25 ± 0.96
Number of inflammatory foci	0~2 3~5 6~9 10~19 >20	0 1 2 3 4	0.25 ± 0.50	3.25 ± 0.96 *,a	2.75 ± 1.26 a	0.50 ± 0.58 b
NAFLD activity score	Sum of score	0~13	2.25 ± 1.89	7.75 ± 2.06 *,a	6.00 ± 2.45 a,b	3.75 ± 1.89 b

Values are presented as the mean ± standard deviation (n = 4). * $p < 0.05$ vs. the normal control (NC) group; significance between two groups was determined using Student's t-test. In the high-fat (HF) diet fed groups, different letters indicate significant differences between groups at $p < 0.05$ by a one-way ANOVA with Fisher's post hoc test. NAFLD, nonalcoholic fatty liver disease; FRB, water extract of fermented rice bran.

2.3.3. Liver Cytokine Levels

There were no significant differences among all groups in hepatic IL-6 and IL-10 levels. Compared to the NC group, the hepatic TNF-α level was significantly higher; however, no change was observed in hepatic TNF-α levels among the HF group and FRB treatment groups. Moreover, the hepatic IL-1β level of the HF group was slightly higher than that of the NC group. Conversely, hepatic IL-1β levels of the HF + 1% FRB and HF + 5% FRB groups were significantly lower than that of the HF group (Table 3).

Table 3. Effects of the water extract of fermented rice bran (FRB) on liver cytokine levels in aged rats with high-fat (HF) diet feeding.

pg/mg Protein	NC	HF	HF + 1% FRB	HF + 5% FRB
TNF-α	7.49 ± 1.40	9.25 ± 0.54 *	8.25 ± 1.32	8.45 ± 1.24
IL-1β	46.21 ± 9.01	55.27 ± 7.68 [a]	44.03 ± 5.11 [b]	35.53 ± 4.64 [c]
IL-6	7.92 ± 2.80	8.24 ± 0.43	8.72 ± 2.10	8.82 ± 2.67
IL-10	2.54 ± 0.55	1.95 ± 0.37	1.49 ± 0.29	1.46 ± 0.61

Values are presented as the mean ± standard deviation (n = 6). * $p < 0.05$ vs. the normal control (NC) group; significance between two groups was determined using Student's t-test. In the HF diet-fed groups, different letters indicate significant differences between groups at $p < 0.05$ by a one-way ANOVA and Fisher's post hoc test. TNF, tumor necrosis factor; IL, interleukin.

2.3.4. Plasma and Liver Lipid Peroxidation

No significant difference was found in liver thiobarbituric acid-reactive substance (TBARS) levels. Compared to the NC group, a slight increase was seen in the plasma TBARS level in the HF group, but the FRB treatment groups only showed a trend of lower plasma TBARS levels compared to the HF group (Table 4).

Table 4. Effects of the water extract of fermented rice bran (FRB) on plasma and liver thiobarbituric acid-reactive substances (TBARSs) in aged rats with high-fat (HF) diet feeding.

	NC	HF	HF + 1% FRB	HF + 5% FRB
Plasma TBARS (ng/μL)	2.51 ± 0.98	4.85 ± 2.67	4.26 ± 2.81	3.07 ± 0.87
Liver TBARS (ng/mg)	44.04 ± 21.57	42.34 ± 9.54	59.43 ± 23.16	47.59 ± 32.50

Values are presented as the mean ± standard deviation (n = 6). Significance between the normal control (NC) and HF groups was determined using Student's t-test. In the HF diet-fed groups, significant differences between groups at $p < 0.05$ by a one-way ANOVA with Fisher's post hoc test.

2.4. *Lipid Metabolism Indicators*

2.4.1. Hepatic Total Cholesterol (TC) and TG Concentrations

As shown in Table 5, significant differences were not observed in hepatic TC concentrations among all groups. The HF group displayed a slightly higher liver TG level compared to the NC group, while FRB-treated groups showed a tendency of decreased liver TG levels compared to the HF group.

Table 5. Effects of the water extract of fermented rice bran (FRB) on hepatic total cholesterol (TC) and triglyceride (TG) levels in aged rats with high-fat (HF) diet feeding.

mg/g	NC	HF	HF + 1% FRB	HF + 5% FRB
Liver TC	3.82 ± 1.34	3.50 ± 0.52	3.48 ± 1.58	3.10 ± 0.68
Liver TG	60.68 ± 23.64	71.22 ± 15.26	67.30 ± 20.06	59.43 ± 4.57

Values are presented as the mean ± standard deviation (n = 6). Significance between the normal control (NC) and HF groups was determined using Student's t-test. In the HF diet-fed groups, significant differences between groups were determined by a one-way ANOVA with Fisher's post hoc test.

2.4.2. Hepatic Lipid Metabolism-Related Protein Expressions

There were no significant changes in liver AMPKα, phosphorylated (p)-AMPKα/AMPKα, leptin receptor, or adipoR2 protein expressions among all groups (Figure 5B,D–F). SIRT1 and p-AMPKα protein expressions were significantly elevated in the HF group compared to the NC group, while there were no differences between the HF group and FRB treatment groups (Figure 5A,C).

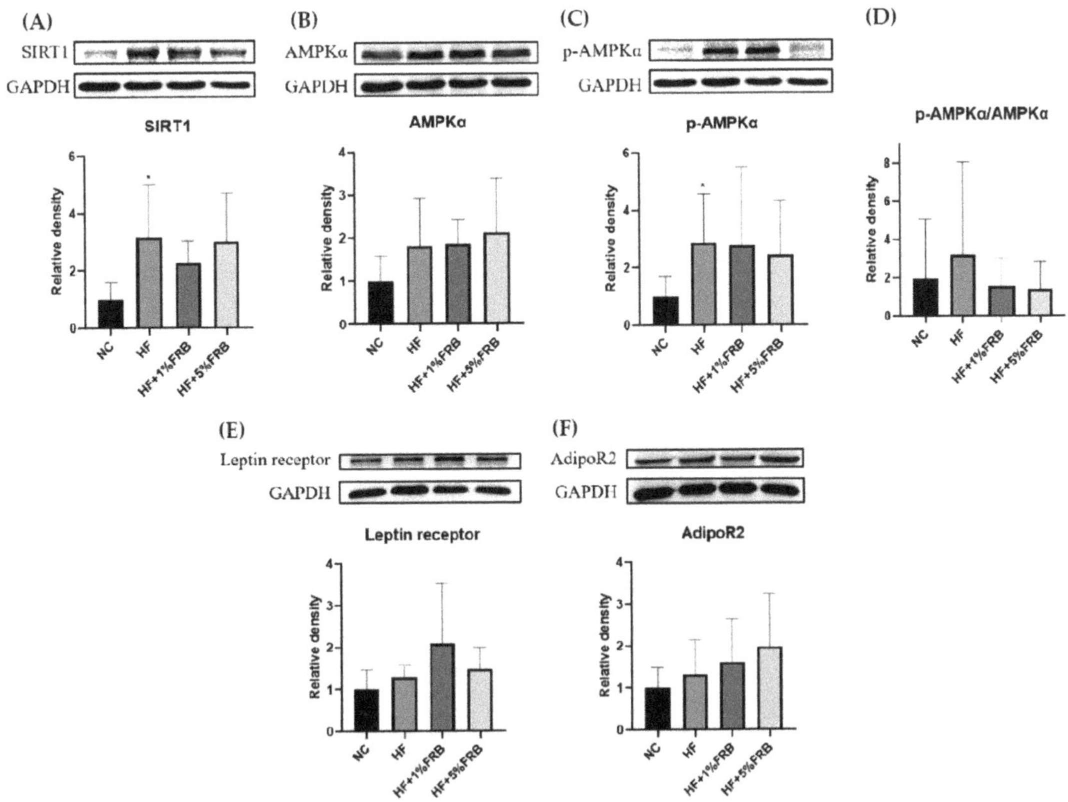

Figure 5. Effects of the water extract of fermented rice bran (FRB) on hepatic lipid metabolism-related protein expressions in aged rats with high-fat (HF) diet feeding. Values are presented as the mean ± standard deviation ($n = 6$). * $p < 0.05$ vs. the normal control (NC) group; significance between two groups was determined using Student's *t*-test. In the HF diet-fed groups, significant differences between groups were determined by a one-way ANOVA with Fisher's post hoc test. Western blot analysis of (**A**) NAD-dependent sirtuin-1 (SIRT1), (**B**) adenosine monophosphate-activated protein kinase-α (AMPKα), (**C**) phosphorylated (p)-AMPKα, (**D**) p-AMPKα/AMPKα, (**E**) leptin receptor, and (**F**) adiponectin receptor 2 (AdipoR2) protein expressions. Glyceraldehyde 3 phosphate dehydrogenase (GAPDH) was used as an internal control. Quantitative analysis of protein levels and the ratio of each internal control were calculated by setting the value of the mean of the NC group.

2.4.3. Hepatic Fatty Acid Metabolism-Related Gene Messenger (m)RNA Levels

There were no significant differences in mRNA levels of SREBP-1c, ACC, SCD1, FAS or PPARα among these groups. In addition, CPT-1 mRNA levels showed no difference between the NC and HF groups, while it was significantly lower in the HF + 1% FRB group than that of the HF group. On the other hand, mRNA levels of SREBP-1c and ACC were slightly higher in the HF group compared to the NC group, but the HF + 1% FRB and HF + 5% FRB groups had a decreasing trend of SREBP-1c and ACC levels compared to the HF group (Table 6).

Table 6. Effects of the water extract of fermented rice bran (FRB) on hepatic fatty acid metabolism-related gene mRNA levels in aged rats with high-fat (HF) diet feeding.

mRNA Levels	NC	HF	HF + 1% FRB	HF + 5% FRB
SREBP-1c	1.00 ± 0.56	1.04 ± 0.85	0.85 ± 0.66	0.59 ± 0.27
ACC	1.00 ± 0.36	1.61 ± 1.32	1.43 ± 1.10	1.03 ± 0.49
SCD1	1.00 ± 0.22	0.59 ± 0.53	0.39 ± 0.31	0.21 ± 0.16
FAS	1.00 ± 0.51	0.69 ± 0.76	0.28 ± 0.19	0.24 ± 0.12
PPARα	1.00 ± 0.42	0.86 ± 0.20	0.57 ± 0.41	0.88 ± 0.64
CPT-1	1.00 ± 0.59	0.84 ± 0.15 [a]	0.45 ± 0.27 [b]	0.69 ± 0.36 [a,b]

Values are presented as the mean ± standard deviation (n = 6). Significance between the normal control (NC) and HF groups was determined using Student's t-test. In the HF diet-fed groups, different letters indicate significant differences between groups at $p < 0.05$ by a one-way ANOVA with Fisher's post hoc test. Comparative quantification of each gene was normalized to β-actin using the $2^{-\Delta\Delta Ct}$ method and the ratio of each internal control was calculated by setting the value of the mean of the NC group. SREBP-1c, sterol response element-binding protein-1c; ACC, acetyl CoA carboxylase; SCD1, stearoyl coenzyme A desaturase 1; FAS, fatty acid synthase; PPARα, peroxisome proliferator-activated receptor α; CPT-1, carnitine palmitoyl transferase-1.

2.4.4. Plasma Adipokine Levels

Compared to the NC group, plasma leptin levels were significantly higher in the HF group. However, FRB-treated groups showed a decreasing trend of plasma leptin levels compared to the HF group (Figure 6A). No differences were found in plasma adiponectin levels or adiponectin/leptin ratios (Figure 6B,C).

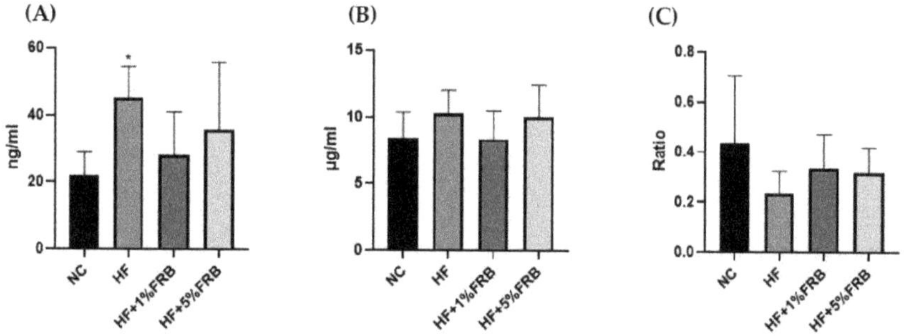

Figure 6. Effects of the water extract of fermented rice bran (FRB) on plasma adipokine levels in aged rats with high-fat (HF) diet feeding. Plasma levels of (**A**) leptin, (**B**) adiponectin, and (**C**) the adiponectin/leptin ratio. Values are presented as the mean ± standard deviation (n = 6). * $p < 0.05$ vs. the normal control (NC) group; significance between two groups was determined using Student's t-test. In the HF diet-fed groups, significant differences between groups were determined by a one-way ANOVA with Fisher's post hoc test.

2.5. Blood Glucose Regulators

According to the results of the IR analysis presented in Figure 7, there were no differences in fasting blood glucose levels, fasting insulin levels, or homeostasis model assessment of the IR index (HOMA-IRI) among all groups.

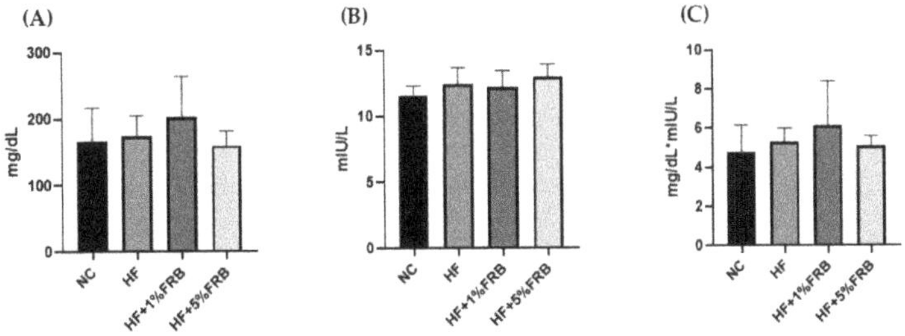

Figure 7. Effects of the water extract of fermented rice bran (FRB) on the insulin resistance analysis in aged rats with high-fat (HF) diet feeding. (**A**) Fasting blood glucose level, (**B**) fasting plasma insulin level, and (**C**) homeostasis model assessment of the insulin resistance index (HOMA-IRI). Values are presented as the mean ± standard deviation (n = 6). Significance between the normal control (NC) and HF groups was determined using Student's t-test. In the HF diet-fed groups, significant differences between groups were determined by a one-way ANOVA with Fisher's post hoc test.

2.6. Intestinal Damage Indicators

2.6.1. Intestinal Tight Junction Protein mRNA Levels

As shown in Figure 8, the intestinal mRNA levels of zonula occludens (ZO)-1 and occludin presented no differences among all groups. There was also no change in intestinal claudin-1 mRNA levels between the NC and HF groups; however, the intestinal claudin-1 mRNA level was significantly lower in the HF + 1% FRB group compared to the HF group.

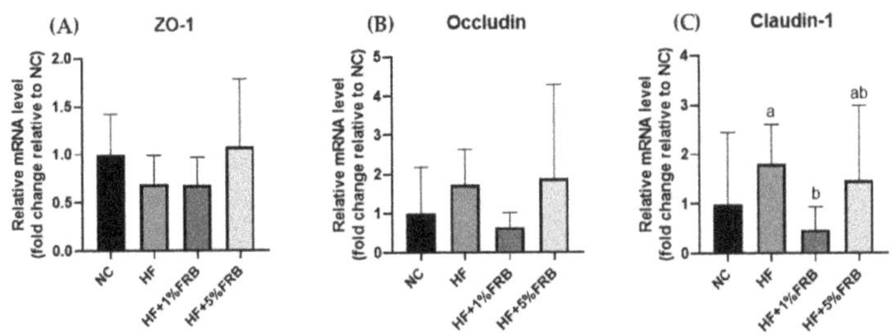

Figure 8. Effects of the water extract of fermented rice bran (FRB) on intestinal zonula occludens (ZO)-1, occludin, and claudin-1 mRNA levels in aged rats with high-fat (HF) diet feeding. Values are presented as the mean ± standard deviation (n = 6). * $p < 0.05$ vs. the normal control (NC) group; significance between two groups was determined using Student's t-test. In the HF diet-fed groups, different letters indicate significant differences between groups at $p < 0.05$ by a one-way ANOVA with Fisher's post hoc test. Comparative quantification of each gene was normalized to β-actin using the $2^{-\Delta\Delta Ct}$ method, and the ratio of each internal control was calculated by setting the value of the mean of the NC group.

2.6.2. Fecal Microbiotic Analysis

The Firmicutes-to-Bacteroidetes (F/B) Ratio and Alpha-Diversity Index

There were no significant differences in the F/B ratio among all groups (Figure 9A). In order to determine alterations in the fecal microbiotic community structure, an alpha-diversity analysis was conducted. The Chao1 and ACE indices commonly describe the species richness, while Shannon and Simpson indices refer to species diversity. There were no changes in the fecal microbiotic richness among all groups (Figure 9B). In addition, no differences were found in the Shannon and Simpson indices between the NC and HF groups. However, the HF + 1% FRB group showed significantly higher levels of the Shannon and Simpson indices compared to the HF group (Figure 9C).

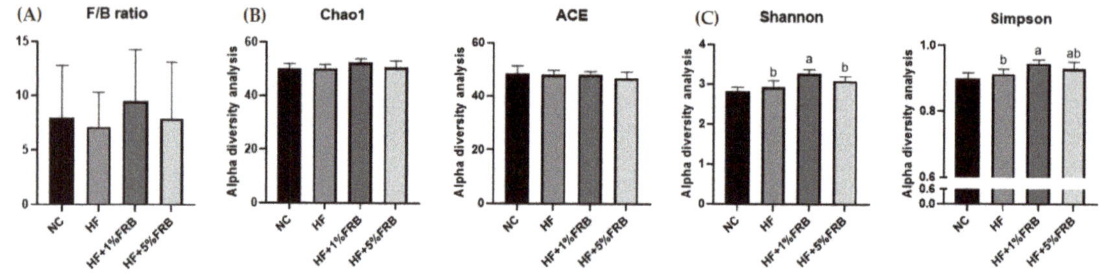

Figure 9. Effects of the water extract of fermented rice bran (FRB) on the Firmicutes-to-Bacteroidetes (F/B) ratio and α-diversity of the fecal microbiota in aged rats with high-fat (HF) diet feeding. (**A**) F/B ratio. (**B**) Community richness of the fecal microbiota. (**C**) Community diversity of the fecal microbiota. Values are presented as the mean ± standard deviation (n = 5). * $p < 0.05$ vs. the normal control (NC) group; significance between two groups was determined using Student's t-test. In the HF diet-fed groups, different letters indicate significant differences between groups at $p < 0.05$ by a one-way ANOVA with Fisher's post hoc test.

Beta-Diversity Index

To assess the variations in the fecal microbiota, a beta-diversity analysis was performed using a principal coordinate analysis (PCoA) plot. As shown in Figure 10, the PCoA showed different microbiotic distributions between the HF and FRB treatment groups.

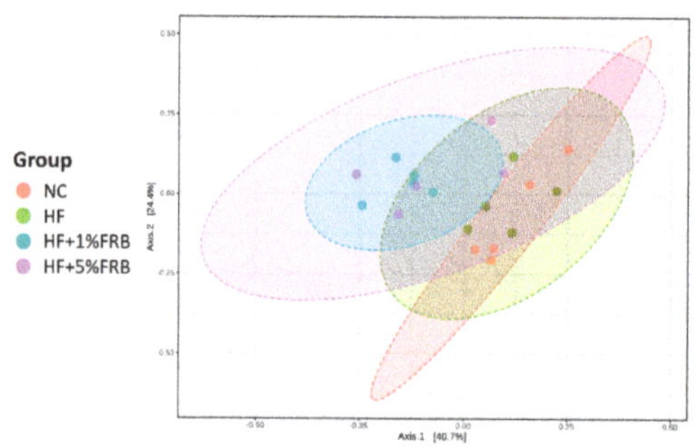

Figure 10. Effects of the water extract of fermented rice bran (FRB) on a principal coordinate analysis (PCoA) of the fecal microbiota in aged rats with high-fat (HF) diet feeding. Values are presented as the mean ± standard deviation (n = 5).

Linear Discriminant Analysis of the Effect Size (LEfSe)

The LEfSe approach and linear discriminant analysis (LDA) score were used to identify changes in bacterial taxonomical abundances among groups. It was found that the RF39 (order) and *Anaeroplasma* (genus) of the Tenericutes phylum and the Ruminococcaceae (family) and *Lactococcus garvieae* (species) of the Firmicutes phylum were abundant in the NC group (Figure 11). When comparing differences between the NC and HF groups, pathogenic bacteria such as the Proteobacteria (phylum), Peptococcaceae (family), *rc4_4* (genus), *Oscillospira* (genus), and *Ruminococcus gnavus* (species), which belong to the Firmicutes phylum, were overrepresented in the HF group (Figure 12). When comparing differences between the HF group and FRB-treated groups, SCFA-producing bacteria such as *Lactobacillales* (order), *Lactobacillaceae* (family), Lachnospiraceae (family), *Blautia* (genus), and *Blautia producta* (species) of the Firmicutes phylum were overexpressed in the HF + 1% FRB group (Figure 13). Moreover, the dominant bacteria in the HF + 5% FRB group also contained the SCFA-producing bacteria, including Lactobacillales (order), *Lactobacillaceae* (family), *Lachnospiraceae* (family), *Coprococcus* (genus) and *Lactobacillus* (genus) of the

Firmicutes phylum as well as *Bacteroides uniformis* (species) of the Bacteroidetes phylum (Figure 14).

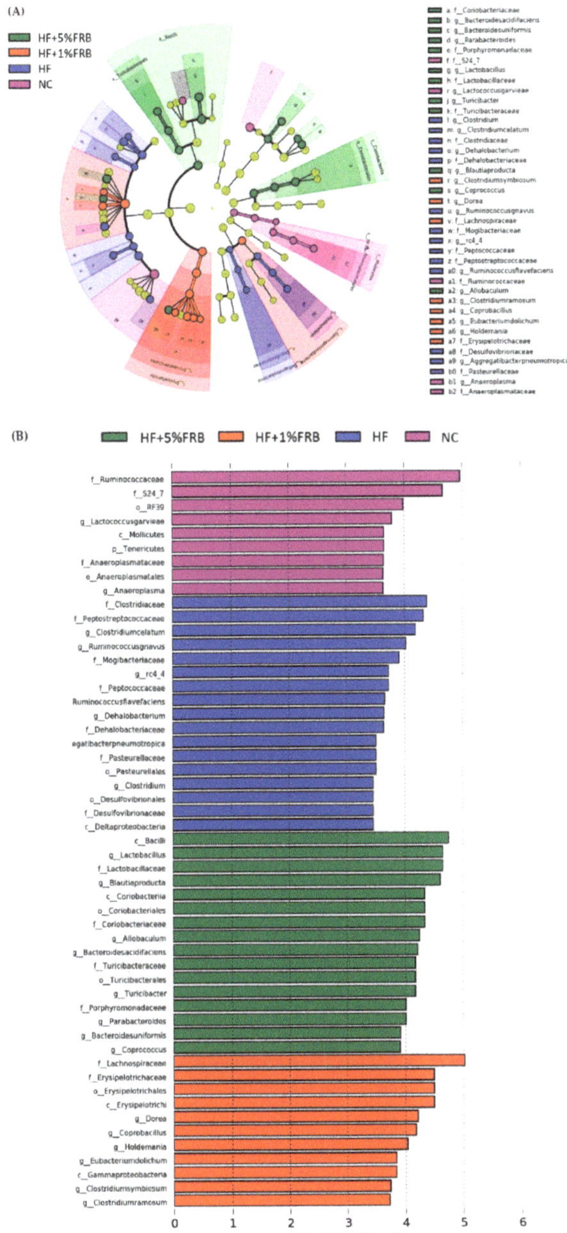

Figure 11. Effects of the water extract of fermented rice bran (FRB) on taxonomies of fecal microbiotic compositions in aged rats with high-fat (HF) diet feeding. (**A**) A linear discriminant analysis of the effect size (LEfSe) of the most significant abundance differences in the fecal microbiota among all groups (n = 5). (**B**) Bacteria meeting the LDA threshold ($\geq$2) differed among all groups (n = 5).

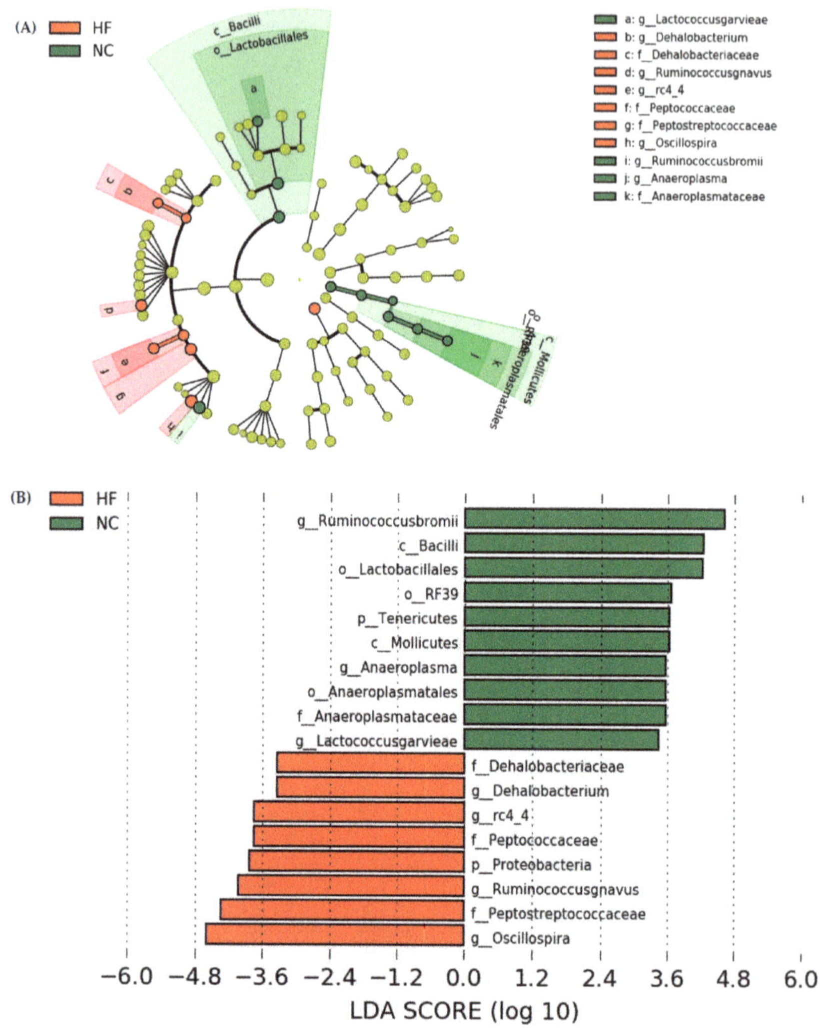

Figure 12. Effects of the water extract of fermented rice bran (FRB) on taxonomies of fecal microbiotic compositions in aged rats with high-fat (HF) diet feeding. (**A**) A linear discriminant analysis of the effect size (LEfSe) of the most significant abundance differences in the fecal microbiota in the NC and HF groups ($n = 5$). (**B**) Bacteria meeting the LDA threshold ($\geq$2) differed in the NC and HF groups ($n = 5$).

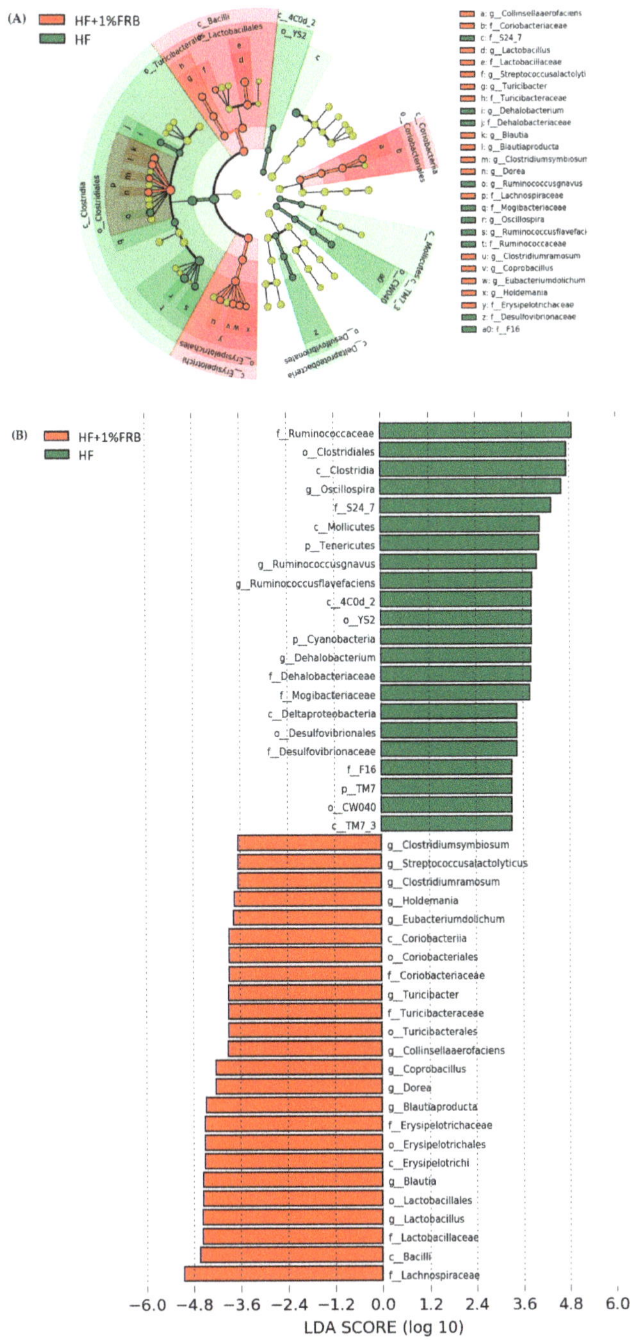

Figure 13. Effects of the water extract of fermented rice bran (FRB) on taxonomies of fecal microbiotic compositions in aged rats with high-fat (HF) diet feeding. (**A**) A linear discriminant analysis of the effect size (LEfSe) of the most significant abundance differences in the fecal microbiota in the HF and HF + 1% FRB groups (n = 5). (**B**) Bacteria meeting the LDA threshold ($\geq$2) differed in the HF and HF + 1% FRB groups (n = 5).

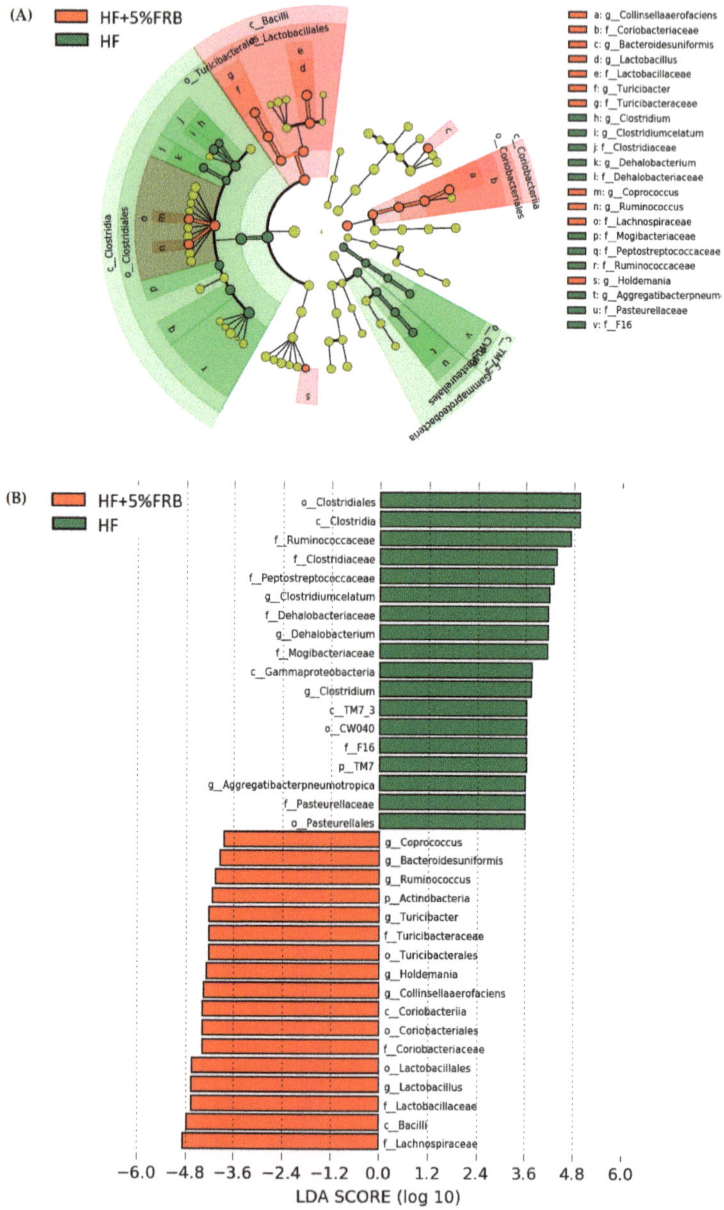

Figure 14. Effects of the water extract of fermented rice bran (FRB) on taxonomies of fecal microbiotic compositions in aged rats with high-fat (HF) diet feeding. (**A**) A linear discriminant analysis of the effect size (LEfSe) of the most significant abundance differences in the fecal microbiota in the HF and HF + 5% FRB groups ($n = 5$). (**B**) Bacteria meeting the LDA threshold (≥ 2) differed in the HF and HF + 5% FRB groups ($n = 5$).

2.6.3. Fecal SCFA Concentrations

There were no significant differences in isobutyric acid, butyric acid, isovaleric acid, valeric acid, 4-methylvaleric acid, or hexanoic acid levels among all groups. The heptanoic acid level was significantly lower in the HF group than in the NC group, whereas no change was seen between the HF group and FRB-treated groups. Moreover, compared to the NC group, the HF group showed a trend of a lower propionic acid level; however, only an increasing trend was observed in the HF + 1% FRB and HF + 5% FRB groups compared to the HF group (Table 7).

Table 7. Effects of the water extract of fermented rice bran (FRB) on fecal short-chain fatty acid (SCFA) concentrations in aged rats with high-fat (HF) diet feeding.

SCFAs (μM)	NC	HF	HF + 1% FRB	HF + 5% FRB
Propionic acid	55.28 ± 34.73	27.50 ± 24.33	32.33 ± 11.02	43.34 ± 27.99
Isobutyric acid	116.43 ± 4.45	120.84 ± 10.61	122.40 ± 14.67	124.08 ± 8.86
Butyric acid	91.22 ± 23.26	74.47 ± 62.19	52.00 ± 15.29	64.25 ± 22.99
Isovaleric acid	14.30 ± 7.36	13.63 ± 11.76	12.39 ± 2.13	15.16 ± 7.11
Valeric acid	29.44 ± 30.24	13.69 ± 10.47	11.43 ± 3.20	13.60 ± 4.66
4-Methylvaleric acid	1.79 ± 0.20	1.77 ± 0.53	1.81 ± 0.55	2.42 ± 1.27
Hexanoic acid	1.43 ± 0.55	1.16 ± 0.41	0.87 ± 0.42	1.40 ± 1.35
Heptanoic acid	1.23 ± 0.32	0.48 ± 0.26 *	0.50 ± 0.39	0.53 ± 0.15

Values are presented as the mean ± standard deviation (n = 4). * $p < 0.05$ vs. the normal control (NC) group; significance between two groups was determined using Student's t-test. In the HF diet-fed groups, significant differences between groups were determined by a one-way ANOVA with Fisher's post hoc test.

3. Discussion

3.1. Food Intake and BWs

HF diets are widely used in animal models to induce the development of NAFLD, simulating clinical conditions of METS, hyperlipidemia, obesity, and IR [30]. It was reported that patients over 50 years of age with diabetes or obesity had a 66% risk of experiencing non-alcoholic steatohepatitis (NASH) with advanced fibrosis [31]. In the present study, as shown in Figure 2, HF-diet-fed old rats had significantly higher caloric intake and final BWs than normal-diet-fed old rats, while FRB supplementation showed a trend to decrease the final BWs.

3.2. Liver Damage

After feeding the HF diet for 8 weeks, hepatic steatosis and inflammatory cell infiltration as well as the higher TNF-α level were observed (Figure 4, Tables 2 and 3). Plasma AST, ALT, and TBARS levels and the hepatic IL-1β level were only slightly increased (Figure 3, Tables 3 and 4). However, based on the liver histopathological examinations, the NAFLD score was significantly elevated. Thus, these results showed that HF-diet-induced liver damage was induced [32] although the liver damage might be mild in this study. As reported previously, TNF-α, IL-6, and IL-1 cytokine family members are stimulated by an HF diet; however, anti-inflammatory adipokines such as adiponectin and IL-10 were shown to have decreased [14]. On the other hand, Liu et al. reported that old rats fed a HF diet for 12 weeks presented a significant increase in serum ALT levels and NAFLD activity scores [33]. Nunes-Souza et al. also found that plasma AST and ALT levels, hepatic fat deposition as well as plasma and hepatic lipid peroxidation were significantly elevated by the 14-week HF diet feeding in aged mice [34]. In this study, the HF diet feeding period was only 8 weeks. The shorter feeding period might be the reason why plasma AST and ALT levels, and plasma and hepatic lipid peroxidation did not significantly change.

When supplemented with FRB, liver damage was alleviated as described by significantly lower NAFLD scores and liver IL-1β levels, and slightly lower levels of liver TNF-α and plasma TBARSs (Figure 4, Tables 2–4). Likewise, Ai et al. found that FRB could mitigate hepatic steatosis in mice with type 2 diabetes mellitus (T2DM) [35]. Lipid accumulation was also improved by FRB treatment in IR-HepG2 cells [35]. In a mouse model of obesity, it was reported that supplementation with FRB could decrease fatty depositions in liver tissues [36]. Another study showed that the inflammatory reactions were inhibited after FRB treatment by suppressing the mRNA levels of TNF-α and IL-6 in RBL-2H3 cells [37]. Consequently, the present results supported the protective effects of FRB on HF diet-induced liver damage possibly due to ameliorating hepatic steatosis and cytokines in aged rats.

3.3. Lipid and Glucose Metabolism

In this study, no differences were found in hepatic TC or TG levels among the groups. The hepatic TG level only showed an increasing trend in the HF group compared to the NC group (Table 5). It was found that hepatic lipids were significantly elevated, and IR was observed in young rats fed the HF diet for 8 weeks [38]. Thus, age might be a factor that caused no change in hepatic TC and TG levels. In future studies, a young control group should be designed as a control group due to the age-related differences.

This study demonstrated that plasma leptin levels and hepatic SIRT1 and p-AMPKα protein expressions were significantly higher in rats fed the HF diet for 8 weeks, while the plasma adiponectin/leptin ratio exhibited a downward trend (Figures 5 and 6). It was pointed out that hypoadiponectinemia and higher levels of serum leptin commonly occurred in NASH patients [9]. Increased serum leptin levels and decreased serum adiponectin levels were also observed in mice fed a HF diet for 10 weeks [36]. In addition, SIRT1 and AMPK were indicated to suppress lipogenesis through inhibition of SREBP-1 levels and also activation of FA oxidation via PPARα in the liver [39]. Chronic HF diet consumption reduced AMPK activity in liver tissues, which in turn led to inhibition of FA oxidation, followed by metabolic inflammation, oxidative stress, and IR [40]. However, results showed a reverse trend of hepatic SIRT1 and p-AMPKα protein expressions compared to previous studies. It was suspected that the aged rats may increase the activities of catabolic enzymes after HF diet intake in a compensatory manner, but the underlying mechanisms should be explored in future studies. Furthermore, mRNA levels of hepatic FA metabolism-related genes did not show a difference after 8 weeks of HF dietary intake in the present study (Table 6). Although hepatic lipogenic genes such as *SREBP-1c* and *ACC* only had upward trends in the HF group, hepatic steatosis was found in the HF group according to hepatic histopathology. Expressions of these hepatic FA metabolism-related proteins must be measured in the future study.

Previous studies suggested that FRB exerted anti-obesity and anti-dyslipidemic effects on HF diet-fed mice [36] and rats [41]. Alauddin et al. reported that FRB could improve IR, decrease the serum leptin/adiponectin ratio, and activate the liver AMPK signaling pathway in stroke-prone spontaneously hypertensive rats [25]. In the present study, after the FRB intervention, a decreasing trend was shown in plasma leptin and an increasing trend was shown in the adiponectin/leptin ratio (Figure 6), which indicated that FRB might have the potential for attenuating the imbalanced adipokines if the feeding period was extended. On the other hand, small-molecule compounds found in FRB are given in Supplementary Table S1. The results indicated that the fermentation process might result in activation of some components such as ferulic acid and folic acid, which are involved in regulating lipid and glucose metabolism [42,43], demonstrating that fermentation may have the potential to release activated substances. Nevertheless, the present study indicated that it had no impact on the fasting glucose, insulin levels and HOMA-IRI among the groups (Figure 7). It was speculated that the therapeutic effect of FRB on NAFLD was inconsistent with past studies, which might have been due to the age of the animals, the fat composition in the HF diet, bacterial species used for fermentation, or the feeding period. Elucidating the exact mechanism requires further study. However, these results still showed that FRB may be beneficial for regulating lipid metabolism homeostasis in aged rats fed a HF diet.

3.4. Intestinal Damage

In a previous study, HF diet intake increased the intestinal permeability, decreased tight junction expressions, and activated liver Toll-like receptor 4 (TLR4)/nuclear factor (NF)-κB inflammation [44]. However, in this study, the HF diet did not change mRNA levels of intestinal tight junction proteins (Figure 8). The protein expressions of the intestinal tight junction protein should be determined in the future. According to the gut microbiota composition, compared to normal diet-fed rats, fecal samples from HF diet-fed rats were more enriched in Proteobacteria (phylum), Peptococcaceae (family), *rc4_4* (genus), *Dehalobacterium* (genus), and *Oscillospira* (genus) (Figure 12), which were similar to past studies in a HF diet-induced obesity rodent model [45,46]. At the species level, *Ruminococcus gnavus* was also abundant in the HF group, which was reported to alter the gut community in IBD patients [47].

On the other hand, the species diversity increased after supplementation with 1% FRB (Figure 9C), and some dominant bacteria, including the probiotic Lactobacillaceae (family) and SCFA-producing Lachnospiraceae (family), were observed in the FRB-treated groups (Figures 13 and 14). A previous study revealed that FRB modulated the composition of the gut microbiota, especially enriching SCFA-producing bacteria such as *Dubosiella* and *Lactobacillus* and also increasing the SCFA levels in type 2 diabetic mice [35]. In a colitis mouse model, FRB enhanced SCFAs and tryptamine production, which promoted the tight junction barrier integrity [28]. Studies indicated that *Bacteroides uniformis* and *Bacteriodes acidifaciens*, which were overexpressed in the HF + 5% FRB group (Figure 11), could alleviate obesity-related metabolic dysfunction in mice [48,49]. In addition, the Coriobacteriaceae (family) was correlated with bile salts, steroid metabolism, and dietary polyphenol activation [50], which increased in the HF + 5% FRB group (Figure 14). Although some SCFA-producing bacteria were enriched in the FRB-treated groups, fecal SCFA levels were not significantly elevated in this study. The fecal propionic acid level only had an increasing trend after FRB supplementation (Table 7). The reason may be related to differences in the intervention dosage and periods compared with the previous studies. However, FRB still showed the potential to regulate the intestinal homeostasis by altering the composition of the gut microbiota in aged rats with HF diet feeding.

3.5. Gut-Liver Axis

Altered tight junction proteins may cause impairment of intestinal epithelial integrity, which would subsequently lead to bacteria translocation [51]. As mentioned previously, increased intestinal permeability and higher LPS levels were shown in NASH patients [51]. Changes in bacterial abundances in the gut microbiota could play a modulatory role in the gut-liver axis and pathogenesis of NAFLD [52]. In NAFLD patients, decreased bacterial diversity and lower levels of SCFA-producing members of the Lachnospiraceae, Lactobacillaceae, and Ruminococcaceae of the Firmicutes were observed [53]. In this study, rats fed a HF diet showed higher pathogenic bacterial abundances in fecal samples (intestinal damage), higher TNF-α levels and NAFLD score in the liver (liver injury), which might be the link between the intestinal and liver damage under HF diet feeding.

Dietary supplementation was thought to normalize the gut microbiome, which may be a beneficial strategy to regulate the NAFLD progression [54]. In this study, abundances of the beneficial Lachnospiraceae and Lactobacillaceae in the gut microbiota increased when HF diet-fed rats were supplemented with FRB (Figures 13 and 14). Additionally, the hepatic IL-1β level and NAFLD score was inhibited in the FRB-treated groups (Table 3). As mentioned above, these results supported FRB supplementation possibly having a potential effect of maintaining a healthier gut microbiotic composition which was connected with the lower hepatic cytokine levels in rats fed the HF diet and FRB. On the other hand, the amino acids composition such as tryptophan, isoleucine and GABA (γ-Aminobutyric acid) of FRB were enhanced than that of RB (Supplementary Table S2). A previous study showed that tryptophan intervention supported intestinal integrity and ameliorated hepatic steatosis in NAFLD mouse model [55]. Therefore, higher level of functional amino acids in FRB seemed to play an important role in management of NAFLD by maintaining the gut and liver health.

3.6. Comparison between 1% and 5% of FRB Supplementation

Previous studies showed that 5% of FRB supplementation could attenuate metabolic syndrome [25] and obesity [35]. Muscle atrophy [26] and intestinal inflammatory disorders [28] were also prevented after 10% of RB fermented by *A. kawachii* and *Lactobacillus* sp. In the present study, rats in the HF+1% FRB and 5% FRB groups showed significantly lower hepatic IL-1β levels, a decreasing trend of final BWs, plasma leptin levels, and hepatic TNF-α levels compared to the HF group. Beneficial bacteria in the gut microbiota also increased in both HF+1% FRB and 5% FRB groups. Moreover, the NAFLD score was significantly reduced in the HF + 5% FRB group. Taken together, 5% of FRB treatment might have distinct effects in preventing liver damage, which is possibly due to the higher contents of nutrients and bioactive substances in FRB, including polyphenol compounds, ferulic acid and functional amino acids [56].

3.7. Study Limitations

This study still has several limitations. Hepatic protein expressions of leptin, adiponectin, lipogenesis-, and lipolysis-related pathways as well as intestinal tight junction proteins must be measured in order to clarify the protective effects of FRB on NAFLD in aged rats fed a HF diet through the mechanism of the gut-liver axis. Additionally, a young control group is required to discuss the relationship among age, HF diet consumption, and FRB supplementation. Lastly, extending the experimental period should be considered in future studies.

4. Materials and Methods

4.1. Animals

Male Sprague Dawley (SD) rats at 36 weeks of age (BioLasco Taiwan, Ilan, Taiwan) were used in the present study. All rats were housed in individual cages in an animal room maintained at 23 ± 2 °C with 55% ± 10% relative humidity and a 12 h light–dark cycle. All rats were allowed free access to a standard rodent diet (LabDiet 5001 Rodent Diet; PMI Nutrition International, St. Louis, MO, USA) and water for 8 weeks of acclimation before the study. All procedures were approved by the Institutional Animal Care and Use Committee of Taipei Medical University (with an identification code of LAC-2020-0143).

4.2. Study Protocol

After 8 weeks of acclimatization, all rats were randomly divided into four groups (six rats/group) based on their BWs, including a normal control (NC) group (normal diet), high-fat (HF) group (HF diet; 60% of total calories from fat), HF + 1% FRB group (HF diet + 1%FRB *w/w*), and HF + 5% FRB group (HF diet + 5%FRB *w/w*). The diet formula was designed according to the method of Islam et al. [28]. Details of the experimental diets are given in Table 8. All rats were fed their diet and water *ad libitum*, and the BW change and diet consumption were measured every week. The experiment was carried out for 8 weeks. At the end of the 7th week of the study, feces were collected for fecal SCFA and microbiotic analyses. After 8 weeks of the experimental period, rats were anesthetized and sacrificed. Blood samples were collected in heparin-containing tubes and centrifuged (3000 rpm for 20 min at 4 °C) to obtain plasma. All of the plasma samples were stored at −80 °C until being assayed. Liver and intestinal tissues were rapidly excised and stored at −80 °C for further analysis.

Table 8. Composition of the experimental diets.

Ingredient (g/kg)	NC	HF	HF + 1% FRB	HF + 5% FRB
Cornstarch [1]	465	0	0	0
Maltodextrin [2]	155	125	122.65	113.258
Sucrose [3]	100	68.8	67.81	63.86
Casein [4]	140	200	198.82	194.11
L-cysteine [5]	2	3	3	3
Fresh soybean oil [6]	40	25	24.62	23.1
Lard [7]	0	245	245	245
Cellulose [8]	50	50	50	50
Mineral mixture (AIN-93M-MIX) [9]	35	35	33.404	27.02
Vitamin mixture (AIN-93M-MIX) [10]	10	10	10	10
Choline bitartrate [11]	3	3	3	3
Tert-butylhydroquinone [12]	0.008	0.008	0.008	38
Water extract of fermented rice bran (FRB)	0	0	7.6	38
kcal/g	3.808	5.25	5.25	5.22

NC, normal control group; HF, high-fat diet group; HF + 1% FRB, high-fat diet+1% FRB *w/w* group; HF + 5% FRB, high-fat diet+5% FRB *w/w* group. [1] Cornstarch: 902956, MP Biomedicals, Irvine, CA, USA; [2] maltodextrin: 960429, MP Biochemicals; [3] sucrose: Taiwan Sugar Corporation, Taipei, Taiwan; [4] casein: 901293, MP Biochemicals; [5] l-cysteine: 101454, MP Biochemicals; [6] fresh soybean oil: Taiwan Sugar Corporation; [7] lard: 902140, MP Biochemicals; [8] cellulose: 900453, MP Biochemicals; [9] mineral mixture (AIN-93M-MIX): 960401, MP Biochemicals; [10] vitamin mixture (AIN-93M-MIX): 2960402, MP Biochemicals; [11] choline bitartrate: 101384, MP Biochemicals; [12] tert-Butylhydroquinone: 195590, MP Biochemicals.

4.3. *Fermentation Processes and Analysis of FRB Components*

4.3.1. Production Procedures of FRB

Rice bran was fermented with *Aspergillus kawachii* (Bioresource Collection and Research Center, Hsinchu, Taiwan), and FRB was produced based on procedures described in a previous study [25]. Rice bran (100g; YuanShun, Yunlin, Taiwan) and 100 g of fat-free rice bran (Guanshan, Taitung, Taiwan) were initially mixed in a stainless-steel container and steamed for 15 min, followed by the addition of 50 g of purified water to mix the ingredients well. Then, the mixture was sterilized (1.2 atm for 30 min at 121 °C) and cooled to about 30 °C, after which a spore solution of *A. kawachii* was inoculated and placed in a microcentrifuge tube containing 1 mL of distilled water (dH_2O) to wash out the spores until the solution turned light-proof black. The spore solution was then diluted with 1000 mL of dH_2O and evenly shaken. The spore solution was dropped into the culture medium and incubated at 25 °C in a fermentation chamber overnight, then the water was poured from the medium, and then poured again once at 08:00 and 16:00. After 4 days of fermentation, the FRB product was obtained. The FRB solution was stirred with equal proportions of purified water and centrifuged (3000 rpm for 15 min at 4 °C), condensed for 30~40 min, and lyophilized at −40 to −60 °C for 3 days to collect FRB extracts. The lyophilized powder was kept at 4 °C until being used.

4.3.2. Antioxidative Status of FRB

Total Antioxidant Capacity (TAC)

The TAC of FRB was determined according to the manufacturer's instruction (Cayman Chemical, Ann Arbor, MI, USA). The optical density (OD) was read at 750 nm with a microplate reader (Molecular Devices, Sunnyvale, CA, USA).

Di(Phenyl)-(2,4,6-trinitrophenyl) Iminoazanium (DPPH) Antioxidant Assay

Radical-scavenging activity was quantified by a DPPH antioxidant assay using a commercial kit (D678-01, Dojindo, Rockville, MD, USA) and measuring the absorbance at 517 nm.

4.4. Measurements and Analytical Procedures

4.4.1. Determination of Liver Damage

Liver Function Index

Activities of plasma aspartate aminotransferase (AST) and alanine aminotransferase (ALT) were detected with the ADVIA® Chemistry XPT System (Siemens Healthcare Diagnostics, Eschborn, Germany).

Liver Histopathological Examination

The caudate lobe of liver tissues was fixed in a 10% formaldehyde solution. Sections of liver tissues were then stained with hematoxylin A- and eosin (H&E) and evaluated by a veterinarian. Images of the tissues were captured with a digital camera at 200× magnification. Histopathological evaluations of macrovesicular steatosis, microvesicular steatosis, hypertrophy, and the number of inflammatory foci were separately scored, and the severity was graded as described by Liang et al. [57] with minor modifications. Adding the scores of the above four parameters was used to calculate the NAFLD score.

Liver Cytokine Levels

The extraction method of liver tissues was described by Chen et al. [58]: 0.5 g of liver tissues was homogenized in 1.5 mL of ice-cold buffer containing 50 mM Tris (pH 7.2), 150 mM NaCl, 1% Triton X-100, and 0.1% protease inhibitor (PI) (HYK0010, MedChemExpress, Monmouth Junction, NJ, USA). The homogenized solution was then centrifuged at 3000 rpm for 15 min at 4 °C, and the supernatant was collected. Concentrations of hepatic TNF-α, IL-1β, IL-6, and IL-10 were determined by corresponding enzyme-linked immunosorbent assay (ELISA) kits, including rat TNF-α (BioLegend Systems, San Diego, CA, USA), IL-1β (Rat IL-1, R&D Systems, Minneapolis, MN, USA), IL-6 (Rat IL-6, R&D Systems) and IL-10 (Rat IL-10, R&D Systems, Minneapolis, MN, USA). The OD was read at 450 nm with a microplate reader (Molecular Devices, Sunnyvale, CA, USA) for all cytokines.

Plasma and Liver Lipid Peroxidation

Liver tissues (0.1 g) were homogenized as described for the method of liver cytokines. Plasma and the supernatants of liver homogenates were used for the lipid peroxidation analysis. Lipid peroxidation in plasma and the liver was measured by TBARS levels with a TBARS kit (TBARS 10009055 (TCA method) Assay Kit, Cayman Chemical, MI, USA) according to the assay kit instructions.

4.4.2. Determination of Lipid Metabolism

Hepatic TC and TG Concentrations

For the liver TC determination, 0.01 g of liver tissue was homogenized in 200 μL of solvent (chloroform: isopropanol: nonyl phenoxypolyethoxylethanol (NP-40) of 7: 11: 0.1) and centrifuged at 8876 rpm for 10 min at 4 °C. Avoiding the pellet, the liquid (organic phase) was transferred to a new tube, air dried at 50 °C to remove the chloroform, and samples were placed under a vacuum for 30 min to remove trace amounts of the organic solvent. Dried lipids were then dissolved in 200 μL of 1× assay diluent with vortexing until the solution turned homogenous and then was stored at −80 °C. To measure the concentrations of liver TGs, 0.1 g of liver tissue was homogenized in 500 μL of diluted NP-40 reagent containing protease inhibitors. The mixture was centrifuged at 7247 rpm for 10 min at 4 °C, and the supernatant (including a layer of insoluble fat) was collected and stored at −80 °C. TC and TG contents were further analyzed with a cholesterol colorimetric assay kit (Cell Biolabs, San Diego, CA, USA) and colorimetric TG assay kit (Cayman Chemical), and results were expressed as milligrams per gram (mg/g) of liver tissue.

Hepatic Lipid Metabolism-Related Protein Expressions

A Western blot analysis was performed to determine expressions of SIRT1, AMPKα, p-AMPKα, leptin receptor and AdipoR2. The method of crude extraction preparation of liver tissues was assessed according to previously described protocols [59]. Liver tissue (0.1 g) was homogenized in 0.5 mL of RIPA buffer containing 1% of a PI (HYK0010, MedChemExpress) and phosphatase inhibitor (PPI) (HYK0022, MedChemExpress). After being placed in an ice bath for 30 min, the homogenate was centrifuged at 7939 rpm for 15 min at 4 °C, followed by collection of the supernatant. Liver proteins (50 μg) were separated by 10% sodium dodecylsulfate polyacrylamide gel electrophoresis (SDS-PAGE), and then electroblotted onto polyvinylidene difluoride (PVDF) transfer membranes (Pall, Port Washington, NY, USA) and incubated with 5% bovine serum albumin (BSA). These blots were incubated with primary antibodies listed in Table 9, and glyceraldehyde 3-phosphate dehydrogenase (GAPDH) was used as an internal control. The blots were finally treated with goat anti-rabbit immunoglobulin G (IgG) horseradish peroxidase (HRP; Croyez Bioscience, Taipei, Taiwan), and specific bindings of anti-bodies were assayed by the UVP Chemidoc it 515 Imaging System (UVP, Upland, CA, USA) using a T Western Lightning kit (PerkinElmer Lifesciences, Boston, MA, USA). Bands were quantified using Image-Pro Plus software (Media Cybernetics, Rockville, MD, USA).

Table 9. Antibodies used for Western blotting.

	Antibody (Ab)	Ab Type	Product No.	Source
Primary antibody	SIRT1	monoclonal	#9475	Cell Signaling Technology
	AMPKα	polyclonal	#2532	Cell Signaling Technology
	p-AMPKα	monoclonal	#2535	Cell Signaling Technology
	Leptin receptor	polyclonal	DF7139	Affinity Biosciences
	AdipoR2	polyclonal	DF12811	Affinity Biosciences
Internal control	GAPDH	monoclonal	HRP-60004	Proteintech
Secondary antibody	anti-rabbit IgG		C04003	Croyez Bioscience

SIRT1, NAD-dependent sirtuin-1; AMPKα, adenosine monophosphate-activated protein kinase-α; p-AMPKα, phosphorylated-AMPKα; AdipoR2, adiponectin receptor 2.

Hepatic FA Metabolism-Related Gene mRNA Levels

Total RNA of the liver was extracted with the TRI Reagent® (Sigma-Aldrich, St. Louis, MO, USA), according to instructions from the manufacturer. The quality and quantity of total RNA were evaluated by measuring the OD 260/280 ratio on a BioTek epoch reader with the Gen5TM Take3 Module (BioTek Instruments, Winooski, VT, USA). The concentration of total RNA was adjusted to 4000 ng/μL and then reverse-transcribed with a RevertAid First Strand cDNA Synthesis kit (#K1621, ThermoFisher Scientific, Waltham, MA, USA). The concentration of complementary (c)DNA was calculated by the BioTek epoch reader with the Gen5TM Take3 Module system and adjusted to 50 ng/μL. The resulting cDNA was amplified in a 96-well polymerase chain reaction (PCR) plate using SYBR Green/ROX qPCR Master Mix (2X) (ThermoFisher Scientific) on a QuantStudio 1 Real-Time PCR System (ThermoFisher Scientific). Gene levels were normalized to β-actin, and the ratio to β-actin was calculated by setting the value of the NC group to 1. Information on primers is given in Table 10.

Plasma Adipokine Levels

A Leptin Mouse/Rat ELISA kit (Biovendor, Brno, Czech Republic) and Rat Adiponectin ELISA kit (Assaypro, Charles, MO, USA) were used to detect plasma leptin and adiponectin levels, respectively, following the manufacturer's instructions. The absorbance was read on a microplate reader (Molecular Devices) at 450 nm for all adipokines.

Table 10. Primers used for the quantitative polymerase chain reaction.

	Forward 5′→3′	Reverse 5′→3′
SREBP-1c	AGGAGGCCATCTTGTTGCTT	GTTTTGACCCTTAGGGCAGC
ACC	GGAAGACCTGGTCAAGAAGAAAAT	CACCAGATCCTTATTATTGT
SCD1	GTTGGGTGCCTTATCGCTTTCC	CTCCAGCCAGCCTCTTGTCTAC
FAS	CGGCGTGTGATGGGGCTGGTA	AGGAGTAGTAGGCGGTGGTGTAGA
PPARα	CGGGTCATACTCGCAGGAAA	AAGCGTCTTCTCAGCCATGC
CPT-1	GCATCCCAGGCAAAGAGACA	CGAGCCCTCATAGAGCCAGA
ZO-1	CTTGCCACACTGTGACCCTA	ACAGTTGGCTCCAACAAGGT
Occludin	CTGTCTATGCTCGTCATCG	CATTCCCGATCTAATGACGC
Claudin-1	AAACTCCGCTTTCTGCACCT	TTTGCGAAACGCAGGACATC
β-Actin	CACCAGTTCGCCATGGATGACGA	CCATCACACCCTGGTGCCTAGGGC

SREBP-1c, sterol response element-binding protein-1c; ACC, acetyl CoA carboxylase; SCD1, stearoyl coenzyme A desaturase 1; FAS, fatty acid synthase; PPARα, peroxisome proliferator-activated receptor α; CPT-1, carnitine palmitoyl transferase-1; ZO-1, zonula occludens-1.

4.4.3. IR Analysis

Plasma insulin levels were assayed with a Rat Insulin ELISA kit (Mercodia, Uppsala, Sweden). The fasting blood glucose level was detected via the Glucose Monitoring System (Abbott Diabetes Care, Oyl, UK). The HOMA-IRI was calculated with the following formula: HOMA-IRI = fasting blood glucose (mg/dL) × fasting insulin (mIU/L)/405.

4.4.4. Determination of Intestinal Damage

Intestinal Tight Junction Protein mRNA Levels

The intestinal tight junction protein mRNA levels were measured by quantitative (q)PCR methods. The method of ileum sample preparation was the same as that for the measurement of hepatic FA metabolism-related genes. Information on primers is given in Table 10.

Fecal Microbiotic Analysis

The fecal microbiotic composition was analyzed using a 16S ribosomal (r)RNA Next Generation Sequencing (NGS) system. Fresh feces were collected into sterilized 2-mL Eppendorf tubes and stored at −80 °C for analysis. The amplified fecal DNA was purified with Agencourt AMPure XP Reagent beads (Beckman Coulter, Brea, CA, USA). A qPCR (KAPA SYBR FAST qPCR Master Mix) was used to quantify each library with the Roche LightCycler 480 system, and then pooled equally to 4 nM for the Illumina MiSeq NGS system (illumina, San Diego, CA, USA). More than 80,000 reads with paired-end sequencing (>250 bp*2) were generated, and the QIIME2 workflow [60] classified organisms from the amplicons using the GreenGenes taxonomy database. MicrobiomeAnalyst [61] was used to perform statistical and visual analyses of the microbiome data.

Fecal SCFA Concentrations

Fecal SCFAs were extracted according to the method described by García-Villalba et al. [62] with slight modification. Fresh feces were weighed and suspended in 1 mL of water with 0.5% phosphoric acid per 0.1 g of sample into sterilized 2-mL Eppendorf tubes and stored at −80 °C until analysis. The SCFA analysis was performed by gas chromatographic-mass spectrometric (GC-MS) system of Agilent 5977B coupled with a 7820A autoinjector (Agilent Technologies, Santa Clara, CA, USA). For GC-MS, a Nukol™ 30 m × 0.25 × 0.25 μm capillary column (24107, Supelco, Bellefonte, PA, USA) was used. Injection was made in the pulsed split mode with an injection volume of 1 μL and an injector temperature of 250 °C. GC separation was carried out at 90 °C initially, then increased to 150 °C at 15 °C /min, to 170 °C at 3 °C/min and finally to 200 °C at 50 °C/min (total time 16.267 min). The solvent delay was 3.2 min. The identification of SCFAs was based on the retention times of standard compounds using a commercial standard solution (46975-

U, Supelco). Data were quantitated with MassHunter Quantitative Analysis software (Agilent Technologies).

4.4.5. Statistical Analysis

Data are expressed as the mean ± standard deviation (SD). Statistical analysis was performed using GraphPad Prism vers. 8.0.1 (GraphPad Software, San Diego, CA, USA). Student's *t*-test was used to determine statistical differences between the NC and HF groups. A one-way analysis of variance (ANOVA), followed by Fisher's post hoc test was used to determine statistical differences among the HF, HF + 1% FRB, and HF + 5% FRB groups. Statistical significance was assigned at the $p < 0.05$ level.

5. Conclusions

It was found that the water extract of fermented rice bran (by *Aspergillus kawachii*) had a higher anti-oxidative ability. The current results demonstrated that after 8 weeks of FRB treatment, rats in the HF + 5% FRB group had a significantly lower NAFLD score and hepatic IL-1β level, a decreasing trend of final BW and plasma leptin level, as well as an increase in beneficial bacteria in the gut microbiota. In summary, it was suggested that FRB showed the potential for alleviating liver damage induced by a HF diet, possibly through regulating the imbalanced adipokines and altering the gut microbiotic composition (Figure 15).

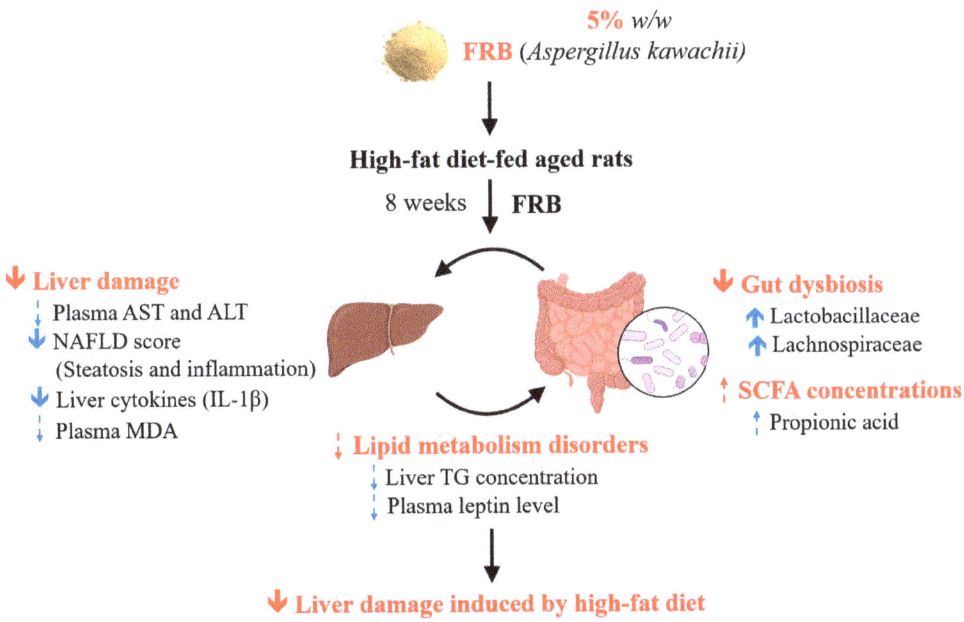

Figure 15. Effects of the water extract of fermented rice bran (FRB) on liver damage and intestinal injury in aged rats with high-fat (HF) diet feeding. In this study, it was indicated that FRB ameliorated liver damage induced by the HF diet which was represented as a lower non-alcoholic fatty liver disease (NAFLD) score and hepatic interleukin (IL)-1β level in rats. The protective effects of FRB against liver damage may have been due to regulating the plasma adipokines and maintaining homeostasis of the gut microbiota. Solid blue arrow: levels of analytical items were significantly increased or decreased in the HF + 5% FRB group than in the HF group; dotted blue arrow: levels of analytical items in the HF + 5% FRB group showed an increasing or decreasing trend compared to the HF group.

Supplementary Materials: The following supporting information can be downloaded at: https://www.mdpi.com/article/10.3390/plants11050607/s1, Table S1: LC/MS analysis of the water extract of fermented rice bran (FRB), Table S2: Amino acid compositions of the water extracts of rice bran (RB) and fermented RB (FRB), Table S3: Fatty acid compositions of the water extracts of rice bran (RB) and fermented RB (FRB), Table S4: Components in the water extracts of rice bran (RB) and fermented RB (FRB).

Author Contributions: Conceptualization, S.-C.Y. and H.S.; formal analysis, T.-Y.C., Y.-L.C. and W.-T.L.; data curation, T.-Y.C. and Y.-L.C.; writing—original draft preparation, T.-Y.C.; writing—review and editing, S.-C.Y.; supervision, W.-C.C., C.-L.Y., Y.-T.T., Y.-L.C. and S.-C.Y. All authors have read and agreed to the published version of the manuscript.

Funding: This research was funded by Innovative Translational research, Ministry of Education (DP2-109-21121-01-O-03-03) and Formosa Produce Corporation, Taipei, Taiwan (109-6202-029-300).

Institutional Review Board Statement: The study was approved by the Institutional Animal Care and Use Committee of Taipei Medical University (LAC-2020-0143).

Informed Consent Statement: Not applicable.

Data Availability Statement: Not applicable.

Acknowledgments: We thank the Laboratory Animal Center at Taipei Medical University, Taiwan for technical support in the animal experiment.

Conflicts of Interest: The authors declare that there are no conflict of interest regarding the publication of this paper.

Abbreviations

ACC	acetyl CoA carboxylase
AdipoR2	adiponectin receptor 2
ALT	alanine aminotransferase
AMPKα	adenosine monophosphate-activated protein kinase-α
ANOVA	analysis of variance
AST	aspartate aminotransferase
BMI	body-mass index
BSA	bovine serum albumin
BWs	body weights
CPT-1	carnitine palmitoyl transferase-1
dH_2O	distilled water
DPPH	di(phenyl)-(2,4,6-trinitrophenyl) iminoazanium
ELISA	enzyme-linked immunosorbent assay
ER	endoplasmic reticulum
FAs;	FFAs free fatty acids
FAS	fatty acid synthase
F/B	Firmicutes-to-Bacteroidetes
FER	food efficiency ratio
GABA	γ-aminobutyric acid
GAPDH	glyceraldehyde 3-phosphate dehydrogenase
GC-MS	gas chromatographic-mass spectrometric
H&E	hematoxylin A- and eosin
HF	high-fat
HOMA-IRI	homeostasis model assessment of the IR index
IBD	inflammatory bowel disease
IgG	immunoglobulin G
IL	interleukin
IR	insulin resistance
LDA	linear discriminant analysis
LEfSe	linear discriminant analysis of the effect size
METS	metabolic syndrome

NAFLD	non-alcoholic fatty liver disease
NAHSIT	nutrition and health survey in Taiwan
NASH	non-alcoholic steatohepatitis
NC	normal control
NF	nuclear factor
NGS	next generation sequencing
NP	nonyl phenoxypolyethoxylethanol
OD	optical density
P	phosphorylated
PCoA	principal coordinate analysis
PCR	polymerase chain reaction
PI	protease inhibitor
PPARα	peroxisome proliferator-activated receptor α
PPI	phosphatase inhibitor
PVDF	polyvinylidene difluoride
R	ribosomal
RB;	FRB water extract of fermented rice bran
SCD1	stearoyl coenzyme A desaturase 1
SCFA	short-chain fatty acid
SD	standard deviation
SDS-PAGE	sodium dodecylsulfate polyacrylamide gel electrophoresis
SIRT1	nicotinamide adenine dinucleotide-dependent sirtuin-1
SREBP-1c	sterol response element-binding protein-1c
TAC	total antioxidant capacity
TBARS	thiobarbituric acid-reactive substance
TC	total cholesterol
TG	triglyceride
TLR4	Toll-like receptor 4
TNF	tumor necrosis factor
T2DM	type 2 diabetes mellitus
WHO	world health organization
ZO	zonula occludens

References

1. World Health Organization. *Decade of Healthy Ageing: Baseline Report*; WHO: Geneva, Switzerland, 2020.
2. World Health Organization. Obesity and Overweight. Available online: https://www.who.int/news-room/fact-sheets/detail/obesity-and-overweight (accessed on 15 February 2022).
3. Taiwan National Nutrition and Health Survey 2013–2016. Health Promotion Administration, Ministry of Health and Welfare, Taiwan. Available online: https://www.hpa.gov.tw/ (accessed on 30 November 2021).
4. 2019 Cause of Death Statistics. Ministry of Health and Welfare, Taiwan. Available online: https://www.mohw.gov.tw/ (accessed on 30 November 2021).
5. Younossi, Z.; Tacke, F.; Arrese, M.; Chander Sharma, B.; Mostafa, I.; Bugianesi, E.; Wai-Sun Wong, V.; Yilmaz, Y.; George, J.; Fan, J.; et al. Global perspectives on nonalcoholic fatty liver disease and nonalcoholic steatohepatitis. *Hepatology* **2019**, *69*, 2672–2682. [CrossRef] [PubMed]
6. Hsu, C.S.; Kao, J.H. Non-alcoholic fatty liver disease: An emerging liver disease in taiwan. *J. Formos. Med. Assoc. Taiwan Yi Zhi* **2012**, *111*, 527–535. [CrossRef] [PubMed]
7. Manne, V.; Handa, P.; Kowdley, K.V. Pathophysiology of nonalcoholic fatty liver disease/nonalcoholic steatohepatitis. *Clin. Liver Dis.* **2018**, *22*, 23–37. [CrossRef]
8. Bettermann, K.; Hohensee, T.; Haybaeck, J. Steatosis and steatohepatitis: Complex disorders. *Int. J. Mol. Sci.* **2014**, *15*, 9924–9944. [CrossRef]
9. Adolph, T.E.; Grander, C.; Grabherr, F.; Tilg, H. Adipokines and non-alcoholic fatty liver disease. *Mult. Interact.* **2017**, *18*, 1649.
10. Iwabu, M.; Okada-Iwabu, M.; Yamauchi, T.; Kadowaki, T. Adiponectin/adiponectin receptor in disease and aging. *NPJ Aging Mech. Dis.* **2015**, *1*, 15013. [CrossRef] [PubMed]
11. Handa, P.; Maliken, B.D.; Nelson, J.E.; Morgan-Stevenson, V.; Messner, D.J.; Dhillon, B.K.; Klintworth, H.M.; Beauchamp, M.; Yeh, M.M.; Elfers, C.T.; et al. Reduced adiponectin signaling due to weight gain results in nonalcoholic steatohepatitis through impaired mitochondrial biogenesis. *Hepatology* **2014**, *60*, 133–145. [CrossRef] [PubMed]
12. Baselga-Escudero, L.; Souza-Mello, V.; Pascual-Serrano, A.; Rachid, T.; Voci, A.; Demori, I.; Grasselli, E. Beneficial effects of the mediterranean spices and aromas on non-alcoholic fatty liver disease. *Trends Food Sci. Technol.* **2017**, *61*, 141–159. [CrossRef]

13. Kohjima, M.; Higuchi, N.; Kato, M.; Kotoh, K.; Yoshimoto, T.; Fujino, T.; Yada, M.; Yada, R.; Harada, N.; Enjoji, M.; et al. Srebp-1c, regulated by the insulin and ampk signaling pathways, plays a role in nonalcoholic fatty liver disease. *Int. J. Mol. Med.* **2008**, *21*, 507–511.
14. Shabalala, S.C.; Dludla, P.V.; Mabasa, L.; Kappo, A.P.; Basson, A.K.; Pheiffer, C.; Johnson, R. The effect of adiponectin in the pathogenesis of non-alcoholic fatty liver disease (nafld) and the potential role of polyphenols in the modulation of adiponectin signaling. *Biomed. Pharmacother.* **2020**, *131*, 110785. [CrossRef]
15. Safari, Z.; Gérard, P. The links between the gut microbiome and non-alcoholic fatty liver disease (nafld). *Cell. Mol. Life Sci. CMLS* **2019**, *76*, 1541–1558. [CrossRef]
16. Salazar, N.; Valdés-Varela, L.; González, S.; Gueimonde, M.; de Los Reyes-Gavilán, C.G. Nutrition and the gut microbiome in the elderly. *Gut Microbes* **2017**, *8*, 82–97. [CrossRef]
17. Albenberg, L.G.; Wu, G.D. Diet and the intestinal microbiome: Associations, functions, and implications for health and disease. *Gastroenterology* **2014**, *146*, 1564–1572. [CrossRef] [PubMed]
18. Makki, K.; Deehan, E.C.; Walter, J.; Bäckhed, F. The impact of dietary fiber on gut microbiota in host health and disease. *Cell Host Microbe* **2018**, *23*, 705–715. [CrossRef] [PubMed]
19. Graf, D.; Di Cagno, R.; Fåk, F.; Flint, H.J.; Nyman, M.; Saarela, M.; Watzl, B. Contribution of diet to the composition of the human gut microbiota. *Microb. Ecol. Health Dis.* **2015**, *26*, 26164. [CrossRef] [PubMed]
20. Clarke, S.F.; Murphy, E.F.; Nilaweera, K.; Ross, P.R.; Shanahan, F.; O'Toole, P.W.; Cotter, P.D. The gut microbiota and its relationship to diet and obesity: New insights. *Gut Microbes* **2012**, *3*, 186–202. [CrossRef]
21. Mn, L.; Venkatachalapathy, N.; Manickavasagan, A. Physicochemical characteristics of rice bran. In *Brown Rice*; Springer: Cham, Switzerland, 2017; pp. 79–90.
22. Bodie, A.R.; Micciche, A.C.; Atungulu, G.G.; Rothrock, M.J.; Ricke, S.C. Current trends of rice milling byproducts for agricultural applications and alternative food production systems. *Front. Sustain. Food Syst.* **2019**, *3*, 47. [CrossRef]
23. Jariwalla, R.J. Rice-bran products: Phytonutrients with potential applications in preventive and clinical medicine. *Drugs Exp. Clin. Res.* **2001**, *27*, 17–26.
24. Sohail, M.; Rakha, A.; Butt, M.S.; Iqbal, M.J.; Rashid, S. Rice bran nutraceutics: A comprehensive review. *Crit. Rev. Food Sci. Nutr.* **2017**, *57*, 3771–3780. [CrossRef]
25. Alauddin, M.; Shirakawa, H.; Koseki, T.; Kijima, N.; Ardiansyah; Budijanto, S.; Islam, J.; Goto, T.; Komai, M. Fermented rice bran supplementation mitigates metabolic syndrome in stroke-prone spontaneously hypertensive rats. *BMC Complement. Altern. Med.* **2016**, *16*, 442. [CrossRef]
26. Rusbana, T.B.; Agista, A.Z.; Saputra, W.D.; Ohsaki, Y.; Watanabe, K.; Ardiansyah, A.; Budijanto, S.; Koseki, T.; Aso, H.; Komai, M.; et al. Supplementation with fermented rice bran attenuates muscle atrophy in a diabetic rat model. *Nutrients* **2020**, *12*, 2409. [CrossRef] [PubMed]
27. Shibayama, J.; Kuda, T.; Shikano, A.; Fukunaga, M.; Takahashi, H.; Kimura, B.; Ishizaki, S. Effects of rice bran and fermented rice bran suspensions on caecal microbiota in dextran sodium sulphate-induced inflammatory bowel disease model mice. *Food Biosci.* **2018**, *25*, 8–14. [CrossRef]
28. Islam, J.; Koseki, T.; Watanabe, K.; Budijanto, S.; Oikawa, A.; Alauddin, M.; Goto, T.; Aso, H.; Komai, M.; Shirakawa, H. Dietary supplementation of fermented rice bran effectively alleviates dextran sodium sulfate-induced colitis in mice. *Nutrients* **2017**, *9*, 747. [CrossRef] [PubMed]
29. Yu, Y.; Zhang, J.; Wang, J.; Sun, B. The anti-cancer activity and potential clinical application of rice bran extracts and fermentation products. *RSC Adv.* **2019**, *9*, 18060–18069. [CrossRef]
30. Fellmann, L.; Nascimento, A.R.; Tibiriça, E.; Bousquet, P. Murine models for pharmacological studies of the metabolic syndrome. *Pharmacol. Ther.* **2013**, *137*, 331–340. [CrossRef]
31. Rinella, M.E. Nonalcoholic fatty liver disease: A systematic review. *JAMA* **2015**, *313*, 2263–2273. [CrossRef] [PubMed]
32. Leoni, S.; Tovoli, F.; Napoli, L.; Serio, I.; Ferri, S.; Bolondi, L. Current guidelines for the management of non-alcoholic fatty liver disease: A systematic review with comparative analysis. *World J. Gastroenterol.* **2018**, *24*, 3361–3373. [CrossRef]
33. Liu, C.-Y.; Chang, C.-W.; Lee, H.-C.; Chen, Y.-J.; Tsai, T.-H.; Chiau, J.-S.C.; Wang, T.-E.; Tsai, M.-C.; Yeung, C.-Y.; Shih, S.-C. Metabolic damage presents differently in young and early-aged c57bl/6 mice fed a high-fat diet. *Int. J. Gerontol.* **2016**, *10*, 105–111. [CrossRef]
34. Nunes-Souza, V.; César-Gomes, C.J.; Da Fonseca, L.J.S.; Guedes, G.D.S.; Smaniotto, S.; Rabelo, L.A. Aging increases susceptibility to high fat diet-induced metabolic syndrome in c57bl/6 mice: Improvement in glycemic and lipid profile after antioxidant therapy. *Oxidative Med. Cell. Longev.* **2016**, *2016*, 1987960. [CrossRef]
35. Ai, X.; Wu, C.; Yin, T.; Zhur, O.; Liu, C.; Yan, X.; Yi, C.; Liu, D.; Xiao, L.; Li, W.; et al. Antidiabetic function of lactobacillus fermentum mf423-fermented rice bran and its effect on gut microbiota structure in type 2 diabetic mice. *Front. Microbiol.* **2021**, *12*, 1427. [CrossRef]
36. Park, S.; Chang, H.C.; Lee, J.J. Rice bran fermented with kimchi-derived lactic acid bacteria prevents metabolic complications in mice on a high-fat and -cholesterol diet. *Foods* **2021**, *10*, 1501. [CrossRef]
37. Fan, J.; Choi, K.; Han, G.D. Inhibitory effects of water extracts of fermented rice bran on allergic response. *Food Sci. Biotechnol.* **2010**, *19*, 1573–1578. [CrossRef]

38. Xia, H.M.; Wang, J.; Xie, X.J.; Xu, L.J.; Tang, S.Q. Green tea polyphenols attenuate hepatic steatosis, and reduce insulin resistance and inflammation in high-fat diet-induced rats. *Int. J. Mol. Med.* **2019**, *44*, 1523–1530. [CrossRef] [PubMed]
39. Fulco, M.; Sartorelli, V. Comparing and contrasting the roles of ampk and sirt1 in metabolic tissues. *Cell Cycle* **2008**, *7*, 3669–3679. [CrossRef] [PubMed]
40. Crispino, M.; Trinchese, G.; Penna, E.; Cimmino, F.; Catapano, A.; Villano, I.; Perrone-Capano, C.; Mollica, M.P. Interplay between peripheral and central inflammation in obesity-promoted disorders: The impact on synaptic mitochondrial functions. *Int. J. Mol. Sci.* **2020**, *21*, 5964. [CrossRef] [PubMed]
41. Lim, S.-I.; Seong, K.-S.; Song, S.-M.; Hwang, J.-T.; Lee, B.-Y. Effect of rice bran and soybean fermented by *bacillus* spp. On lipid profiles of liver, serum, and feces in rats fed high fat diet. *J. Korean Soc. Appl. Biol. Chem.* **2011**, *54*, 237–242. [CrossRef]
42. Naowaboot, J.; Piyabhan, P.; Munkong, N.; Parklak, W.; Pannangpetch, P. Ferulic acid improves lipid and glucose homeostasis in high-fat diet-induced obese mice. *Clin. Exp. Pharmacol. Physiol.* **2016**, *43*, 242–250. [CrossRef]
43. da Silva, R.P.; Kelly, K.B.; Al Rajabi, A.; Jacobs, R.L. Novel insights on interactions between folate and lipid metabolism. *Biofactors* **2014**, *40*, 277–283. [CrossRef]
44. Feng, D.; Zou, J.; Su, D.; Mai, H.; Zhang, S.; Li, P.; Zheng, X. Curcumin prevents high-fat diet-induced hepatic steatosis in apoe (−/−) mice by improving intestinal barrier function and reducing endotoxin and liver tlr4/nf-κb inflammation. *Nutr. Metab.* **2019**, *16*, 79. [CrossRef]
45. Crawford, M.S.; Mohr, A.E.; Sweazea, K.L. Novel organic mineral complex prevents high-fat diet-induced changes in the gut and liver of male sprague-dawley rats. *J. Nutr. Metab.* **2020**, *2020*, 8846401. [CrossRef]
46. Ding, Y.; Song, Z.; Li, H.; Chang, L.; Pan, T.; Gu, X.; He, X.; Fan, Z. Honokiol ameliorates high-fat-diet-induced obesity of different sexes of mice by modulating the composition of the gut microbiota. *Front. Immunol.* **2019**, *10*, 2800. [CrossRef] [PubMed]
47. Hall, A.B.; Yassour, M.; Sauk, J.; Garner, A.; Jiang, X.; Arthur, T.; Lagoudas, G.K.; Vatanen, T.; Fornelos, N.; Wilson, R.; et al. A novel ruminococcus gnavus clade enriched in inflammatory bowel disease patients. *Genome Med.* **2017**, *9*, 103. [CrossRef] [PubMed]
48. Cano, P.G.; Santacruz, A.; Moya, Á.; Sanz, Y. Bacteroides uniformis cect 7771 ameliorates metabolic and immunological dysfunction in mice with high-fat-diet induced obesity. *PLoS ONE* **2012**, *7*, e41079.
49. Yang, J.Y.; Lee, Y.S.; Kim, Y.; Lee, S.H.; Ryu, S.; Fukuda, S.; Hase, K.; Yang, C.S.; Lim, H.S.; Kim, M.S.; et al. Gut commensal bacteroides acidifaciens prevents obesity and improves insulin sensitivity in mice. *Mucosal Immunol.* **2017**, *10*, 104–116. [CrossRef]
50. Zhao, X.; Zhang, Z.; Hu, B.; Huang, W.; Yuan, C.; Zou, L. Response of gut microbiota to metabolite changes induced by endurance exercise. *Front. Microbiol.* **2018**, *9*, 765. [CrossRef] [PubMed]
51. Ilan, Y. Leaky gut and the liver: A role for bacterial translocation in nonalcoholic steatohepatitis. *World J. Gastroenterol.* **2012**, *18*, 2609–2618. [CrossRef]
52. Milosevic, I.; Vujovic, A.; Barac, A.; Djelic, M.; Korac, M.; Radovanovic Spurnic, A.; Gmizic, I.; Stevanovic, O.; Djordjevic, V.; Lekic, N.; et al. Gut-liver axis, gut microbiota, and its modulation in the management of liver diseases: A review of the literature. *Int. J. Mol. Sci.* **2019**, *20*, 395. [CrossRef]
53. Wang, B.; Jiang, X.; Cao, M.; Ge, J.; Bao, Q.; Tang, L.; Chen, Y.; Li, L. Altered fecal microbiota correlates with liver biochemistry in nonobese patients with non-alcoholic fatty liver disease. *Sci. Rep.* **2016**, *6*, 32002. [CrossRef]
54. Ji, Y.; Yin, Y.; Sun, L.; Zhang, W. The molecular and mechanistic insights based on gut-liver axis: Nutritional target for non-alcoholic fatty liver disease (nafld) improvement. *Int. J. Mol. Sci.* **2020**, *21*, 3066. [CrossRef]
55. Ritze, Y.; Bárdos, G.; Hubert, A.; Böhle, M.; Bischoff, S.C. Effect of tryptophan supplementation on diet-induced non-alcoholic fatty liver disease in mice. *Br. J. Nutr.* **2014**, *112*, 1–7. [CrossRef]
56. Ardiansyah. A short review: Bioactivity of fermented rice bran. *J. Oleo Sci.* **2021**, *70*, 1565–1574. [CrossRef] [PubMed]
57. Liang, W.; Menke, A.L.; Driessen, A.; Koek, G.H.; Lindeman, J.H.; Stoop, R.; Havekes, L.M.; Kleemann, R.; van den Hoek, A.M. Establishment of a general nafld scoring system for rodent models and comparison to human liver pathology. *PLoS ONE* **2014**, *9*, e115922. [CrossRef] [PubMed]
58. Chen, Y.L.; Shirakawa, H.; Lu, N.S.; Peng, H.C.; Xiao, Q.; Yang, S.C. Impacts of fish oil on the gut microbiota of rats with alcoholic liver damage. *J. Nutr. Biochem.* **2020**, *86*, 108491. [CrossRef] [PubMed]
59. Chen, Y.L.; Peng, H.C.; Wang, X.D.; Yang, S.C. Dietary saturated fatty acids reduce hepatic lipid accumulation but induce fibrotic change in alcohol-fed rats. *Hepatobiliary Surg. Nutr.* **2015**, *4*, 172–183.
60. Bolyen, E.; Rideout, J.R.; Dillon, M.R.; Bokulich, N.A.; Abnet, C.C.; Al-Ghalith, G.A.; Alexander, H.; Alm, E.J.; Arumugam, M.; Asnicar, F.; et al. Reproducible, interactive, scalable and extensible microbiome data science using qiime 2. *Nat. Biotechnol.* **2019**, *37*, 852–857. [CrossRef]
61. Chong, J.; Liu, P.; Zhou, G.; Xia, J. Using microbiomeanalyst for comprehensive statistical, functional, and meta-analysis of microbiome data. *Nat. Protoc.* **2020**, *15*, 799–821. [CrossRef]
62. García-Villalba, R.; Giménez-Bastida, J.A.; García-Conesa, M.T.; Tomás-Barberán, F.A.; Carlos Espín, J.; Larrosa, M. Alternative method for gas chromatography-mass spectrometry analysis of short-chain fatty acids in faecal samples. *J. Sep. Sci.* **2012**, *35*, 1906–1913. [CrossRef]

plants

MDPI

Article

Neurotherapy of Yi-Gan-San, a Traditional Herbal Medicine, in an Alzheimer's Disease Model of *Drosophila melanogaster* by Alleviating Aβ_{42} Expression

Ming-Tsan Su [1], Yong-Sin Jheng [1], Chen-Wen Lu [1], Wen-Jhen Wu [1], Shieh-Yueh Yang [2], Wu-Chang Chuang [3], Ming-Chung Lee [4] and Chung-Hsin Wu [1,*]

[1] School of Life Science, National Taiwan Normal University, Taipei 11677, Taiwan; mtsu@ntnu.edu.tw (M.-T.S.); zxx159753@gmail.com (Y.-S.J.); kumo.lu@gmail.com (C.-W.L.); efgy78@gmail.com (W.-J.W.)
[2] MagQu Co., Ltd., New Taipei City 11231, Taiwan; syyang@magqu.com
[3] Sun Ten Pharmaceutical Co., Ltd., Taipei 23143, Taiwan; cwctd331@sunten.com.tw
[4] Brion Research Institute of Taiwan, Taipei 23143, Taiwan; mileslee@sunten.com.tw
* Correspondence: megawu@ntnu.edu.tw

Abstract: Alzheimer's disease (AD), a main cause of dementia, is the most common neurodegenerative disease that is related to the abnormal accumulation of amyloid β (Aβ) proteins. Yi-Gan-San (YGS), a traditional herbal medicine, has been used for the management of neurodegenerative disorders and for the treatment of neurosis, insomnia and dementia. The aim of this study was to examine antioxidant capacity and cytotoxicity of YGS treatment by using 2,2-Diphenyl-1-picrylhydrazyl (DPPH) and 3-(4,5-Dimethylthiazol-2-yl)-2,5-diphenyltetrazolium bromide (MTT) assays in vitro. We explored neuroprotective effects of YGS treatment in alleviating Aβ neurotoxicity of Drosophila melanogaster in vivo by comparing survival rate, climbing index, and Aβ expressions through retinal green fluorescent protein (GFP) expression, highly sensitive immunomagnetic reduction (IMR) and Western blotting assays. In the in vitro study, our results showed that scavenging activities of free radical and SH-SY5Y nerve cell viability were increased significantly ($p < 0.01$–0.05). In the in vivo study, Aβ_{42}-expressing flies (Aβ_{42}-GFP flies) and their WT flies (mCD8-GFP flies) were used as an animal model to examine the neurotherapeutic effects of YGS treatment. Our results showed that, in comparison with those Aβ_{42} flies under sham treatments, Aβ_{42} flies under YGS treatments showed a greater survival rate, better climbing speed, and lower Aβ_{42} aggregation in *Drosophila* brain tissue ($p < 0.01$). Our findings suggest that YGS should have a beneficial alternative therapy for AD and dementia via alleviating Aβ neurotoxicity in the brain tissue.

Keywords: Alzheimer's disease; amyloid β; immunomagnetic reduction; *Drosophila melanogaster*

Citation: Su, M.-T.; Jheng, Y.-S.; Lu, C.-W.; Wu, W.-J.; Yang, S.-Y.; Chuang, W.-C.; Lee, M.-C.; Wu, C.-H. Neurotherapy of Yi-Gan-San, a Traditional Herbal Medicine, in an Alzheimer's Disease Model of *Drosophila melanogaster* by Alleviating Aβ_{42} Expression. *Plants* **2022**, *11*, 572. https://doi.org/10.3390/plants11040572

Academic Editors: Juei-Tang Cheng, I-Min Liu and Szu-Chuan Shen

Received: 26 January 2022
Accepted: 19 February 2022
Published: 21 February 2022

1. Introduction

Dementia is the most frequent age-related neurocognitive disorder. Patients with dementia are known to frequently experience disturbing behavioral and psychological symptoms, such as excitement, aggression, hallucinations, insomnia, anxiety, wandering, and depression [1–3]. Alzheimer's disease (AD), a main cause of dementia, is the most common neurodegenerative disease that is related to the abnormal accumulation of amyloid β (Aβ) proteins [4]. Pathological indicators of AD include the presence of Aβ plaques, which damage neurons, particularly those surrounding the hippocampus [5]. Aβ plaques are neuropathological biomarkers for AD. The challenge with assaying AD biomarkers is ascribed to the ultralow concentrations of Aβ_{42} proteins in the cerebrospinal fluid and the blood [6].

Yi-Gan-San (YGS, Shun-Ning-Yi OTC medicine), a traditional Chinese (Kampo) herbal medicine, is composed of Atractylodes, Poria, Chuanxiong, Angelica, Bupleurum, Licorice and Uncaria. Since ancient times, YGS has been used to treat patients who have symptoms, such as nervousness, short temper, irritability and sleeplessness. Recently, several studies have shown that the administration of YGS is effective for treating the behavioral and psychological symptoms of dementia (BPSD) [7,8]. In Japan, YGS is also called yokukansan and served as a remedy for restlessness and agitation in children [9]. Yokukansan has been approved by the Japanese Ministry of Health, Labor and Welfare as a remedy for neurosis, insomnia, irritability and night crying in children. YGS is the extract of multiple crude drugs containing a large number of ingredients. YGS can protect against the cytotoxic effect of a low concentration of corticosterone on hippocampal neurons [10]. To date, over 70 basic research articles have been published on the pharmacological efficacy and mechanisms of YGS about the pharmacokinetics, metabolism and brain distribution of their active ingredients. Among these research articles, YGS has been confirmed as a new potential therapeutic agent for the management of neurodegenerative disorders such as AD, and for the treatment of neurosis, insomnia and dementia [7,8,11–14]. Although molecular complexity is the main obstacle to studying the mechanism of YGS, it is an advantage for various pharmacological effects.

The argument that recapitulation of human AD by using transgenic animal models has improved the understanding of its pathological mechanisms is indisputable. The *Drosophila melanogaster* model has made its mark as an effective tool for the study of human AD [15–17], including its cellular aspects and associated physiological and behavioral traits, through the use of both conventional and innovative genetic tools. It is undeniable that human genetics research has improved the understanding of genes related to neurodegenerative diseases. However, the inspection of human subjects is hampered by moral and technical limitations. Therefore, many studies turn to AD animal models, such as the fruit fly (*Drosophila melanogaster*), mouse (*Mus musculus*), zebrafish (*Danio rerio*), and nematode (*Caenorhabditis elegans*); with each mirroring differing aspects of AD to generalize human diseases. Among these animal models, the *Drosophila melanogaster* AD model has been selected as an ideal tool to study AD disorders because targeted expression of Aβ proteins in adult *Drosophila* can result in changes in the appearance of the structure, including a reduction in external eye size, and a loss of ommatidia organization. Thus, this study selected *Drosophila* as an AD model as a drug discovery tool for AD. Pathological evidence of AD includes the presence of Aβ protein in the brain tissue [18–20]. It is very important to find suitable analytical methods that can detect Aβ expression in *Drosophila* brain tissue. The novel techniques that have been successfully developed to detect Aβ in cerebrospinal fluid (CSF) are clearly suitable for detecting Aβ expression in *Drosophila* brain tissue [21,22]. For example, immunomagnetic reduction (IMR) can detect ultralow concentrations of Aβ protein in human CSF and blood for early diagnosis of AD through the use of antibody-functionalized magnetic nanoparticles dispersed in an aqueous solution [23–26].

In this study, we shed light on how *Drosophila melanogaster* became an animal model for AD, as well as its contribution as a tool for discovering therapeutic drugs for AD. We used highly sensitive IMR assay technology that is capable of detecting ultralow concentrations of Aβ in *Drosophila* brain tissue. Our results demonstrated that YGS treatments had better antioxidant capacity, and low cytotoxicity for SH-SY5Y nerve cells for the in vitro study, and had a greater survival rate, better climbing speed, and lower $A\beta_{42}$ aggregation in the brain tissue. On the basis of our data, YGS treatment may be a beneficial therapy for alleviating neurodegenerative disorders in AD.

2. Results

2.1. Chromatographic Fingerprints of YGS

YGS is widely used as a traditional herbal medicine that is composed of seven dried medicinal herbs: *Uncis ramulus*, *Cnidii rhizoma*, *Bupleuri radix*, *Atratylodis Lanceae rhizoma*, Poria, *Angelicae radix* and *Glycyrrhizae radix* in specific ratios. Each plant material was identified by external morphology, and the marker compound of the plant specimen according to the Taiwan Pharmacopoeia standard. In Figure 1, we used 3D high-performance liquid chromatography (HPLC) and UV detection methods to analyze individual active substances and confirm which medicinal material corresponds to the chromatographic peak of YGS. We adopted chlorogenic acid, ferulic acid, liquiritin, glycyrrhizic acid, atractylenolide III and ligustilide, and other standard products to prepare a standard solution, and analyzed and compared the standard and sample solution with the same analytical method. Our 3D HPLC data showed the bioactive substances of YGS were chlorogenic acid, ferulic acid, liquiritin, glycyrrhizin, atractylenolide III, ligustilide, and were determined qualitatively within 60 min. The possible function of these bioactive substances was described in detail separately in the discussion section.

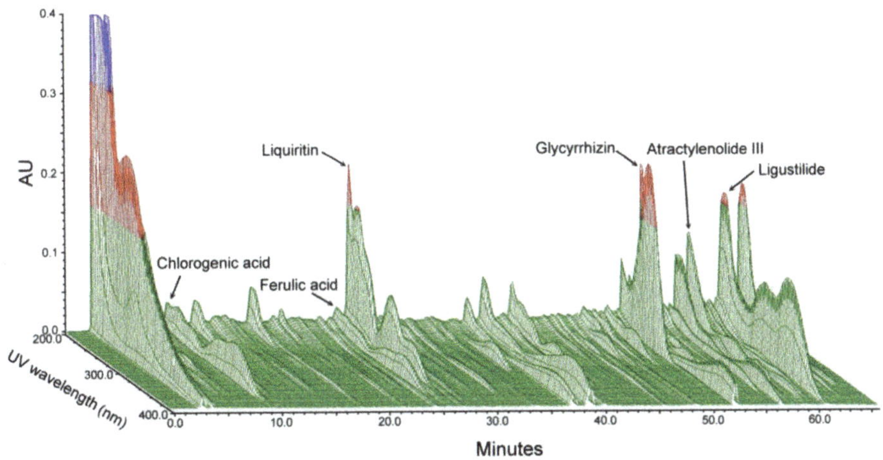

Figure 1. Chromatographic fingerprints of YGS from 3D HPLC. The bioactive marker compounds, namely chlorogenic acid, ferulic acid, liquiritin, glycyrrhizin, atractylenolide III, and ligustilide, were determined qualitatively within 60 min under the selected HPLC condition. Abbreviations: YGS, Yi-Gan-San; AU, arbitrary perfusion units; 3D, three-dimension; HPLC, high-performance liquid chromatography.

2.2. YGS Treatment Shows Better DPPH Free Radical Scavenging Activity

Figure 2A shows quantified DPPH free radical scavenging activities of YGS extracts at varying concentrations. By using the radical scavenging activity of L-ascorbic acid as the reference standard, we measured 56.4–91.7% of radical scavenging activity under 0.1–100 mg/mL YGS extract treatments. Quantified DPPH radical scavenging activities of YGS extract treatments at 0.1 and 1.0 mg/mL had similar free radical scavenging activity (56.4–61.2%), while there were significant differences ($p < 0.01$, Figure 2A) for those YGS extract treatments at 10–100 mg/mL (78.3–91.7%) from YGS extract treatments at 0.1 and 1.0 mg/mL.

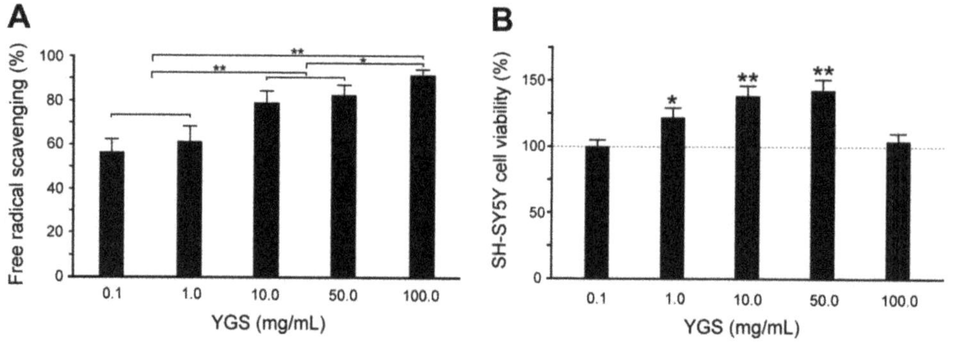

Figure 2. Antioxidant capacity and cytotoxicity of YGS treatment. (**A**) Quantified scavenging activities of free radicals are significantly greater with concentrations of YGS treatments by DPPH assay ($N = 3$ for each group). (**B**) Quantified relative SH-SY5Y cell viability was significantly greater with concentrations of YGS treatments by MTT assay ($N = 3$ for each group). Values are mean ± SEM (** $p < 0.01$, * $p < 0.05$, one-way ANOVA followed by a Student–Newman–Keuls multiple comparisons post-test). Abbreviations: YGS, Yi-Gan-San; AU, DPPH, 1,1-diphenyl-2-picrylhydrazyl; L-AA, L-Ascorbic acid; MTT, 3-(4,5-Dimethylthiazol-2-yl)-2,5-diphenyltetrazolium bromide; SEM, standard error of the mean; ANOVA, analysis of variance.

2.3. YGS Treatment Shows Higher Cell Viability of SH-SY5Y Cells

Figure 2B shows the quantified cell viability of human neuroblastoma SH-SY5Y cells at different concentrations of YGS extracts treatment. Significant cell viability of SH-SY5Y cells was observed under YGS treatment at 0.5–10 mg/mL. When compared to sham treatment, the cell viability of SH-SY5Y cells was significantly enhanced from 120.9–140.8% under YGS treatment at 0.5–10 mg/mL ($p < 0.01$–0.05, Figure 2B). However, the cell viability of SH-SY5Y cells under YGS treatment at 20 mg/mL has no significant difference from those with sham treatment ($p > 0.05$, Figure 2B). Our results reveal that SH-SY5Y cells show better cell viability only under YGS treatment from 0.5–10 mg/mL.

2.4. YGS Treatment Shows Higher Survival Rate for $A\beta_{42}$ Flies

Figure 3 shows the survival rate of mCD8 and Aβ42 flies under sham and YGS treatments. Both days after eclosion of mCD8 flies (as normal control) with sham and YGS treatment are longer than those of $A\beta_{42}$ flies with sham and YGS treatment, and days after eclosion of $A\beta_{42}$ flies with YGS treatment are longer than those of $A\beta_{42}$ flies with sham treatment. Our results reveal that YGS treatment should be able to prolong survival duration for $A\beta_{42}$ flies.

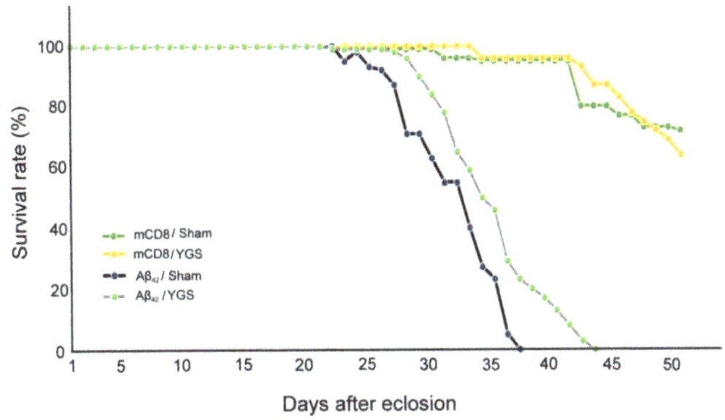

Figure 3. The survival rate of mCD8 and Aβ_{42} flies under sham and YGS treatments. Days after eclosion of mCD8 flies (normal control) with sham ($N = 300$) and YGS treatment ($N = 300$) are longer than those of Aβ_{42} flies with sham ($N = 300$) and YGS treatment ($N = 300$), and days after eclosion of Aβ_{42} flies with YGS treatment are longer than those of Aβ_{42} flies with sham treatment. In other words, YGS treatment should be able to prolong survival duration for Aβ_{42} flies. Abbreviations: YGS, Yi-Gan-San; Aβ, amyloid-beta.

2.5. YGS Treatment Shows Better Climbing Index for Aβ_{42} Flies

Figure 4 shows a quantified climbing index of mCD8 and Aβ42 flies at 1, 5, and 10 days after eclosion under sham and YGS treatments. Our results show a quantified climbing index of Aβ_{42} flies at 5 and 10 days after eclosion, and under YGS treatment was significantly greater than those Aβ_{42} flies under sham treatment ($N = 30$ for each group, $p < 0.01$–0.05, Figure 4). In other words, YGS treatment should be able to enhance the climbing index for Aβ_{42} flies.

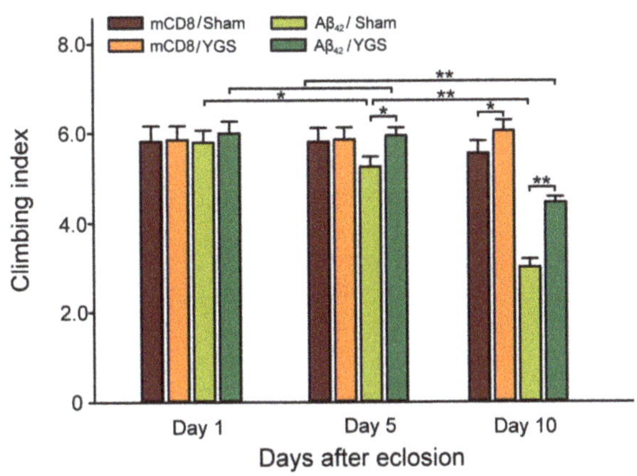

Figure 4. Climbing index of mCD8 and Aβ_{42} flies at 1, 5, 10 days after eclosion under sham and YGS treatments. The quantified climbing index of Aβ_{42} flies at 5 and 10 days after eclosion under YGS treatment was significantly greater than those Aβ_{42} flies under sham treatment ($N = 30$ for each group). In other words, YGS treatment should be able to enhance the climbing index for Aβ_{42} flies. Values are mean ± SEM (** $p < 0.01$, * $p < 0.05$, two-way ANOVA followed by a Student–Newman–Keuls multiple comparisons post-test). Abbreviations: YGS, Yi-Gan-San; Aβ, amyloid-beta; SEM, standard error of the mean; ANOVA, analysis of variance.

2.6. YGS Treatment Shows Greater Green Fluorescent Protein (GFP) Fluorescence in the External Eyes for GFP-Aβ_{42} Flies

We modeled GFP fluorescence in Aβ_{42}-expressing flies and observed that it was more sensitive and suitable for analyzing Aβ_{42} toxicity. As shown in Figure 5, GFP fluorescence in the external eyes of GFP-Aβ_{42} flies was weaker than those of GFP-WT flies. When GFP-Aβ_{42} flies were treated with 0.1% and 1% YGS treatments, GFP fluorescence in the external eyes was stronger than those of GFP-Aβ_{42} flies with sham treatment (Figure 5A). We quantified and then compared that GFP fluorescence in the external eyes of GFP-mCD8 flies was significantly stronger than those of GFP-Aβ_{42} flies with sham, 0.1% and 1% YGS treatments ($N = 30$ for each group, $p < 0.01$, Figure 5B); while GFP fluorescence in the external eyes of GFP-Aβ_{42} flies with 0.1% and 1% YGS treatments was significantly stronger than those of GFP-Aβ_{42} flies with sham treatment ($N = 30$ for each group, $p < 0.01$, Figure 5B). In addition, GFP fluorescence in the external eyes of GFP-Aβ_{42} flies with 1% YGS treatment was significantly stronger than those of GFP-Aβ_{42} flies with 0.1% YGS treatment ($N = 30$ for each group, $p < 0.05$, Figure 5B). The results revealed that GFP-Aβ_{42} flies showed increasing GFP fluorescence in the external eyes with an increasing dose of YGS treatment.

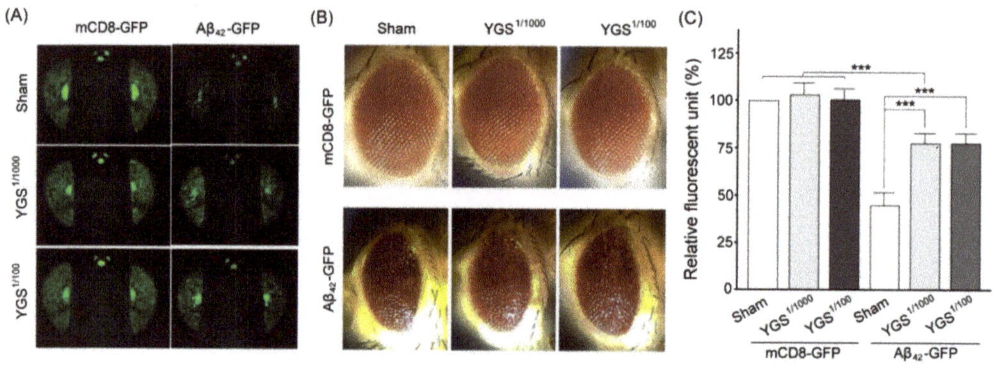

Figure 5. Eye GFP expressions of mCD8-GFP and Aβ_{42}-GFP flies under sham and YGS treatments. (**A**) GFP expressions of external eyes in mCD8-GFP (normal control) flies under sham and YGS treatments were obviously greater than those of Aβ_{42}-GFP under sham and YGS treatments, while GFP expressions of external eyes in the Aβ_{42}-GFP flies under YGS treatment were obviously greater than those of Aβ_{42}-GFP flies under sham treatment. (**B**) Completeness of external eyes in mCD8-GFP flies under sham and YGS treatments was obviously better than those of Aβ_{42}-GFP under sham and YGS treatments, while completeness of external eyes in the Aβ_{42}-GFP flies under YGS treatment was obviously better than those of Aβ_{42}-GFP flies under sham treatment. (**C**) Quantified GFP fluorescence of external eyes of mCD8-GFP flies under sham and YGS treatments were significantly greater than those of Aβ_{42}-GFP under sham and YGS treatments, while quantified GFP expressions of external eyes in the Aβ_{42}-GFP flies under YGS treatment was significantly greater than those of Aβ_{42}-GFP flies under sham treatment ($N = 30$ for each group). Values are mean ± SEM (*** $p < 0.001$, two-way ANOVA followed by a Student–Newman–Keuls multiple comparisons post-test). Abbreviations: YGS, Yi-Gan-San; Aβ, amyloid-beta, GFP, green fluorescent protein; SEM, standard error of the mean; ANOVA, analysis of variance.

2.7. YGS Treatment Shows Reduced Aβ Expression for Aβ_{42} Flies by IMR Assay

We used a highly sensitive IMR assay that can detect ultralow concentrations of Aβ protein in human blood for early diagnosis of AD [23–25]. We established a curve of the relationship between Aβ_{42} expression and the number of GFP-Aβ_{42} flies by using an IMR assay. Our results showed that the increase in the IMR signal as Aβ_{42} concentration increased was linked to the number of Aβ_{42}-expressing flies (Figure 6A, Right). In addition, we plotted a relationship curve and then found that the relationship between IMR signals and Aβ_{42} expression is positively correlated (Figure 6A, Left). IMR signals (in percentages) for Aβ_{42} concentrations between 0.1 and 10,000 pg/mL were used to explore the analytical relationship, which followed a logistic function. According to the relationship curve of Figure 6A, we quantified and then compared that the Aβ_{42} concentration of GFP-Aβ_{42} flies with sham treatment was significantly stronger than those of GFP-Aβ_{42} flies with 1% YGS treatment (sham, 26.4 pg/mL vs. YGS, 22.8 pg/mL, $p < 0.05$, Figure 6B). The results revealed down-regulation of Aβ_{42} expression for GFP-Aβ_{42} flies under YGS treatment.

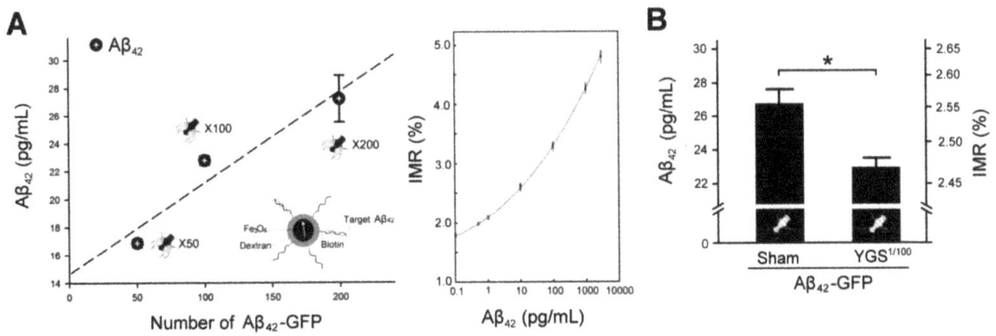

Figure 6. IMR assay of Aβ_{42} expressions for Aβ_{42}-GFP flies under sham and YGS treatments. (**A**) Left shows a relationship between Aβ_{42} concentrations and the number of Aβ_{42}-GFP flies by IMR assay. A magnetic nanoparticle coated with Aβ_{42} bioprobes was shown. Right shows the IMR value was increased with the Aβ_{42} concentration. (**B**) Quantified Aβ_{42} concentration of Aβ_{42}-GFP flies under sham treatment was significantly greater than those of Aβ_{42}-GFP flies under YGS treatment (N = 100 for each group). Values are mean ± SEM (* $p < 0.05$, one-way ANOVA followed by a Student–Neman–Keuls multiple comparisons post-test). Abbreviations: YGS, Yi-Gan-San; Aβ, amyloid-beta, GFP, green fluorescent protein; IMR, immunomagnetic reduction; SEM, standard error of the mean; ANOVA, analysis of variance.

2.8. *YGS Treatment Shows Reduced Aβ Expression for GFP-Aβ_{42} Flies by Western Blotting Assay*

For further verification, we used Western blotting to examine the Aβ_{42} expressions of mCD8 and Aβ_{42} flies under sham and YGS treatments, shown in Figure 7. Our results showed that Aβ_{42} expressions of mCD8 flies under sham and YGS treatments are quite low, while the Aβ_{42} expression of Aβ_{42} flies under sham treatment was greater than those of GFP-Aβ_{42} flies under YGS treatment (Figure 7A). We quantified and then compared that Aβ_{42} concentrations of Aβ_{42} concentrations of Aβ_{42} flies under sham treatment are significantly greater than those of GFP-Aβ_{42} flies under YGS treatment (sham, 0.96 vs. YGS, 0.61, $p < 0.01$, Figure 7B). The results further confirmed the down-regulation of Aβ_{42} expression for Aβ_{42} flies under YGS treatment.

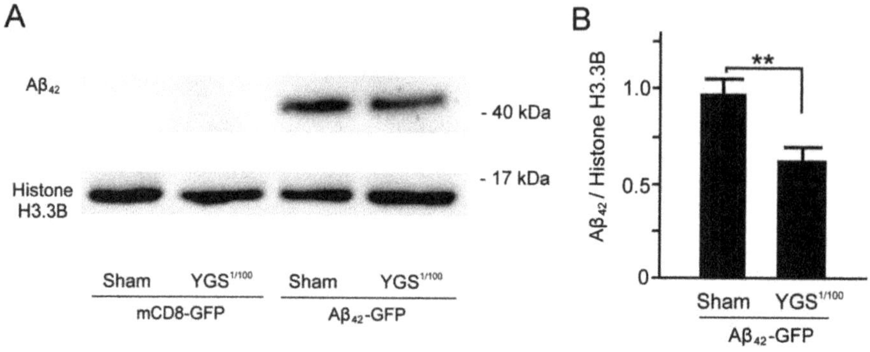

Figure 7. Western blotting analysis of Aβ_{42} expressions of Aβ_{42}-GFP flies under sham and YGS treatments. (**A**) An example of Western blotting of Aβ_{42} expressions of Aβ_{42}-GFP flies under sham and YGS treatments. (**B**) Quantified Aβ_{42} expressions of the Aβ_{42}-GFP flies under YGS treatment were significantly weaker than those of Aβ_{42}-GFP flies under sham treatment (N = 100 for each group). Values are mean ± SEM (** $p < 0.01$, two-way ANOVA followed by a Student–Newman–Keuls multiple comparisons post-test). Abbreviations: YGS, Yi-Gan-San; Aβ, amyloid-beta, GFP, green fluorescent protein; kDa, kilodaltons; WT, wild-type; SEM, standard error of the mean; ANOVA, analysis of variance.

3. Discussion

Oxidative stress has been implicated in the progression of a number of neurodegenerative diseases, including Alzheimer's disease (AD), Parkinson's disease (PD) and amyotrophic lateral sclerosis (ALS) [27]. These diseases are characterized by extensive oxidative damage to lipids, proteins and DNA. Oxidative stress is the result of an imbalance in the pro-oxidant/antioxidant homeostasis, leading to a generation of toxic reactive oxygen species (ROS). Aβ can generate H_2O_2 with the further generation of ROS through Fenton chemistry. To prove YGS has a neuroprotective function, we first examine whether YGS has the antioxidant ability by using the DPPH assay. In this study, we found that YGS has a good antioxidant ability because YGS extracts show significant DPPH radical scavenging activity at concentrations of 0.1–100 mg/mL. Bioactive markers of YGS were chlorogenic acid, ferulic acid, liquiritin, glycyrrhizin, atractylenolide III, and ligustilide that were determined qualitatively. It has been reported that YGS has neuroprotective effects and rescues neurons, possibly via the PI3K/Akt pathway [28]. Chlorogenic acid, an active compound of Uncaria, could protect against neurodegeneration with supplementation via inhibiting oxidative stress in the brain [29–32]. In addition, chlorogenic acid treatment can protect Aβ-induced injury in SH-SY5Y neurons and alleviate cognitive impairments in AD transgenic mice via enhancing the activation of the mTOR/TFEB signaling pathway [33]. Ferulic acid, an active compound of Angelica, can delay Aβ-induced pathological symptoms [34]. Liquiritin, an active compound of Licorice, can attenuate rheumatoid arthritis via reducing inflammation, inhibiting the MAPK signal pathway [35], and ameliorating Aβ-induced spatial learning and memory impairment by inhibiting oxidative stress and neural apoptosis [36]. Glycyrrhizin, another active compound of Licorice, can ameliorate inflammatory pain via blockage of the HMGB1-TLR4- NF-kB pathway [37], and prevent cognitive impairment in aged mice by reducing neuroinflammation and AD-related pathology [38]. Atractylenolide III, an active compound of Atractylodes, has anti-inflammatory and neuroprotective effects that may serve as a therapeutic agent in the treatment of depression [39]. Ligustilide, an active compound of Chuanxiong, can improve aging-induced memory deficit by regulating mitochondrial related inflammation and inhibiting oxidative stress in SAMP8 mice [40].

In this study, we modeled GFP fluorescence in Aβ_{42}-expressing flies that were more sensitive and suitable for analyzing Aβ_{42} toxicity and identified relevant therapeutic compounds. *Drosophila* Aβ models can help us to approach AD studies in uncovering crucial mechanisms and pathways. In addition, *Drosophila* models can be developed as an excellent tool for various drug testing purposes. As suggested from our results, YGS treatment could significantly reduce GFP fluorescence in the external eyes. These results should provide supporting evidence for the neurotherapy of YGS in an Alzheimer's disease model of *Drosophila melanogaster* by alleviating $A\beta_{42}$ expression. Although *Drosophila* was an excellent tool for studying AD mechanisms and pathways, there are still risks when using it as a disease model because the pathology may be unique to vertebrates and cannot be transformed into the invertebrate *Drosophila*.

Since the expression of Aβ in the *Drosophila* brain tissue is extremely low, it is very important to find a suitable analytical method to detect the Aβ change. We found that IMR, an ultra-high-sensitivity technology, should be very suitable for detecting ultralow concentrations of Aβ protein in the *Drosophila* brain tissue for diagnosis of AD through the use of antibody-functionalized magnetic nanoparticles dispersed in aqueous solution and the superconducting-quantum-interference-device (SQUID) [23–25]. Most studies utilizing SQUID IMR have focused on exploring the relationship between Aβ protein in plasma and in CSF [25,26]. We have achieved promising results in terms of the feasibility of detecting AD in *Drosophila* brain tissue. Our Western blotting data confirmed the results that down-regulation of $A\beta_{42}$ expression for $A\beta_{42}$ flies under YGS treatment by IMR assay is credible. As far as we know, this study should be the first application of IMR technology to detect Aβ protein expressions in the *Drosophila* brain tissue for the diagnosis of AD.

The exact mechanism by which YGS treatment alleviates $A\beta_{42}$ expression remains unclear. Therefore, further clinical trials are warranted to verify the benefits of YGS treatment in AD patients.

4. Materials and Methods

4.1. Yi-Gan-San (YGS) Preparation

The YGS, an over-the-counter drug called Shun-Ning-Yi, in Taiwan, was manufactured by the Sun-Ten Pharmaceutical company, New Taipei City, Taiwan. The Ministry of Health, Labor and Welfare of Japan has approved this as a remedy for neurosis, insomnia and irritability in children. YGS is a traditional herbal medicine consisting of seven herbs: *Uncis ramulus, Cnidii rhizoma, Bupleuri radix, Atratylodis Lanceae rhizoma*, Poria, *Angelicae radix* and *Glycyrrhizae radix* in specific ratios. As suggested by the Sun-Ten Pharmaceutical company, YGS is a Chinese herbal formula and its mass fraction of constituents was composed of Bupleurum 2 g, Licorice 2 g, Chuanxiong 3.2 g, Angelica 4 g, Atractylodes 4 g, Poria 4 g, and Uncaria 4 g per 100 g of the final product. In Taiwan, YGS was traditionally used as powders. In the prescription of YGS powder, such as *Angelica sinensis* and Chuanxiong, it is rich in volatile oil, which can retain a high content of volatile oil when used in powder, while the volatile oil can be easily volatilized when used in soup. YGS follows the traditional concept of using powder as medicine, mixing crude drug powder according to the proportion of traditional Chinese medicine formula, and then adding a small amount of starch for granulation, with the proportion of starch being 3.3%.

4.2. Phytochemical Screening

Before the phytochemical screening, precisely weigh 0.5 g of the YGS powder, ultrasonically shake with 20 mL of 70% methanol at room temperature for 15 min, then shake at 160 rpm for 20 min in a 40 °C water bath, and then centrifuge to take the supernatant. Another 20 mL of 70% methanol was added to the precipitate after centrifugation, ultrasonically shaken for 15 min at room temperature, and then centrifuged again at 160 rpm for 20 min in a 40 °C water bath, and the supernatant was taken. The two supernatants were combined and made up to 50 mL with 70% methanol, mixed evenly, and filtered through a 0.45 μm filter to obtain the test solution. The chromatographic fingerprint analysis of YGS was conducted using 3D HPLC (Burdick & Jackson, Gyeonggi-do, Korea). To confirm which bioactive substances of YGS were in the chromatographic peaks, we used chlorogenic acid, ferulic acid, liquiritin, glycyrrhizic acid, atractylenolide III and ligustilide and other standard products to prepare a standard solution and analyze and compare the standard and sample solution with the same analytical method. HPLC analysis systems regarding column type and identification are described: the model of controller and pump pressurized system was a Waters 600, the degasser was a Waters In-Line Degasser AF, the autosampler was a Waters 717 plus, the photodiode array detector was a Waters 2996, the pre-column was a Lichrospher RP-18 endcapped (5 μm, ID × L = 4.0 × 10 mm, Merck), and the analytical column was a Cosmosil 5C18-MS-II (5 μm, ID × L = 4.6 × 250 mm, Nacalai tesque). HPLC analysis conditions regarding mobile phase, flow rate and detector time are described: column temperature was 35 °C, the flow rate was 1.0 mL/min, and analysis time was 65 min.

4.3. DPPH Assay

Antioxidant activities of the YGS treatment were assessed by DPPH assay. YGS extracts that were diluted in distilled water in a concentration range of 0.1 to 100 mg/mL were mixed with 100 μL of 1.5mM/mL DPPH (D9132, Sigma-Aldrich Co., St. Louis, MO, USA) in methanol in a 96-well plate. After 30 min at room temperature, the absorbance of the samples was recorded. The color changes were recorded spectrophotometrically at 517 nm using a microplate spectrophotometer (μQuant™, BioTek Intruments, Inc., Winooski, VT, USA). Appropriate blanks (methanol) and standards (L-ascorbic acid in water, L-AA; A5960, Sigma-Aldrich Co., St. Louis, MO, USA) were recorded simultaneously. Each assay was carried out in triplicate. The DPPH scavenging was calculated by using the following expression:

$$\text{DPPH scavenging (\%)} = 100 \times [(\text{absorbance of sample + DPPH}) - (\text{absorbance of sample blank})]/[(\text{absorbance of DPPH}) - (\text{absorbance of methanol})]$$

Concentrations of YGS that cause 50% scavenging (IC_{50}) were calculated from the graph in which scavenging activity was plotted against the corresponding YGS concentration.

4.4. MTT Assay under YGS Treatment

Triplicate cultures of 1×10^5 SH-SY5Y cells per well for each 24-well plate. After the YGS or sham treatment, we added 0.5 mg/mL 3-(4,5-dimethylthiazol-2-yl)-2,5-diphenyltetrazolium bromide (MTT, M5655, Sigma-Aldrich Co., St. Louis, MO, USA) to the culture media. SH-SY5Y cells were incubated for 1 h at 37 °C in a humidified atmosphere (95% air and 5% CO_2), and then the MTT solution was discarded, and 100 μL of dimethyl sulfoxide (DMSO, Sigma-Aldrich Co., St. Louis, MO, USA) was added. The absorbance was read at an optical density (OD) of 570 nm with a microplate spectrophotometer (μQuant™, BioTek Intruments, Inc., Winooski, VT, USA).

4.5. Animals and Experimental Design

$A\beta_{42}$-expressing flies were generated by the Goldstein Laboratory (stock No. 32038). In this study, $A\beta_{42}$-expressing flies ($A\beta_{42}$-GFP flies) and their WT (mCD8-GFP flies) were used to examine the effect of YGS treatment by comparing retinal GFP expression without the need for histological assessment. The stronger $A\beta_{42}$ expressions in GFP-$A\beta_{42}$ flies and the weaker retinal GFP expression was monitored [15]. In this study, unless otherwise specified, both GFP-$A\beta_{42}$ and their WT flies were cultured and maintained on a standard cornmeal-yeast-agar medium at 25 °C and 60% humidity, in which the standard medium of GFP-$A\beta_{42}$ was uniformly mixed with 1% and 0.1% YGS by weight of the medium.

4.6. Survival Rate and Behavior Analysis in Drosophila

$A\beta_{42}$ flies and mCD8 flies (N = 300 for each group) were exposed to sham and 1.0% YGS treatment. The flies were observed daily for the incidence of mortality, and the survival rate was determined by counting the number of dead flies for 50 days. The data were subsequently analyzed and plotted as cumulative mortality and percentage survival after the treatment period.

The negative geotaxis assay was used to evaluate the locomotor performance of flies ($N = 30$ for each group). In brief, after the treatment period of 5 days, the flies from each group were briefly immobilized in ice and transferred into a clean tube (11 cm in length, 3.5 cm in diameter) and labeled accordingly. The flies were initially allowed to recover from immobilization for 10 min and thereafter were tapped at the bottom of the tubes. Observations were made for the total number of flies that crossed the 6 cm line within a period of 6 s and recorded. The results were expressed as a percentage of flies that escaped beyond a minimum distance of 6 cm in 6 s during three independent experiments.

4.7. Retinal GFP Expression Assay in Drosophila

We took an image of the outer Drosophila eye with an Olympus-BH2 microscope. Tissue sections and fluorescence were imaged using a Leica DM IRB fluorescence microscope. For examination with fluorescence microscopy, we removed the Drosophila head with spring scissors between the head and the thorax that was arranged in pairs so that individual experimental flies could be imaged and compared directly to individual flies from their respective control groups. Fluorescence was quantified by using publicly available NIH Image J software. The mean retinal fluorescence from $A\beta_{42}$ flies was normalized to the mean fluorescence from mCD8 flies (30 files for each group).

4.8. IMR Assay in Drosophila

Drosophila (100 files for each group) was homogenized and mixed with a reagent, which consisted of magnetic nanoparticles that were functionalized with monoclonal antibodies against a target protein, and then dispersed in PBS of pH 7.2 (MagQu Co., Ltd.) at room temperature. The magnetic nanoparticles used were dextran-coated Fe_3O_4 particles (MF-DEX-0060, MagQu Co., Ltd.). For each sample at each target protein concentration, IMR signal measurements were performed in duplicate. The signals were converted to biomarker concentrations using standard curves. All plasma samples were blinded for IMR measurements. The tau reagent (MF-TAU-0060, MagQu Co., Ltd.) contained magnetic nanoparticles immobilized with a monoclonal antibody (T9450, Sigma) against human tau protein. The $A\beta_{1-42}$ reagent (MF-AB2-0060, MagQu Co., Ltd.) contained magnetic nanoparticles coated with a monoclonal antibody against human $A\beta_{1-42}$ protein. These reagents were superparamagnetic, with a saturated magnetization of 0.3 emu/g. A SQUID-based AC magnetic susceptometer (XacPro-S, MagQu Co., Ltd.) was used to determine the time-dependent AC magnetic susceptibility, which approximates the association between magnetic nanoparticles and target protein molecules in the plasma [24] of each mixture. The IMR signal, which refers to the reduction in magnetic susceptibility caused by the association between magnetic nanoparticles and the target protein molecule, as detected by the magnetic susceptometer, represents the concentration of the target protein.

4.9. Western Blotting in Drosophila

Total proteins were extracted from the head tissue of *Drosophila* following the treatment described (100 files for each group). The removed tissue was homogenized in a buffer solution that was placed on ice for one hour and then centrifuged at 4 °C for 13,000 rpm for another 20 min. The separated solution was quantified by using a BCA protein assay kit (Thermo Fisher Scientific Inc. Waltham, MA, USA). Proteins were separated on 12.5% or 15% SDS polyacrylamide gels (Bionovas Pharmaceuticals Inc., Washington, DC, USA), and proteins were transferred to polyvinylidene difluoride membranes (GE Healthcare Life Sciences, Barrington, IL, USA). The antibodies used in this study were anti-Histone H3.3B (Thermo Fisher Scientific Inc.), and anti-amyloid-beta (anti-$A\beta$) (Covance Cat#SIG-39220, BioLegend, Dedham, MA, USA). Antibodies were detected by suitable horseradish peroxidase (HRP)-conjugated secondary antibody (Santa Cruz Biotechnology Inc.), and then proteins' immunoreactive bands were visualized by the enhanced chemiluminescence

(ECL) substrate (Millipore, Billerica, MA, USA), and the band intensities were quantified with the Image J analysis software (version 1.48t, Wayne Rasnabd, USA).

4.10. Statistical Analysis

All data are presented as means ± standard errors of the mean. One-way or two-way analysis of variance was performed, followed by the Student–Newman–Keuls post hoc test. The p-values of at least < 0.05 were considered significant. All data are obtained in at least three independent experiments.

5. Conclusions

Transgenic Drosophila Aβ models have successfully provided valuable information for studying AD mechanisms and pathways. In addition, Drosophila provided valuable drug testing in vivo experiments for YGS treatment. For in vitro experiments, our results showed that YGS treatment has a good antioxidant ability and low cytotoxicity. For in vivo experiments, our results showed that YGS treatment can reduce Aβ and Tau expressions in Drosophila melanogaster by IMR assay and Western blotting that were quite consistent with the change in appearance traits. To the best of our knowledge, this study was the first to conduct an evidence-based investigation of the effectiveness of alternative therapy with the traditional Chinese medicine YGS treatment in alleviating Aβ neurotoxicity of *Drosophila melanogaster* assessed by highly sensitive IMR assay and evaluated from appearance traits in GFP expression of external eyes.

Author Contributions: Conceived and designed the experiments: C.-H.W. and M.-T.S.; analyzed the data, S.-Y.Y., C.-W.L. and W.-J.W.; contributed reagents/materials/analysis tools, S.-Y.Y., M.-C.L. and W.-C.C.; wrote the paper, C.-H.W. and M.-T.S.; surgical procedures, Y.-S.J.; behavioral tests, Y.-S.J. and M.-T.S.; Western blotting assay, Y.-S.J. and W.-J.W.; immunohistochemistry, Y.-S.J. and M.-T.S. All authors have read and agreed to the published version of the manuscript.

Funding: This study was supported by grants from the Ministry of Science and Technology of Taiwan under grant numbers MOST 107-2321-B-003-001-, MOST 108-2321-B-003-001-, MOST 109-2321-B-003-001-, and MOST 107-2320-B-003-003-MY3.

Institutional Review Board Statement: This study was conducted according to the guidelines of the Care and Use of Laboratory Animals of the National Institutes of Health.

Informed Consent Statement: Not applicable.

Data Availability Statement: The data is confidential.

Conflicts of Interest: The authors declare no conflict of interest.

References

1. Lawlor, B.A. Behavioral and psychological symptoms in dementia: The role of atypical antipsychotics. *J. Clin. Psychiatry* **2004**, *65*, 5–10. [PubMed]
2. Ballard, C.G.; Gauthier, S.; Cummings, J.L.; Brodaty, H.; Grossberg, G.T.; Robert, P.; Lyketsos, C.G. Management of agitation and aggression associated with Alzheimer disease. *Nat. Rev. Neurol.* **2009**, *5*, 245–255. [CrossRef] [PubMed]
3. Cerejeira, J.; Lagarto, L.; Mukaetova-Ladinska, E. Behavioral and Psychological Symptoms of Dementia. *Front. Neurol.* **2012**, *3*, 73. [CrossRef] [PubMed]
4. Leinonen, V.; Koivisto, A.M.; Savolainen, S.; Rummukainen, J.; Tamminen, J.N.; Tillgren, T.; Vainikka, S.; Bm, O.T.P.; Mölsä, J.; Fraunberg, M.; et al. Amyloid and tau proteins in cortical brain biopsy and Alzheimer's disease. *Ann. Neurol.* **2010**, *68*, 446–453. [CrossRef] [PubMed]
5. Hardy, J.A.; Higgins, G.A. Alzheimer's disease: The amyloid cascade hypothesis. *Science* **1992**, *256*, 184–185. [CrossRef] [PubMed]
6. Chiu, M.-J.; Yang, S.-Y.; Horng, H.-E.; Yang, C.-C.; Chen, T.-F.; Chieh, J.-J.; Chen, H.-H.; Chen, T.-C.; Ho, C.-S.; Chang, S.-F.; et al. Combined Plasma Biomarkers for Diagnosing Mild Cognition Impairment and Alzheimer's Disease. *ACS Chem. Neurosci.* **2013**, *4*, 1530–1536. [CrossRef]
7. Mizukami, K.; Asada, T.; Kinoshita, T.; Tanaka, K.; Sonohara, K.; Nakai, R.; Yamaguchi, K.; Hanyu, H.; Kanaya, K.; Takao, T.; et al. A randomized cross-over study of a traditional Japanese medicine (kampo), yokukansan, in the treatment of the behavioural and psychological symptoms of dementia. *Int. J. Neuropsychopharmacol.* **2009**, *12*, 191–199. [CrossRef]

8. Teranishi, M.; Kurita, M.; Nishino, S.; Takeyoshi, K.; Numata, Y.; Sato, T.; Tateno, A.; Okubo, Y. Efficacy and Tolerability of Risperidone, Yokukansan, and Fluvoxamine for the Treatment of Behavioral and Psychological Symptoms of Dementia: A Blinded, Randomized Trial. *J. Clin. Psychopharmacol.* **2013**, *33*, 600–607. [CrossRef]
9. Ikarashi, Y.; Mizoguchi, K. Neuropharmacological efficacy of the traditional Japanese Kampo medicine yokukansan and its active ingredients. *Pharmacol. Ther.* **2016**, *166*, 84–95. [CrossRef]
10. Nakatani, Y.; Tsuji, M.; Amano, T.; Miyagawa, K.; Miyagishi, H.; Saito, A.; Imai, T.; Takeda, K.; Ishii, D.; Takeda, H. Neuroprotective effect of yokukansan against cytotoxicity induced by corticosterone on mouse hippocampal neurons. *Phytomedicine* **2014**, *21*, 1458–1465. [CrossRef]
11. Iwasaki, K.; Satoh-Nakagawa, T.; Maruyama, M.; Monma, Y.; Nemoto, M.; Tomita, N.; Tanji, H.; Fujiwara, H.; Seki, T.; Fujii, M.; et al. A Randomized, Observer-Blind, Controlled Trial of the Traditional Chinese Medicine Yi-Gan San for Improvement of Behavioral and Psychological Symptoms and Activities of Daily Living in Dementia Patients. *J. Clin. Psychiatry* **2005**, *66*, 248–252. [CrossRef]
12. Kawanabe, T.; Yoritaka, A.; Shimura, H.; Oizumi, H.; Tanaka, S.; Hattori, N. Successful Treatment with Yokukansan for Behavioral and Psychological Symptoms of Parkinsonian Dementia. *Prog. Neuropsychopharmacol. Biol. Psychiatry* **2010**, *34*, 284–287. [CrossRef]
13. Nagata, K.; Yokoyama, E.; Yamazaki, T.; Takano, D.; Maeda, T.; Takahashi, S.; Terayama, Y. Effects of Yokukansan on Behavioral and Psychological Symptoms of Vascular Dementia: An Open-Label Trial. *Phytomedicine* **2012**, *19*, 524–528. [CrossRef]
14. Okahara, K.; Ishida, Y.; Hayashi, Y.; Inoue, T.; Tsuruta, K.; Takeuchi, K.; Yoshimuta, H.; Kiue, K.; Ninomiya, Y.; Kawano, J.; et al. Effects of Yokukansan on Behavioral and Psychological Symptoms of Dementia in Regular Treatment for Alzheimer's Disease. *Prog. Neuropsychopharmacol. Biol. Psychiatry* **2010**, *34*, 532–536. [CrossRef]
15. Burr, A.A.; Tsou, W.-L.; Ristic, G.; Todi, S.V. Using membrane-targeted green fluorescent protein to monitor neurotoxic protein-dependent degeneration ofDrosophilaeyes. *J. Neurosci. Res.* **2014**, *92*, 1100–1109. [CrossRef]
16. Yeh, P.-A.; Chien, J.-Y.; Chou, C.-C.; Huang, Y.-F.; Tang, C.-Y.; Wang, H.-Y.; Su, M.-T. Drosophila notal bristle as a novel assessment tool for pathogenic study of Tau toxicity and screening of therapeutic compounds. *Biochem. Biophys. Res. Commun.* **2010**, *391*, 510–516. [CrossRef]
17. Tan, F.H.P.; Azzam, G. Drosophila melanogaster: Deciphering Alzheimer's Disease. *Malays. J. Med. Sci.* **2017**, *24*, 6–20. [CrossRef]
18. Edwards, P.; Wischik, C.M. Monitoring pathological assembly of tau and ?-amyloid proteins in Alzheimer's disease. *Acta Neuropathol.* **1995**, *89*, 50–56. [CrossRef]
19. Braak, H.; Del Tredici, K. The preclinical phase of the pathological process underlying sporadic Alzheimer's disease. *Brain* **2015**, *138*, 2814–2833. [CrossRef]
20. He, Z.; Guo, J.L.; McBride, J.D.; Narasimhan, S.; Kim, H.; Changolkar, L.; Zhang, B.; Gathagan, R.; Yue, C.; Dengler, C.; et al. Amyloid-β plaques enhance Alzheimer's brain tau-seeded pathologies by facilitating neuritic plaque tau aggregation. *Nat. Med.* **2018**, *24*, 29–38. [CrossRef]
21. Olsson, B.; Lautner, R.; Andreasson, U.; Öhrfelt, A.; Portelius, E.; Bjerke, M.; Hölttä, M.; Rosén, C.; Olsson, C.; Strobel, G.; et al. CSF and blood biomarkers for the diagnosis of Alzheimer's disease: A systematic review and meta-analysis. *Lancet Neurol.* **2016**, *15*, 673–684. [CrossRef]
22. Bjerke, M.; Engelborghs, S. Cerebrospinal Fluid Biomarkers for Early and Differential Alzheimer's Disease Diagnosis. *J. Alzheimer's Dis.* **2018**, *62*, 1199–1209. [CrossRef]
23. Chieh, J.-J.; Huang, K.-W.; Chuang, C.P.; Wei, W.C.; Dong, J.J.; Lee, Y.Y. Immunomagnetic Reduction Assay on Des-Gamma-Carboxy Prothrombin for Screening of Hepatocellular Carcinoma. *IEEE Trans. Biomed. Eng.* **2015**, *63*, 1681–1686. [CrossRef]
24. Yang, S.-Y.; Chiu, M.-J.; Chen, T.-F.; Horng, H.-E. Detection of Plasma Biomarkers Using Immunomagnetic Reduction: A Promising Method for the Early Diagnosis of Alzheimer's Disease. *Neurol. Ther.* **2017**, *6*, 37–56. [CrossRef] [PubMed]
25. Yang, S.-Y.; Liu, H.-C.; Chen, W.-P. Immunomagnetic Reduction Detects Plasma Aβ1–42 Levels as a Potential Dominant Indicator Predicting Cognitive Decline. *Neurol. Ther.* **2020**, *9*, 435–442. [CrossRef] [PubMed]
26. Teunissen, C.E.; Chiu, M.-J.; Yang, C.-C.; Yang, S.-Y.; Scheltens, P.; Zetterberg, H.; Blennow, K. Plasma Amyloid-β (Aβ42) Correlates with Cerebrospinal Fluid Aβ42 in Alzheimer's Disease. *J. Alzheimer's Dis.* **2018**, *62*, 1857–1863. [CrossRef] [PubMed]
27. Barnham, K.J.; Masters, C.L.; Bush, A.I. Neurodegenerative diseases and oxidatives stress. *Nat. Rev. Drug Discov.* **2004**, *3*, 205–214. [CrossRef]
28. Doo, A.-R.; Kim, S.-N.; Park, J.-Y.; Cho, K.H.; Hong, J.; Eun-Kyung, K.; Moon, S.K.; Jung, W.S.; Lee, H.; Jung, J.H.; et al. Neuroprotective effects of an herbal medicine, Yi-Gan San on MPP+/MPTP-induced cytotoxicity in vitro and in vivo. *J. Ethnopharmacol.* **2010**, *131*, 433–442. [CrossRef]
29. Heitman, E.; Ingram, D.K. Cognitive and neuroprotective effects of chlorogenic acid. *Nutr. Neurosci.* **2017**, *20*, 32–39. [CrossRef]
30. Zhang, J.-G.; Geng, C.-A.; Huang, X.-Y.; Chen, X.-L.; Ma, Y.-B.; Zhang, X.-M.; Chen, J.-J. Chemical and biological comparison of different sections of Uncaria rhynchophylla (Gou-Teng). *Eur. J. Mass Spectrom.* **2017**, *23*, 11–21. [CrossRef]
31. Lin, C.-M.; Lin, Y.-T.; Lee, T.-L.; Imtiyaz, Z.; Hou, W.-C.; Lee, M.-H. *In vitro* and in vivo evaluation of the neuroprotective activity of Uncaria hirsuta Haviland. *J. Food Drug Anal.* **2020**, *28*, 147–158. [CrossRef]
32. Lee, T.-K.; Kang, I.-J.; Kim, B.; Sim, H.J.; Kim, D.-W.; Ahn, J.H.; Lee, J.-C.; Ryoo, S.; Shin, M.C.; Cho, J.H.; et al. Experimental Pretreatment with Chlorogenic Acid Prevents Transient Ischemia-Induced Cognitive Decline and Neuronal Damage in the Hippocampus through Anti-Oxidative and Anti-Inflammatory Effects. *Molecules* **2020**, *25*, 3578. [CrossRef]

33. Gao, L.; Li, X.; Meng, S.; Ma, T.; Wan, L.; Xu, S. Chlorogenic Acid Alleviates Aβ_{25-35}-Induced Autophagy and Cognitive Impairment via the mTOR/TFEB Signaling Pathway. *Drug Des. Dev. Ther.* **2020**, *14*, 1705–1716. [CrossRef]
34. Wang, N.; Zhou, Y.; Zhao, L.; Wang, C.; Ma, W.; Ge, G.; Wang, Y.; Ullah, I.; Muhammad, F.; Alwayli, D.; et al. Ferulic acid delayed amyloid β-induced pathological symptoms by autophagy pathway via a fasting-like effect in Caenorhabditis elegans. *Food Chem. Toxicol.* **2020**, *146*, 111808. [CrossRef]
35. Zhai, K.-F.; Duan, H.; Cui, C.-Y.; Cao, Y.-Y.; Si, J.-L.; Yang, H.-J.; Wang, Y.-C.; Cao, W.-G.; Gao, G.-Z.; Wei, Z.-J. Liquiritin from Glycyrrhiza uralensis Attenuating Rheumatoid Arthritis via Reducing Inflammation, Suppressing Angiogenesis, and Inhibiting MAPK Signaling Pathway. *J. Agric. Food Chem.* **2019**, *67*, 2856–2864. [CrossRef]
36. Jia, S.-L.; Wu, X.-L.; Li, X.-X.; Dai, X.-L.; Gao, Z.-L.; Lu, Z.; Zheng, Q.; Sun, Y.-X. Neuroprotective effects of liquiritin on cognitive deficits induced by soluble amyloid-β1–42 oligomers injected into the hippocampus. *J. Asian Nat. Prod. Res.* **2016**, *18*, 1186–1199. [CrossRef]
37. Sun, X.; Zeng, H.; Wang, Q.; Yu, Q.; Wu, J.; Feng, Y.; Deng, P.; Zhang, H. Glycyrrhizin ameliorates inflammatory pain by inhibiting microglial activation-mediated inflammatory response via blockage of the HMGB1-TLR4-NF-kB pathway. *Exp. Cell Res.* **2018**, *369*, 112–119. [CrossRef]
38. Kong, Z.-H.; Chen, X.; Hua, H.-P.; Liang, L.; Liu, L.-J. The Oral Pretreatment of Glycyrrhizin Prevents Surgery-Induced Cognitive Impairment in Aged Mice by Reducing Neuroinflammation and Alzheimer's-Related Pathology via HMGB1 Inhibition. *J. Mol. Neurosci.* **2017**, *63*, 385–395. [CrossRef]
39. Gong, W.-X.; Zhou, Y.-Z.; Qin, X.-M.; DU, G.-H. Involvement of mitochondrial apoptotic pathway and MAPKs/NF-κ B inflammatory pathway in the neuroprotective effect of atractylenolide III in corticosterone-induced PC12 cells. *Chin. J. Nat. Med.* **2019**, *17*, 264–274. [CrossRef]
40. Zhu, W.-L.; Zheng, J.-Y.; Cai, W.-W.; Dai, Z.; Li, B.-Y.; Xu, T.-T.; Liu, H.-F.; Liu, X.-Q.; Wei, S.-F.; Luo, Y.; et al. Ligustilide improves aging-induced memory deficit by regulating mitochondrial related inflammation in SAMP8 mice. *Aging* **2020**, *12*, 3175–3189. [CrossRef]

plants

MDPI

Article

Major Plant in Herbal Mixture Gan-Mai-Da-Zao for the Alleviation of Depression in Rat Models

Ying-Xiao Li [1], Kai-Chun Cheng [2], Chao-Tien Hsu [3], Juei-Tang Cheng [4,*] and Ting-Ting Yang [5,*]

1 Department of Nursing, Tzu Chi University of Science and Technology, Hualien 970302, Taiwan; yxli@ems.tcust.edu.tw
2 Department of Pharmacy, College of Pharmacy and Health Care, Tajen University, Pingtung 90741, Taiwan; kc-cheng@tajen.edu.tw
3 Department of Pathology, E-Da Hospital, I-Shou University, Kaohsiung 82445, Taiwan; ed103797@edah.org.tw
4 Department of Medical Research, Chi-Mei Medical Center, Tainan City 71004, Taiwan
5 School of Chinese Medicine for Post-Baccalaureate, I-Shou University, Kaohsiung 82445, Taiwan
* Correspondence: jtcheng5503@gmail.com (J.-T.C.); tingting@isu.edu.tw (T.-T.Y.)

Abstract: Gan-Mai-Da-Zao (GMDZ) is a well-known product in Chinese traditional medicine and includes three major plants: blighted wheat (Fu Mai), licorice (Gan Cao), and jujube (Da Zao). GMDZ is widely used as an efficacious and well-tolerated prescription for depression in clinics. The present study was designed to investigate the main plant of GMDZ for its antidepressant-like effect using the unpredictable chronic mild stress (UCMS) model on rats who received an injection with p-chlorophenylalanine (PCPA) to produce the chemical model. In rats subjected to the UCMS model, forced swim tests, open field tests, and sucrose preference tests were applied to estimate the chronic effect of GMDZ. We found that the oral administration of GMDZ for 21 days significantly alleviated the behavior in rats with depression induced by either UCMS or PCPA. The expression levels of the serotonin transporter (5-HTT) and brain-derived neurotrophic factor (BDNF) in the hippocampus of the rats with depression were markedly increased by GMDZ. Additionally, rats that received the herbal mixture without licorice showed a markedly lower response than GMDZ. These results suggest that GMDZ may alleviate the depressive-like behaviors in depressive rats, possibly via licorice (Gan Cao), to increase 5-HTT and BDNF signals in the hippocampus. The present study confirmed the antidepressant-like effects of GMDZ. Additionally, licorice (Gan Cao) may play a key role in the effectiveness of GMDZ.

Keywords: Gan-Mai-Da-Zao; depression; brain-derived neurotrophic factor; serotonin transporter; unpredictable chronic mild stress

Citation: Li, Y.-X.; Cheng, K.-C.; Hsu, C.-T.; Cheng, J.-T.; Yang, T.-T. Major Plant in Herbal Mixture Gan-Mai-Da-Zao for the Alleviation of Depression in Rat Models. *Plants* **2022**, *11*, 258. https://doi.org/10.3390/plants11030258

Academic Editors: Domenico Trombetta and Suresh Awale

Received: 21 November 2021
Accepted: 17 January 2022
Published: 19 January 2022

1. Introduction

Major depression is a common psychiatric disorder and it may lead to emotional depression, suicidal tendencies, and a recurrence of morbidity [1]. Dysfunction of the serotonin (5-HT) system is commonly considered to be the cause of depression. Depression has been shown as the negative factor that affects the rate of adult hippocampal neurogenesis. The serotonin transporter (5-HTT) responsible for the reuptake of 5-HT is related to the role of 5-HT in neurodevelopmental processes [2]. The evidence from animal models and human studies indicates that reduced function of 5-HTT is associated with the decreased expression of brain-derived neurotrophic factor (BDNF) [3]. Additionally, BDNF is highly expressed in the adult hippocampus and hypothalamus [4], and is involved in the etiopathology of mood disorders [5]. In depressive patients, serum BDNF levels are markedly decreased [6] that can be restored by antidepressant treatment [7]. Although today's treatments for depression have greatly improved, it is still necessary to find more safe and effective agents to prevent depression.

Traditional Chinese medicine (TCM) has shown the therapeutic effects of depression [8,9]. GMDZ is one of the well-known products in TCM and it has widely been used to treat depressive patients in Asia. Despite the large variety of TCM patterns among participants, dozens of herbal formulas for depression were GMDZ-based [10]. This was first documented in the Chinese medical book Jin-Gui-Yao-Lue (Synopsis of Prescriptions of the Golden Chamber) written by Dr. Zongjing Zhang (AD 152-219) [11]. GMDZ is believed to be effective for depression [12]. Clinical studies indicated that GMDZ decoction is an efficacious and well-tolerated antidepressant prescription for depressive disorders, even postpartum depression [13–15]. Moreover, GMDZ could protect hippocampal neurons against glutamate toxicity in depression-like rats [16,17]. The composition of GMDZ includes three major plants: blighted wheat (FuMai, M), licorice (GanCao, G), and jujube (DaZao, D) [16]. The combination of three plants may enhance the efficiency and/or reduce toxicity in clinical applications. However, the role of these components in GMDZ is still vague.

To investigate the antidepressant-like effects of GMDZ, unpredictable chronic mild stress (UCMS) was used in the present study. The behavior tests, including a forced swimming test (FST), open field test (OFT), and sucrose preference test (SPT), were then performed. Moreover, we compared the variations between one herb-deleted mixture and GMDZ to understand the main plant in GMDZ using the rats with depression induced by pretreatment with a serotonin synthesis inhibitor, p-chlorophenylalanine (PCPA). The levels of 5-HTT and BDNF in the hippocampus of rats were also assessed to obtain further insight into the mechanism(s) regarding the antidepressive effects of GMDZ.

2. Results

2.1. *Chronic GMDZ Treatment Ameliorated Depression-Like Behaviors in UCMS Rats and PCPA Treated Rats*

FST shows a high predictive validity for antidepressant activity. OFT is classically used to assess anxiety in rodents. In the present study, the UCMS group significantly prolonged the immobility time in the FST (Figure 1a); and also reduced the time spent at the center and decreased the total distance traveled in the OFT compared with the control group (Figure 1b,c). Similar results were observed in depressive rats induced by PCPA in FST (Figure 1e) and OFT (Figure 1f,g). To assess the antidepressant-like effects of GMDZ, rats were orally administrated GMDZ for 21 days while fluoxetine (10 mg/kg) was included as the positive control. The results showed that GMDZ produced a significant reduction in the duration of immobility as well as the time in the center and distance of traveling in OFT. It indicated that GMDZ significantly ameliorated depression-like behaviors compared with the vehicle-treated animals. Additionally, fluoxetine showed marked effects on FST and OFT in the UCMS group which were not observed in the PCPA-induced model.

The SPT is a procedure which is used to measure the hedonic value of sucrose, typically reduced in animals with depression-like disorders. Both the UCMS and PCPA groups (Figure 1h) showed a significant decrease in sucrose consumption as compared with the control. However, the sucrose consumption in both models was significantly restored by the chronic administration of GMDZ. PCPA-induced stressed rats treated with fluoxetine did not differ in sucrose preference from the vehicle-treated group. Therefore, GMDZ produces antidepressant effects through different mechanisms to fluoxetine.

2.2. *Effects of Two-Herb Mixture of GMDZ on Behavioral Tests in PCPA-Induced Rats*

To understand the major plant in GMDZ, a one-plant deletion from the mixture was prepared. Then, five groups of PCPA-induced rats were administered with GMDZ, G and M, M and DZ, G and DZ, and the vehicle, respectively, for 21 days. In both G and M and G and DZ groups, rats showed a shorter immobility time in FST (Figure 2a), a longer central zone duration and a greater total distance in OFT (Figure 2b,c), and a higher sucrose intake in SPT (Figure 2d) which ameliorated the depression-like behaviors. However, M and DZ did not show an obvious effect in FST, OFT, or SPT, respectively. These results suggest that

G seems to play a crucial role in the effect of GMDZ using PCPA-induced rats showing depression-like behavior.

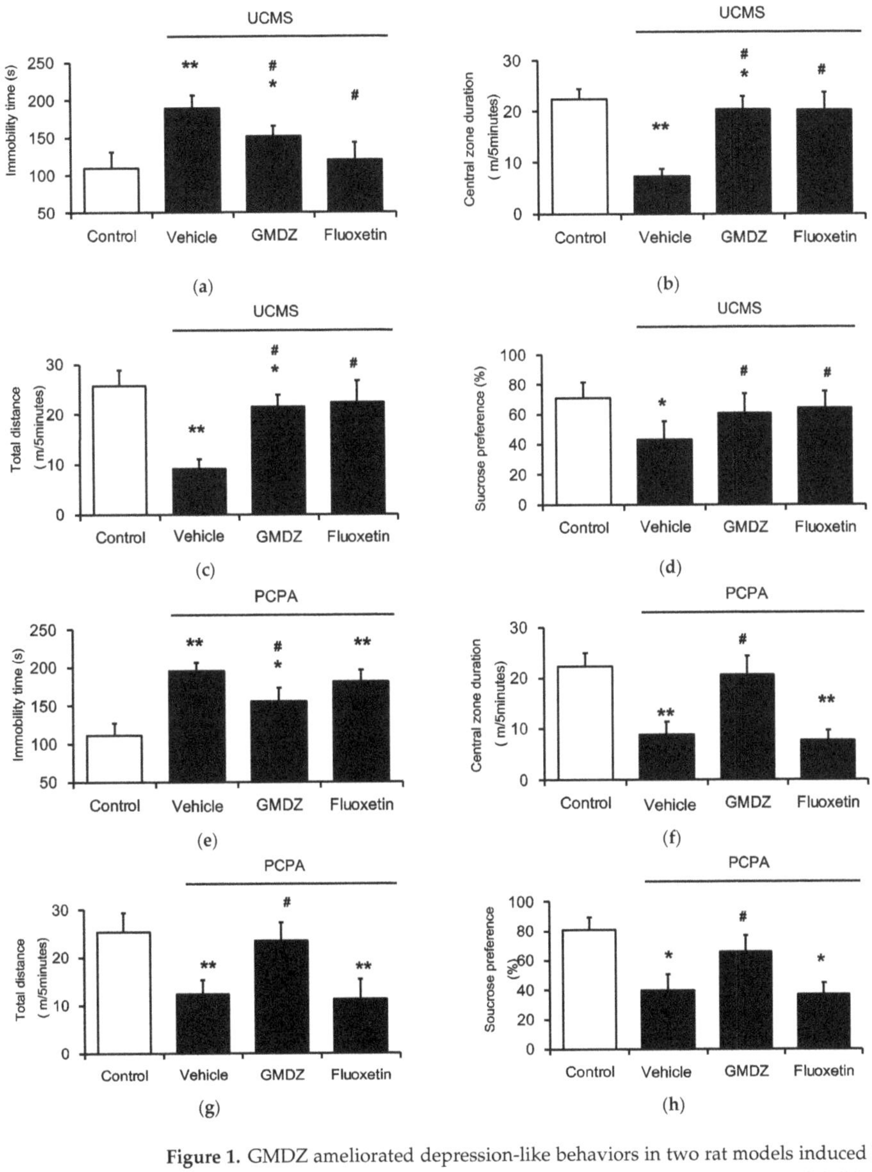

Figure 1. GMDZ ameliorated depression-like behaviors in two rat models induced by the UCMS or PCPA injection. (**a**) The changes of immobility time in FST in UCMS model; (**b**) the changes of time in central in OFT in UCMS model; (**c**) the changes of traveling distance in OFT in UCMS model; (**d**) the changes of sucrose solution consumption (%) in SPT in UCMS model; (**e**) the changes of immobility time in FST in PCPA treated groups; (**f**) the changes of time in central in OFT in PCPA treated groups; (**g**) the changes of traveling distance in OFT in PCPA treated groups; (**h**) the changes of sucrose solution consumption (%) in SPT in PCPA treated groups. Responses to fluoxetine used as the positive control. Data are expressed as mean ± SE ($n = 8$). * $p < 0.05$, ** $p < 0.01$ compared with the normal control group; # $p < 0.05$ compared with vehicle-treated group.

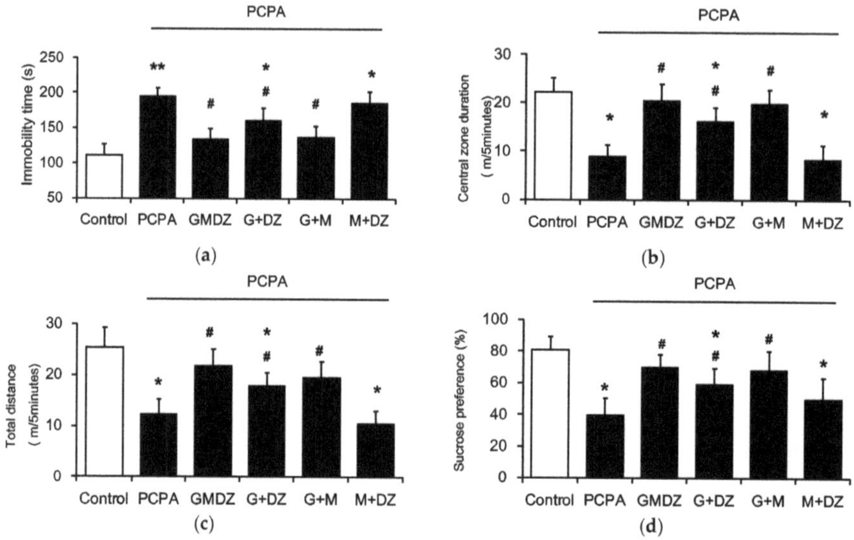

Figure 2. Effects of the one herb-deleted mixture of GMDZ in depression-like behaviors using the PCPA treated rats. (**a**) The changes of immobility time in FST; (**b**) the changes of time in central in OFT; (**c**) the changes of traveling distance in OFT; (**d**) the changes of sucrose solution consumption (%) in SPT. Data are expressed as mean ± SE ($n = 8$). * $p < 0.05$, ** $p < 0.01$ compared with the normal control; # $p < 0.05$ compared with vehicle-treated group.

2.3. *Chronic GMDZ Treatment Restores the 5-HTT and BDNF Levels in the Hippocampus of PCPA Treated Rats*

To investigate the potential mechanism(s) underlying GMDZ-induced antidepressant effects, we determined the gene expressions and protein levels of 5-HTT (Figure 3a,b) and BDNF in the hippocampus of PCPA-treated rats (Figure 3c,d). The mRNA and protein levels of 5-HTT were significantly decreased in the PCPA-treated group, which was restored by GMDZ. Additionally, chronic treatment with GMDZ also increased the protein and mRNA levels of BDNF in the hippocampus of PCPA-treated rats.

Same as the results in behavior experiments, G and M or G and DZ treatment significantly restored the mRNA and protein levels of 5-HTT. Moreover, as shown in Figure 3, G and M or G and DZ treatment also restored the protein level and mRNA level of BDNF. However, the treatment of M and DZ did not modify the expressions of 5-HTT and BDNF in the hippocampus. It supports that G played an important role in GMDZ for depression improvement.

2.4. *Effects of Glycyrrhizic Acid on 5-HTT and BDNF Expression in the Corticosterone-Treated H19-7 Cell Line*

We hypothesized that licorice played a major role in GMDZ mixture, since the antidepressant-like effect in G and M and G and DZ treatments seemed to be more significant than that in the M and DZ treatment group. In addition, long-term exposure to stress or high glucocorticoid levels leads to depression-like behavior in rodents [18]. Therefore, we investigated the potential mechanism of glycyrrhizic acid (the active component of licorice) for corticosterone-induced stress injury in cells. Our results show that the in vitro findings correlate with the vivo results. It showed that chronic exposure of H19-7 cells to corticosterone markedly decreased the gene expressions of 5-HTT (Figure 4a) and BDNF (Figure 4b). Interestingly, glycyrrhizic acid significantly reversed these expressions in a dose-dependent manner.

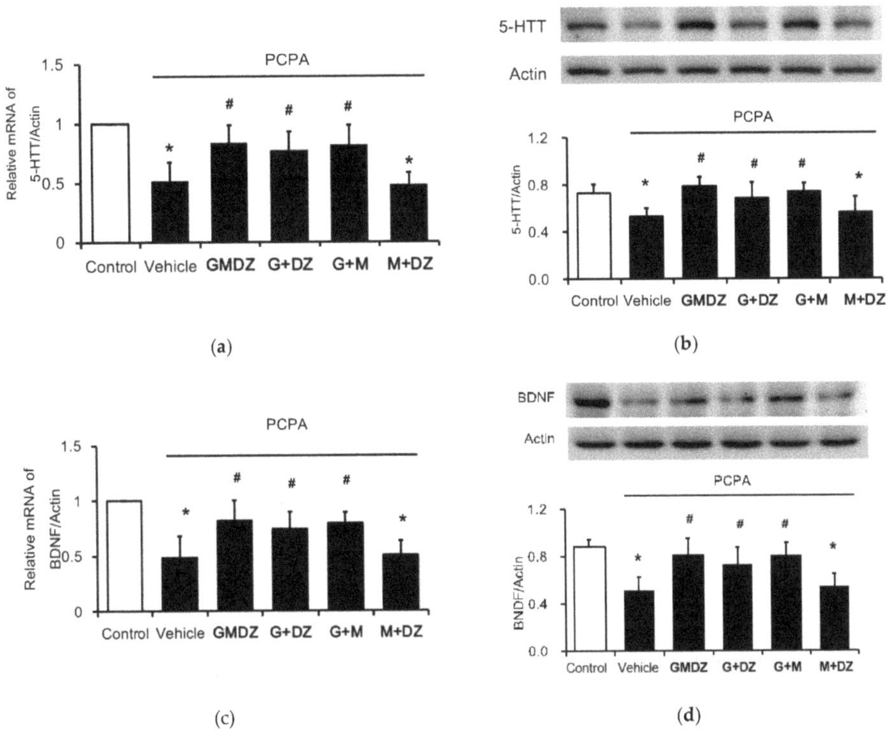

Figure 3. Effects of GMDZ and the one plant-deleted mixture of GMDZ on the expressions of 5-HTT and BDNF in the hippocampus. (**a**) The mRNA levels of 5-HTT; (**b**) The protein levels of 5-HTT; (**c**) the mRNA levels of BDNF; (**d**) the protein levels of BDNF. Data are expressed as mean ± SE (n = 8). * $p < 0.05$ compared with the normal control; # $p < 0.05$ compared with vehicle-treated group.

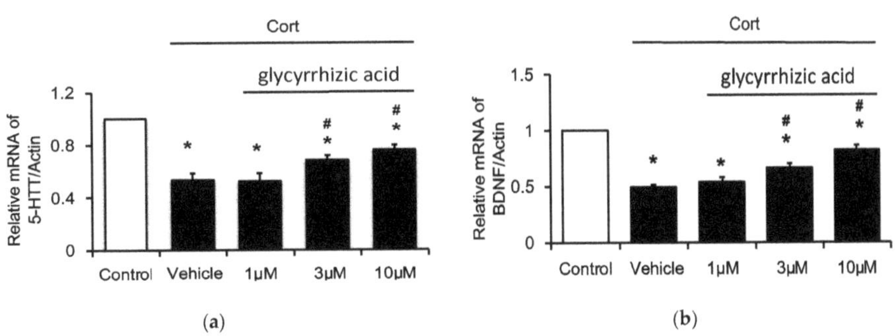

Figure 4. Effects of glycyrrhizic acid (GA) on 5-HTT and BDNF expressions in the corticosterone-treated H19-7 cells. H19-7 cells were incubated under normal differentiating conditions in the presence or absence of corticosterone (Cort) at concentration of 1 μM for five days. Then, cells were respectively incubated with glycyrrhizic acid at 1 μM, 3 μM, or 10 μM for 24 h. (**a**) The mRNA levels of 5-HTT; (**b**) the mRNA levels of BDNF. Data are expressed as mean ± SE (n = 8). * $p < 0.05$ compared to the cells without treatment of corticosterone (control). # $p < 0.05$ compared with the vehicle-treated group.

3. Discussion

UCMS rats exhibited a significantly increased immobility time in the FST, decreased locomotor activity in OFT, and reduced sucrose intake in the SPT, as described previously [19]. In the present study, we demonstrated that chronic administration of GMDZ ameliorated depression-like behaviors in UCMS rats. Moreover, depletion of 5-HT by PCPA failed to block the antidepressant action of GMDZ, indicating that the action mechanisms of GMDZ varied from fluoxetine which is one of SSRI. Otherwise, we found that GMDZ may promote the BDNF signaling pathway and enhance 5-HTT expression in the hippocampus in the depressive rats.

Serotonin (5-HT) released from the axon terminalis is selectively taken up from the synaptic cleft into these terminals via the 5-HTT [2]. The decrease of 5-HTT expressions is associated with stress-induced anxiety and depression-like behaviors [20]. It demonstrates an interaction between 5-HT and BDNF [21]. The loss of BDNF appears to exacerbate neurochemical and behavioral abnormalities in 5-HTT mutant mice [22]. BDNF modulated the 5-HTT gene promoter to influence the function of 5-HTT [23]. Our results are consistent with previous findings that 5-HTT-gene and protein expressions were suppressed in PCPA rats [16]. Administration of GMDZ reversed the expression level of 5-HTT compared with the vehicle-treated UCMS group. Additionally, chronic GMDZ administration the stress-induced reversed the decrease of BDNF level [24] in the hippocampus. Therefore, GMDZ may play an important role in the 5-HTT and BDNF regulation. GMDZ promoted the increase of 5-HTT and BDNF expressions, which may contribute to the antidepressant effect [7].

The effects of GMDZ may include improved cerebral microcirculatory regulation, mood stabilization, and the alleviation of impatience, as noted in a previous report [15]. As the herbal mixture of GMDZ seems highly complex, the present study aimed to understand the main plant that played a major role in the therapeutic effects of GMDZ. Using the deletion of one plant from the original mixture in GMDZ, three products (G and M, G and DZ, and M and DZ) were administrated to PCPA-treated rats, respectively. This showed that G and M and G and DZ treatments, but not M and DZ treatment, produced significant antidepressant-like effects in the PCPA-induced model, same as the effects of GMDZ. Additionally, G and M or G and DZ administration also reversed the decreased 5-HTT and BDNF levels in depression. As the antidepressant-like effect in the G and M and G and DZ treatments seemed more significant than that in the M and DZ treatment group, licorice (G) might play a major role in the effectiveness of GMDZ. It has been documented that licorice induces an antidepressant-like effect in animals [25,26]. GanCao (licorice) contains triterpenoid glycosides and flavonoid glycosides. Glycyrrhizin or glycyrrhizic acid as the active ingredient in licorice has been extensively studied [27]. Glycyrrhizic acid has been reported to show anti-inflammatory and anti-nociceptive activities in mice [28]. The present study found that glycyrrhizic acid increases the 5-HTT and BDNF expressions, both were reduced in H19-7 cells treated with corticosterone [23]. Therefore, glycyrrhizic acid as one of the active principles in GanCao (licorice) seems to have participated in the alleviation of depressive disorders in animals. However, the real action mechanism(s) shall be clarified in the future.

According to the traditional TCM theory, the blighted wheat seems to have a predominant function in GMDZ. It seems possible that the active constituent from blighted wheat is through the conversion of gut microbiota in animals. Stress leads to anxiety/depression by altering the gut microbiota, and it is possible to improve anxiety and depression by probiotics modulation [29]. The blighted wheat seems beneficial for the protection of exhaustive physical exercise, oxidative stress injury on brain tissues [30]. However, it needs more investigations for the blighted wheat in the future. Otherwise, Dazao contains various triterpenoids (e.g., betulinic acid and oleanolic acid) and glycosides [31]. But it seems only plays a supporting role in the regulation of depression. However, the systemic administration of betulinic acid and the oral administration of oleanolic acid show analgesic effects on acetic acid-induced writhing in the animal model [32,33].

4. Materials and Methods

4.1. Preparation of Extracts of the GMDZ Decoction

GMDZ is mainly prepared from three dried raw plants: licorice (Glycyrrhiza uralensis fisch, GanCao, G), blighted wheat (Triticum aestivum L, FuMai, M), and jujuba (Ziziphus jujuba Mill, DaZao, DZ). In the present study, the used GMDZ was a commercial product purchased from Sun Ten Pharmaceutical Co., Ltd. (Taipei, Taiwan). The specification of the GMDZ preparation is as follows: Glycyrrhiza uralensis (root and rhizome): 18.75%; Triticum aestivum (fruit): 62.50%; Ziziphus jujuba (fruit): 18.75%. After extraction in distilled water (ratio 1:10), the product of GMDZ was similar to that in a previous report [34]. To identify the major plant in GMDZ, the commercial product of each plant in GMDZ was also purchased. Then, we used the single deletion method to mimic the original prescription and new mixtures in three, such as GanCao + FuMai (G and M), FuMai + DaZao (M and DZ), GanCao + DaZao (G and DZ), were obtained. Each product has the same ratio as that in GMDZ decoction. The amount of each deleted plant was replaced by the starch to reach the same amount of GMDZ. Then, the treated doses were expressed as the dried weight of each product per bodyweight of the animals (g/kg body weight).

4.2. Experimental Animals

Male Sprague-Dawley rats (250–300 g) obtained from National Animal Center (Taipei, Taiwan) were maintained in the animal center of Chi-Mei Medical Center (Tainan, Taiwan). In brief, they were housed in pairs on a 12/12-hr light/dark cycle (light beginning at 7:00 am)) with ad libitum access to food and water except during behavioral tests. The project was approved by the Institutional Animal Care and Use Committee of Chi-Mei Medical Center (No. 105122622). All the animal procedures were performed according to the Guide for the Care and Use of Laboratory Animals published by the US National Institutes of Health (NIH Publication No. 85-23, revised 1996).

4.3. Experimental Design

To investigate the antidepressant-like effect of GMDZ, rats were randomly divided into four groups (eight rats in each group): control group, UCMS model group, UCMS + fluoxetine group, and UCMS + GMDZ group. Additionally, to find the major herb, rats induced by PCPA were also used as PCPA + fluoxetine group, PCPA + GMDZ group, PCPA + G and M group, PCPA + M and DZ group, and PCPA + G and DZ group.

PCPA (Sigma, St. Louis, MO, USA), a specific inhibitor of serotonin (5-HT) biosynthesis, was administered (100 mg/kg) once a day for seven days, as described previously [35]. PCPA has shown a high degree of 5-HT depletion (>90%) yielded by similar treatment [36]. On the seventh day, changes in 5-HTT expression in the hippocampus were confirmed by Western blots. Two days after the model induced by PCPA, rats were administered with GMDZ (2.5 g/kg) [34], G and M mixture (2.5 g/kg), M and D mixture (2.5 g/kg), G and D mixture (2.5 g/kg) and fluoxetine (10 mg/kg) once daily by oral gavage for a three-week period.

4.4. UCMS Procedure

According to the previous report [37], the UCMS-induced depressive animal model was induced. Experimental rats (n = 8 per group) were exposed to unpredictable mild stressors randomly every day in four weeks. The stressors applied included the following: physical restraint (1h), 1 min tail pinch (2.5 cm from the end of the tail), reversed light/dark cycle (24 h), overnight illumination (12 h), soiled cage (12 h), and cage tilt (18 h, 45°). Each stressor was randomly assigned two or three times over the experimental period. The non-stressed control rats were normally housed in groups (three to four per cage) in the other room, and the stressed rats were singly housed [37]. At least 12 h of rest was provided between a stressor and a test to avoid effects of acute stress [38].

After the first week, the animals were administered with GMDZ (2.5 g/kg) or fluoxetine (10 mg/kg) by oral gavage once a day for three weeks. The dosage was applied according to a previous study [34]. Behavioral tests were performed 2 h after the last treatment.

4.5. Behavioral Tests

The sequence of the behavioral test was SPT, OPT, and FST. There was a three-day time interval between these tests.

SPT: To evaluate the anhedonia response [39], rats were exposed to two identical bottles (one containing tap water and the other 1% sucrose solution) for 1 h, followed by 12 h of tap water and food deprivation [40]. The bottles were weighed before and after the 1 h test period; the sucrose preference (%) was then determined. Animals were habituated three days to the two-bottle choice (both bottles were filled with tap water and placed through the top of the cage lid) before the test.

OPT: To measure locomotion and anxious behaviors, rats were placed in an open field area made of a 70 × 70 × 40 cm wooden box and equipped with an infrared floor to measure locomotor activity [41]. The arena was subdivided into a central and a peripheral zone. Rats were placed in the open field boxes for 5 min under normal light conditions, and the locomotor activity and time stay in central was tracked with a video system (Viewpoint, Lyon, France). Individual animals were gently placed in the same corner of the apparatus in all trials.

FST: To assess learned-helplessness, rats were gently placed in a clear plastic cylinder (height = 40 cm; diameter = 30 cm) filled with water to 30 cm high at 24 °C ± 0.5 °C for 6 min [42]. Immobility time was calculated by subtracting active time from the total time. The behavior experiments were recorded using a side-mounted camera and assessed using a video tracking software.

4.6. Tissue Preparation

Rats were sacrificed 24 h after the behavior experiments. The treatments continued during these days. Rats were euthanized by intraperitoneal injection (IP) of a lethal dose of pentobarbital. The brain was quickly removed and immediately frozen in liquid nitrogen for further analysis.

4.7. Cell Cultures

Rat-derived hippocampus H19-7 cell line cells (CRL-2526; American Type Culture Collection, Manassas, VA, USA) were maintained at 37 °C and 5% CO_2 in Dulbecco's modified Eagle's medium (DMEM; HyClone, South Logan, UT, USA) with 4 mM l-glutamine that was adjusted with sodium bicarbonate (1.5 g/L), glucose (4.5 g/L), G418 (200 μg/mL), and puromycin (1 μg/mL) and supplemented with 10% fetal bovine serum [43]. Cells (1×10^6) were plated on 60-mm culture dishes, and at 80% confluence. H19-7 cells were incubated with or without the rat stress hormone, corticosterone, at the concentration of 1 μM, in differentiating medium for five days, as described previously [44]. Then, glycyrrhizic acid at various concentrations was pretreated with the corticosterone-incubated H19-7 cells for 24 h. Finally, the H19-7 cells were collected for assay as described below.

4.8. Western Blotting Analysis

The protein concentration of the total protein lysates was determined using the BCA protein assay kit (Pierce Biotechnology, Rockford, IL, USA). The protein lysates (30 μg) were separated by electrophoresis and transferred to a polyvinylidene difluoride membrane (Millipore, Billerica, MA, USA). After blocking and washings, the following primary antibodies were incubated at 4 °C overnight: anti-BDNF (1:1000; Abcam, Cambridge, UK), anti-5-HTT (1:1000; Millipore, Billerica, MA, USA), anti-β-actin (1:5000, Sigma-Aldrich, St. Louis, MO, USA). The protein bands were visualized using the enhanced chemiluminescence kit (PerkinElmer, Boston, MA, USA). The quantification was determined using software (Gel-Pro Analyzer version 4.0 software (Media Cybernetics Inc., Silver Spring, MD, USA).

4.9. Real-Time Reverse Transcription-Polymerase Chain Reaction

According to our previous method [45], the mRNA expression levels of each signal were determined. In brief, total RNA was extracted from the cell lysates with TRIzol reagent (Carlsbad, CA, USA). Total RNA (200 ng) was reverse-transcribed into cDNA with random hexamer primers (Roche Diagnostics, Mannheim, Germany). All PCR experiments were performed using a LightCycler (Roche Diagnostics GmbH, Mannheim, Germany). The concentration of each PCR product was calculated relative to a corresponding standard curve. The relative gene expression was subsequently indicated as the ratio of the target gene level to that of β-actin. The primers for BDNF, 5-HTT and β-actin are listed as follows:

BDNF F: 5′-GCAGTCAAGTGCCTTTGGAG-3′;
BDNF R: 5′-CGGCATCCAGGTAATTTTG -3′;
5-HTT F: 5′-CATCAGCCCTCTGTTTCTCC -3′;
5-HTT R: 5′-CGGACGACATCCCTATGC -3′;
β-actin F: 5′-CTAAGGCCAACCGTGAAAAG-3′;
β-actin R: 5′-GCCTGGATGGCTACGTACA-3′.

4.10. Statistical Analysis

Data are presented as the mean ± standard errors (SE). Statistical analysis was conducted by using SPSS Version 21. Comparisons between groups were performed using the one-way analysis of variance (ANOVA) with Bonferroni's posthoc method. The p values of less than 0.05 were considered statistically significant.

5. Conclusions

The present study demonstrated that GMDZ ameliorated depression-like behaviors in rats. GMDZ regulated the 5-HTT and BDNF expression in the depression model. In addition, GanCao (licorice) played a crucial function in GMDZ decoction for antidepression.

Author Contributions: Conceptualization, T.-T.Y. and J.-T.C.; methodology, K.-C.C.; investigation, Y.-X.L.; resources, C.-T.H.; writing—original draft preparation, Y.-X.L.; writing—review and editing, K.-C.C. and J.-T.C.; supervision, T.-T.Y. All authors have read and agreed to the published version of the manuscript.

Funding: This research received no external funding.

Institutional Review Board Statement: The study was conducted according to the guidelines of the Care and Use of Laboratory Animals of the National Institutes of Health, and approved by the Institutional Animal Ethics Committee of Chi-Mei Medical Center (No. 105122622).

Data Availability Statement: The data is confidentiality.

Conflicts of Interest: The authors declare no conflict of interest.

References

1. Lepine, J.P.; Briley, M. The increasing burden of depression. *Neuropsychiatr. Dis. Treat.* **2011**, *7*, 3–7. [CrossRef] [PubMed]
2. Hansson, S.R.; Mezey, E.; Hoffman, B.J. Serotonin transporter messenger RNA in the developing rat brain: Early expression in serotonergic neurons and transient expression in non-serotonergic neurons. *Neuroscience* **1998**, *83*, 1185–1201. [CrossRef]
3. Molteni, R.; Cattaneo, A.; Calabrese, F.; Macchi, F.; Olivier, J.D.; Racagni, G.; Ellenbroek, B.A.; Gennarelli, M.; Riva, M.A. Reduced function of the serotonin transporter is associated with decreased expression of BDNF in rodents as well as in humans. *Neurobiol. Dis.* **2010**, *37*, 747–755. [CrossRef] [PubMed]
4. Schmidt-Kastner, R.; Wetmore, C.; Olson, L. Comparative study of brain-derived neurotrophic factor messenger RNA and protein at the cellular level suggests multiple roles in hippocampus, striatum and cortex. *Neuroscience* **1996**, *74*, 161–183. [CrossRef]
5. Lesch, K.P.; Bengel, D.; Heils, A.; Sabol, S.Z.; Greenberg, B.D.; Petri, S.; Benjamin, J.; Muller, C.R.; Hamer, D.H.; Murphy, D.L. Association of anxiety-related traits with a polymorphism in the serotonin transporter gene regulatory region. *Science* **1996**, *274*, 1527–1531. [CrossRef] [PubMed]

6. Karege, F.; Perret, G.; Bondolfi, G.; Schwald, M.; Bertschy, G.; Aubry, J.M. Decreased serum brain-derived neurotrophic factor levels in major depressed patients. *Psychiatry Res.* **2002**, *109*, 143–148. [CrossRef]
7. Shirayama, Y.; Chen, A.C.; Nakagawa, S.; Russell, D.S.; Duman, R.S. Brain-derived neurotrophic factor produces antidepressant effects in behavioral models of depression. *J. Neurosci.* **2002**, *22*, 3251–3261. [CrossRef] [PubMed]
8. Qin, F.; Wu, X.A.; Tang, Y.; Huang, Q.; Zhang, Z.J.; Yuan, J.H. Meta-analysis of randomized controlled trials to assess the effectiveness and safety of Free and Easy Wanderer Plus, a polyherbal preparation for depressive disorders. *J. Psychiatr. Res.* **2011**, *45*, 1518–1524. [CrossRef]
9. Wang, Y.; Fan, R.; Huang, X. Meta-analysis of the clinical effectiveness of traditional Chinese medicine formula Chaihu-Shugan-San in depression. *J. Ethnopharmacol.* **2012**, *141*, 571–577. [CrossRef]
10. Yeung, W.F.; Chung, K.F.; Ng, K.Y.; Yu, Y.M.; Ziea, E.T.; Ng, B.F. A meta-analysis of the efficacy and safety of traditional Chinese medicine formula Ganmai Dazao decoction for depression. *J. Ethnopharmacol.* **2014**, *153*, 309–317. [CrossRef]
11. Jun, J.H.; Lee, J.A.; Choi, T.Y.; Yun, K.J.; Lim, H.J.; Lee, M.S. Herbal medicine (Gan Mai Da Zao decoction) for depression: A systematic review protocol. *BMJ Open* **2014**, *4*, e003690. [CrossRef]
12. Spitzer, R.L.; Robins, E.; Endicott, J. *Research Diagnostic Criteria (RDC) for a Selected Group of Functional Disorders*; New York State Psychiatric Institute: New York, NY, USA, 1978.
13. Yang, F.; Qiao, Q.; Gong, Y. Antidepressant effect of gan-maida-zao tang on 30 women with postpartum depression. *Shanxi J TCM* **2009**, *30*, 851–852.
14. Wu, J. Antidepressant effect of gan-mai-da-zao tang on depression compared with fluoxetine. *J. Chin. Physician* **2002**, *11*, 18–19.
15. Kurebayashi, L.F.; Turrini, R.N.; Kuba, G.; Shimizu, M.H.; Takiguch, R.S. Chinese phytotherapy to reduce stress, anxiety and improve quality of life: Randomized controlled trial. *Rev. Esc. Enferm. USP* **2016**, *50*, 853–860. [CrossRef] [PubMed]
16. Lou, J.S.; Li, C.Y.; Yang, X.C.; Fang, J.; Yang, Y.X.; Guo, J.Y. Protective effect of gan mai da zao decoction in unpredictable chronic mild stress-induced behavioral and biochemical alterations. *Pharm. Biol.* **2010**, *48*, 1328–1336. [CrossRef] [PubMed]
17. Almeida, R.D.; Manadas, B.J.; Melo, C.V.; Gomes, J.R.; Mendes, C.S.; Graos, M.M.; Carvalho, R.F.; Carvalho, A.P.; Duarte, C.B. Neuroprotection by BDNF against glutamate-induced apoptotic cell death is mediated by ERK and PI3-kinase pathways. *Cell Death Differ.* **2005**, *12*, 1329–1343. [CrossRef] [PubMed]
18. McEwen, B.S. Stress and hippocampal plasticity. *Annu. Rev. Neurosci.* **1999**, *22*, 105–122. [CrossRef]
19. Willner, P.; Towell, A.; Sampson, D.; Sophokleous, S.; Muscat, R. Reduction of sucrose preference by chronic unpredictable mild stress, and its restoration by a tricyclic antidepressant. *Psychopharmacology* **1987**, *93*, 358–364. [CrossRef] [PubMed]
20. Tang, M.; Lei, J.; Sun, X.; Liu, G.; Zhao, S. Stress-induced anhedonia correlates with lower hippocampal serotonin transporter protein expression. *Brain Res.* **2013**, *1513*, 127–134. [CrossRef]
21. McClung, C.A.; Nestler, E.J. Neuroplasticity mediated by altered gene expression. *Neuropsychopharmacology* **2008**, *33*, 3–17. [CrossRef]
22. Ren-Patterson, R.F.; Cochran, L.W.; Holmes, A.; Sherrill, S.; Huang, S.J.; Tolliver, T.; Lesch, K.P.; Lu, B.; Murphy, D.L. Loss of brain-derived neurotrophic factor gene allele exacerbates brain monoamine deficiencies and increases stress abnormalities of serotonin transporter knockout mice. *J. Neurosci. Res.* **2005**, *79*, 756–771. [CrossRef]
23. Mossner, R.; Daniel, S.; Albert, D.; Heils, A.; Okladnova, O.; Schmitt, A.; Lesch, K.P. Serotonin transporter function is modulated by brain-derived neurotrophic factor (BDNF) but not nerve growth factor (NGF). *Neurochem. Int.* **2000**, *36*, 197–202. [CrossRef]
24. Zetterstrom, T.S.; Pei, Q.; Madhav, T.R.; Coppell, A.L.; Lewis, L.; Grahame-Smith, D.G. Manipulations of brain 5-HT levels affect gene expression for BDNF in rat brain. *Neuropharmacology* **1999**, *38*, 1063–1073. [CrossRef]
25. Dhingra, D.; Sharma, A. Antidepressant-like activity of Glycyrrhiza glabra L. in mouse models of immobility tests. *Prog. Neuro-Psychopharmacol. Biol. Psychiatry* **2006**, *30*, 449–454. [CrossRef]
26. Hatano, T.; Fukuda, T.; Miyase, T.; Noro, T.; Okuda, T. Phenolic constituents of licorice. III. Structures of glicoricone and licofuranone, and inhibitory effects of licorice constituents on monoamine oxidase. *Chem. Pharm. Bull.* **1991**, *39*, 1238–1243. [CrossRef]
27. Dhingra, D.; Sharma, A. Evaluation of antidepressant-like activity of glycyrrhizin in mice. *Indian J. Pharmacol.* **2005**, *37*, 390–395. [CrossRef]
28. Wang, H.L.; Li, Y.X.; Niu, Y.T.; Zheng, J.; Wu, J.; Shi, G.J.; Ma, L.; Niu, Y.; Sun, T.; Yu, J.Q. Observing anti-inflammatory and anti-nociceptive activities of glycyrrhizin through regulating COX-2 and pro-inflammatory cytokines expressions in mice. *Inflammation* **2015**, *38*, 2269–2278. [CrossRef] [PubMed]
29. Liu, X.; Cao, S.; Zhang, X. Modulation of gut microbiota-brain axis by probiotics, prebiotics, and diet. *J. Agric. Food Chem.* **2015**, *63*, 7885–7895. [CrossRef]
30. Borah, M.; Sarma, P.; Das, S. A study of the protective effect of triticum aestivum L. in an experimental animal model of chronic fatigue syndrome. *Pharmacogn. Res.* **2014**, *6*, 285–291. [CrossRef] [PubMed]
31. Lee, S.M.; Park, J.G.; Lee, Y.H.; Lee, C.G.; Min, B.S.; Kim, J.H.; Lee, H.K. Anti-complementary activity of triterpenoides from fruits of Zizyphus jujuba. *Biol. Pharm. Bull.* **2004**, *27*, 1883–1886. [CrossRef]
32. Park, S.H.; Sim, Y.B.; Kang, Y.J.; Kim, S.S.; Kim, C.H.; Kim, S.J.; Suh, H.W. Mechanisms involved in the antinociceptive effects of orally administered oleanolic acid in the mouse. *Arch. Pharmacal Res.* **2013**, *36*, 905–911. [CrossRef] [PubMed]
33. Oyebanji, B.O.; Saba, A.B.; Oridupa, O.A. Studies on the anti-inflammatory, analgesic and antipyrexic activities of betulinic acid derived from Tetracera potatoria. *Afr. J. Tradit. Complement. Altern. Med.* **2014**, *11*, 30–33. [CrossRef] [PubMed]

34. Huang, H.L.; Lim, S.L.; Lu, K.H.; Sheen, L.Y. Antidepressant-like effects of Gan-Mai-Dazao-Tang via monoamine regulatory pathways on forced swimming test in rats. *J. Tradit. Complement. Med.* **2018**, *8*, 53–59. [CrossRef] [PubMed]
35. Garcia-Garcia, L.; Shiha, A.A.; Bascunana, P.; de Cristobal, J.; Fernandez de la Rosa, R.; Delgado, M.; Pozo, M.A. Serotonin depletion does not modify the short-term brain hypometabolism and hippocampal neurodegeneration induced by the lithium-pilocarpine model of status epilepticus in rats. *Cell. Mol. Neurobiol.* **2016**, *36*, 513–519. [CrossRef]
36. Kornum, B.R.; Licht, C.L.; Weikop, P.; Knudsen, G.M.; Aznar, S. Central serotonin depletion affects rat brain areas differently: A qualitative and quantitative comparison between different treatment schemes. *Neurosci. Lett.* **2006**, *392*, 129–134. [CrossRef]
37. Tianzhu, Z.; Shihai, Y.; Juan, D. Antidepressant-like effects of cordycepin in a mice model of chronic unpredictable mild stress. *Evid.-Based Complement. Altern. Med.* **2014**, *2014*, 438506. [CrossRef]
38. Rowland, N.E. Food or fluid restriction in common laboratory animals: Balancing welfare considerations with scientific inquiry. *Comp. Med.* **2007**, *57*, 149–160. [PubMed]
39. Bessa, J.M.; Ferreira, D.; Melo, I.; Marques, F.; Cerqueira, J.J.; Palha, J.A.; Almeida, O.F.; Sousa, N. The mood-improving actions of antidepressants do not depend on neurogenesis but are associated with neuronal remodeling. *Mol. Psychiatry* **2009**, *14*, 739, 764–773. [CrossRef]
40. Nielsen, C.K.; Arnt, J.; Sanchez, C. Intracranial self-stimulation and sucrose intake differ as hedonic measures following chronic mild stress: Interstrain and interindividual differences. *Behav. Brain Res.* **2000**, *107*, 21–33. [CrossRef]
41. Prut, L.; Belzung, C. The open field as a paradigm to measure the effects of drugs on anxiety-like behaviors: A review. *Eur. J. Pharmacol.* **2003**, *463*, 3–33. [CrossRef]
42. Dunn, A.J.; Swiergiel, A.H. Effects of interleukin-1 and endotoxin in the forced swim and tail suspension tests in mice. *Pharmacol. Biochem. Behav.* **2005**, *81*, 688–693. [CrossRef] [PubMed]
43. Huang, C.W.; Lai, M.C.; Cheng, J.T.; Tsai, J.J.; Huang, C.C.; Wu, S.N. Pregabalin attenuates excitotoxicity in diabetes. *PLoS ONE* **2013**, *8*, e65154. [CrossRef] [PubMed]
44. Piron, M.; Villereal, M.L. Chronic exposure to stress hormones alters the subtype of store-operated channels expressed in H19-7 hippocampal neuronal cells. *J. Cell. Physiol.* **2013**, *228*, 1332–1343. [CrossRef] [PubMed]
45. Lo, S.H.; Hsu, C.T.; Niu, H.S.; Niu, C.S.; Cheng, J.T.; Chen, Z.C. Ginsenoside Rh2 improves cardiac fibrosis via PPARdelta-STAT3 signaling in type 1-Like diabetic rats. *Int. J. Mol. Sci.* **2017**, *18*, 1364. [CrossRef] [PubMed]

plants

MDPI

Article

Vescalagin from Pink Wax Apple (*Syzygium samarangense* (Blume) Merrill and Perry) Protects Pancreatic β-Cells against Methylglyoxal-Induced Inflammation in Rats

Wen-Chang Chang [1,2], James Swi-Bea Wu [2] and Szu-Chuan Shen [3,*]

[1] Department of Food Science, National Chiayi University, Chiayi 600355, Taiwan; wcchang@mail.ncyu.edu.tw
[2] Graduate Institute of Food Science and Technology, National Taiwan University, Taipei 10617, Taiwan; jsbwu@ntu.edu.tw
[3] Graduate Program of Nutrition Science, National Taiwan Normal University, Taipei 10610, Taiwan
* Correspondence: scs@ntnu.edu.tw; Tel.: +886-2-77491437; Fax: +886-2-23639635

Abstract: Methylglyoxal (MG) is the primary precursor of advanced glycation end products involved in the pathogenesis of inflammation and diabetes. A previous study in our laboratory found anti-inflammatory and anti-hyperglycemic effects of the polyphenol vescalagin (VES) in rats with MG-induced carbohydrate metabolic disorder. The present study further investigated the occurrence of inflammation in pancreatic β-cells in MG-induced diabetic rats and the mechanism by which VES prevents it. The results showed that VES downregulates the protein expression levels of advanced glycation end product receptors and CCAAT/enhancer binding protein-β and upregulates the protein expression levels of pancreatic duodenal homeobox-1, nuclear factor erythroid 2-related factor 2 and glyoxalase I from the pancreatic cells. The results also revealed that VES elevates glutathione and antioxidant enzyme contents and then downregulates c-Jun N-terminal kinase and p38 mitogen-activated protein kinases pathways to protect pancreatic β-cells in MG-administered rats.

Keywords: vescalagin; methylglyoxal; inflammation; antioxidant; insulin secretion

Citation: Chang, W.-C.; Wu, J.S.-B.; Shen, S.-C. Vescalagin from Pink Wax Apple (*Syzygium samarangense* (Blume) Merrill and Perry) Protects Pancreatic β-Cells against Methylglyoxal-Induced Inflammation in Rats. *Plants* **2021**, *10*, 1448. https://doi.org/10.3390/plants10071448

Academic Editor: Suresh Awale

Received: 8 June 2021
Accepted: 12 July 2021
Published: 15 July 2021

1. Introduction

Diabetes mellitus (DM) is a chronic disease associated with carbohydrate metabolism and caused by a deficiency in insulin secretion or by the ineffectiveness in insulin action [1]. The prevalence of DM, obesity, and many other metabolic syndromes has been linked to the increased consumption of methylglyoxal (MG)-containing foods [2,3]. MG is a major precursor of advanced glycation end products (AGEs) which in turn lead to oxidative stress [4]. Oxidative stress plays an important role in the pathophysiology of inflammation, insulin resistance, atherogenesis, and diabetes [5]. Cell studies indicated that MG and AGEs may promote the production of inflammatory cytokines that cause the damage of pancreatic β-cells [4–6]. MG has also been found from animal studies to cause inflammation in Sprague–Dawley rats, to induce their pancreatic impairment, and to affect insulin secretion as a consequence [5]. It appears that the protection of pancreatic β-cells can be an effective way for the maintenance of insulin secretion and the prevention of DM.

DM is associated with protein glycation [6]. AGEs may activate advanced glycation end product receptors (RAGE) on the membrane of pancreatic cells and then upregulate the two key inflammatory transcription factors, namely early growth response-1 and nuclear factor kappa B (NF-κB) [4,7]. NF-κB may promote the release of cytokines, including tumor necrosis factor alpha (TNF-α), interleukin-1 beta (IL-1β), and interleukin-6 (IL-6) [4]. High levels of these cytokines and reactive oxygen species (ROS) reduce the release of translocation-specific transcription factor pancreatic duodenal homeobox-1 (PDX-1) from the nucleus of a β-cell to the cytoplasm and hinder the synthesis of insulin [8]. Exposure to MG may decrease insulin secretion and the survival rate of pancreatic β-cells via redox-independent inhibition of phosphatidylinositol 3-kinase insulin signal pathway [9]. There are positive correlations between MG content and the survival rates of pancreatic β-cells both in vivo and in vitro [5,9].

Serum MG levels are less than 1 μM in healthy humans but can be elevated to 2–6 μM in diabetic patients, with a positive correlation with the degree of hyperglycemia [10]. Recently, many foods, including steak, wine, and beer, were found to be associated with high MG levels in the serum of human blood. Broiled steak has been reported to contain 12.73 μg/g of MG in a study in the United States [11] Certain samples of wine and beer were found to contain 21.59 and 13.88 μM MG, respectively, in Europe [12,13]. Under physiological conditions, glutathione (GSH) may activate glyoxalase to convert MG into D-lactate [6]. Glyoxalase I (GLO-1) and glyoxalase II (GLO-2) may also catalyze the detoxifying conversion of MG into D-lactoylglutathione and D-lactate, suggesting that GLO-1 may retard MG-induced formation of AGEs [14]. Nuclear factor erythroid 2-related factor 2 (Nrf2) has been reported to promote the expression of GLO-1 and the conversion of MG to D-lactate [15]. Many antioxidants, including quercetin and phenolic acids, have been found to attenuate oxidative damage by activating Nrf2 [16].

Ellagitannins are bioactive polyphenols with antioxidant and anti-inflammatory activities [17]. Pink wax apple (*Syzygium samarangense* (Blume) Merrill and Perry cv. Pink) fruit contains the ellagitannin vescalagin (VES). VES has been reported to be antitumor, cardiovascular disease preventive, insulin-resistance alleviative, and dyslipidemia mediative [18,19]. In a previous study, our laboratory found that VES may reduce serum glucose content and in the meanwhile increase serum insulin and C-peptide levels [20]. No studies with regard to the protective effect of VES on pancreatic β-cells have been reported yet. The aim of the present study was to elucidate the mechanism by which VES ameliorates the MG-induced inflammation and insulin secretion reduction by assessing the activities of oxidation enzymes, inflammatory proteins, and the insulin secretion-related proteins in pancreatic β-cells in rats.

2. Results

2.1. Diet Intake and Body Weight in Rats Orally Administered with MG

Table 1 shows the diet intake and body weight in rats orally administered with MG. In the present study, the rats administered orally once a day with MG at 300 mg/kg followed with another chemical (PIO, AG, or VES) at 30 mg/kg for a period of 8 weeks showed no significant difference in diet intake, drink intake, or body weight as compared with the normal group (Table 1), indicating no occurrence of acute-phase response in rats in the feeding period.

Table 1. Diet intakes, drink intakes, and body weights of rats administered with methylglyoxal and pioglitazone, aminoguanidine, or vescalagin.

Items/Groups	Normal	MG	MG+PIO	MG+AG	MG+VES
Diet intake (g/rat/day)	28.93 ± 5.66 [a]	29.60 ± 5.42 [a]	30.48 ± 5.41 [a]	27.60 ± 5.76 [a]	28.99 ± 4.84 [a]
Drink intake (mL/rat/day)	42.40 ± 3.29 [a]	44.23 ± 3.67 [a]	49.67 ± 5.09 [a]	43.42 ± 4.76 [a]	46.63 ± 4.78 [a]
Body weight (g)	416.11 ± 9.58 [a]	444.50 ± 19.0 [a]	420.64 ± 17.6 [a]	411.01 ± 36.3 [a]	424.37 ± 21.6 [a]

Normal: rats fed with normal diet and deionized water. MG: rats fed with normal diet with the addition of methylglyoxal (300 mg/kg body weight/day). MG+PIO: rats fed with normal diet with the addition of methylglyoxal (300 mg/kg body weight/day) and pioglitazone (30 mg/kg body weight/day). MG+AG: rats fed with normal diet with the addition of methylglyoxal (300 mg/kg body weight/day) and aminoguanidine (30 mg/kg body weight/day). MG+VES: rats fed with normal diet with the addition of methylglyoxal (300 mg/kg body weight/day) and vescalagin (30 mg/kg body weight/day). Feeding period was 8 weeks. Different letters ([a]) on the same line signify a statistically significant difference at $p < 0.05$. Results are from 8 repetitions and expressed as mean ± SD.

2.2. Amylase and Lipase Activities in the Rats

Figure 1 shows the effect of VES on amylase and lipase activities in MG-administered rats. When the contents of amylase and lipase are increased in the blood, this is recognized as a phenomenon of pancreatic cell inflammation in mammals. However, if the contents of amylase and lipase are significantly increased, it is recognized as acute pancreatitis. There were no significant differences in serum lipase activity among all groups. MG group (2259 ± 124 U/L) showed higher serum amylase activity as compared with the normal group (1869 ± 142 U/L) ($p < 0.05$), indicating that MG may induce hyperglycemia in rats. Our results show significantly lower activities of serum amylase in MG+PIO, MG+AG, and MG+VES treatment groups, by 19%, 35.8%, and 27.6%, respectively, as compared with the MG group ($p < 0.05$) (Figure 1). We proposed that VES may ameliorate hyperglycemia in MG-administered rats via the inhibition of carbohydrate digestion enzymes.

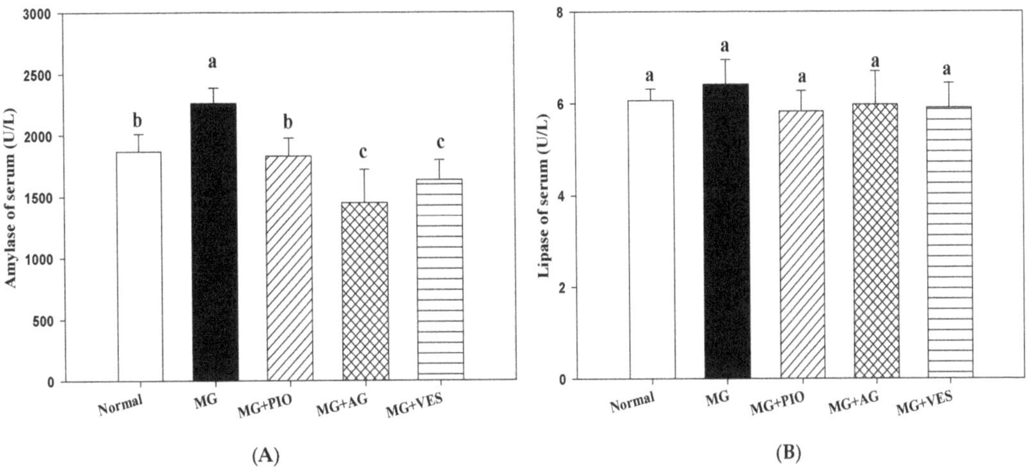

Figure 1. Effects of pioglitazone, aminoguanidine, and vescalagin on pancreatic inflammation indices (**A**) amylase concentration and (**B**) lipase concentration in the serum of MG-administered rats. Normal: rats fed with normal diet and deionized water. MG: rats fed with normal diet with the addition of methylglyoxal (300 mg/kg body weight/day). MG+PIO: rats fed with normal diet with the addition of methylglyoxal (300 mg/kg body weight/day) and pioglitazone (30 mg/kg body weight/day). MG+AG: rats fed with normal diet with the addition of methylglyoxal (300 mg/kg body weight/day) and aminoguanidine (30 mg/kg body weight/day). MG+VES: rats fed with normal diet with the addition of methylglyoxal (300 mg/kg body weight/day) and vescalagin (30 mg/kg body weight/day). Feeding period was 8 weeks. Different letters (a–c) signify a statistically significant difference at $p < 0.05$. Results are from 8 repetitions and expressed as mean ± SD.

2.3. *RAGE and Insulin Secretion-Related Protein Expression Levels in Pancreatic β-Cells of the Rats*

Figure 2 shows the effect of VES on the expression levels of RAGE and insulin secretion-related proteins in pancreatic β-cells of MG-administered rats. The results indicate that the expression of RAGE and C/EBPβ proteins in pancreatic β-cells can be elevated by MG and the elevation can be offset by PIO, AG, and VES ($p < 0.05$) (Figure 2A,B). The expression levels of pancreatic PDX-1 and GLO-1 proteins were increased in rats administered with MG and PIO, AG, or VES (Figure 2C,D). An increase in PDX-1 expression and a decrease in C/EBPβ expression correspond to an increase in insulin secretion and glucose tolerance [5]. We speculate that VES inhibits AGE formation and inflammation reaction in pancreatic cells. The present study also found significant reductions in pancreatic GSH content in the MG group as compared with normal group ($p < 0.05$). Furthermore, those groups with PIO, AG, and VES in the diet showed 202%, 154%, and 311% elevations in pancreatic GSH content, respectively, in comparison with the MG group ($p < 0.05$) (Figure 2E). The results suggest that VES elevates GSH content to protects pancreatic cells in MG-administered rats.

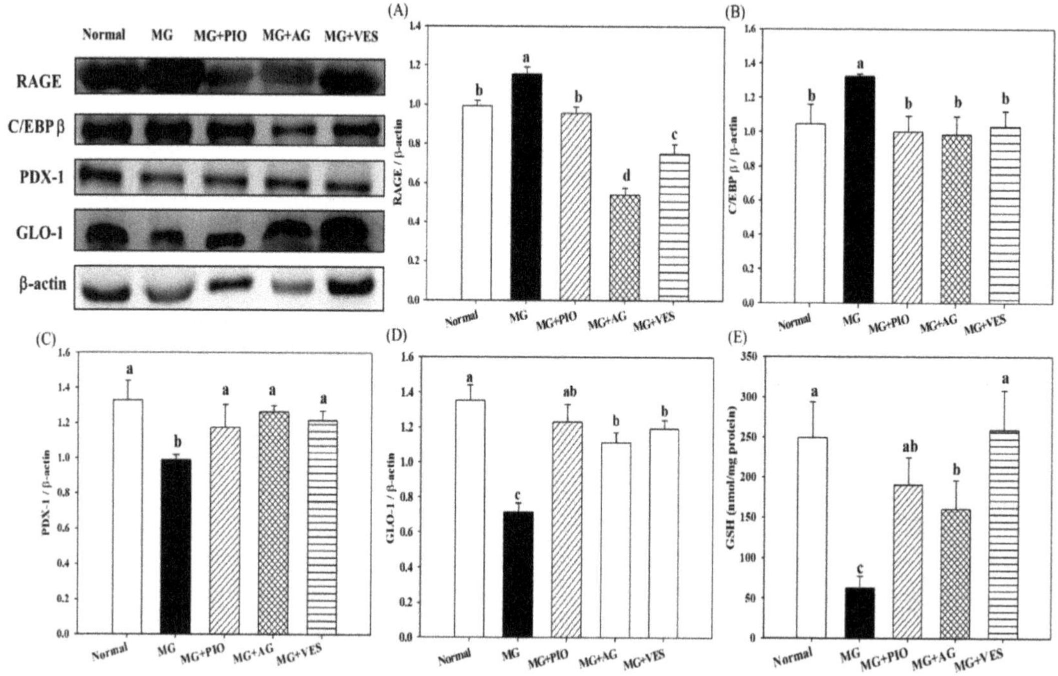

Figure 2. Effects of pioglitazone, aminoguanidine, and vescalagin on (**A**) RAGE expression, (**B**) C/EBP β expression, (**C**) PDX-1 expression, (**D**) GLO-1 expression, and (**E**) GSH concentration in the pancreatic β-cells of MG-administered rats. Normal: rats fed with normal diet and deionized water. MG: rats fed with normal diet with the addition of methylglyoxal (300 mg/kg body weight/day). MG+PIO: rats fed with normal diet with the addition of methylglyoxal (300 mg/kg body weight/day) and pioglitazone (30 mg/kg body weight/day). MG+AG: rats fed with normal diet with the addition of methylglyoxal (300 mg/kg body weight/day) and aminoguanidine (30 mg/kg body weight/day). MG+VES: rats fed with normal diet with the addition of methylglyoxal (300 mg/kg body weight/day) and vescalagin (30 mg/kg body weight/day). Feeding period was 8 weeks. Different letters (a–c) signify a statistically significant difference at $p < 0.05$. Results are from 8 repetitions and expressed as mean ± SD.

2.4. Expression Levels of Inflammation Proteins in Pancreatic β-Cells of the Rats

Figure 3 shows the effects of VES on the expression levels of inflammation proteins in pancreatic β-cells of MG-administered rats. MG and AGEs can increase oxidative stress and promote the generation of inflammatory cytokines, as found in our previous study [20]. The present study shows that pancreatic levels of the inflammatory proteins NF-κB, ICAM-1, and TNF-α were increased in MG-administered rats ($p < 0.05$). The pancreatic levels of NF-κB, ICAM-1, and TNF-α protein were reduced in MG-administered rats by feeding with VES (Figure 3A,B,D). It was found that PIO, AG, and VES activated pancreatic Nrf2 in MG-administered rats (Figure 3E). Many antioxidants, such as VES and PIO, have been evaluated for their ability to activate Nrf2 and to attenuate oxidative damage and inflammatory reactions [6].

2.5. Activities of Antioxidant Enzymes in Pancreatic β-Cells of the Rats

Figure 4 shows the effect of VES on antioxidative enzymes in pancreatic β-cells of MG-administered rats. The present study showed that MG downregulated the expression levels of Nrf2 and antioxidant enzymes, including GSH, SOD, and catalase, indicating the occurrence of redox imbalance after the administration of MG (Figures 2E and 4A,B). The present study also found that GSH and catalase contents were elevated in pancreatic β-cells in MG+PIO, MG+AG, and MG+VES groups as compared with the MG group

(Figures 2E and 4B). In comparison with the MG group, those groups administered with MG and PIO, AG, or VES in the diet showed significant reduction of 47%, 52%, or 57%, respectively, in pancreatic MDA content ($p < 0.05$) (Figure 4C). The present study showed that activation of Nrf2 in pancreatic β-cells reduces inflammatory reaction, which may ameliorate β-cell damage (Figure 3). We speculated that PIO, AG, and VES could effectively activate Nrf2 by phosphorylation, thereby increasing antioxidant enzyme contents and suppressing lipid peroxidation in the pancreas of the MG-administered rat.

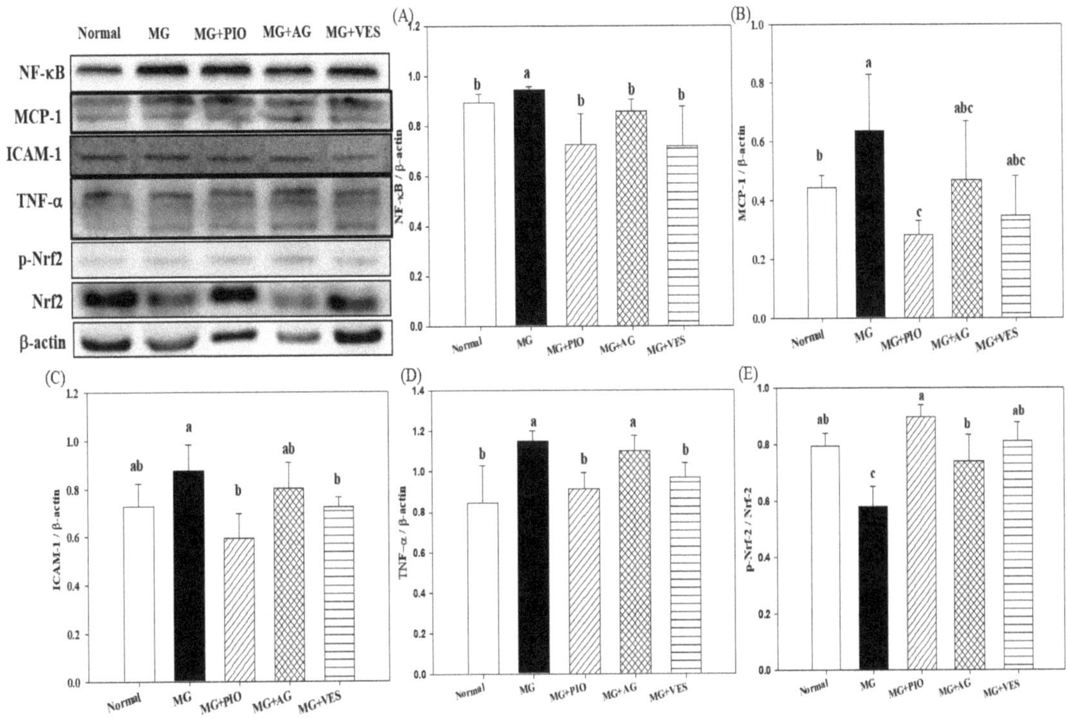

Figure 3. Effects of pioglitazone, aminoguanidine, and vescalagin on the expression levels of inflammation proteins (**A**) NF-κB, (**B**) MCP-1, (**C**) ICAM-1, (**D**) TNF-α, and (**E**) Nrf2 in the pancreatic β-cells of MG-administered rats. MG: rats fed with normal diet with the addition of methylglyoxal (300 mg/kg body weight/day). MG+PIO: rats fed with normal diet with the addition of methylglyoxal (300 mg/kg body weight/day) and pioglitazone (30 mg/kg body weight/day). MG+AG: rats fed with normal diet with the addition of methylglyoxal (300 mg/kg body weight/day) and aminoguanidine (30 mg/kg body weight/day). MG+VES: rats fed with normal diet with the addition of methylglyoxal (300 mg/kg body weight/day) and vescalagin (30 mg/kg body weight/day). Feeding period was 8 weeks. Different letters (a–c) signify a statistically significant difference at $p < 0.05$. Results are from 8 repetitions and expressed as mean ± SD.

2.6. Mitogen-Activated Protein Kinases (MAPKs) and Inflammation in Pancreatic β-Cells of the Rats

Finally, according to the above observations, we will confirm whether VES can ameliorate serum glucose by downregulating the inflammatory proteins of the common inflammation pathway from pancreatic cells. Figure 5 shows the effect of VES on the expression levels of MARKs in pancreatic β-cells of MG-administered rats. The pancreatic levels of JNK and p38 phosphorated proteins were reduced in rats administered with VES in addition to MG, suggesting that VES may downregulate the phosphorylation of JNK and p38 proteins in MAPK pathways and protect pancreatic β-cells in MG-administered rats.

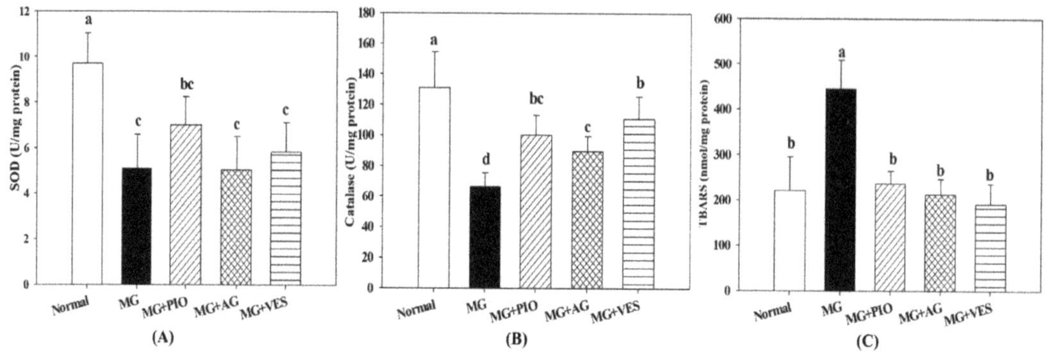

Figure 4. Effects of pioglitazone, aminoguanidine, and vescalagin on the concentrations of antioxidant indicators (**A**) SOD, (**B**) catalase, and (**C**) TBARS in the pancreatic β-cells of MG-administered rats. MG: rats fed with normal diet with the addition of methylglyoxal (300 mg/kg body weight/day). MG+PIO: rats fed with normal diet with the addition of methylglyoxal (300 mg/kg body weight/day) and pioglitazone (30 mg/kg body weight/day). MG+AG: rats fed with normal diet with the addition of methylglyoxal (300 mg/kg body weight/day) and aminoguanidine (30 mg/kg body weight/day). MG+VES: rats fed with normal diet with the addition of methylglyoxal (300 mg/kg body weight/day) and vescalagin (30 mg/kg body weight/day). Feeding period was 8 weeks. Different letters (a–c) signify a statistically significant difference at $p < 0.05$. Results are from 8 repetitions and expressed as mean ± SD.

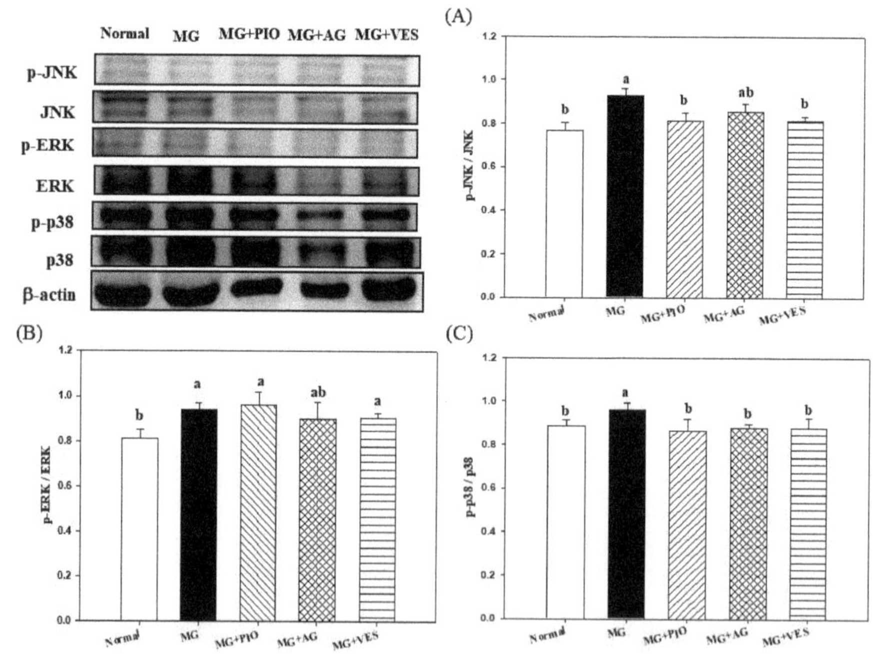

Figure 5. Effects of pioglitazone, aminoguanidine, and vescalagin on the expression levels of inflammation proteins (**A**) JNK, (**B**) ERK, and (**C**) p38 in the pancreatic β-cells of MG-administered rats. MG: rats fed with normal diet with the addition of methylglyoxal (300 mg/kg body weight/day). MG+PIO: rats fed with normal diet with the addition of methylglyoxal (300 mg/kg body weight/day) and pioglitazone (30 mg/kg body weight/day). MG+AG: rats fed with normal diet with the addition of methylglyoxal (300 mg/kg body weight/day) and aminoguanidine (30 mg/kg body weight/day). MG+VES: rats fed with normal diet with the addition of methylglyoxal (300 mg/kg body weight/day) and vescalagin (30 mg/kg body weight/day). Feeding period was 8 weeks. Different letters (a, b) signify a statistically significant difference at $p < 0.05$. Results are from 8 repetitions and expressed as mean ± SD.

3. Discussion

Recent studies showed that MG injection of Sprague–Dawley rats may cause glucose intolerance, reduce the insulin-stimulated glucose uptake in adipose tissue, and result in pancreatic dysfunction and that oral administration of MG at 300 mg/kg/day may induce inflammation in the pancreatic β-cells [5,10,20]. These animal studies also showed that there was no significant difference in diet intake, drink intake, and body weight among all groups. The aforementioned studies are consistent with our results.

Some chemicals, such as MG and AGEs, may damage the pancreatic tissue and even lead to pancreatitis and diabetes [2,5,21]. Recent studies demonstrated that amylase and lipase may enhance carbohydrate and lipid digestion in the induction of hyperglycemia and hyperlipidemia in rats [22,23]. Mnafgui et al. found elevated amylase activities in the serum and pancreas of diabetic rats [24]. The increase in α-amylase activity also occurred in the rats orally administered with MG, indicating that oral administration is a feasible method to induce hyperglycemia in the animal [20]. The elevated serum amylase activity may be associated with pancreatic β-cell inflammation in rats [24]. A range of berry polyphenols, for example, flavonols, anthocyanidins, resveratrol, ellagitannins, and proanthocyanidins, can inhibit amylase to a level that affects starch degradation in rats [25]. Liu et al. proposed that polyphenols and flavonoids act as antioxidants to suppress the activity of amylase in vitro [23]. Our results show a significantly lower activity of serum amylase in the MG+VES treatment group as compared with the MG group ($p < 0.05$) (Figure 1). We reported that the effect of MG in reducing the insulin and C-peptide contents in serum, upregulating inflammatory cytokines, and inducing hyperglycemia in MG-administered rats previously [20]. Both amylase and lipase levels have been used as indices for the inflammation of the pancreas in rats [21]. Previous studies demonstrated that VES could trap MGO through a conjugating reaction and thus inhibit further glycation reaction. VES was able to inhibit both α-amylase and α-glucosidase [20,25]. We speculate that MG promotes AGEs production and inflammation and hurts pancreatic β-cells in rats. The following assessments with regards to the effects of VES on inflammation and the insulin secretion of pancreatic β-cells in rats administered with MG were performed to support the postulation.

RAGE downregulates the expression levels of insulin secretion-related proteins and induces inflammation in pancreatic β-cells [4,6]. Pioglitazone is an insulin sensitizer used for controlling the serum glucose of type 2 diabetes mellitus. The antidiabetic drug PIO is a peroxisome proliferator-activated receptor-γ (PPARγ) ligand. The activation of PPARγ is known to affect pancreatic β-cell functions, including insulin production [26]. AG is a nucleophilic hydrazine derivative and is well known for its action in blocking advanced glycation, and it has been shown to prevent diabetic complications. AG was reported to block glycation, and it also inhibits the production of toxic aldehydes by quinone enzymes [27]. AG, PIO, and VES are effective agents in scavenging MG molecules, downregulating RAGE expression, and retarding AGE formation [20,28]. Reduced MG intake may alleviate the production of the pancreatic serum proteins that are promotive for RAGE expression, glycation, and inflammation [2,4,29]. The results in the present study reconfirm the above-mentioned findings. RAGE may downregulate PDX-1 protein expression in pancreatic RIN-m5F cells [16]. In β-cells, PDX-1 protein is a positive regulator of insulin synthesis, whereas C/EBPβ protein is a repressor of insulin gene transcription [5]. MG was found to downregulate PDX-1, to upregulate C/EBPβ, and to inhibit insulin secretion from isolated β-cells from rats [5]. The expression of pancreatic PDX-1 protein was increased in rats administered with MG and PIO, AG, or VES. The expression of pancreatic C/EBPβ protein was increased by the administration with MG, and the increment could be offset by PIO, AG, and VES (Figure 2). The present study indicates that VES is promotive of insulin secretion and glucose tolerance in rats.

Dhar et al. proposed that the reduction in MG content may ameliorate insulin resistance and β-cell damage [5]. Reflecting the impact of MG on biological systems, cells developed a GLO enzyme system that is dedicated to MG degradation and was well conserved during evolution [4]. The GLO system is present in the cytosol of all animal cells and comprises two enzymes, GLO-1 and GLO-2, and a catalytic amount of GSH [30]. GLO-1 and GLO-2 catalyze MG to D-lactoylglutathione and D-lactate and thereby suppress dicarbonyl-mediated glycation reactions [30]. The present study found that the pancreatic GLO-1 protein expression was elevated in MG-administered rats administered with PIO, AG, or VES ($p < 0.05$) (Figure 2D), suggesting that the increase in GLO-1 expression is involved in the mechanism for MG reduction. A high expression of GLO-1 prevented the increase of MG and AGEs in streptozotocin-induced diabetic rats [31]. The activity of the GLO enzyme system is vital to mitigate the dicarbonyl components, such as MG and AGEs, oxidative stress, and pancreatic β-cell damage [30]. In a previous study, we found the effectiveness of VES in ameliorating the serum glucose level in the rat via the reduction in MG and AGE contents [20]. GLO-1 activity is GSH-dependent [4,16,32]. We propose that VES elevates GSH content (Figure 2D), promotes GLO-1 expression (Figure 2E) in MG-administered rats, accelerates metabolism of MG, attenuates AGE formation in pancreatic β-cells, and thereby protects these cells.

The elevation in MG and inflammation cytokines may induce abnormality in β-cells [20]. The molecular mechanism underlying MG toxicity appears to be complex and may involve the glycation of cellular proteins and induction of cell death [4,16]. MG and AGEs promote the release of proinflammatory cytokines, increase the activity of many pro-oxidant enzymes such as NADPH oxidase and JNK, and upregulate the expression of NF-κB in pancreatic cells in rats [4,20]. The contents of inflammatory cytokines TNF-α and IL-6 are elevated in the serum of MG-administered rats [20]. The present study shows the pancreatic levels of inflammatory proteins were reduced in MG-administered rats by feeding with VES (Figure 3), suggesting that VES may reduce the expression levels of inflammation protein to protect pancreatic β-cells. Previous studies showed that antioxidants, including PIO and VES, may reduce serum contents of TNF-α and IL-6 in MG-administered rats [6,16,20]. We found that VES activated pancreatic Nrf2 in MG-administered rats (Figure 3E). Nrf2 has been shown to improve glucose tolerance, insulin sensitivity and metabolic syndrome in rats and to inhibit oxidative stress and inflammatory reactions in the pancreas [33–35]. We propose that VES downregulates the expression levels of inflammation proteins, elevates Nrf2 expression, and protects β-cells in MG-administered rats (Figure 3).

Cytokine-induced β-cell dysfunction can be reduced through inhibition of ROS production, alleviation of inflammatory reaction, or increase in the activities of antioxidative enzymes [8]. Nrf2 is a crucial regulator of the cellular redox homeostasis and plays a pivotal role, through the promotion of the expression levels of antioxidant enzymes and other antioxidant proteins, in protecting cells against ROS damage and inflammation [36]. The antioxidant enzymes controlled by Nrf2 include catalase, glutathione peroxidase, GSH, and superoxide dismutase (SOD) [36]. Previous studies showed that PPARγ agonists, such as PIO, could exhibit a regulatory effect upon Nrf2 [16,36]. Previous studies demonstrated that MG promotes the generation of MDA, which is a product of lipid peroxidation and a marker of dysfunctional insulin expression and diabetes [6,37]. GSH is well known to serve diverse biological functions, including the reduction in MDA content, the alleviation of inflammation, and the protection of pancreatic β-cells [37,38]. The present study suggested that PIO, AG, and VES elevated GSH content in MG-administered rats via the activation of Nrf2 protein expression (Figures 2E and 3E).

Studies indicated that MAPKs play a pivotal role in the development of insulin resistance induced by various factors, including inflammatory cytokines, ROS, chemicals, and oxidants [4,36,39]. MG was observed to activate MAPK pathways of inflammation in osteoblasts, endothelial cells, and Jurkat leukemia cells in a redox-dependent manner [4]. The present study showed that the pancreatic levels of JNK and p38 phosphorated proteins were reduced in rats administered with VES in addition to MG (Figure 5), suggesting that VES may downregulate the phosphorylation of JNK and p38 proteins in MAPK pathways and protect pancreatic β-cells in MG-administered rats.

MG may promote the phosphorylation of p38 in pancreatic islets via the elevation in expression levels of proinflammation proteins such as NF-κB [8]. Reduction in the MG content often alleviates inflammation [4,6,10,14]. In a previous study we found VES to reduce the MG content in the serum of MG-administered rats [20]. VES is a polyphenol with antioxidant and anti-inflammatory bioactivities [17]. The present study showed that VES promotes the activities of antioxidant enzymes, downregulates the expression of inflammation factor NF-κB and the phosphorylation and expression levels of inflammation proteins JNK and p38 MAPKs in pancreatic β-cells (Figures 3 and 4), and increases insulin secretion in MG-administered rats (Figure 5). Based on the above-described results, we propose that VES downregulates the expression levels of inflammation proteins and protects the pancreatic β-cells in MG-administered rats via the inactivation of JNK and p38 MAPK pathways.

4. Materials and Methods

4.1. Chemicals

VES was extracted and purified from unripe pink wax apple fruit (*Syzygium samarangense* (Blume) Merrill and Perry) following the reported procedure [40]. 1,4-Dithiothreit, acrylamide, aminoguanidine (AG), ammonium peroxydisulfate, D-glucose, ethyl alcohol, ethyl ether, glycine, methanol, MG, pioglitazone hydrochloride (PIO), sodium chloride, sodium dodecyl sulfate (SDS), sodium phosphate dibasic, sulfuric acid, thiourea, tris base, Triton X-100, Tween-20, and urea were purchased from Sigma (St. Louis, MO, USA).

4.2. Animals and Diets

Male Wistar rats (5-week old) were supplied by National Laboratory Animal Breeding and Research Center, Taipei, Taiwan. The rats were maintained in standard laboratory conditions at 22 ± 1 °C on 12-h light/12-h dark cycle with free access to food and water for the duration of the study. The room conditions and treatment procedures were in accordance with the National Institutes of Health (NIH) Guide for the Care and Use of Laboratory Animals, and all protocols were approved by the Institutional Animal Care and Use Committee of National Taiwan Normal University, Taipei, Taiwan (approval number 103042). The rats were fed a normal diet and deionized water for 1 week and then divided into 5 groups, 8 animals each, to feed for 8 more weeks. Among them, 1 group was fed a normal diet and 30 mg/mL deionized water (Normal group), while the other 4 groups were fed with the normal diet supplemented with MG at 300 mg/kg b.w./day and orally administered, once a day, with 30 mg/mL deionized water (MG group) or the solution of PIO (MG+PIO group), AG (MG+AG group), or VES (MG+VES group) at 30 mg/kg body weight/day. The dosage of MG followed the experimental designs reported by Dhar et al. and Lee et al. [5,6]. All animals were sacrificed by ethyl ether asphyxia at the end of the feeding period. The following operations were then performed.

4.3. Blood and Pancrease Tissue Sample Collection

Blood samples were taken from venter vein of the sacrificed rat, allowed to clot for 30 min at room temperature, and then centrifuged at 3000× *g* for 20 min to obtain the serum, which was stored at −80 °C before use. The pancreas tissue was picked out and rinsed with cold modified Tyrode calcium-free solution. The tissue was cut into small pieces (2–3 mm) in Tyrode buffer and washed several times with this solution to remove blood and possible fat tissue contamination and stored at −80 °C before use.

4.4. Biochemical Analyses

ELISA kits for rat α-amylase, catalase, GSH, lipase, superoxide dismutase (SOD), and thiobarbituric acid reactive substances (TBARS) were purchased from Randox Laboratories (Crumlin, Antrim, UK). Analyses were performed following the supplier's protocols. Briefly, pancreas tissues were homogenized on ice with RIPA buffer (Cell Signaling Technology, Beverly, MA, USA) using a homogenizer and centrifuged at 12,000× *g* (4 °C,

60 min) to collect supernatant as the pancreas extract. The protein concentration in the extract was determined using Bio-Rad Protein Assay Kit (Bio-Rad, Hercules, CA, USA) with bovine serum albumin as the standard. Pancreas extract was mixed with the GSH commercial reaction reagent, and the absorbance at OD450 nm was measured in kinetic mode for 40–60 min at room temperature by ELISA reader. Data were further analyzed following the GSH kit formula. Pancreas extract was mixed with the SOD commercial reaction reagent (WST-1 and enzyme working solution) and then incubated at 37 °C for 20 min. Absorbance was then measured at 450 nm with a spectrophotometer at room temperature. Data were analyzed following the SOD kit formula. Pancreas extract was mixed the catalase commercial reaction reagent, and then the supplier's protocols were followed (1. add H_2O_2 to samples and incubate 30 min at 25 °C; 2. add stop solution to samples; 3. add development mix and incubate 10 min at 25 °C). Absorbance was then measured at 570 nm with a spectrophotometer. Data were analyzed following the catalase kit formula. Malondialdehyde (MDA), other aldehydes, and lipid hydroperoxides are able to form adducts with TBA. The MDA concentration was used as an index of lipid peroxidation using the thiobarbituric acid reactive substances (TBARS) method. Pancreas extract was added to a mixture containing 0.5% TBA and 3.75% BHT in methanol. Extracts were heated in a boiling water bath for 30 min and then cooled on ice. The MDA–TBA adduct can be easily quantified colorimetrically at 532 nm. Data were analyzed following the MDA kit formula.

4.5. Western Blot Analysis

Aliquots of the pancreas extract, containing 100 μg protein in each, were used to evaluate the expression levels of CCAAT/enhancer binding protein-β (C/EBPβ), extracellular signal-regulated kinase (ERK), phosphorylated ERK (p-ERK), GLO-1, intercellular adhesion molecule 1 (ICAM-1), c-Jun N-terminal kinase (JNK), phosphorylated p-JNK (JNK), monocyte chemoattractant protein-1 (MCP-1), NF-κB, Nrf2, phosphorylated Nrf2 (p-Nrf2), p38 MAPK (p38), phosphorylated p38 (p-p38), PDX-1, RAGE, and TNF-α. The protein extract was separated by using dodecyl sulfate polyacrylamide gel electrophoresis on 10% polyacrylamide gel and transferred to a polyvinylidene difluoride membrane (Millipore, Billerica, MA, USA). The membrane was incubated with block buffer (PBS containing 0.05% Tween-20 and 5% *w*/*v* nonfat dry milk) for 1 h and washed with PBS containing 0.05% Tween-20 (PBST) 3 times. Then, it was probed with each of the 1:1000 diluted solutions of anti-NF-κB, anti-MCP-1, and anti-ICAM-1 (GeneTex, Irvine, CA, USA); 1:1000 diluted solutions of anti-ERK, anti-p-ERK, anti-Nrf2, anti-PDX-1, anti-RAGE, and anti-TNF-α (Cell Signaling Technology, Beverly, MA, USA); and 1:1000 diluted solutions of anti-C/EBPβ, anti-GLO-1, anti-JNK, anti-p-JNK, anti-p-Nrf2, anti-p38, and anti-p-p38 (Epitomics, Burlingame, CA, USA) overnight at 4 °C. The intensity of the blots probed with 1:2000 diluted solution of rabbit monoclonal antibody to bind actin (Gene Tex, Irvine, CA, USA) was used as the control to ensure that a constant amount of protein was loaded into each lane of the gel. The membrane was washed with PBST 3 times, 5 min each; shaken in a solution of horseradish peroxidase-linked anti-mouse IgG or anti-rabbit IgG secondary antibody (Gene Tex, Irvine, CA, USA); washed with PBST 3 more times, 5 min each; and then exposed to the enhanced chemiluminescence reagent (Millipore, Bedford, MA, USA) following the manufacturer's instructions. The films were scanned and analyzed using the UVP Biospectrum image system (Level, Cambridge, UK).

4.6. Statistical Analysis

Results expressed as means ± SD were analyzed by one-way ANOVA and Duncan's new multiple range tests. All the p values less than 0.05 were considered to be significant.

5. Conclusions

The present study elucidates the mechanism by which VES alleviates MG-caused inflammation in pancreatic β-cells by evaluating the activities of antioxidant enzymes and the expression levels of anti-inflammation proteins in rats. Our experiment results revealed that VES elevates GSH content; upregulates the protein expression levels of Nrf2, GLO-1, and antioxidant enzymes; downregulates JNK and p38 MAPK pathways; and thereby protects pancreatic β-cells and improves insulin secretion in MG-administered rats. These findings support the potential for VES to become a health food ingredient in the prevention of diabetes.

Author Contributions: W.-C.C. and S.-C.S. designed the research; S.-C.S. acquired funding; W.-C.C. performed the experimental work and data curation; W.-C.C. and J.S.-B.W. wrote the manuscript. All authors discussed, edited, and approved the final version. All authors have read and agreed to the published version of the manuscript.

Funding: The authors would like to thank the Ministry of Science and Technology of the Republic of China (ROC), Taiwan, for financially supporting this research under contract No. MOST 107-2320-B-003-004-MY3 and No. MOST 108-2221-E-415-015-MY2.

Institutional Review Board Statement: The study was conducted according to the guidelines of the Declaration of Helsinki, and approved by the Ethics Committee of National Taiwan Normal University, Taipei, Taiwan (approval number 103042, approval date 23 December 2015).

Informed Consent Statement: Not applicable.

Data Availability Statement: Data available in a publicly accessible repository.

Acknowledgments: The authors would like to thank the Ministry of Science and Technology of the Republic of China (ROC), Taiwan, for financially supporting this research.

Conflicts of Interest: The authors declare no conflict of interest.

References

1. Huang, D.W.; Shen, S.C.; Wu, J.S.B. Effects of caffeic acid and cinnamic acid on glucose uptake in insulin-resistant mouse hepatocytes. *J. Agric. Food Chem.* **2009**, *57*, 7687–7692. [CrossRef]
2. Nemet, I.; Varga-Defterdarovic, L.; Turk, Z. Methylglyoxal in food and living organisms. *Mol. Nutr. Food Res.* **2006**, *50*, 1105–1117. [CrossRef]
3. Lin, S.; Yang, Z.; Liu, H.; Tang, L.; Cai, Z. Beyond glucose: Metabolic shifts in responses to the effects of oral glucose tolerance test and the high-fructose diet in rats. *Mol. BioSyst.* **2011**, *7*, 1537–1548. [CrossRef]
4. Matafome, P.; Sena, C.; Seiça, R. Methylglyoxal, obesity, and diabetes. *Endocrine* **2013**, *43*, 472–484. [CrossRef]
5. Dhar, A.; Dhar, I.; Jiang, B.; Desai, K.M.; Wu, L. Chronic methylglyoxal infusion by minipump causes pancreatic β-cell dysfunction and induces type 2 diabetes in sprague-dawley rats. *Diabetes* **2011**, *60*, 899–908. [CrossRef]
6. Lee, B.H.; Hsu, W.H.; Chang, Y.Y.; Kuo, H.F.; Hsu, Y.W.; Pan, T.M. Ankaflavin: A natural novel PPARγ agonist upregulates Nrf2 to attenuate methylglyoxal-induced diabetes in vivo. *Free Radic. Biol. Med.* **2012**, *53*, 2008–2016. [CrossRef]
7. Ramasamy, R.; Yan, S.; Schmidt, A. Receptor for AGE (RAGE): Signaling mechanisms in the pathogenesis of diabetes and its complications. *Ann. N. Y. Acad. Sci.* **2011**, *1243*, 88–102. [CrossRef]
8. Novotny, G.W.; Lundh, M.; Backe, M.B.; Christensen, D.P.; Hansen, J.B.; Dahllof, M.S.; Pallesen, E.M.H.; Mandrup-Poulsen, T. Transcriptional and translational regulation of cytokine signaling in inflammatory β-cell dysfunction and apoptosis. *Arch. Biochem. Biophys.* **2012**, *528*, 171–184. [CrossRef] [PubMed]
9. Fiory, F.; Lombardi, A.; Miele, C.; Giudicelli, J.; Beguinot, F.; Van Obberghen, E. Methylglyoxal impairs insulin signalling and insulin action on glucose-induced insulin secretion in the pancreatic beta cell line INS-1E. *Diabetologia* **2011**, *54*, 2941–2952. [CrossRef]
10. Dhar, A.; Desai, K.M.; Wu, L. Alagebrium attenuates acute methylglyoxal-induced glucose intolerance in sprague-dawley rats. *Brit. J. Pharmacol.* **2010**, *159*, 166–175. [CrossRef]
11. Uribarri, J.; Woodruff, S.; Goodman, S.; Cai, W.; Chen, X.; Pyzik, R.; Yong, A.; Striker, G.E.; Vlassara, H. Advanced glycation end products in foods and a practical guide to their reduction in the diet. *J. Am. Diet. Assoc.* **2010**, *110*, 911–916. [CrossRef] [PubMed]
12. De Revel, G.; Bertrand, A. A method for the detection of carbonyl compounds in wine: Glyoxal and methylglyoxal. *J. Sci. Food Agric.* **1993**, *61*, 267–272. [CrossRef]
13. Barros, A.; Rodrigues, J.A.; Almeida, P.J.; Oliva-Teles, M.T. Determination of glyoxal, methylglyoxal and diacetyl in selected beer and wine by HPLC with UV spectrophotometric detection after derivatization with o-phenylendiamine. *J. Liq. Chromatogr. Relat. Technol.* **1999**, *22*, 2061–2069. [CrossRef]
14. Desai, K.; Wu, L. Methylglyoxal and advanced glycation endproducts: New therapeutic horizons? *Recent Pat. Cardiovasc. Drug Discov.* **2007**, *2*, 89–99. [CrossRef]

15. Vander Jagt, D.L.; Hunsaker, L.A. Methylglyoxal metabolism and diabetic complications: Roles of aldose reductase, glyoxalase-I, betaine aldehyde dehydrogenase and 2-oxoaldehyde dehydrogenase. *Chem. Biol. Interact.* **2003**, *143*, 341–351. [CrossRef]
16. Lee, B.H.; Hsu, W.H.; Hsu, Y.W.; Pan, T.M. Dimerumic acid protects pancreas damage and elevates insulin production in methylglyoxal-treated pancreatic RINm5F cells. *J. Funct. Foods* **2013**, *5*, 642–650. [CrossRef]
17. Heber, D. Multitargeted therapy of cancer by ellagitannins. *Cancer Lett.* **2008**, *269*, 262–268. [CrossRef] [PubMed]
18. Fridrich, D.; Glabasnia, A.; Fritz, J.; Esselen, M.; Pahlke, G.; Hofmann, T.; Marko, D. Oak ellagitannins suppress the phosphorylation of the epidermal growth factor receptor in human colon carcinoma cells. *J. Agric. Food Chem.* **2008**, *56*, 3010–3015. [CrossRef] [PubMed]
19. Shen, S.C.; Chang, W.C. Hypotriglyceridemic and hypoglycemic effects of vescalagin from Pink wax apple [*Syzygium samarangense (Blume)* Merrill and Perry cv. Pink] in high-fructose diet-induced diabetic rats. *Food Chem.* **2013**, *136*, 858–863. [CrossRef]
20. Chang, W.C.; Shen, S.C.; Wu, J.S.B. Protective effects of vescalagin from pink wax apple [*Syzygium samarangense* (Blume) Merrill and Perry] fruit against methylglyoxal-induced inflammation and carbohydrate metabolic disorder in rats. *J. Agric. Food Chem.* **2013**, *61*, 7102–7109. [CrossRef]
21. Koyu, A.; Gokalp, O.; Gumral, N.; Oktem, F.; Karahan, N.; Yilmaz, N.; Saygin, M. Impact of caffeic acid phenethyl ester treatment on vancomycin-induced pancreatic damage in rats. *Toxicol. Ind. Health* **2013**, *32*, 306–312. [CrossRef]
22. Malloy, J.; Gurney, K.; Shan, K.; Yan, P.; Chen, S. Increased variability and abnormalities in pancreatic enzyme concentrations in otherwise asymptomatic subjects with type 2 diabetes. *Diabetes Metab. Syndr. Obes.* **2012**, *5*, 419–424.
23. Liu, S.; Li, D.; Huang, B.; Chen, Y.; Lu, X.; Wang, Y. Inhibition of pancreatic lipase, α-glucosidase, α-amylase, and hypolipidemic effects of the total flavonoids from *Nelumbo nucifera* leaves. *J. Ethnopharmacol.* **2013**, *149*, 263–269. [CrossRef]
24. Mnafgui, K.; Kchaou, M.; Hamden, K.; Derbali, F.; Slama, S.; Nasri, M.; Salah, H.B.; Allouche, N.; Elfeki, A. Inhibition of carbohydrate and lipid digestive enzymes activities by *Zygophyllum album* extracts: Effect on blood and pancreas inflammatory biomarkers in alloxan-induced diabetic rats. *J. Physiol. Biochem.* **2014**, *70*, 93–106. [CrossRef]
25. McDougall, G.J.; Stewart, D. The inhibitory effects of berry polyphenols on digestive enzymes. *BioFactors* **2005**, *23*, 189–195. [CrossRef]
26. Irwin, N.; McKinney, J.M.; Bailey, C.J.; McClenaghan, N.H.; Flatt, P.R. Acute and long-term effects of peroxisome proliferator-activated receptor-γ activation on the function and insulin secretory responsiveness of clonal beta-cells. *Horm. Metab. Res.* **2011**, *43*, 244–249. [CrossRef] [PubMed]
27. Yu, P.H.; Zuo, D.M. Aminoguanidine inhibits semicarbazide- sensitive amine oxidase activity; implication for advanced glycation and angiopathy in diabetes. *Diabetologia* **1997**, *363*, 1243–1250. [CrossRef]
28. Yildiz, G.; Demiryürek, A.T.; Sahin-Erdemli, I.; Kanzik, I. Comparison of antioxidant activities of aminoguanidine, methylguanidine and guanidine by luminol-enhanced chemiluminescence. *Birt. J. Pharmacol.* **1998**, *124*, 905–910. [CrossRef]
29. Singh, R.; Barden, A.; Mori, T.; Beilin, L. Advanced glycation end-products: A review. *Diabetologia* **2001**, *44*, 129–146. [CrossRef]
30. Rabbani, N.; Thornalley, P.J. Methylglyoxal, glyoxalase 1 and dicarbonyl proteome. *Amino Acids* **2012**, *42*, 1133–1142. [CrossRef]
31. More, S.S.; Vartak, A.P.; Vince, R. Restoration of glyoxalase enzyme activity precludes cognitive dysfunction in a mouse model of Alzheimer's disease. *ACS Chem. Neurosci.* **2013**, *4*, 330–338. [CrossRef] [PubMed]
32. Jack, M.; Ryals, J.; Wright, D. Protection from diabetes-induced peripheral sensory neuropathy-a role for elevated glyoxalase I? *Exp. Neurol.* **2012**, *234*, 62–69. [CrossRef] [PubMed]
33. Chartoumpekis, D.V.; Ziors, P.G.; Psyrogiannis, A.I.; Papavassiliou, A.G.; Kyriazopoulou, V.E.; Sykiotis, G.P.; Habeos, I.G. Nrf2 represses FGF21 during long-term high-fat diet-induced obesity in mice. *Diabetes* **2011**, *60*, 2465–2473. [CrossRef]
34. Yu, Z.W.; Li, D.; Ling, W.H.; Jin, T.R. Role of nuclear factor (erythroid-derived 2)-like 2 in metabolic homeostasis and insulin action: A novel opportunity for diabetes treatment? *World J. Diabetes* **2012**, *3*, 19–28. [CrossRef]
35. Pi, J.; Zhang, Q.; Fu, J.; Woods, C.G.; Hou, Y.; Corkey, B.E.; Collins, S.; Andersen, M.E. ROS signaling oxidative stress and Nrf2 in pancreatic beta-cell function. *Toxicol. Appl. Pharm.* **2010**, *244*, 77–83. [CrossRef]
36. Wang, X.; Gu, C.; He, W.; Ye, X.; Chen, H.; Zhang, X.; Hai, C. Glucose oxidase induces insulin resistance via influencing multiple targets in vitro and *in vivo*: The central role of oxidative stress. *Biochimie* **2012**, *94*, 1705–1717. [CrossRef]
37. Nain, P.; Saini, V.; Sharma, S.; Nain, J. Antidiabetic and antioxidant potential of *Emblica officinalis Gaertn.* leaves extract in streptozotocin-induced type-2 diabetes mellitus (T2DM) rats. *J. Ethnopharmacol.* **2012**, *142*, 65–71. [CrossRef]
38. Duh, P.D.; Lin, S.L.; Wu, S.C. Hepatoprotection of *Graptopetalum paraguayense* E. Walther on CCl4-induced liver damage and inflammation. *J. Ethnopharmacol.* **2011**, *134*, 379–385. [CrossRef]
39. Hirosumi, J.; Tuncman, G.; Chang, L.; Görgün, C.Z.; Uysal, K.T.; Maeda, K.; Hotamisligil, G.S. A central role for JNK in obesity and insulin resistance. *Nature* **2002**, *420*, 333–336. [CrossRef]
40. Chang, W.C.; Shen, S.C. Effect of water extracts from edible Myrtaceae plants on uptake of 2-(N-(7-nitrobenz-2-oxa-1,3-diazol-4-yl)amino)-2-deoxyglucose (2-NBDG) in tumor necrosis factor-α-treated FL83B mouse hepatocytes. *Phytother. Res.* **2013**, *27*, 236–243. [CrossRef]

plants

Article

Comparative Phytochemical Profile and Biological Activity of Four Major Medicinal Halophytes from Qassim Flora

Hamdoon A. Mohammed [1,2,*], Hussein M. Ali [3,4], Kamal A. Qureshi [5], Mansour Alsharidah [6], Yasser I. Kandil [7], Rana Said [8], Salman A. A. Mohammed [3], Mohsen S. Al-Omar [9], Osamah Al Rugaie [10], Ahmed A. H. Abdellatif [11,12], Essam Abd-Elmoniem [13], Manal M. Abbas [14], Khalid M. Mohany [15] and Riaz A. Khan [1,*]

1 Department of Medicinal Chemistry and Pharmacognosy, College of Pharmacy, Qassim University, Buraydah 51452, Saudi Arabia
2 Department of Pharmacognosy, Faculty of Pharmacy, Al-Azhar University, Cairo 11371, Egypt
3 Department of Pharmacology and Toxicology, College of Pharmacy, Qassim University, Buraydah 51452, Saudi Arabia; hu.ali@qu.edu.sa (H.M.A.); m.azmi@qu.edu.sa (S.A.A.M.)
4 Department of Biochemistry, Faculty of Medicine, Al-Azhar University, Assiut 71524, Egypt
5 Department of Pharmaceutics, Unaizah College of Pharmacy, Qassim University, Unaizah 51911, Saudi Arabia; ka.qurishe@qu.edu.sa
6 Department of Physiology, College of Medicine, Qassim University, Buraydah 51452, Saudi Arabia; malsharidah@qu.edu.sa
7 Biochemistry and Molecular Biology Department, Faculty of Pharmacy, Al-Azhar University, Cairo 11371, Egypt; kandil.yasser@azhar.edu.eg
8 Pharmacological and Diagnostic Research Centre, Faculty of Pharmacy, Al-Ahliyya Amman University, Amman 19328, Jordan; rana.said2@gmail.com
9 Department of Medicinal Chemistry and Pharmacognosy, Faculty of Pharmacy, JUST, Irbid 22110, Jordan; m.omar@qu.edu.sa
10 Department of Basic Medical Sciences, College of Medicine and Medical Sciences, Qassim University, Unaizah 51911, Saudi Arabia; o.alrugaie@qu.edu.sa
11 Department of Pharmaceutics, College of Pharmacy, Qassim University, Buraydah 51452, Saudi Arabia; a.abdellatif@qu.edu.sa
12 Department of Pharmaceutics and Industrial Pharmacy, Faculty of Pharmacy, Al-Azhar University, Assiut 71524, Egypt
13 Department of Plant Production and Protection, College of Agriculture and Veterinary Medicine, Qassim University, Buraydah 51452, Saudi Arabia; ruessam2@yahoo.com
14 Department of Medical Laboratory Sciences, Faculty of Allied Medical Sciences, Al-Ahliyya Amman University, Amman 19328, Jordan; m.abbas2@ammanu.edu.jo
15 Department of Medical Biochemistry and Molecular Biology, Faculty of Medicine, Assiut University, El Fateh 71515, Egypt; khalidmohany@aun.edu.eg
* Correspondence: ham.mohammed@qu.edu.sa (H.A.M.); ri.khan@qu.edu.sa (R.A.K.)

Citation: Mohammed, H.A.; Ali, H.M.; Qureshi, K.A.; Alsharidah, M.; Kandil, Y.I.; Said, R.; Mohammed, S.A.A.; Al-Omar, M.S.; Rugaie, O.A.; Abdellatif, A.A.H.; et al. Comparative Phytochemical Profile and Biological Activity of Four Major Medicinal Halophytes from Qassim Flora. *Plants* **2021**, *10*, 2208. https://doi.org/10.3390/plants10102208

Academic Editors: Juei-Tang Cheng, Ki Hyun Kim, I-Min Liu and Szu-Chuan Shen

Received: 9 September 2021
Accepted: 11 October 2021
Published: 18 October 2021

Abstract: Four halophytic plants, *Lycium shawii*, *Anabasis articulata*, *Rumex vesicarius*, and *Zilla spinosa*, growing in the central Qassim area, Saudi Arabia, were phytochemically and biologically investigated. Their hydroalcoholic extracts' UPLC-ESIQ-TOF analyses demonstrated the presence of 44 compounds of phenolic acids, flavonoids, saponins, carbohydrates, and fatty acids chemical classes. Among all the plants' extracts, *L. shawii* showed the highest quantities of total phenolics, and flavonoids contents (52.72 and 13.01 mg/gm of the gallic acid and quercetin equivalents, respectively), along with the antioxidant activity in the TAA (total antioxidant activity), FRAP (ferric reducing antioxidant power), and DPPH-SA (2,2-diphenyl-1-picryl-hydrazyl-scavenging activity) assays with 25.6, 56.68, and 19.76 mg/gm, respectively, as Trolox equivalents. The hydroalcoholic extract of the *L. shawii* also demonstrated the best chelating activity at 21.84 mg/gm EDTA equivalents. Among all the four halophytes, the hydroalcoholic extract of *L. shawii* exhibited the highest antiproliferative activity against MCF7 and K562 cell lines with IC_{50} values at 194.5 μg/mL and 464.9 μg/mL, respectively. The hydroalcoholic extract of *A. articulata* demonstrated better cytotoxic activity amongst all the tested plants' extracts against the human pancreatic cancer cell lines (PANC1) with an IC_{50} value of 998.5 μg/mL. The *L. shawii* induced apoptosis in the MCF7 cell lines, and the percentage of the necrotic cells changed to 28.1% and 36.5% for the IC_{50} and double-IC_{50} values at 22.9% compared with the untreated groups. The hydroalcoholic extract of *L. shawii* showed substantial antibacterial activity against *Bacillus cereus* ATCC 10876 with a MIC value of 12.5 mg/mL. By contrast, the *A. articulata* and *Z. spinosa* exhibited antifungal activities against *Aspergillus niger* ATCC 6275 with MIC values at 12.5 and 50 mg/mL, respectively. These findings suggested that the *L. shawii* is a potential halophyte with remarkable biological properties, attributed to its contents of phenolics and flavonoid classes of compounds in its extract.

Keywords: *Lycium shawii; Anabasis articulata; Rumex vesicarius; Zilla spinosa;* anticancer; antimicrobial; antioxidant; biogenetic interrelationship; flavonoid contents; trace elements

1. Introduction

Secondary metabolite-derived compounds from plants serve the basic needs of humans and animals as medicaments [1–3]. Natural products have global acceptability and use due to their diversity, ease of access, sustainability, procurements, efficacy, safety, and widespread occurrence [4]. Nevertheless, some plants have been identified as toxic botanicals, including digitalis, belladonna, and ephedra, etc. [5]. However, these plants are widely used for specific purposes by people to treat certain critical disorders [6–8]. Since ancient times, the necessity to explore plants' activity against various diseases has remained continued, and the approaches to the scientific confirmations of phytochemicals' biological activity are a well-established tactic for new drug discovery, and drug development in modern times [6,9,10].

The environmental factors, and habitat-related effects, associated with high salinity, desert climate, and water and nutrient scarcity have emphasized the importance of halophytic plants in the fields of drug discovery, and alternative medicines [11,12]. These harsh environmental conditions compel plants to maintain higher levels of compounds with defensive roles which are produced as part of their survival mechanism against excessive oxidative stress, bacterial infection, and animal grazing encroachments [13]. Secondary metabolites, such as phenolics, flavonoids, alkaloids, and saponins comparatively higher presence make these plants more intriguing for chemical and biological evaluations [14–16]. Additionally, the presence of these compounds also confer significant nutritional and health-promoting benefits, including the medicinal properties in these plants [14,15].

Generally, the central region of Saudi Arabia has a high-salinity ecosystem, which affects plants' growth, and is a significant challenge behind the slowed development of agriculture in the area [17,18]. The potential of some of these plants of halophytic nature, such as *Lycium shawii, Anabasis articulata, Zilla spinosa*, and *Rumex vesicarius*, to flourish in this harsh habitat demonstrates their species' capacity to adapt to and survive in unfavorable environmental conditions and makes these plants noticeable for their presence and traditional medicinal uses, including their uses in livestock feeds, and human nutrition. The environmental adaptation of these plants is based on the presence of certain enzymes, and their productions of enzymatic and non-enzymatic compounds to protect these plants against intercellular oxidative stress caused by dryness of the habitat's atmosphere, water scarcity in the soil, and soil salinity [14,19]. These plants also produce phenolics and flavonoid compounds to neutralize the reactive oxygen species (ROS) as part of their antioxidant defense systems [13,20]. Locally, these plants are also used as livestock feed [21,22] and are employed to treat various medical conditions of the ailing population. *Lycium shawii* has a rich history of use in central Saudi Arabia for infection control [23], and for allergy treatments in Wadi Hagul, Egypt [21]. *Rumex vesicarius* is used as a diuretic and for treatments of GIT (Gastro Intestinal Tract) disorders, including dysentery and dyspepsia [23]. The plant is also part of the human diet [24]. In Algeria, the aerial parts of the *Anabasis articulata* are decocted and used as a remedy for diabetes [25]. The entire plant has been used to treat hypertension [26]. *Zilla spinosa* has a long history of treating urinary and gallbladder stones [27] in this area of the region and has a purgative action [26].

The four plants, i.e., *Lycium shawii*, *Anabasis articulata*, *Zilla spinosa*, and *Rumex vesicarius*, were selected based on defined criteria [28], specifically including their high distribution in the Qassim region, similar halophytic character, soil type, the environmental similarity of growing conditions, and their common folklore uses in different ailments. The current study investigated the phytochemical contents, anticancer, antimicrobial, and antioxidant activities of the hydroalcoholic extracts of these four major halophytic plants growing in central Saudi Arabia. The study aimed to provide a comparative chemical contents status, and antioxidant potentials of the plants' extracts, together with their different biological activity levels' as examined in the in vitro conditions.

2. Results and Discussion

2.1. Trace Elements Analysis

Trace elements have various functions in maintaining the general health status of living organisms. Consequently, any disturbances in the trace elements' presence levels, by either deficiency or their increments up to the toxic levels due to systemic accumulation, forms the core of the lead causes for several pathological disorders in the biosystems. The trace elements deficiency may result from either a decrease in specific element's intake, or an increase in the levels of these elements due to any biochemical, physiological, or environmental causes, whereby both may lead to impairment, regression, and higher activity of the biochemical pathways to increase the risks for diseases generation over the time [29,30].

The edible halophytes are a good natural source of trace elements for the local population [12,15,31]. The high contents of trace elements in these halophytic plants are also considered responsible for supplementing their antioxidants and supporting several other biological actions [32]. Magnesium (Mg) is one of the important trace elements which acts as a cofactor for about ~300 enzymes, including the regulation of blood glucose, protein synthesis, regulation of blood pressure, and functions of both nerves, and the muscles [33]. A previous study also demonstrated that magnesium decreases the risk of ischemic heart disease, caused by the reduced blood supply to cardiac muscles [34]. Among all the four plants, magnesium was measured at higher levels in *Anabasis articulata* and *Rumex vesicarius* (1272 $\pm$ 18.52 µg/kg and 1250 $\pm$ 20.0 µg/kg); however, its contents in the *Lycium shawii* (1163 $\pm$ 9.17 µg/kg), and *Zilla spinosa* (1191 $\pm$ 10.14 µg/kg) (Table 1) were measured in parallel to other halophytes found in this region. Another abundant element, Manganese (Mn), has important functions in the activating and synthesizing many enzymes, e.g., isomerases, ligases, transferases, hydrolases, pyruvate decarboxylase, arginase, glutamine synthetase, lysates, and oxidoreductases. It also has a role in regulating blood glucose levels, improving immunity, and mineralizing the bones [35]. Moreover, manganese also supports intracellular metabolic energy production, and protects the cells from free radicals led damages [36]. Among these four halophytes, the manganese was detected at the highest concentration in *Rumex vesicarius* at 96.03 $\pm$ 1.04 µg/kg levels, and the lowest concentration was found in *Lycium shawii* at 28.53 $\pm$ 0.42 µg/kg occurrence, as detected in their respective plants' dry powders. Iron (Fe) is part of the structures of multiple proteins, including mitochondrial cytochrome enzymes responsible for energy production, several cellular functions, and also for cell differentiation [37]. In addition, iron is also a component of myoglobin and hemoglobin and functions as a carrier of oxygen [38]. *Anabasis articulata* contained the highest levels of iron concentrations at 243.33 $\pm$ 2.08 µg/kg, while the *Zilla spinosa* showed the lowest levels of iron at 93.47 $\pm$ 1.16 µg/kg concentrations from the respective plants' dry powders. Copper (Cu) represents one of the fundamental trace elements, also part of the structures of several enzymes, i.e., tyrosinase, cytochrome oxidase [39], and the antioxidant superoxide dismutase [40]. In addition, it also forms a component of copper-containing ceruloplasmin protein that is associated with red blood cells formation, and its deficiency leads to the progression of anemia [41]. All the studied halophytic plants in the current work contained copper at concentrations ranging from 10.34 $\pm$ 2.11 µg/kg to 13.67 $\pm$ 0.50 µg/kg of the plants' dry powders. *Anabasis articulata* showed the highest levels of copper at 13.67 ± 0.50 µg/kg, while *Lycium shawii* had the lowest concentration levels of copper at 10.34 ± 2.11 µg/kg of the plants' powder. Zinc (Zn), associated with several molecular, cellular, metabolic, and immunological functions, including antioxidant, anti-inflammatory, and anti-apoptotic responses [42], was also detected in all four halophytes. The element is essential for normal spermatogenesis, taste sensation, and gastric enzymes secretions [43]. A previous study reported that zinc is related to insulin secretion, and it increases insulin sensitivity for the tissue, and provides an improved glucose utilization for the diabetic conditions [44]. Cobalt (Co), another element present in these plants, represents a key constituent of vitamin B12 [45], and is part of the structure of methyl malonyl-CoA-mutase enzyme, which has a role in amino acid [46] and purine, as well as, pyrimidine metabolisms [47].

Table 1. Trace elements contents of the four plants (µg/kg).

Elements	*Lycium shawii*	*Anabasis articulata*	*Rumex vesicarius*	*Zilla spinosa*
Fe	12.33 ± 4.16	24.33 ± 2.08	16.67 ± 2.31	93.47 ± 1.16
Cu	10.34 ± 2.11	13.67 ± 0.50	10.96 ± 0.93	10.44 ± 0.95
Mn	28.53 ± 0.42	39.1 ± 0.85	96.03 ± 1.04	34.8 ± 0.36
Co	15.67 ± 4.07	14.07 ± 1.31	18.0 ± 3.55	12.23 ± 2.07
Zn	12.93 ± 1.34	44.63 ± 0.40	109 ± 0.00	10.52 ± 1.10
Mg	1163 ± 9.17	1272 ± 18.52	1250 ± 20.0	1191 ± 10.14

Moreover, cobalt also helps to normalize blood glucose levels, and increases adiponectin secretion, consequently improving the conditions of diabetes mellitus, obesity, hypertension, and cardiovascular diseases, and thus helps to treat them [48,49]. Zinc and cobalt contents of the plants play important roles in the biological activity and enhancement of the general health of humans and animals. Zinc and cobalt were detected in all the four plants with the highest levels detected in the *Rumex vesicarius* at 109 ± 0.0 µg/kg and 18.0 ± 3.55 µg/kg concentrations, respectively, while the *Zilla spinosa* contained the lowest concentrations of these two elements at 10.52 ± 1.10 µg/kg and 12.23 ± 2.07 µg/kg levels of the dried plants' powders, respectively (Table 1). The overall mineral contents in these four plants are supportive indications of the traditional uses of these plants as part of food, livestock feeds, nutritional supplements, and medicinal herbs. The higher levels of the trace elements in these plants is also an indication of the presence of these elements in higher concentrations in the Qassim soil, and the plants' adaptation and intake of these elements for various purposes of the plants' physiology, biomechanism, metabolism, defense, and environmental factors.

2.2. LC-MS Profiles of the Plants

The LC-MS analyses of the plants' hydroalcoholic extracts' results are summarized in Table 2 (for details see the Supplementary File).

Table 2. LC-MS result analyses of the four plants.

Sr	RT (min)	Observed Mass (m/z)	Calcd. Mass (m/z)	Ion	Molecular Formula	Identity *	Relative % of the Identified Compounds **			
							R. vesicarius	*L. shawii*	*A. articulata*	*Z. spinosa*
1	0.54	181.0721	182.0794	[M-H]-	Sorbitol	$C_6H_{14}O_6$	0.0859	0.0358	0.0008	0.0013
2	0.55	341.1070	342.1138	[M-H]-	Hexose-based disaccharide	$C_{12}H_{22}O_{11}$	2.7116	4.0265	0.1345	0.2199
3	0.55	683.2218	684.2291	[M-H]-	Galabiose	$C_{24}H_{44}O_{22}$	0.0262	0.0818		
4	1.00	202.1069	203.1142	[M-H]-	L-Acetyl carnitine	$C_9H_{17}NO_4$	0.0260	0.0053	0.0003	0.0006
5	2.21	165.0533	166.0606	[M-H]-	3-Phenyl lactic acid	$C_9H_{10}O_3$		0.1023		0.0013
6	2.95	353.0859	354.0931	[M-H]-	Chlorogenic acid	$C_{16}H_{18}O_9$	0.4393	0.0056		
7	4.72	387.1985	388.2058	[M-H]-	5-Methoxy-7,8-diprenylflavone	$C_{26}H_{28}O_3$		0.0005	0.0007	0.0004
8	4.76	471.1875	472.1947	[M-H]-	Eugenol rutinoside	$C_{22}H_{32}O_{11}$		0.0010	0.0014	0.0006

Table 2. *Cont.*

Sr	RT (min)	Observed Mass (m/z)	Calcd. Mass (m/z)	Ion	Molecular Formula	Identity *	Relative % of the Identified Compounds **			
							R. vesicarius	*L. shawii*	*A. articulata*	*Z. spinosa*
9	4.89	463.0886	464.0959	[M-H]-	Spiraeoside	$C_{21}H_{20}O_{12}$	0.0005			
10	4.95	447.0890	448.0963	[M-H]-	Orientin	$C_{21}H_{20}O_{11}$		0.0021	0.0003	0.0002
11	5.54	593.1488	594.1560	[M-H]-	Quercetin-3,7-dirhamnosyl	$C_{27}H_{30}O_{15}$	0.0311	0.0118	0.0006	0.0011
12	5.60	609.1427	610.1500	[M-H]-	3-Gluco-7-rhamnosyl quercetin	$C_{27}H_{30}O_{16}$	0.0050	1.3642	0.0069	0.0346
13	5.73	463.0861	464.0934	[M-H]-	Hyperoside	$C_{21}H_{20}O_{12}$	0.2741	0.0087	0.0003	
14	5.92	447.0906	448.0978	[M-H]-	Luteolin 7-O-glucoside	$C_{21}H_{20}O_{11}$	0.0054	0.1352		
15	6.32	577.1539	578.1625	[M-H]-	Isorhoifolin	$C_{27}H_{30}O_{14}$	0.0406	0.0123		
16	6.36	593.1472	594.1545	[M-H]-	Kaempferol 3-neohesperidosid	$C_{27}H_{30}O_{15}$	0.0008	1.5347	0.0328	0.2967
17	6.57	447.0918	448.0991	[M-H]-	Kaempferol-3-O-glucoside	$C_{21}H_{20}O_{11}$	0.2531	0.0081		
18	6.81	285.0402	286.0475	[M-H]-	3,6,2′,4′-Tetrahydroxyflavone	$C_{15}H_{10}O_{6}$		0.0053		
19	6.81	447.0935	448.1008	[M-H]-	Luteolin-4′-O-glucoside	$C_{21}H_{20}O_{11}$	0.0010	0.0211		
20	6.82	431.0962	432.1035	[M-H]-	Vitexin	$C_{21}H_{20}O_{10}$	0.0031	0.0351	0.0013	
21	6.90	431.0985	432.1058	[M-H]-	Isovitexin	$C_{21}H_{20}O_{10}$	0.0006	0.0543	0.0002	
22	7.76	829.2216	830.2289	[M-H]-	Alatanin	$C_{39}H_{42}O_{20}$	0.0052	0.0007		
23	7.77	289.1097	290.1169	[M-H]-	5-O-Methylvisamminol	$C_{16}H_{18}O_{5}$		0.0307	0.8107	0.0879
24	8.06	312.1217	313.1290	[M-H]-	Acetylcaranine	$C_{18}H_{19}NO_{4}$	0.0054	0.0011	0.0258	0.0033
25	8.59	301.0335	302.0407	[M-H]-	Quercetin	$C_{15}H_{10}O_{7}$	0.9066	0.0009		
26	9.00	431.0985	432.1058	[M-H]-	Apigenin-7-O-glucoside	$C_{21}H_{20}O_{10}$	0.0069	0.0003		
27	9.17	809.4290	810.4363	[M-H]-	Azukisaponin III	$C_{42}H_{66}O_{15}$		0.4169		0.0582
28	9.24	315.0506	316.0579	[M-H]-	6-Methoxy luteolin	$C_{16}H_{12}O_{7}$	0.5725	0.0041		
29	9.99	582.2586	583.2659	[M-H]-	Tricoumaroyl spermidine	$C_{34}H_{37}N_{3}O_{6}$	0.0116	0.3624	0.0034	0.0214
30	10.35	315.0508	316.0580	[M-H]-	Rhamnetin	$C_{16}H_{12}O_{7}$	1.2297	0.0015		
31	11.70	329.2309	330.2382	[M-H]-	9,10,11-Trihydroxy-(12Z)-12- octadecenoic acid	$C_{18}H_{34}O_{5}$	0.2259	0.0448	0.6680	0.3237
32	14.24	247.1339	248.1412	[M-H]-	3-Hydroxy-14-calamenoic acid	$C_{15}H_{20}O_{3}$	5.1008	0.0380	0.0034	0.005
33	14.25	293.1737	294.1810	[M-H]-	Gingerol	$C_{17}H_{26}O_{4}$	0.2933	0.3180	0.2920	0.3002
34	20.49	293.2101	294.2174	[M-H]-	Hydroxyoctadectrienoic acid	$C_{18}H_{30}O_{3}$	0.0406	0.0096	0.0258	0.0806
35	25.68	253.2149	254.2222	[M-H]-	9-Hexadecenoic acid	$C_{16}H_{30}O_{2}$	0.0819	0.1653	0.1796	0.0264
36	26.16	279.2304	280.23773	[M-H]-	Linoleic acid	$C_{18}H_{32}O_{2}$	0.2374	0.4492	0.7048	2.0156
37	27.42	621.4395	622.4468	[M-H]-	Ginsenoside	$C_{36}H_{62}O_{8}$	2.4514	0.2084		
38	27.91	255.2308	256.2381	[M-H]-	Palmitic acid	$C_{16}H_{32}O_{2}$	1.5057	2.0933	2.8938	4.1485
39	28.23	281.2464	282.2536	[M-H]-	Oleic acid	$C_{18}H_{34}O_{2}$	1.2639	1.5673	2.0126	3.0127
40	29.63	311.2242	312.2315	[M-H]-	Octadecenedioic acid	$C_{18}H_{32}O_{4}$	2.6267	0.0293	0.0002	0.0011
41	29.72	575.4705	576.4778	[M-H]-	*cis*-Epoxy octadecenoate	$C_{36}H_{64}O_{5}$	3.1036	0.0091		
42	29.96	283.2620	284.2693	[M-H]-	Stearic acid	$C_{18}H_{36}O_{2}$	6.0177	8.1601	10.6431	11.1964
43	30.16	409.3085	410.3157	[M-H]-	γ-Tocotrienol	$C_{28}H_{42}O_{2}$	0.4789	0.6317	0.7026	0.5591
44	30.30	423.4227	424.4299	[M-H]-	Octacosanoic acid	$C_{28}H_{56}O_{2}$	0.0020		0.0003	0.0126
Total relative percentages of the identified compounds							30.07%	21.99%	19.15%	22.41%

** Compounds were tentatively identified; * relative % (percentages) of the occurrence levels of the identified compounds were calculated in comparison to the area of all the peaks in the LC-chromatogram for each plant.

The tentatively identified compounds were arranged in an ascending elution order of the chromatographic analysis with the specific retention time of each compound. The compounds' structures were tentatively assigned based on the mass spectral pattern, fragment ions peaks, and their abundances in the corresponding MS spectrum. The relative percentage of the identified compounds' presence was calculated by considering their peak area concerning the area of all the peaks of the chromatogram. Forty-four compounds were identified, representing the primary and secondary plant metabolites, e.g., carbohydrates as sorbitol, galabiose, and hexose-based disaccharide. The presence of secondary metabolites, e.g., flavonoids, saponins, and phenolic acids were also confirmed. The LC-MS analyses results (Table 2) showed certain levels of similarities between the four halophytic plants' constituents, which were detected in different proportions. For instance, the compound at a retention time of 5.6 min with the corresponding molecular weight (MW) at m/z 609.1427 [M-H] was tentatively identified as quercetin-3-glucosyl-7-rhamnosyl, which was detected in all the four plants in different concentrations but at the same retention time (Figure 1 and Table 2). Some fatty acids, i.e., oleic, linoleic, palmitic, and stearic acids, were also commonly present in all these plants, together with other long-chain saturated and unsaturated fatty acids. Only 27 of the 44 compounds identified in the current analyses were detected in *A. articulata* and *Z. spinosa*. However, the majority of the identified compounds were found in the *R. vesicarius* and *L. shawii* extracts, which were at 38 and 43 compounds, respectively, as found populated in the extracts through the LC-MS analysis.

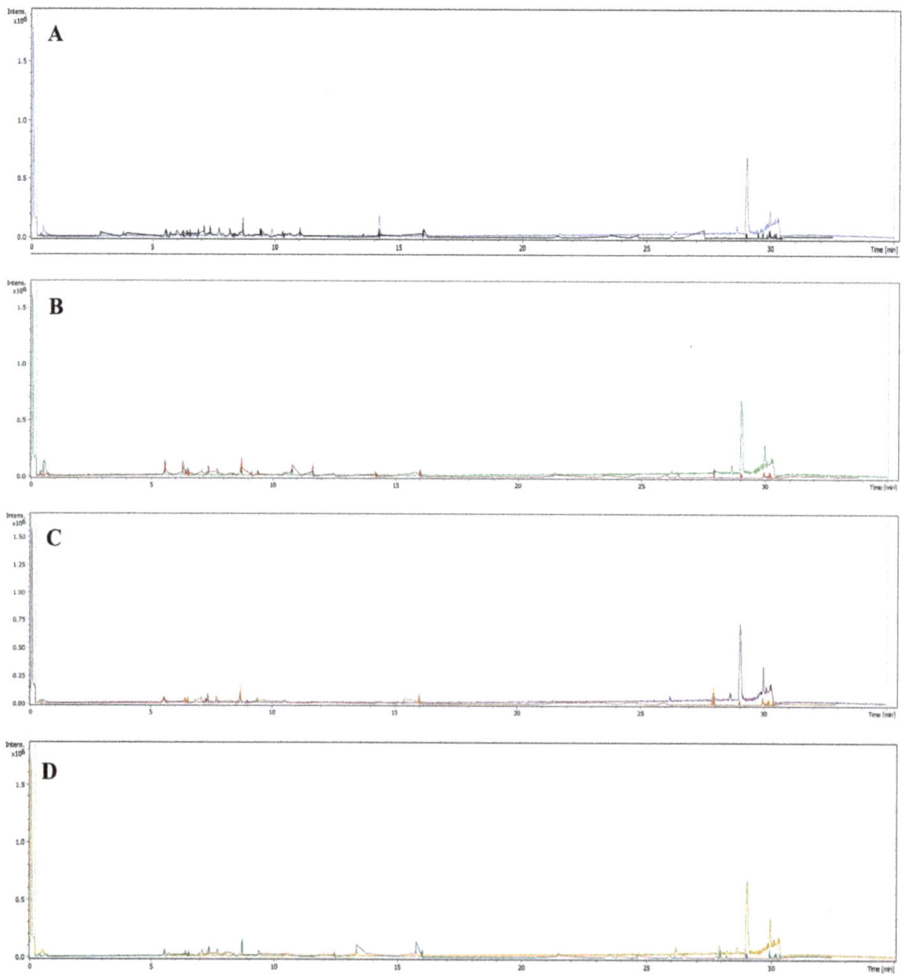

Figure 1. LC-MS chromatograms of the four halophytic plants; (**A**) *R. vesicarius*, (**B**) *L. shawii*, (**C**) *A. articulata*, and (**D**) *Z. spinosa*.

Chromatographic analyses also exhibited that the total percentages of the identified compounds from each plant were 30.07 %, 21.99 %, 19.15 %, and 22.41 % for *R. vesicarius*, *L. shawii*, *A. articulata*, and *Z. spinosa*, respectively, based on the compounds' discernible peaks as available in their respective LC-chromatograms. The presence of these compounds in the plants also revealed their health benefits as saccharides and fatty acids, together with the phenolics and flavonoids classes of compounds, which were present in abundance in these plants' extracts. The presence of the saturated fatty acids, e.g., stearic acid, at higher proportions in the extracts of all the plants (from 6% to 11% of the identified compounds), is noticeable from the health-benefit view-point that makes the part of the human diet, probably also as the thrombogenic and atherogenic risk factors improvement entity, as reported earlier that the intake of 19 g of the stearic acid in the diet is effective [50]. The four plants also contained variable amounts of other dietary fatty acid constituents, i.e., oleic, palmitic, and linoleic acids, which are dominant in the halophytes [51,52] and are known to play roles with their nutritive values. Moreover, the highest percentages of the fatty acids were found in the *Z. spinosa* in comparison to other plants under the current study. In addition, sorbitol, hexose-based disaccharide, and L-acetyl carnitine were also identified in the extracts of all four plants at variable concentrations. The nutritional and medicinal values of these plants were also demonstrated by the presence of γ-tocotrienol, and gingerol in their extracts (Table 2).

Secondary metabolites, flavonoids, and phenolics are also reported for their health benefits [53,54] and are known to possess pharmacological activities of different kinds, including antimicrobial [55,56], anticancer [55,57,58], and several other [59–61]. The results in Table 2 showed noticeable variations in the phenolics and flavonoids distributions among these investigated four plants. Among the flavonoids and phenolics compounds identified in the plants under current investigation, hyperoside, quercetin, apigenin-7-O-glucoside, rhamnetin, and chlorogenic acid [62–65] have been reported for their antimicrobial activity, whereas spiraeoside, orientin, luteolin-7-O-glucoside, vitexin, isovitexin, quercetin, and apigenin-7-O-glucoside [62–64,66–69] have been reported for their anticancer activities. The results in Table 2 also demonstrated the presence of three flavonoids, i.e., quercetin-3,7-dirhamnosyl, quercetin-3-glucosyl-7-rhamnosyl, and 3-O-neohesperidoside kaempferol, which were identified in all the four plants; however, their representative occurrence (percentages) in these plants varied. The highest number of identified flavonoids were represented in the *L. shawii* extract (18 compounds representing 3.20% of the total peaks area of the plant extract's LC-MS chromatogram) followed by the *R. vesicarius* extract (16 compounds representing 3.34% of the total peaks area of the plant's extract's LC-MS chromatogram). The lowest numbers, and percentages of the flavonoids were identified in *A. articulata* and *Z. spinosa*, which showed eight (0.0431% of the area of all the peaks of the plant's extract's LC-MS chromatogram) and five (0.33% of the total area of all the peaks in the plant's extract LC-MS chromatogram) of the identified flavonoids, respectively. The structures of the flavonoid contents, varied in C_6-C_3-C_6 flavonoid basic skeletal substitutions of hydroxyl, methylation, prenylation, and glycosylations of mono- and di-glycosidic nature, were encountered; their structures are provided in Figure 2.

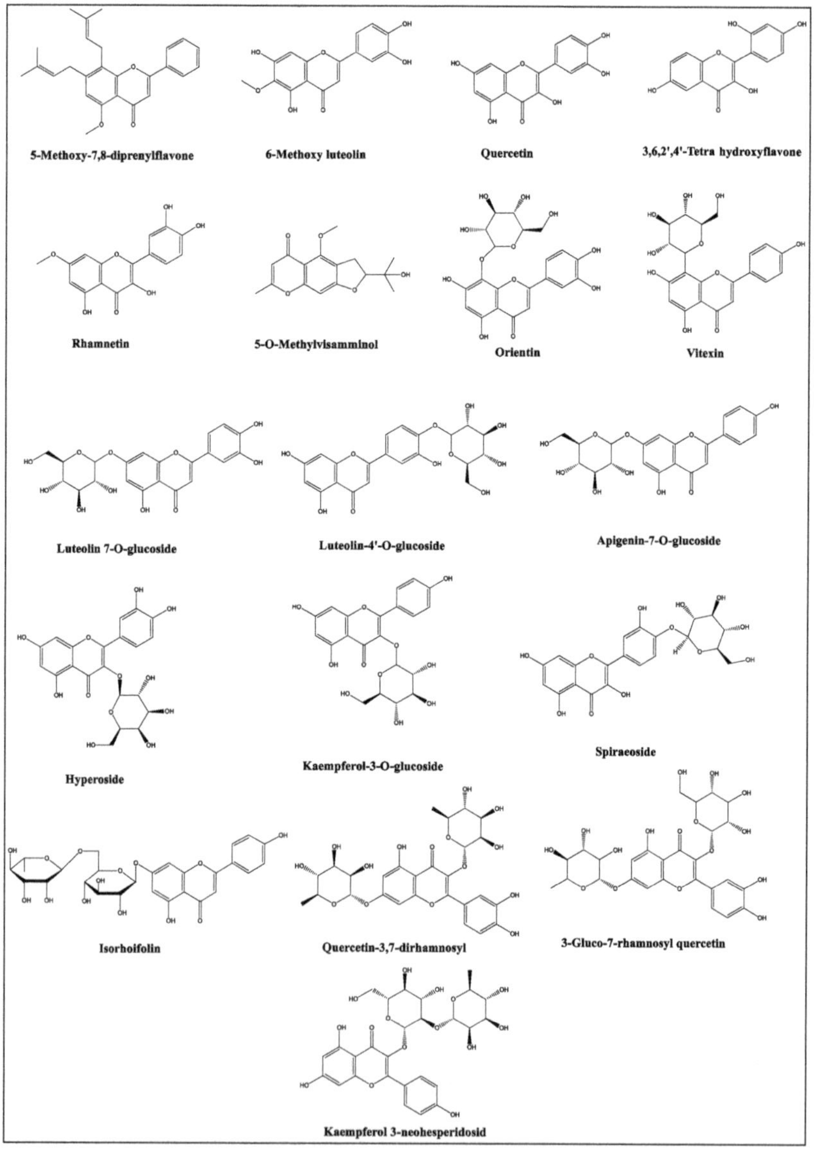

Figure 2. Structures of the flavonoid contents of the four halophytic plants.

The reported biological activities of these plants and their common use in traditional medicine are mostly attributed, but maybe not limited, to their contents of flavonoids and phenolics compounds. Furthermore, the variations in the phenolics and flavonoids constituents' representations, and concentrations could be among the major reasons for variations observed in their antioxidant, antimicrobial, and anticancer activities as encountered in these plants in the current study (Tables 3–6). The presence of major phenolics and flavonoid compounds also indicated the roles of the antioxidant compounds in these four plants in relation to exhibiting the combined antioxidant potential of each plant, and their treatment efficacy vis-a-vis their uses in containing, and curing of various disease, especially, of the oxidative stress origins. The complex flavonoids structures of mono- and di-glycosidal patterns have lent credence to the known roles and receptor interactions of flavonoids and their glycosides in various molecular modeling based studies [70]. This approach also established the attempt of comparison study of these four halophytes in relation to their constituents, and their biological activities.

Table 3. Quantitative analysis of the total phenolics, flavonoids, and antioxidant activity in mg/gm of the dried plants' extracts.

Quantitative Tests Plants	TPC	TFC	TAA	FRAP	DPPH-SA	MCA
Lycium shawii	52.72 ± 3.17	13.01 ± 0.63	25.60 ± 4.61	56.68 ± 0.62	19.76 ± 0.04	21.84 ± 0.22
Anabasis articulata	21.13 ± 0.32	11.48 ± 1.52	12.43 ± 0.46	19.67 ± 0.40	7.15 ± 0.46	11.89 ± 0.31
Zilla spinosa	22.36 ± 0.67	7.29 ± 0.26	14.36 ± 0.38	23.68 ± 0.93	7.22 ± 0.13	13.32 ± 0.58
Rumex vesicarius	28.54 ± 1.13	12.64 ± 0.28	10.79 ± 0.46	33.09 ± 2.10	14.22 ± 0.29	13.01 ± 0.09

All the measurements were conducted in triplicate; mean and standard deviations were calculated. TPC, total phenolic contents calculated in mg/gm gallic acid equivalent; TFC, total flavonoid contents calculated in mg/gm quercetin equivalent; TAA, total antioxidant activity in mg Trolox equivalents per gm (gram) of the extract; FRAP, ferric reducing antioxidant power in mg Trolox equivalent per gm of the dry extract; DPPH-SA, 2,2-diphenyl-1-picrylhydrazyl-scavenging activity in mg Trolox equivalent per gm of the dry extract; MCA, metal chelating activity in mg; and EDTA, equivalents per gm of the extract.

Table 4. Antiproliferative IC_{50} values for the plants' hydroalcoholic extracts.

Cell Lines	*Lycium shawii* (µg/mL)	*Anabasis articulata* (µg/mL)	*Rumex vesicarius* (µg/mL)	*Zilla spinosa* (µg/mL)
MCF-7, IC_{50}	194.5	2030	1759	1077
(95% CI)	(153.2 to 246.9)	(944.7 to 4363)	(661.2 to 4680)	(661.7 to 1754)
K562, IC_{50}	464.9	2729	1319	736.9
(95% CI)	(326.0 to 662.9)	(1170 to 6366)	(766.2 to 2270)	(475.6 to 1142)
PANC-1, IC_{50}	2619	998.5	Not converged	Not converged
(95% CI)	(1246 to 5506)	(740.9 to 1346)		
Fibroblast, IC_{50}	3109	3659	3139	1888
(95% CI)	(1225 to 7894)	(1201 to 11,151)	(1408 to 6999)	(1105 to 3225)

Table 5. Annexin V-staining FACS analysis of the hydroalcoholic extract of *L. shawii*.

	(A) Media Untreated	(B) Half IC_{50}	(C) IC_{50}	(D) Double IC_{50}
Viable Q3	64%	64.3%	36.2%	9.5%
Early apoptosis Q4	0.2%	1.2%	2.3%	0.4%
Late apoptosis Q2	12.8%	13.9%	33.4%	53.6%
Necrosis Q1	22.9%	20.6%	28.1%	36.5%

Table 6. Results of MIC and MBC of *Lycium shawii* (L), *Anabasis articulata* (A), and *Zilla spinosa* (Z) hydroalcoholic plant extracts and positive controls (PC).

Hydroalcoholic Plant Extracts	*B. cereus* ATCC 10876		*A. niger* ATCC 6275	
	MIC (mg/mL)	MBC (mg/mL)	MIC (mg/mL)	MBC (mg/mL)
Lycium shawii	12.5	25	-	-
Anabasis articulata	-	-	12.5	25
Rumex vesicarius	-	-	50	100

2.3. Total Phenolic and Flavonoid Contents

The quantitative analysis of the total phenolics and flavonoids were also conducted since the percentages of the identified constituents obtained by LC-MS analysis only showed 30.07%, 21.99%, 19.15%, and 22.41% as the identified constituents in these plants, which meant that the percentages of the unidentified constituents by the LC-MS technique were at 69.92%, 78.01%, 80.85%, and 77.59% of the plants' constituents from these four plants, i.e., *R. vesicarius*, *L. shawii*, *A. articulata*, and *Z. spinosa*, respectively (Table 2). Although, all the four plants under investigation are growing in a similar environment and locality with similar salinity levels and in the marshy area, the results in Table 3 for phenolics and flavonoid quantities revealed distinct variations among these plants' species.

For instance, the phenolics contents in *A. articulata*, and *Z. spinosa* were at the lowest levels of 21.13 mg/gm and 22.36 mg/gm GAE (Gallic Acid Equivalent) as compared to the *R. vesicarius* contents at 28.54 mg/gm GAE, respectively, of the dried extracts of the respective plants. In addition, nearly two folds of the *R. vesicarius*, and more than two-folds of the *A. articulata* and *Z. spinosa* phenolics contents levels, the phenolics contents were measured in the *L. shawii* extract which was found at 52.72 mg/gm of the GAE of the dried plants' extracts. Despite these differences in the phenolics contents among these species, the assays revealed that the flavonoids quantity in three out of these four plant species, i.e., *L. shawii, A. articulata*, and *R. vesicarius*, were at higher levels of 11.48 to 13.01 mg/gm QE (Quercetin Equivalent) of the dry extracts of the respective plants. By contrast, *Z. spinosa* extract accounted for the lowest flavonoids contents at 7.29 mg/gm QE of the plant's dried extract. The results obtained for the phenolics and flavonoids contents analysis in these species were consistent with the LC-MS chromatographic profiling of the respective plants, and they exhibited similar patterns of the presence of these constituents which validated the current findings. For instance, the highest numbers of the identified phenolics and flavonoids by the LC-MS were recorded for the *L. shawii* extract (18 flavonoids representing 3.20 % of the total peaks' area of the chromatogram), and *R. vesicarius* (16 compounds representing 3.34 % of the area of all the peaks of the chromatogram). The lowest identified phenolics and flavonoid contents were observed in the *A. articulata*, and *Z. spinosa* (8 compounds, 0.0431%, and 5 compounds, 0.33%, respectively, of the total peaks areas of their respective chromatograms).

The influence of environmental conditions on the plant constituents can be postulated by comparing the levels of phenolics and flavonoids contents in *R. vesicarius* against the similar species of the plant growing in Algeria that contained 43.28 and 19.72 mg/gm catechin equivalents of phenolics and flavonoids contents, respectively [71]. However, for the plant species growing under similar conditions, the total phenolics and flavonoid contents in *L. shawii* revealed 101.70 mg/gm GAE and 59.8 mg/gm of QE of the phenolics and flavonoids, respectively [72]. The current levels of *Z. spinosa* phenolics and flavonoids contents were nearly similar to that recorded for the species growing in the southern part of Saudi Arabia [73]. The presence of phenolics and flavonoid contents in these plants also supported the use of these plants as part of foods and livestock feed, as well as their use for different biological activities where the roles of antioxidants are involved [74–76].

2.4. Antioxidant Activity

The antioxidant potentials of the plants were measured with different methods to evaluate the free radicals scavenging potentials (2,2-diphenyl-1-picrylhydrazyl-scavenging activity, DPPH-SA), metal chelation potentials (metal chelating activity, MCA), and reducing-power of the plants' extracts (total antioxidant power, TAP, and ferric reducing antioxidant power, FRAP). The antioxidant estimations results (Table 3) confirmed the positive relationship between the antioxidant potentials of the plants and their phenolics and flavonoids contents. The *L. shawii* extract contained the highest phenolics and flavonoids contents, and the highest level of antioxidant activity was observed for this plant as compared to all the other plants investigated under the current study. The extract of *L. shawii* reduced the molybdenum (VI) to molybdenum (V) in the TAA, and ferric ions in the FRAP estimation by 26.60 mg and 56.68 mg of Trolox equivalent (TE), respectively. The next higher levels of reducing activity were recorded for *R. vesicarius*, followed by *Z. spinosa*, and *A. articulata* with FRAP activities equal to 33.09, 23.68, and 19.67 TE/gm of the plants' extracts. The TAA of *R. vesicarius* was less than the *Z. spinosa* and *A. articulata*, thereby suggesting that these extracts reduced the ferric ions more than the molybdenum (VI) ions as compared between the two methods results. The scavenging activity of *L. shawii* extract against DPPH free radicals scavenging (19.76 TE) was significantly higher than those of the other plants' extracts. However, the *R. vesicarius* DPPH-SA (14.22 TE) activity was as high as twice to that of the *Z. spinosa*, and *A. articulata* plants at 7.22 and 7.15 TE, respectively. The results displayed in Table 3 also confirmed higher iron-chelating power of the extracts of *L. shawii* of MCA at 21.84 mg EDTA-equivalents, and *R. vesicarius* and *Z. spinosa* showed the MCA at 13.01 mg and 13.32 mg of EDTA-equivalents, respectively. A lower MCA potential was recorded for *A. articulata* extract. The overall antioxidant potential of all the four plants revealed that the phenolics and flavonoid contents have an essential role in these plants' antioxidant capacity.

2.5. Cytotoxicity Analysis

The cytotoxic activity evaluations were conducted for all the plants' hydroalcoholic extracts' against the three cancer cell lines, i.e., breast cancer cell line (MCF-7), chronic myeloid leukemia cell line (K562), and the human pancreatic cancer cell line (PANC-1) (Supplementary File). The fibroblast cell lines were also cultured with the plant extracts to measure their toxicity on the normal cells. Among all the four halophytes, the hydroalcoholic extract of *L. shawii* showed the highest antiproliferative effects against MCF-7 and K562 cell lines with IC_{50} values at 194.5 μg/mL and 464.9 μg/mL, respectively. The hydroalcoholic extract of *A. articulata* demonstrated the highest cytotoxic activity against the human pancreatic cancer cell line (PANC-1) with IC_{50} values at 998.5 μg/mL (Table 4). The plant extracts' cytotoxic effects against MCF-7 and K562 cell lines were dose-dependent (Figure 3). The results in Table 4 also demonstrated the plants' safety toward the normal fibroblast cells, as the IC_{50} were measured above 3000 μg/mL for all the plants' hydroalcoholic extracts, except for the *Z spinosa* hydroalcoholic extract, which showed an IC_{50} value at 1888 μg/mL. The overall results indicated a higher selectivity of *L. shawii* extract towards the MCF-7 and K562 cell lines as compared to the PANC-1 cell lines, and the fibroblast's normal cell lines.

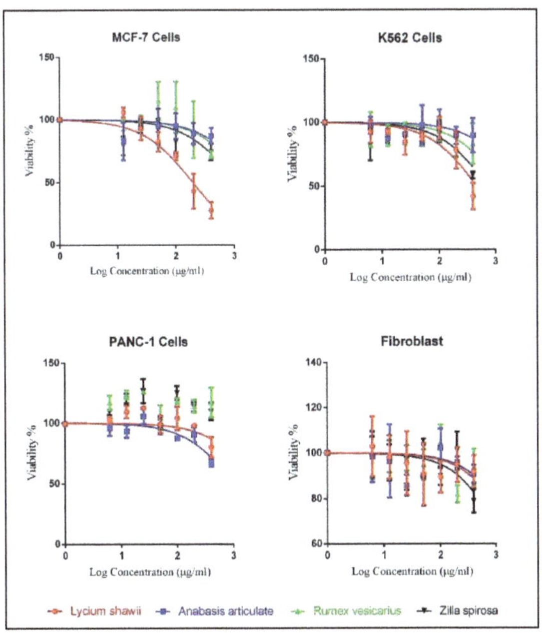

Figure 3. The log dose-response curve of cancer cell lines compared to the fibroblasts. Values are represented as mean ± SD.

The flow cytometry analysis demonstrated that the *L. shawii* could induce apoptosis in MCF-7 cell lines (Figure 4). Annexin-V conjugated-FITC represented the apoptotic cells, while the PI dyes represented membrane damage due to the necrosis and late apoptosis. The untreated cells expressing the negative group viability were at 94%, while the viable cells decreased, for both the IC_{50} and double IC_{50}, at 36.2% and 9.5% for the *L. shawii*. However, half IC_{50} did not show any significant changes. The necrotic cells percentage for *L. shawii* increased to 28.1% and 36.5% for the IC_{50} and 2 × IC_{50} as compared to the 22.9% of the untreated group (Table 5). In addition, using the concentrations equal to IC_{50} and 2 × IC_{50} for *L. shawii* extract, the late apoptotic cells viability increased to 33.4% and 53.6%, respectively. These results demonstrated that the *L. shawii* cytotoxic effects were dose-dependent, as, once the concentration was increased from IC_{50} to double IC_{50}, the percentage of the necrosis increased in a directly proportional relationship. These data also confirmed that the *L. shawii* extract could inhibit the growths of human cancer cell lines, MCF-7, effectively.

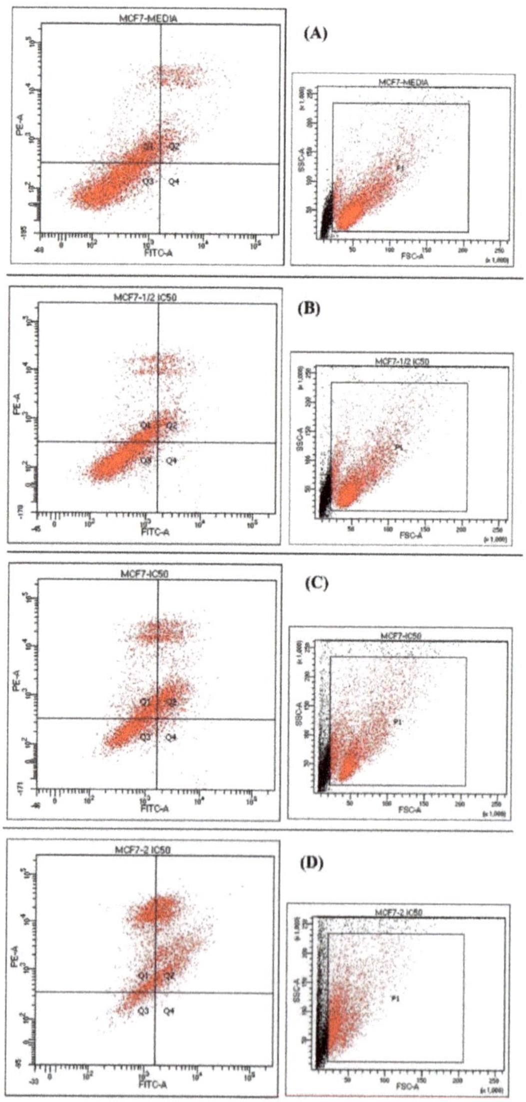

Figure 4. Effects of *L. shawii* extract treatment on apoptosis in MCF-7 cell lines. (**A**) Untreated control, (**B**) 0.5 % DMSO, (**C**) IC_{50} of *L. shawii* extract (519.2 μg/mL), and (**D**) double IC_{50} of *L. shawii* extract (1038.4 μg/mL).

2.6. Antimicrobial Activity

2.6.1. Preliminary Antimicrobial Screening

Results of the preliminary antimicrobial screenings of the hydroalcoholic extracts of *L. shawii*, *A. articulata*, *R. vesicarius*, and *Z. spinosa* revealed that *L. shawii* possesses significant antimicrobial activity against the tested *Bacillus cereus* ATCC 10876, whereas the hydroalcoholic extracts of the *A. articulata* and *Z. spinosa* exhibited substantial antifungal activity against the tested *A. niger* ATCC 6275 (Figure 5). The results demonstrated that the hydroalcoholic extract of *L. shawii* inhibited *B. cereus* ATCC 10876 with a diameter of 9.0 $\pm$ 0.1 mm, while *A. articulata* and *Z. spinosa* inhibited *A. niger* ATCC 6275 with diameters of 7.5 $\pm$ 0.2 mm, and 7.7 $\pm$ 0.2 mm, respectively. Additionally, Table 6 also displayed that the *L. shawii*, *A. articulata*, *R. vesicarius*, and *Z. spinosa* have no antimicrobial activity against other tested strains, whereas *R. vesicarius* exhibited no antimicrobial potential against all of the tested microorganisms. The currently observed weak antimicrobial activity of some of these plants' extracts in comparison to their reported antimicrobial activity, from other plants' counterparts found in the non-halophytic environment, could be attributed to their environment and habitat's variations effects on these plants, which also seemingly have affected their constituents in concentrations and types, which were produced as a result of the non-halophytic conditions of these plants. The desert climate effects on the anti-microbial activity of the currently studied four plants demonstrated the effects of the plant environment on the constituents and their biological activity. For example, *L. shawii* plant species growing in Yemen and Tunisia have been reported to have antimicrobial activity against several microbes, including, *S. aureus*, *K. pneumoniae*, and *E. faecalis*, whereas the *L. shawii* species growing here in this region of Saudi Arabia was found to be inactive against these microbes [77,78]. Moreover, the current findings for the antimicrobial activity of *L. shawii*, *A. articulata*, and *R. vesicarius* were consistent with the previous findings regarding the antimicrobial activity of these plant species growing in the similar environmental conditions in Saudi Arabia but at different locations than ours; these plants contained these activities [79].

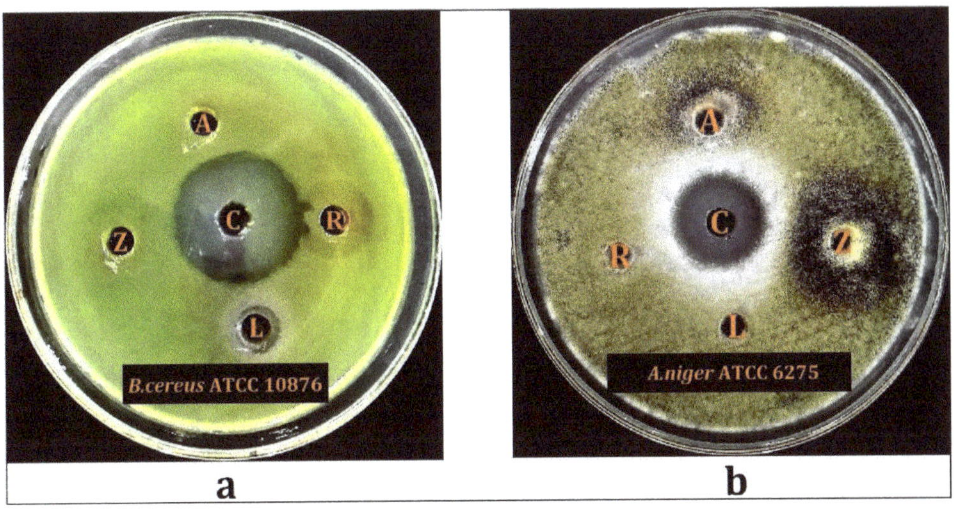

Figure 5. Results of preliminary antimicrobial activity evaluations against, (**a**) *B. cereus* 10876; (**b**) *A. niger* 6275 for *Lycium shawii* (**L**), *Anabasis articulata* (**A**), *Rumex vesicarius* (**R**), *Zilla spinosa* (**Z**) plants' hydroalcoholic extracts, and control antibiotics (**C**).

2.6.2. Minimum Inhibitory Concentration and Minimum Bacterial Concentration

The tests results indicated that the hydroalcoholic extract of *L. shawii* had MIC and MBC values of 12.5 mg/mL, and 25 mg/mL, respectively, against the tested *B. cereus* ATCC 10876. By comparison, the *A. articulata* exhibited MIC and MBC values of 12.5 mg/mL, and 25 mg/mL, respectively, against the *A. niger* ATCC 6275, whereas the *Z. spinosa* showed MIC and MBC values at 50 mg/mL, and 100 mg/mL, respectively, against *A. niger* ATCC 6275. At the same time, the control antibiotics inhibited the growths of all the tested organisms at the given concentrations. The results are summarized in Table 6.

2.7. Plants' Antioxidant Potential, Biological Activity, Flavonoids Plausible Biogenetic Interrelationships, and Molecular Oxygen Proliferations

The current investigation on these four plants exhibited a discernible biogenetic interrelationship in the occurrence of different flavonoid molecules in each plant. The LC-MS analysis revealed that the polyhydroxylated flavonoid aglycones, i.e., quercetin, luteolin, and kaempferol, and their mono- and di-glycosides were the dominant flavonoids among these plants. The progressing oxygen proliferations with the advancing molecular weights in the series of flavonoids present in these plants' extracts showed the increasing anti-oxidative effects of these plant extract in relation to their levels of the, primarily, the flavonoid contents.

The plant's deducible antioxidant levels in relation to their proportions of increasing levels of the antioxidant compounds, especially, the flavonoid contents were observed in these plants which have been observed in the LC-MS analysis and the quantitative estimations of the flavonoid contents and the plant's antioxidant potential, e.g., the *L. shawii* showed the strongest antioxidant activity with the highest concentrations of the phenolics and flavonoids, where 18 flavonoid constituents, highest in numbers and all detected, were present in *L. shawii*. Moreover, the lowest antioxidant level was recorded for the *A. articulata*, and *Z. spinosa*, which have the lowest concentrations of these products. Nonetheless, an observation in the increasing flavonoid contents as a factor of increasing molecular oxygen proliferations in the flavonoids (Figure 6) of these halophytic plants was also found associated with the efficacy of their biological activity, e.g., *L. shawii* showed the strongest anticancer activity (lowest IC_{50} value against MCF-7 and K562 cell lines, Figure 7). The results demonstrated that quantities of phenolics and flavonoids contents in the four plants directly correlate with the plants' antioxidant and anticancer activity levels.

The molecular oxygen proliferations in the flavonoid framework biogenetically produced different, multiple typed, and advancing molecular weights flavonoid contents in alignment with the demands of the halophytic plants to meet their specific requirements of antioxidant potency, disease, and pathogens-fighting capability as part of their defense mechanism [12].

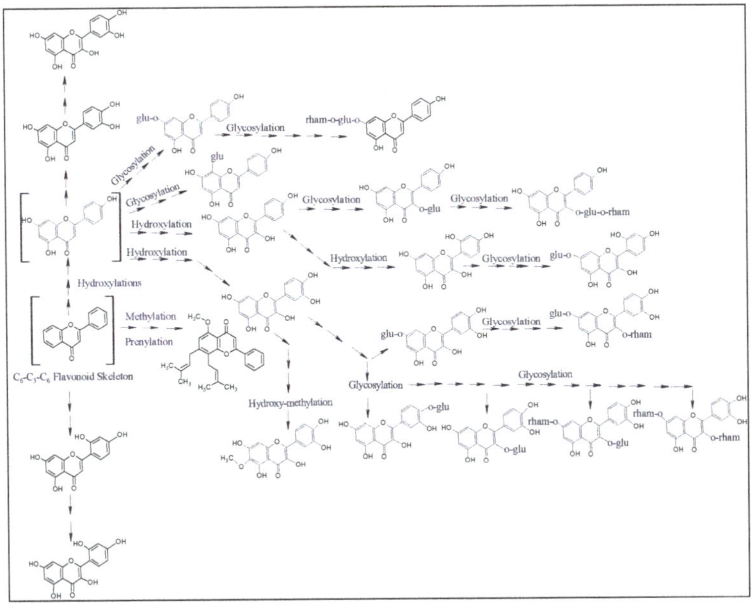

Figure 6. The proposed plausible biogenetic interrelationship and molecular oxygen proliferation in flavonoids from different plants. The products shown in parenthesis have not been detected in the current study.

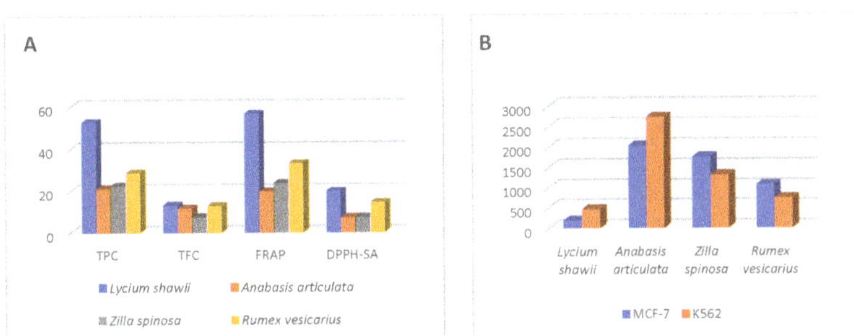

Figure 7. Comparative flavonoids and phenolics contents, antioxidant (**A**), and anticancers (**B**) activities.

3. Materials and Methods

3.1. Plant Materials and Extractions

Aerial parts of the plants were collected in March 2019 from the Qassim University campus surrounding areas. The plants were identified by the institutional botanists at the Department of Plant Production and Protection, College of Agriculture, Qassim University. The plants were dried in shade at room temperature (25 ± 2 °C) for two weeks before grinding. Afterward, 300 gm of plant powders were extracted with 95 % v/v ethanol (hydroalcoholic mixture) (3 × 1 L) for 24 h in stirring conditions. The extracts obtained from each plant were double-filtered through filter papers, and evaporated to dryness on a vacuum rotatory evaporator, Rotavapor®, under a temperature below 40 °C to give a dry gummy mass. The hydroalcoholic plant extracts were stored at −80 °C till further use.

3.2. Liquid Chromatography-Mass Spectroscopy (LC-MS) Analysis

A Bruker Daltonics (Bremen, Germany) Impact II ESI-Q-TOF (electrospray ionization-quadrupole time-of-flight) system equipped with Bruker Daltonics Elute UPLC system

(Bremen, Germany) was used for extracts' scanning under 190 nm and 500 nm range. Specific standards were used to identify the analyte's retention time in the chromatographic analysis. Accurately, 1 mg of the plants' extracts were dissolved in 2.0 mL of DMSO (analytical grade), and the solutions were diluted with acetonitrile to 50 mL. The obtained solutions were centrifuged at 4000 rpm for 2.0 min, 1.0 mL of the clear extracts' solutions were transferred to the autosampler, and the injection volume was adjusted at 3.0 µL. The instrument was operated using Ion Source Apollo II ion Funnel electrospray source. The instrument parameters were adjusted as capillary voltage (2500 V), nebulizer gas (2.0 bar), nitrogen flow (8 L/min), and dry temperature (200 °C). The mass accuracy was 0.1 Da, the mass resolution was 50,000 FSR (full sensitivity resolution), and the TOF repetition rate was up to 20 kHz. Chromatographic separation was performed on 120, C_{18} reverse-phase (RP) column, 100 × 2.1 mm, 1.8 µm (120 Å) from Bruker Daltonics (Bremen, Germany) at 30 °C, and autosampler temperature at 8 °C with a total run time of 35 min using the gradient elutions. The eluent A and B consisted of methanol/5 mM ammonium formate/0.1 % formic acid, and water/methanol (90:10)/5 mM ammonium formate/0.1 % *v*/*v* formic acid, respectively.

3.3. *Quantitative Measurements of the Total Phenolics and Flavonoids Contents*

The method of Quy [80] was used for the measurement of total phenolics and flavonoids contents in the plants' extracts as equivalents to gallic acid and quercetin using Folin–Ciocalteu, and aluminum chloride reagents, respectively. For the phenolics quantification, 0.2 mL of the 10 % *w*/*v* sodium carbonate solution was mixed with 1.6 mL of each plant extract (0.1 mg/mL), and 0.2 mL of the diluted Folin–Ciocalteu reagent (1:5 in distilled water). The mixture was vigorously mixed and kept for 30 min at RT before the absorbances were measured at 760 nm. Three independent measurements were recorded, and total phenolics contents of the plants' extracts were expressed as gallic acid equivalent (GAE) per gm of the dried extract using the slope equation of the gallic acid calibration curve. The total flavonoids were measured by mixing 2 mL of the extracts (0.1 mg/mL), and 0.1 mL of the aluminum chloride (10 % in distilled water) with 0.1 mL of the potassium acetate (0.1 mM) in a test tube. The mixture's absorbance was measured after 30 min incubation at 415 nm, and the quantified total flavonoids were expressed as quercetin equivalent (QE) per gm of the dried extract from three consecutive measurements.

3.4. *Trace Elements Analysis*

The dried plants' powder was used to determine Fe, Cu, Mn, Co, Mg, and Zn trace elements' presence using ICP-OES (Model iCAP 7400 Duo, serial IC 74DC144208, China) instrument according to the reported method of Johnsson [81]. The plants were dried at 70 °C for two days and sifted through a stainless-steel mill < 5 mm pore size. The dried plants' materials (0.5 g) were digested in a mixture of strong acids, including HNO_3, $HClO_4$, and H_2SO_4 (7:2:1) [12], and the trace elements' concentrations were measured from the calibration curve prepared for the individual standard elements. The measurements were conducted in triplicate and were expressed as the mean of the results with their standard deviations.

3.5. *Antioxidant Activity*

3.5.1. Total Antioxidant Capacity

The plant extracts' antioxidant capacity was measured using the method described by Arwa et al. [82]. The molybdate reagent was prepared by mixing sulfuric acid (0.6 M), and ammonium molybdate (4 mM) in sodium phosphate buffer (28 mM). Accurately, 3.6 mL of the molybdate reagent was added to 0.4 mL of the plant extract (containing 200 µg of the extract), and the mixture was vortexed and kept in a warm water bath for 30 min. The mixture was allowed to cool at room temperature, and the absorbance was recorded at 695 nm using a spectrophotometer against a blank, which was prepared in a similar way by mixing 0.4 mL of distilled water with a molybdate reagent. The total antioxidant activity of the extract was calculated as the equivalent of the Trolox using the standard calibration curve from the following equation:

$$y = 0.1954x - 0.1788; R^2 = 0.9646$$

where y is the absorbance of the sample at 695 nm, and x is the concentration of the sample in μg/mL.

3.5.2. DPPH Scavenging Activity

The ability of plant extracts to scavenge the DPPH-free radicals was determined as Trolox equivalents according to the method of Shimada et al. [83]. In brief, 1 mL of the extract's solution in methanol (containing 200 μg of the extract) was added to 1 mL of the DPPH (prepared by dissolving 6 mg of the DPPH in 50 mL of methanol). The mixture was vortexed and kept standing for 30 min in the dark at room temperature. The absorbance of the mixture was measured at 517 nm by a spectrophotometer against methanol as a blank. The method was conducted in triplicate, a standard calibration curve of the Trolox against DPPH was prepared, and the Trolox equivalence of the extract was calculated from the curve slope equation.

3.5.3. Ferric Reducing Antioxidant Power (FRAP) Assay

The FRAP assay was conducted by the method of Benzie and Strain [84] with minor modifications. The working reagent of the FRAP was freshly prepared by mixing one-fold of the TPTZ (2,4,6-Tris(2-pyridyl)-s-triazine, 10 mM prepared in 40 mM HCl), one-fold of the $FeCl_3{\cdot}6H_2O$ (20 mM), and ten-folds of acetate buffer (300 mM, pH 3.6). Accurately, 2 mL of the FRAP reagent was added to 0.1 mL of the extract (containing 200 μg of the dried extract), the mixture was incubated for 30 min at room temperature, and the absorbance was recorded at 593 nm. The procedure was conducted in triplicate, and the prepared FRAP-Trolox calibration curve was used to calculate the extract activity as mg Trolox equivalent per gm of the plant's dried extract.

3.5.4. Metal Chelating Activity Assay

The plant's hydroalcoholic extract's ability to chelate metals compared to the EDTA was estimated using the method of Zengin et al. [85]. In brief, a mixture of the extract solution (2 mL of ethanol containing 200 μg of extract) and ferrous chloride (25 μL, 2 mM) was added to 100 μL of ferrozine to inchoate the color. The mixture's absorbance was recorded at 562 nm against a blank (2 mL of the extract plus 200 μL of the ferrous chloride without ferrozine). The standard calibration curve of EDTA was prepared, and the chelating activity of the extract was calculated as the equivalent of the EDTA.

3.6. *Cytotoxic Assay*

Extracts of *L. shawii*, *R. vesicarius*, *Z. spinosa*, and *A. articulata* were tested for their antiproliferative activity. Toxicity of all the extracts was measured against normal human fibroblast, MCF7, PANC-1, and K562 cell lines using standard MTT assay (Promega, Madison, WI, USA), which measured the ability of the mitochondrial dehydrogenase to reduce MTT to a purple formazan product. The cells were suspended at a density of 12–15 $\times 10^3$ cells/mL in RPMI 1640 media, and 100 μL of each cell type were seeded in each well of a 96-well microtiter plate and incubated for 24 h. The extract was dissolved in DMSO and added to the wells in triplicate to a final concentration ranging from 400 to 12.5 μg/mL in a 2-folds serial dilution (400, 200, 100, 50, 25, and 12.5 μg/mL), and incubated at 37 °C, 5% CO_2 for 24 h. Two controls were used: one contained medium with cells, and the other contained cells plus medium with the vehicle. In addition, the extract was added without cells to check the effects of the background colors. Doxorubicin was used as a positive control. The tests were performed according to the manufacturer's guidelines, and the absorbance was measured at 590 nm using a microplate reader (Biotech, Washington, DC, USA).

3.7. Apoptotic Assay

MCF7 cell lines (5×10^4/well) were plated in 6-well plates 24 h before the experiment. Cells were treated with the inhibitory 1/2 (IC_{50}), (IC_{50}), and double ($2xIC_{50}$) of *L. shawii* for 24 h. A negative control, cells without any treatment, was also used. According to the kit protocol, apoptosis/necrosis was monitored using the TACS Annexin V–FITC Apoptosis Detection Kit (R&D Systems, Minneapolis, MN, USA). The percentage of the apoptotic/necrotic cells was measured by flow cytometry analysis using a FACSCalibur flow cytometer (BD Biosciences, Becton Drive, Franklin Lakes, NJ, USA).

3.8. Antimicrobial Evaluations

3.8.1. Test Organisms

Staphylococcus aureus (*S. aureus*) ATCC 29213, *Staphylococcus saprophyticus* (*S. saprophyticus*) ATCC 43867, *Streptococcus pyogenes* (*S. pyogenes*)-A ATCC 19615, *Streptococcus pneumoniae* (*S. pneumoniae*) ATCC 49619, *Enterococcus faecalis* (*E. faecalis*) ATCC 29212, *Bacillus cereus* (*B. cereus*) ATCC 10876, *Escherichia coli* (*E. coli*) ATCC 25922, *Klebsiella pneumoniae* (*K. pneumoniae*) ATCC 27736, *Pseudomonas aerugenosa* (*P. aerugenosa*) ATCC 9027, *Salmonella typhimurium* (*S. typhimurium*) ATCC 13311, *Shigella flexneri* (*S. flexneri*) ATCC 12022, *Proteus vulgaris* (*P. vulgaris*) ATCC 6380, *Proteus mirabilis* (*P. miribilis*) ATCC 29906, *Candida albicans* (*C. albicans*) ATCC 10231, and *Aspergillus niger* (*A. niger*) ATCC 6275 were used as the test organisms.

3.8.2. Antimicrobial Activity

Preliminary Antimicrobial Activity

The preliminary antimicrobial activity evaluations of the four plants' *L. shawii*, *A. articulata*, *R. vesicarius*, and *Z. spinosa* extracts were performed by the well-diffusion method [86,87]. The antimicrobial activity evaluation was conducted on modified tryptic soy-agar plates. The plant extracts were dissolved in sterile distilled water at a concentration of 100 mg/mL. The levofloxacin (70 µg/mL) and clotrimazole (2.5 mg/mL) were used as antibacterial and antifungal control antibiotics, respectively. First, the suspension of the test organisms (24–48 h old) was adjusted to a turbidity of 0.5 McFarland standard. Next, the 100 µL suspension of each test organism was independently dispensed on the agar plate, and using the sterile cotton swabs the suspensions were evenly distributed across the surfaces of the test plates. Next, the inoculated agar plates were punched, and wells were prepared using a sterile cork borer (06 mm diameter), and an 80 µL sample was used for each well. The inoculated plates were kept at 4–8 °C for 30 min and the samples were allowed to diffuse in the agar plates. Next, the inoculated plates were incubated at 35 ± 2 °C for 24 h for bacteria, and 30 ± 2 °C for 48–72 h for the fungi. Following incubation, the inhibitory zones' diameters were measured on a millimeter scale. Each test was conducted in triplicate, and the findings were recorded in mean ± SD (standard deviation).

MIC and MBC

MIC was measured using the resazurin-based micro broth dilution method, whereas the MBC was measured using the spot inoculation method [86–89]. Two folds serial dilutions of the selected plant extract were prepared with starting concentration of 100 mg/mL with sterile distilled water, and subsequently, various concentrations of the plant extract (0.098–50 mg/mL) were prepared in sterile tryptic soy-broth (TSB). The prepared concentrations of the selected plant extracts were evaluated for their antimicrobial efficacy against selected pathogens. Levofloxacin (10 µg/mL), and clotrimazole (20 µg/mL) were used as control antibiotics.

3.9. Statistical Analysis

The results obtained from the preliminary antimicrobial screening of *Lycium shawii*, *Anabasis articulata*, *Rumex vesicarius*, and *Zilla spinosa* were analyzed to see whether the tested organisms have statistically different mean values of the antimicrobial activity. A one-way ANOVA test combined with Tukey's analysis method (post hoc analysis) was conducted on SPSS version 20.0 SPSS (IBM, Armonk, NY, USA).

4. Conclusions

The study confirmed different phytochemicals representations and trace element concentrations in the four halophytic plants, *L. shawii*, *A. articulata*, *R. vesicarius*, and *Z. spinosa*, which are growing together in the arid area of central Saudi Arabia. The similarities in these plants' biosynthetic ability to produce phenolics and flavonoids at varying concentrations compared to plants growing in normal, non-saline, non-desert environments have been recognized. The high concentrations of phenolics and flavonoids in these halophytic plants were likewise found to be associated with their higher anticancer and antimicrobial activities as compared to their non-halophytic counterparts. The higher levels of antioxidant potential, especially of the *L. shawii*, confirmed the importance of phenolics, and flavonoids contents as biologically active plant constituents. The higher presence of trace elements also supported the prevalent use of *L. shawii* as part of food and animal feeds, and their use in various biological, physiological, and symptomatic reliefs of various ailments by the local population, and the Bedouins. However, extensive pharmacological, bioassay-based activity localization in extracts/fractions, preclinical, and clinical studies are required to approve the plants' safety for long-term, and proven-dose human consumption.

Supplementary Materials: The following are available online at https://www.mdpi.com/article/10.3390/plants10102208/s1.

Author Contributions: Conceptualization, H.A.M.; methodology, H.A.M., H.M.A., K.A.Q., M.M.A., R.S., M.A., E.A.-E., K.M.M. and Y.I.K.; software, R.A.K., M.A., H.M.A. and O.A.R.; validation, H.A.M., H.M.A., M.S.A.-O., M.A., A.A.H.A. and O.A.R.; formal analysis, S.A.A.M., H.A.M. and R.A.K.; investigation, H.A.M., K.A.Q., H.M.A., R.A.K., R.S. and Y.I.K.; resources, H.A.M.; data curation, S.A.A.M., K.A.Q., R.A.K., O.A.R. and M.A.; writing—original draft preparation, H.A.M., H.M.A., S.A.A.M., K.A.Q. and Y.I.K.; writing—review and editing, H.A.M., H.M.A., K.A.Q., R.A.K., R.S. and M.S.A.-O.; supervision, H.A.M.; project administration, H.A.M.; funding acquisition, H.A.M. All authors have read and agreed to the published version of the manuscript.

Funding: This research was funded by the Deputyship for Research & Innovation, Ministry of Education in Saudi Arabia under the project number (QU-IF-1-2-2).

Institutional Review Board Statement: Not applicable.

Informed Consent Statement: Not applicable.

Data Availability Statement: Data are available in the main text and Supplementary File.

Acknowledgments: The authors extend their appreciation to the Deputyship for Research & Innovation, Ministry of Education and, Saudi Arabia for funding this research work through the project number (QU-IF-1-2-2). The authors also thank the technical support of Qassim University.

Conflicts of Interest: The authors declare no conflict of interest.

References

1. Jamshidi-Kia, F.; Lorigooini, Z.; Amini-Khoei, H. Medicinal plants: Past history and future perspective. *J. Herbmed Pharmacol.* **2018**, *7*, 1–7. [CrossRef]
2. Johns, T. *The Origins of Human Diet and Medicine: Chemical Ecology*; University of Arizona Press: Tucson, AZ, USA, 1996; ISBN 0816516871.
3. Farnsworth, N.R.; Soejarto, D.D. Global importance of medicinal plants. *Conserv. Med. plants* **1991**, *26*, 25–51.
4. Gurib-Fakim, A. Medicinal plants: Traditions of yesterday and drugs of tomorrow. *Mol. Aspects Med.* **2006**, *27*, 1–93. [CrossRef]
5. Alamgir, A.N.M. Pharmacognostical Botany: Classification of medicinal and aromatic plants (MAPs), botanical taxonomy, morphology, and anatomy of drug plants. In *Therapeutic Use of Medicinal Plants and Their Extracts*; Springer: Berlin/Heidelberg, Germany, 2017; Volume 1, pp. 177–293.
6. Sasidharan, S.; Saudagar, P. Plant metabolites as new leads to drug discovery. In *Natural Bio-Active Compounds*; Springer: Berlin/Heidelberg, Germany, 2019; pp. 275–295.

7. Anywar, G. Historical Use of Toxic Plants. *Poisonous Plants Phytochem. Drug Discov.* **2020**, 1–17. [CrossRef]
8. Yadesa, T.M.; Ogwang, P.E.; Tolo, C.U. Drugs in Clinical Practice from Toxic Plants and Phytochemicals. *Poisonous Plants Phytochem. Drug Discov.* **2020**, 79–94. [CrossRef]
9. Koparde, A.A.; Doijad, R.C.; Magdum, C.S. Natural products in drug discovery. In *Pharmacognosy: Medicinal Plants*; BoD–Books on Demand: Norderstedt, Germany, 2019; pp. 1–20.
10. Egbuna, C.; Kumar, S.; Ifemeje, J.C.; Ezzat, S.M.; Kaliyaperumal, S. *Phytochemicals as Lead Compounds for New Drug Discovery*; Elsevier: Amsterdam, The Netherlands, 2019; ISBN 0128178914.
11. Nazish, M.; Zafar, M.; Ahmad, M.; Sultana, S. Halophyte Diversity in Pakistan: Wild Resources and Their Ethnobotanical Significance. In *Handbook of Halophytes*; Springer: Cham, Switzerland, 2020; pp. 1–24. [CrossRef]
12. Al-Omar, M.S.; Mohammed, H.A.; Mohammed, S.A.A.; Abd-Elmoniem, E.; Kandil, Y.I.; Eldeeb, H.M.; Chigurupati, S.; Sulaiman, G.M.; Al-Khurayyif, H.K.; Almansour, B.S. Anti-Microbial, Anti-Oxidant, and α-Amylase Inhibitory Activity of Traditionally-Used Medicinal Herbs: A Comparative Analyses of Pharmacology, and Phytoconstituents of Regional Halophytic Plants' Diaspora. *Molecules* **2020**, *25*, 5457. [CrossRef] [PubMed]
13. Ishtiyaq, S.; Kumar, H.; Varun, M.; Ogunkunle, C.O.; Paul, M.S. Role of secondary metabolites in salt and heavy metal stress mitigation by halophytic plants: An overview. *Handb. Bioremediation* **2021**, 307–327. [CrossRef]
14. Mohammed, H.A. The Valuable Impacts of Halophytic Genus Suaeda; Nutritional, Chemical, and Biological Values. *Med. Chem.* **2020**, *16*, 1044–1057. [CrossRef]
15. Mohammed, H.A.; Al-Omar, M.S.; Mohammed, S.A.A.; Alhowail, A.H.; Eldeeb, H.M.; Sajid, M.S.M.; Abd-Elmoniem, E.M.; Alghulayqeh, O.A.; Kandil, Y.I.; Khan, R.A. Phytochemical Analysis, Pharmacological and Safety Evaluations of Halophytic Plant, Salsola cyclophylla. *Molecules* **2021**, *26*, 2384. [CrossRef]
16. Stevanovic, Z.D.; Stankovic, M.S.; Stankovic, J.; Janackovic, P.; Stankovic, M. Use of Halophytes as Medicinal Plants: Phytochemical Diversity and Biological Activity. In *Halophytes and Climate Change: Adaptive Mechanisms and Potential Uses*; CABI: Wallingford, UK, 2019; p. 343.
17. Elhag, M. Evaluation of different soil salinity mapping using remote sensing techniques in arid ecosystems, Saudi Arabia. *J. Sens.* **2016**, *2016*, 7596175. [CrossRef]
18. Mohammed, H.A.; Al-Omar, M.S.; Khan, R.A.; Mohammed, S.A.A.; Qureshi, K.A.; Abbas, M.M.; Al Rugaie, O.; Abd-Elmoniem, E.; Ahmad, A.M.; Kandil, Y.I. Chemical Profile, Antioxidant, Antimicrobial, and Anticancer Activities of the Water-Ethanol Extract of Pulicaria undulata Growing in the Oasis of Central Saudi Arabian Desert. *Plants* **2021**, *10*, 1811. [CrossRef]
19. Jithesh, M.N.; Prashanth, S.R.; Sivaprakash, K.R.; Parida, A.K. Antioxidative response mechanisms in halophytes: Their role in stress defence. *J. Genet.* **2006**, *85*, 237. [CrossRef]
20. Benjamin, J.J.; Lucini, L.; Jothiramshekar, S.; Parida, A. Metabolomic insights into the mechanisms underlying tolerance to salinity in different halophytes. *Plant Physiol. Biochem.* **2019**, *135*, 528–545. [CrossRef]
21. Khdery, G.A.; Farg, E.; Arafat, S.M. Natural vegetation cover analysis in Wadi Hagul, Egypt using hyperspectral remote sensing approach. *Egypt. J. Remote Sens. Sp. Sci.* **2019**, *22*, 253–262. [CrossRef]
22. Baharav, D. Desert habitat partitioning by the dorcas gazelle. *J. Arid Environ.* **1982**, *5*, 323–335. [CrossRef]
23. Rahman, M.A.; Mossa, J.S.; Al-Said, M.S.; Al-Yahya, M.A. Medicinal plant diversity in the flora of Saudi Arabia 1: A report on seven plant families. *Fitoterapia* **2004**, *75*, 149–161. [CrossRef]
24. Mishra, A.P.; Sharifi-Rad, M.; Shariati, M.A.; Mabkhot, Y.N.; Al-Showiman, S.S.; Rauf, A.; Salehi, B.; Župunski, M.; Sharifi-Rad, M.; Gusain, P. Bioactive compounds and health benefits of edible Rumex species-A review. *Cell. Mol. Biol.* **2018**, *64*, 27–34. [CrossRef]
25. Hamza, N.; Berke, B.; Umar, A.; Cheze, C.; Gin, H.; Moore, N. A review of Algerian medicinal plants used in the treatment of diabetes. *J. Ethnopharmacol.* **2019**, *238*, 111841. [CrossRef]
26. El-Shabasy, A. Survey on medicinal plants in the flora of Jizan Region, Saudi Arabia. *Int. J. Bot. Stud.* **2016**, *2*, 38–59.
27. Al-Qahtani, H.; Alfarhan, A.H.; Al-Othman, Z.M. Changes in chemical composition of Zilla spinosa Forssk. medicinal plants grown in Saudi Arabia in response to spatial and seasonal variations. *Saudi J. Biol. Sci.* **2020**, *27*, 2756–2769. [CrossRef]
28. Ahmed, A.S. Biological Activities of Extracts and Isolated Compounds from Bauhinia Galpinii (Fabaceae) and Combretum Vendae (Combretaceae) as Potential Antidiarrhoeal Agents. Ph.D. Thesis, University of Pretoria, Pretoria, South Africa, 2012.
29. Attar, T. A mini-review on importance and role of trace elements in the human organism. *Chem. Rev. Lett.* **2020**, *3*, 117–130.
30. Skalny, A.V. Bioelements and bioelementology in pharmacology and nutrition: Fundamental and practical aspects. In *Pharmacology and Nutritional Intervention in the Treatment of Disease*; InTech: Rijeka, Croatia, 2014; pp. 225–241.
31. Agudelo, A.; Carvajal, M.; del C. Martinez-Ballesta, M. Halophytes of the Mediterranean Basin—Underutilized Species with the Potential to Be Nutritious Crops in the Scenario of the Climate Change. *Foods* **2021**, *10*, 119. [CrossRef] [PubMed]
32. Hanan, Z.; Ibrahim, N.H.; Donia, G.R.; Younis, F.E.; Shaker, Y.M. Scrutinizing of trace elements and antioxidant enzymes changes in Barki ewes fed salt-tolerant plants under South Sinai conditions. *J. Am. Sci.* **2014**, *10*, 54–62.
33. Padayatty, S.; Espey, M.G.; Levine, M.; Vitamin, C. *Encyclopedia of Dietary Supplements*; CRC Press: Boca Raton, FL, USA, 2010.
34. Del Gobbo, L.C.; Imamura, F.; Wu, J.H.Y.; de Oliveira Otto, M.C.; Chiuve, S.E.; Mozaffarian, D. Circulating and dietary magnesium and risk of cardiovascular disease: A systematic review and meta-analysis of prospective studies. *Am. J. Clin. Nutr.* **2013**, *98*, 160–173. [CrossRef]
35. Aschner, J.L.; Aschner, M. Nutritional aspects of manganese homeostasis. *Mol. Aspects Med.* **2005**, *26*, 353–362. [CrossRef]

36. Aschner, M.; Erikson, K.M.; Dorman, D.C. Manganese dosimetry: Species differences and implications for neurotoxicity. *Crit. Rev. Toxicol.* **2005**, *35*, 1–32. [CrossRef] [PubMed]
37. Ganz, T. Systemic iron homeostasis. *Physiol. Rev.* **2013**, *93*, 1721–1741. [CrossRef] [PubMed]
38. Casiday, R.; Frey, R. *Iron Use and Storage in the Body: Ferritin and Molecular Representations*; Department of Chemistry, Washington University: St. Louis, MI, USA, 1998.
39. Lobo, V.; Patil, A.; Phatak, A.; Chandra, N. Free radicals, antioxidants and functional foods: Impact on human health. *Pharmacogn. Rev.* **2010**, *4*, 118. [CrossRef] [PubMed]
40. Malavé, I.; Rodriguez, J.; Araujo, Z.; Rojas, I. Effect of zinc on the proliferation response of human lymphocytes: Mechanism of its mitogenic action. *Immunopharmacology* **1990**, *20*, 1–10. [CrossRef]
41. Attar, T.; Harek, Y.; Larabi, L. Determination of ultra trace levels of copper in whole blood by adsorptive stripping voltammetry. *J. Korean Chem. Soc.* **2013**, *57*, 568–573. [CrossRef]
42. Haase, H.; Rink, L. Multiple impacts of zinc on immune function. *Metallomics* **2014**, *6*, 1175–1180. [CrossRef] [PubMed]
43. Watson, T.D.G. Diet and skin disease in dogs and cats. *J. Nutr.* **1998**, *128*, 2783S–2789S. [CrossRef] [PubMed]
44. Islam, M.R.; Attia, J.; Ali, L.; McEvoy, M.; Selim, S.; Sibbritt, D.; Akhter, A.; Akter, S.; Peel, R.; Faruque, O. Zinc supplementation for improving glucose handling in pre-diabetes: A double blind randomized placebo controlled pilot study. *Diabetes Res. Clin. Pract.* **2016**, *115*, 39–46. [CrossRef]
45. Zhang, Y.; Gladyshev, V.N. Comparative genomics of trace elements: Emerging dynamic view of trace element utilization and function. *Chem. Rev.* **2009**, *109*, 4828–4861. [CrossRef]
46. Takahashi-Iñiguez, T.; García-Hernandez, E.; Arreguín-Espinosa, R.; Flores, M.E. Role of vitamin B 12 on methylmalonyl-CoA mutase activity. *J. Zhejiang Univ. Sci. B* **2012**, *13*, 423–437. [CrossRef] [PubMed]
47. Banerjee, R.; Vlasie, M. Controlling the reactivity of radical intermediates by coenzyme B12-dependent methylmalonyl-CoA mutase. *Biochem. Soc. Trans.* **2002**, *30*, 621–624. [CrossRef]
48. Saker, F.; Ybarra, J.; Leahy, P.; Hanson, R.W.; Kalhan, S.C.; Ismail-Beigi, F. Glycemia-lowering effect of cobalt chloride in the diabetic rat: Role of decreased gluconeogenesis. *Am. J. Physiol. Metab.* **1998**, *274*, E984–E991. [CrossRef]
49. L'Abbate, A.; Neglia, D.; Vecoli, C.; Novelli, M.; Ottaviano, V.; Baldi, S.; Barsacchi, R.; Paolicchi, A.; Masiello, P.; Drummond, G.S. Beneficial effect of heme oxygenase-1 expression on myocardial ischemia-reperfusion involves an increase in adiponectin in mildly diabetic rats. *Am. J. Physiol. Circ. Physiol.* **2007**, *293*, H3532–H3541. [CrossRef]
50. Kelly, F.D.; Sinclair, A.J.; Mann, N.J.; Turner, A.H.; Abedin, L.; Li, D. A stearic acid-rich diet improves thrombogenic and atherogenic risk factor profiles in healthy males. *Eur. J. Clin. Nutr.* **2001**, *55*, 88–96. [CrossRef]
51. Patel, M.K.; Pandey, S.; Brahmbhatt, H.R.; Mishra, A.; Jha, B. Lipid content and fatty acid profile of selected halophytic plants reveal a promising source of renewable energy. *Biomass Bioenergy* **2019**, *124*, 25–32. [CrossRef]
52. Vizetto-Duarte, C.; Figueiredo, F.; Rodrigues, M.J.; Polo, C.; Rešek, E.; Custódio, L. Sustainable Valorization of Halophytes from the Mediterranean Area: A Comprehensive Evaluation of Their Fatty Acid Profile and Implications for Human and Animal Nutrition. *Sustainability* **2019**, *11*, 2197. [CrossRef]
53. Jaganath, I.B.; Crozier, A. Dietary flavonoids and phenolic compounds. *Plant Phenolics Hum. Health Biochem. Nutr. Pharmacol.* **2010**, *1*, 1–50.
54. Fraga, C.G. *Plant Phenolics and Human Health: Biochemistry, Nutrition and Pharmacology*; John Wiley & Sons: Hoboken, NJ, USA, 2010.
55. Mohammed, H.A.; Khan, R.A.; Abdel-Hafez, A.A.; Abdel-Aziz, M.; Ahmed, E.; Enany, S.; Mahgoub, S.; Al-Rugaie, O.; Alsharidah, M.; Aly, M.S.A. Phytochemical Profiling, In Vitro and In Silico Anti-Microbial and Anti-Cancer Activity Evaluations and Staph GyraseB and h-TOP-IIβ Receptor-Docking Studies of Major Constituents of Zygophyllum coccineum L. Aqueous-Ethanolic Extract and Its Subsequent Fra. *Molecules* **2021**, *26*, 577. [CrossRef] [PubMed]
56. Rigane, G.; Ghazghazi, H.; Aouadhi, C.; Ben Salem, R.; Nasr, Z. Phenolic content, antioxidant capacity and antimicrobial activity of leaf extracts from Pistacia atlantica. *Nat. Prod. Res.* **2017**, *31*, 696–699. [CrossRef]
57. Raffa, D.; Maggio, B.; Raimondi, M.V.; Plescia, F.; Daidone, G. Recent discoveries of anticancer flavonoids. *Eur. J. Med. Chem.* **2017**, *142*, 213–228. [CrossRef]
58. Shahidi, F.; Yeo, J. Bioactivities of phenolics by focusing on suppression of chronic diseases: A review. *Int. J. Mol. Sci.* **2018**, *19*, 1573. [CrossRef]
59. Al-Snafi, A.E. Phenolics and flavonoids contents of medicinal plants, as natural ingredients for many therapeutic purposes—A review. *IOSR J. Pharm.* **2020**, *10*, 42–81.
60. Agrawal, A.D. Pharmacological activities of flavonoids: A review. *Int. J. Pharm. Sci. Nanotechnol.* **2011**, *4*, 1394–1398.
61. Croft, K.D. The chemistry and biological effects of flavonoids and phenolic acids a. *Ann. N. Y. Acad. Sci.* **1998**, *854*, 435–442. [CrossRef] [PubMed]
62. Maalik, A.; Khan, F.A.; Mumtaz, A.; Mehmood, A.; Azhar, S.; Atif, M.; Karim, S.; Altaf, Y.; Tariq, I. Pharmacological applications of quercetin and its derivatives: A short review. *Trop. J. Pharm. Res.* **2014**, *13*, 1561–1566. [CrossRef]
63. Sezer, E.D.; Oktay, L.M.; Karadadaş, E.; Memmedov, H.; Selvi Gunel, N.; Sözmen, E. Assessing anticancer potential of blueberry flavonoids, quercetin, kaempferol, and gentisic acid, through oxidative stress and apoptosis parameters on HCT-116 cells. *J. Med. Food* **2019**, *22*, 1118–1126. [CrossRef]

64. Smiljkovic, M.; Stanisavljevic, D.; Stojkovic, D.; Petrovic, I.; Vicentic, J.M.; Popovic, J.; Grdadolnik, S.G.; Markovic, D.; Sankovic-Babice, S.; Glamoclija, J. Apigenin-7-O-glucoside versus apigenin: Insight into the modes of anticandidal and cytotoxic actions. *EXCLI J.* **2017**, *16*, 795. [PubMed]
65. Daglia, M. Polyphenols as antimicrobial agents. *Curr. Opin. Biotechnol.* **2012**, *23*, 174–181. [CrossRef]
66. Nile, A.; Nile, S.H.; Cespedes, C.L.; Oh, J.-W. Spiraeoside extracted from red onion skin ameliorates apoptosis and exerts potent antitumor, antioxidant and enzyme inhibitory effects. *Food Chem. Toxicol.* **2021**, *154*, 112327. [CrossRef] [PubMed]
67. Vaiyapuri, M.; Natesan, K.; Vasamsetti, B.M.K.; Mekapogu, M.; Swamy, M.K.; Thangaraj, K. Orientin: A C-Glycosyl Flavonoid that Mitigates Colorectal Cancer. In *Plant-Derived Bioactives*; Springer: Berlin/Heidelberg, Germany, 2020; pp. 1–19.
68. Goodarzi, S.; Tabatabaei, M.J.; Mohammad Jafari, R.; Shemirani, F.; Tavakoli, S.; Mofasseri, M.; Tofighi, Z. Cuminum cyminum fruits as source of luteolin-7-O-glucoside, potent cytotoxic flavonoid against breast cancer cell lines. *Nat. Prod. Res.* **2020**, *34*, 1602–1606. [CrossRef]
69. Ganesan, K.; Xu, B. Molecular targets of vitexin and isovitexin in cancer therapy: A critical review. *Ann. N. Y. Acad. Sci.* **2017**, *1401*, 102–113. [CrossRef]
70. Panche, A.N.; Diwan, A.D.; Chandra, S.R. Flavonoids: An overview. *J. Nutr. Sci.* **2016**, *5*, E47. [CrossRef] [PubMed]
71. Beddou, F.; Bekhechi, C.; Ksouri, R.; Sari, D.C.; Bekkara, F.A. Potential assessment of Rumex vesicarius L. as a source of natural antioxidants and bioactive compounds. *J. Food Sci. Technol.* **2015**, *52*, 3549–3560. [CrossRef]
72. Alharbi, K.B.; Mousa, H.M.; Ibrahim, Z.H.; El-Ashmawy, I.M. Hepatoprotective effect of methanolic extracts of Prosopis farcta and Lycium shawii against carbon tetrachloride-induced hepatotoxicity in rats. *J. Biol. Sci.* **2017**, *17*, 35–41. [CrossRef]
73. Suleiman, M.H.A.; Ateeg, A.A. Antimicrobial and Antioxidant Activities of Different Extracts from Different Parts of *Zilla spinosa* (L.) Prantl. *Evid.-Based Complement. Altern. Med.* **2020**, *2020*, 6690433. [CrossRef]
74. Zeb, A. Concept, mechanism, and applications of phenolic antioxidants in foods. *J. Food Biochem.* **2020**, *44*, e13394. [CrossRef]
75. Karak, P. Biological activities of flavonoids: An overview. *Int. J. Pharm. Sci. Res.* **2019**, *10*, 1567–1574.
76. Kosanić, M.; Ranković, B.; Stanojković, T.; Rančić, A.; Manojlović, N. Cladonia lichens and their major metabolites as possible natural antioxidant, antimicrobial and anticancer agents. *LWT-Food Sci. Technol.* **2014**, *59*, 518–525. [CrossRef]
77. Dahech, I.; Farah, W.; Trigui, M.; Hssouna, A.B.; Belghith, H.; Belghith, K.S.; Abdallah, F.B. Antioxidant and antimicrobial activities of Lycium shawii fruits extract. *Int. J. Biol. Macromol.* **2013**, *60*, 328–333. [CrossRef]
78. Mothana, R.A.A.; Kriegisch, S.; Harms, M.; Wende, K.; Lindequist, U. Assessment of selected Yemeni medicinal plants for their in vitro antimicrobial, anticancer, and antioxidant activities. *Pharm. Biol.* **2011**, *49*, 200–210. [CrossRef] [PubMed]
79. Akbar, S.; Al-Yahya, M.A. Screening of Saudi plants for phytoconstituents, pharmacological and antimicrobial properties. *Aust. J. Med. Herbal.* **2011**, *23*, 76–87.
80. Do, Q.D.; Angkawijaya, A.E.; Tran-Nguyen, P.L.; Huynh, L.H.; Soetaredjo, F.E.; Ismadji, S.; Ju, Y.-H. Effect of extraction solvent on total phenol content, total flavonoid content, and antioxidant activity of Limnophila aromatica. *J. Food Drug. Anal.* **2014**, *22*, 296–302. [CrossRef] [PubMed]
81. Johnsson, L. Selenium uptake by plants as a function of soil type, organic matter content and pH. *Plant Soil* **1991**, *133*, 57–64. [CrossRef]
82. Aroua, L.M.; Almuhaylan, H.R.; Alminderej, F.M.; Messaoudi, S.; Chigurupati, S.; Al-Mahmoud, S.; Mohammed, H.A. A facile approach synthesis of benzoylaryl benzimidazole as potential α-amylase and α-glucosidase inhibitor with antioxidant activity. *Bioorg. Chem.* **2021**, *114*, 105073. [CrossRef]
83. Shimada, K.; Fujikawa, K.; Yahara, K.; Nakamura, T. Antioxidative properties of xanthan on the autoxidation of soybean oil in cyclodextrin emulsion. *J. Agric. Food Chem.* **1992**, *40*, 945–948. [CrossRef]
84. Benzie, I.F.F.; Strain, J.J. The ferric reducing ability of plasma (FRAP) as a measure of "antioxidant power": The FRAP assay. *Anal. Biochem.* **1996**, *239*, 70–76. [CrossRef] [PubMed]
85. Zengin, G.; Nithiyanantham, S.; Locatelli, M.; Ceylan, R.; Uysal, S.; Aktumsek, A.; Selvi, P.K.; Maskovic, P. Screening of in vitro antioxidant and enzyme inhibitory activities of different extracts from two uninvestigated wild plants: Centranthus longiflorus subsp. longiflorus and Cerinthe minor subsp. auriculata. *Eur. J. Integr. Med.* **2016**, *8*, 286–292. [CrossRef]
86. Balouiri, M.; Sadiki, M.; Ibnsouda, S.K. Methods for in vitro evaluating antimicrobial activity: A review. *J. Pharm. Anal.* **2016**, *6*, 71–79. [CrossRef] [PubMed]
87. Qureshi, K.A.; Elhassan, G.O. Isolation, Purification and Characterization of Antimicrobial Agent Antagonistic to Escherichia coli ATCC 10536 Produced by Bacillus pumilus SAFR-032 Isolated from the Soil of Unaizah, Al Qassim Province of Saudi Arabia. *Pak. J. Biol. Sci.* **2016**, *19*, 191. [CrossRef]
88. Elshikh, M.; Ahmed, S.; Funston, S.; Dunlop, P.; McGaw, M.; Marchant, R.; Banat, I.M. Resazurin-based 96-well plate microdilution method for the determination of minimum inhibitory concentration of biosurfactants. *Biotechnol. Lett.* **2016**, *38*, 1015–1019. [CrossRef]
89. Schwalbe, R.; Steele-Moore, L.; Goodwin, A.C. *Antimicrobial Susceptibility Testing Protocols*; CRC Press: Boca Raton, FL, USA, 2007; ISBN 1420014498.

plants

MDPI

Article

Ellagitannin, Phenols, and Flavonoids as Antibacterials from *Acalypha arvensis* (Euphorbiaceae)

Ever A. Ble-González [1,*], Abraham Gómez-Rivera [1], Alejandro Zamilpa [2], Ricardo López-Rodríguez [1], Carlos Ernesto Lobato-García [1], Patricia Álvarez-Fitz [3], Ana Silvia Gutierrez-Roman [2], Ma Dolores Perez-García [2], Alejandro Bugarin [4] and Manasés González-Cortazar [2,*]

1 División Académica de Ciencias Básicas, Universidad Juárez Autónoma de Tabasco, Carretera Cunduacán-Jalpa Km. 0.5, Cunduacán 86690, Tabasco, Mexico; abgori@gmail.com (A.G.-R.); ricardo.lopezr@ujat.mx (R.L.-R.); carlos.lobato@ujat.mx (C.E.L.-G.)
2 Centro de Investigación Biomédica del Sur, Instituto Mexicano del Seguro Social, Argentina No. 1, Col. Centro, Xochitepec 62790, Morelos, Mexico; azamilpa_2000@yahoo.com.mx (A.Z.); 12gtr.ana@gmail.com (A.S.G.-R.); lola_as@yahoo.com.mx (M.D.P.-G.)
3 Laboratorio de Toxicología, Cátedra CONACyT-Universidad Autónoma de Guerrero, Av. Lázaro Cárdenas s/n. Col. La Haciendita, Chilpancingo 39070, Guerrero, Mexico; paty_fitz@hotmail.com
4 Department of Chemistry and Physics, Florida Gulf Coast University, Fort Myers, FL 33965, USA; abugarin@fgcu.edu
* Correspondence: ble_49@hotmail.com (E.A.B.-G.); gmanases2000@gmail.com (M.G.-C.); Tel.: +52-(993)-581-500 (ext. 4711) (E.A.B.-G.); +52-(777)-3612-155 (M.G.-C.)

Citation: Ble-González, E.A.; Gómez-Rivera, A.; Zamilpa, A.; López-Rodríguez, R.; Lobato-García, C.E.; Álvarez-Fitz, P.; Gutierrez-Roman, A.S.; Perez-García, M.D.; Bugarin, A.; González-Cortazar, M. Ellagitannin, Phenols, and Flavonoids as Antibacterials from *Acalypha arvensis* (Euphorbiaceae). *Plants* **2022**, *11*, 300. https://doi.org/10.3390/plants11030300

Academic Editors: Juei-Tang Cheng, I-Min Liu and Szu-Chuan Shen

Received: 13 December 2021
Accepted: 13 January 2022
Published: 24 January 2022

Abstract: There is a significant need to gain access to new and better antibacterial agents. *Acalypha arvensis*, a plant from the Euphorbiaceae family, has been used in traditional medicine for centuries to treat infectious diseases. This manuscript reports the isolation, characterization, and antibacterial screening of 8 natural products extracted from maceration of aerial parts of *Acalypha arvensis*. Specifically, three extracts were assessed (*n*-hexane, ethyl acetate, and ethanol), in which antibacterial activity was evaluated against diverse bacterial strains. The ethanolic extract showed the best activity against methicillin-sensitive and methicillin-resistant *Staphylococcus aureus*, *Klebsiella pneumoniae*, and *Pseudomonas aeruginosa* strains, which supports the medicinal properties attributed to this plant. The chromatographic fractions AaR4 and AaR5 were the most bioactive, in which the ellagitannin natural product known as corilagin (**1**) was identified for the first time in this plant. Therefore, it can be said that this is the main chemical responsible for the observed antibacterial activity. However, we also identified chlorogenic acid (**2**), rutin (**3**), quercetin-3-*O*-glucoside (**4**), caffeic acid (**5**), among others (**6**–**8**). Hence, this plant can be considered to be a good alternative to treat health-related issues caused by various bacteria.

Keywords: *Acalypha arvensis*; ellagitannin; corilagin; *Staphylococcus aureus*; flavonoids; antibacterial

1. Introduction

The World Health Organization (WHO) has reported that microbial diseases (both from bacteria and fungi) are a leading cause of death in humans [1], which can be transmitted directly or indirectly from one individual to another, and that bacterial resistance causes more than 10 million deaths per year [2–5]. The microorganisms of greatest interest that cause infectious diseases in humans are bacteria such as *Salmonella typhi* (causing typhoid fever), *Staphylococcus aureus* (causing skin infections, sometimes pneumonia, endocarditis, and osteomyelitis), *Streptococcus pneumoniae* (causing pneumonia), among others [6,7]. The cosmopolitan increase of bacterial infections caused mainly by the inappropriate use of antibiotics and/or a deficient control of infections has led to the rise of drug-resistant strains, which represent a major threat to public health and the global economy [1]. Therefore, seeking and development of new generations of antimicrobials to mitigate the spread of antibiotic resistance have become imperative [8]. In fact, a reexamination of traditional medicines has become more common among scientists. This approach has already produced new antibiotics, and it is expected that more novel antibiotics will be discovered/isolated from traditional medicinal plants in the future. [9] Those new bioactive molecules from plant extracts could be essential secondary metabolites that will aid further understand antimicrobial activity, while advancing drug discovery [10].

It is well known that the medicinal properties of plants are related to their ability to synthesize a wide range of bioactive compounds. One of the main families of bioactive molecules found in plant extracts is the phenolic family. These compounds have shown excellent antibacterial activity against resistant pathogens through various reported mechanisms [10–12] and therefore are an important group of potential medicines.

The species *Acalypha arvensis* Poepp Endl., usually known as "spider leaf", "couch grass", "borreguillo", and/or "worm weed", belongs to the family Euphorbiaceae and it can be found from Mexico to tropical South America [13,14]. In traditional medicine, this plant has been used to treat illnesses such as diarrhea, vomiting, scabies, mouth sores (canker sores), spider bites, snake bites, cancer, athlete′s foot, inflammation, fluid retention, headache, and wound healing, among others [14–17]. One of the few pharmacological studies carried out on *A. arvensis* reported that an ethanol extract from leaves and flowers presented antimicrobial activity against enterobacteria pathogenic to humans; *Salmonella typhi* was the most inhibited bacterium (33.73%), whereas *Escherichia coli* (7.35%) was the most resistant [18]. The ethanol extract was also evaluated against three Gram-positive bacteria that cause respiratory infections (*Staphylococcus aureus, Streptococcus pneumoniae and Streptococcus pyogenes*), showing activity only against *Staphylococcus aureus* [19]. In another study, conducted on five species of the genus *Acalypha*, biocidal activity was evaluated, and it was demonstrated that *A. arvensis* has activity against *Pseudomonas aeruginosa* and *Cryptococcus neoformans* at a concentration of 1 mg/mL. On the other hand, it did not show activity against mosquito larvae (*Aedes aegypti* and *Anophles albinamus*). Groups of compounds such as flavonoids, anthocyanins, anthraquinones, coumarins, saponins, and cardenolides were identified in those extracts [20]. However, there are no prior reports indicating the actual bioactive molecules present in this plant species. Thus, the main aims of this work were (1) to evaluate the antimicrobial activity of three extracts and four fractions from *A. arvensis* against microorganisms sensitive and resistant to methicillin and (2) to determine the actual secondary metabolites responsible for the observed bioactivity.

2. Results

2.1. Antibacterial Effect of A. arvensis's Extracts

Maceration of the aerial parts of *Acalypha arvensis* (Aa) species provided the following extracts: hexane (AaHex, 7.4 g, 0.76%), ethyl acetate (AaAcOEt, 19.2 g, 1.98%), and ethanol (AaEtOH, 11.24 g, 1.15%). The three extracts were evaluated against 13 strains of sensitive and methicillin-resistant bacteria (Table 1).

Table 1. MICs of the three extracts obtained from *A. arvensis*.

Extract (mg/mL)					
Bacteria	**AaHex**	**AaAcOEt**	**AaEtOH**	**Control (+)**	**Control (−)**
Sa	>2	2	2	—	*
SaRM1	>2	>2	2	—	*
SaRM2	>2	>2	2	—	*
Se	>2	>2	2	—	*
Sh	>2	>2	>2	—	*
Ef	>2	>2	>2	—	*
Kp1	>2	>2	2	—	*
Kp2	>2	>2	2	—	*
Pa	>2	2	2	—	*

Table 1. *Cont.*

Extract (mg/mL)					
Bacteria	**AaHex**	**AaAcOEt**	**AaEtOH**	**Control (+)**	**Control (−)**
Ec1	>2	>2	>2	—	*
Ec2	>2	>2	>2	—	*
Ec3	>2	>2	>2	—	*
Sd	>2	>2	>2	—	*

(—): no growth; (*): growth.

According to the results in Table 1, the extract that showed the best antibacterial activity in relation to the 13 strains of bacteria evaluated was the ethanolic extract (AaEtOH) with good activity against four strains of Gram-positive bacteria (Sa, SaRM1, SaRM2, and Se) and three strains of Gram-negative bacteria (Kp1, Kp2, and Pa), followed by the ethyl acetate extract (AaAcOEt) that only showed activity against *S. aureus* (Sa) and *P. aeroginosa* (Pa). Lastly, the hexane extract (AaHex) did not exhibit any relevant activity against any of the strains evaluated at the screened concentration. Therefore, the AaEtOH extract was selected for chromatographic fractionation and to carry out the bacterial evaluation.

2.2. Antibacterial Activity of the AaEtOH Fractions (AaR1, AaR2, AaR3, AaR4, and AaR5)

Chromatographic fractionation of the AaEtOH extract produced five clusters, of which only four were evaluated (since AaR1 offered very low yield), against sensitive *S. aureus* (Sa) and methicillin-resistant *S. aureus* (SaRM1) (Table 2).

Table 2. MIC determination of the fractions (AaR2, AaR3, AaR4, and AaR5) obtained with EtOH.

Fractions (mg/mL)						
Bacteria	**AaR2**	**AaR3**	**AaR4**	**AaR5**	**Control (+)**	**Control (−)**
Sa	>2	>2	1	0.5	—	*
SaRM1	>2	>2	<0.5	<0.5	—	*

(—): no growth; (*): growth.

Table 2 shows the significant antibacterial activity from fractions AaR4 and AaR5 against both *S. aureus* sensitive (Sa) and methicillin-resistant (SaRM1) strains.

2.3. HPLC Analysis of Extract and Fractions

Analysis of the AaEtOH extract by high-performance liquid chromatography (HPLC) allowed the identification of four main compounds that have retention times of 8.5, 8.6, 9.3, and 9.4 min and UV-Vis spectra of (λ_{max} = 191.1, 219, 258, 268, and 355 nm), respectively. Those values could correspond to groups of phenolic-type compounds (Figure 1).

Figure 2A,B show the HPLC chemical analysis of two active fractions (AaF4 and AaF5) at 270 nm and Figure 2C shows fraction AaF5 at 350 nm.

In the chromatograms of Figure 2, it can be observed that compound **1** and other derivatives of the same type (ellagitannins) that could not be identified (NI) are present in the two fractions. Therefore, it could be said that the antimicrobial effect produced by the two fractions (AaR4 and AaR5) could be mainly due to the ellagitannin known as corilagin (**1**) and the presence of flavonoids. According to HPLC, the concentration of compound **1** (8.5 min, 270 nm) in the ethanol extract is 18.36 mg of corilagin/g of extract.

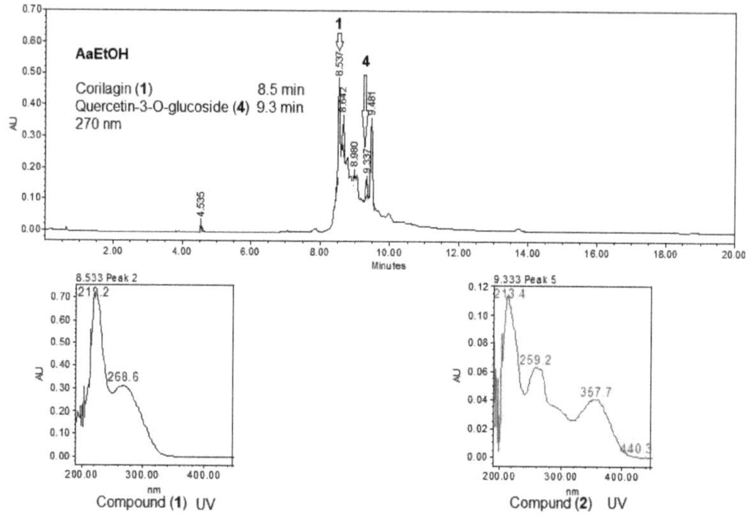

Figure 1. HPLC chromatogram of the AaEtOH extract and UV spectra of compounds (**1**) and (**4**).

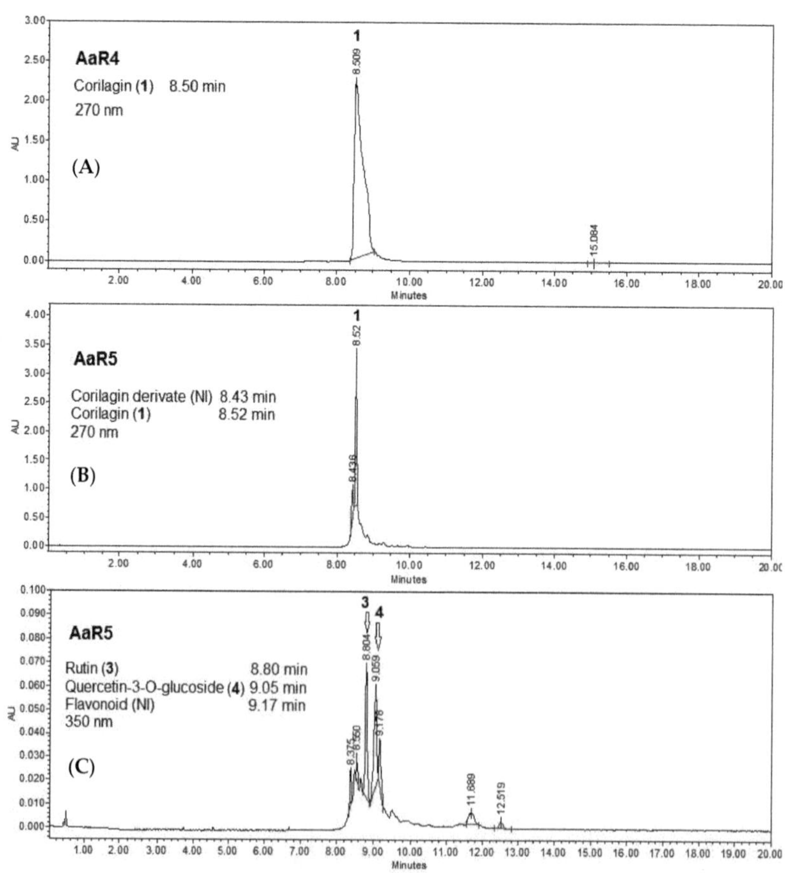

Figure 2. HPLC chromatograms of the fractions: (**A**) AaR4, (**B**) AaR5 at 270 nm and (**C**) AaR5 at 350 nm.

Corilagin (**1**), corilagin derivative (Unidentified, NI), rutin (**3**), quercetin-3-O-glucoside (**4**) and caffeic acid (**5**) are identified.

2.4. Isolation and Structural Elucidation of Compounds **1–3**

Although both fractions (AaR4 and AaR5) were active, it was decided to isolate the compounds from fraction AaR4 since it provided a higher yield.

Chromatographic fractionation of the AaR4 fraction yielded seven compounds. Compound **1** was obtained as an amorphous yellow precipitate. The HPLC chromatogram and UV absorption spectrum (Figure S1) showed a retention time of 8.5 min with absorption lengths of λ_{max} = 221 and 268 nm. Analysis of the ^{1}H NMR spectrum showed three characteristic signals for three aromatic rings at δ 7.06 (2H, s, H-2″ and H-6″), 6.69 (1H, s, H-3′) and 6.67 (1H, s, H-3), assigned to gallic acid. Additionally, a doublet signal is observed at δ 6.36 (1H, d, *J* = 2.2 Hz, H-1⁗) and its ^{13}C-NMR chemical shift at δ 95.0 (CH, C-1⁗) corresponds to an anomeric carbon of a sugar. The COSY experiment analysis allowed us to identify by proton couplings a hexose named α-D-glucose. The anomeric proton (δ 6.36) shows long range correlation (HMBC) with the signal of the carbonyl at δ 166.6 assigned to C-7″ of the gallic acid, so this acid is substituted at C-1‴ position of glucopyranose. Likewise, it is observed a coupling to protons H-6⁗a and H-6⁗b (δ 4.15 and 4.96, respectively) with another carbonyl at δ 170.0. Thus, this was assigned to C-7 of another gallic acid, also for H-3⁗ (δ 4.8) with the carbonyl signal at δ 168.5 from C-7′ of the third gallic acid. Analysis of the obtained one-dimensional (^{1}H, ^{13}C, and DEPT) and two-dimensional (COSY, HMQC, and HMBC) NMR spectra (supplementary data Figures S2–S7), and comparison with data described in the literature [21–23], this compound was identified as an ellagitannin (β-1-*O*-galloyl-3,6-(R)-hexahydroxydiphenoyl-α-D-glucose) known as corilagin (**1**, Figure 3).

Additional analysis of fraction AaR4 was further separated using a second column chromatography. This second separation delivered 3 fractions (fractions 8–10, 13–20, and 23–27). Analysis of these fractions by HPLC and UV-Vis spectra indicated that peaks at 8.36 min (λ_{max} = 211 and 326 nm) co-respond with chlorogenic acid (**2**, Figure 4), 8.9 min (λ_{max} = 213, 255.7, and 355.3 nm) with rutin (**3**, Figure 5), 9.2 min (λ_{max} = 213.4, 255.7 and 355.3 nm) with quercetin-3-*O*-glucoside (**4**, Figure 5), and 9.1 min (λ_{max} = 219.2, 243.9 and 330.3 nm) with caffeic acid (**5**, Figure 6), respectively. These were compared with known standards to stablish their identity.

Fraction 32–37 was analyzed by 1D and 2D NMR spectroscopy (Figures S8–S13) and by comparison with spectroscopic data from the literature [24–27], it was found a mixture of a polyol known as treitol (**6**), a dihydrochalcone known as (*S*,*E*)-1,3-diphenylprop-2-en-1-ol (**7**), and a shikimic acid derivative (1*R*,2*R*,3*R*)-5-(hydroxymethyl)cyclohex-4-ene-1,2,3-triol (**8**) (Figure 3).

Corilagin (**1**): ^{1}H NMR (CD_3OD, 600 MHz) δ 7.06 (2H, s, H-2″ y H-6″), 6.69 (1H, s, H-3′), 6.67 (1H, s, H-3), 6.36 (1H, d, *J* = 2.2 Hz, H-1‴), 3.99 (1H, dd, 2.0, 3.5 Hz, H-2‴), 4.8 (1H, m, H-3‴), 4.46 (1H, dd, *J* = 1.7, 3.3 Hz, H-4‴), 4.52 (1H, dd, *J* = 8.0, 10.9 Hz, H-5‴), 4.15 (1H, dd, *J* = 8.0, 10.9 Hz, H-6a‴), 4.96 (1H, dd, *J* = 10.9, 10.9 Hz, H-6b‴); ^{13}C NMR (CD_3OD, 150 MHz) δ 62.4 (CH, C-4‴), 64.9 (CH_2, C-6‴), 69.4 (CH, C-2‴), 71.5 (CH, C-3‴), 76.1 (CH, C-5‴), 95.0 (CH, C-1‴), 108.3 (CH, C-3), 110.2 (CH, C-3′), 110.9 (2CH, C-2″ y C-6″), 116.6 (C, C-1), 117.2 (C, C-1′), 120.7 (C, C-1″), 125.4 (C, C-2), 125.5 (C, C-2′), 137.6 (C, C-5), 138.1 (C, C-5′), 140.3 (C, C-4″), 145.1 (C, C-6′), 145.2 (C, C-6), 145.6 (C, C-4′), 146.0 (C, C-4), 146.3 (2C, C-3″ y C-5″), 166.6 (C, C-7″), 168.5 (C, C-7′), 170.0 (C, C-7). Treitol *o*

(2*R*,3*R*)-butane-1,2,3,4-tetrol (**6**): ^{1}H NMR (CD_3OD, 600 MHz) δ 3.66 (2H, dd, *J* = 5.7, 5.9 Hz, H-2, H-3), 3.52 (2H, dd, *J* = 5.9, 11.2 Hz, H-1a, H-4a), 3.60 (2H, dd, *J* = 4.9, 11.2 Hz, H-1b, H-4b); ^{13}C NMR (CD_3OD, 150 MHz) δ 73.7 (CH, C-2 y C-3), 64.3 (CH_2, C-1 y C-4).

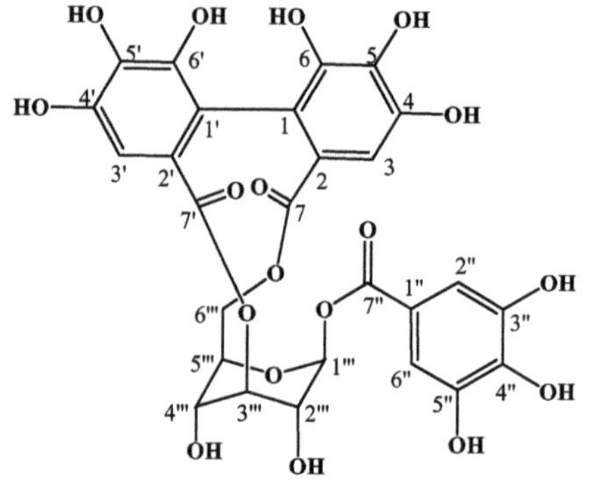

Corilagin (**1**)

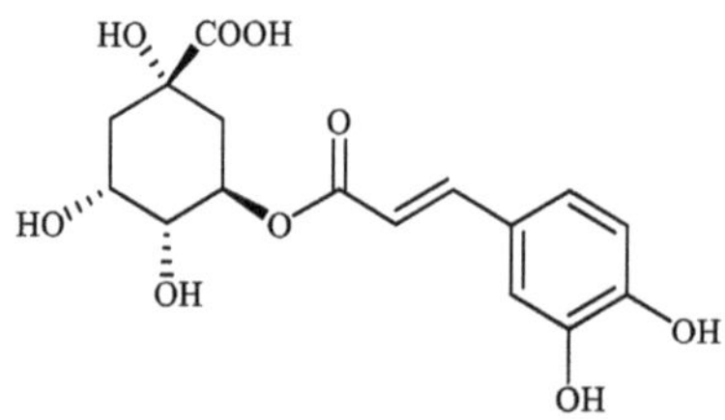

Chlorogenic Acid (**2**)

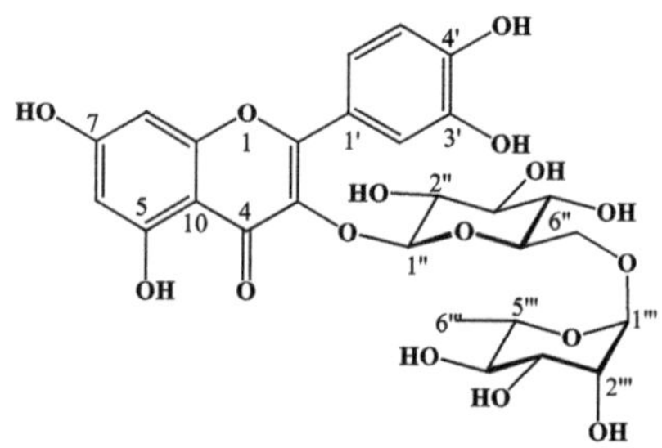

Rutin (**3**)

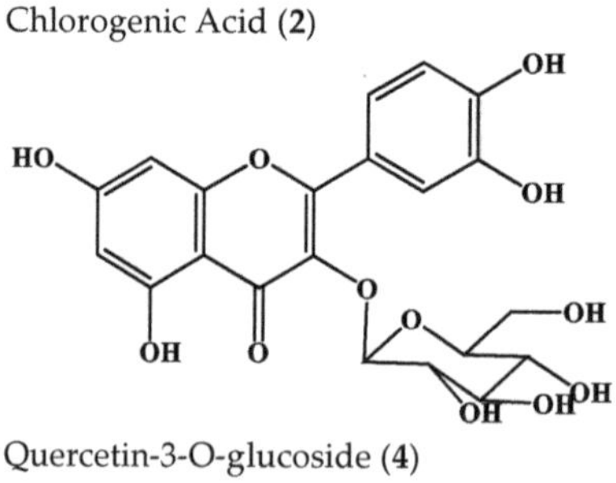

Quercetin-3-O-glucoside (**4**)

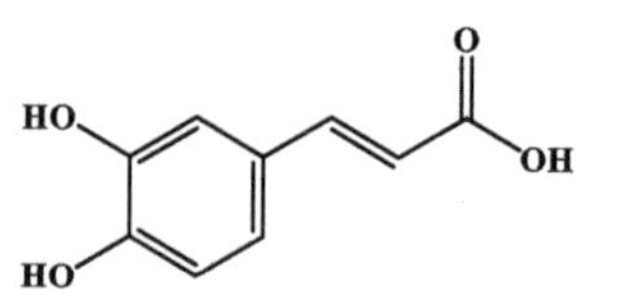

Caffeic acid (**5**)

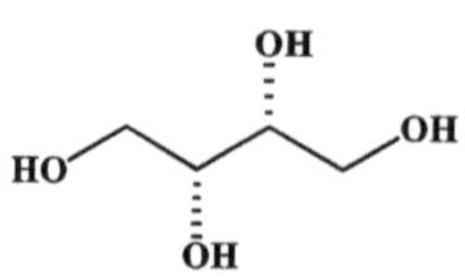

Treitol, (2*R*,3*R*)-butane-1,2,3,4-tetrol (**6**)

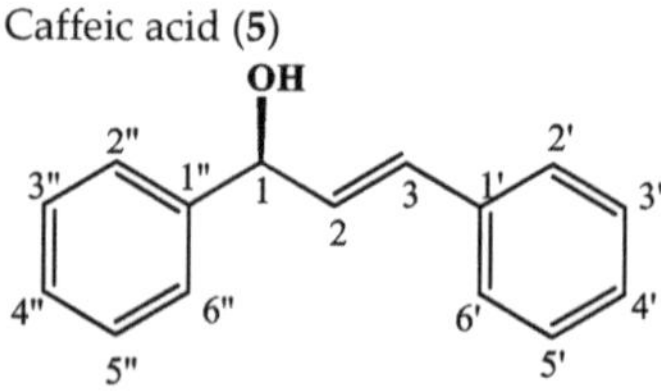

(*S*,*E*)-1,3-Diphenylprop-2-en-1-ol (**7**)

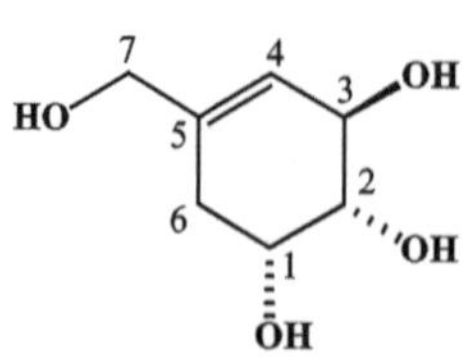

(1*R*,2*R*,3*R*)-5-(hydroxymethyl)cyclohex-4-ene-1,2,3-triol (**8**)

Figure 3. Chemical structures of the compounds isolated from *Acalypha arvensis* (**1**–**8**) in the ethanol extract (AaEtOH).

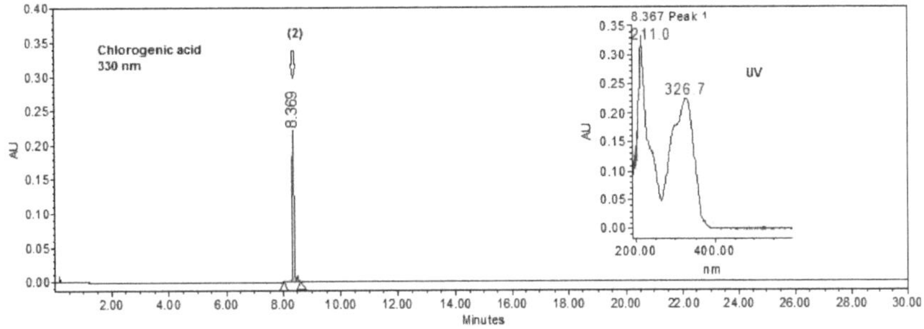

Figure 4. HPLC chromatogram of chlorogenic acid (**2**) and its UV spectrum.

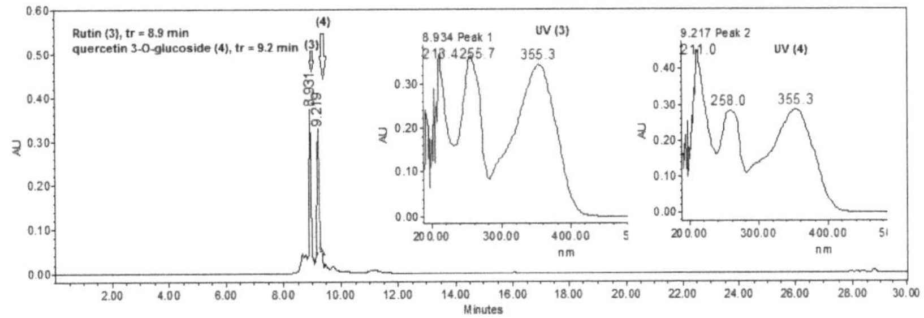

Figure 5. HPLC chromatogram of rutin (**3**), quercetin 3-O-glucoside (**4**), and their UV spectra.

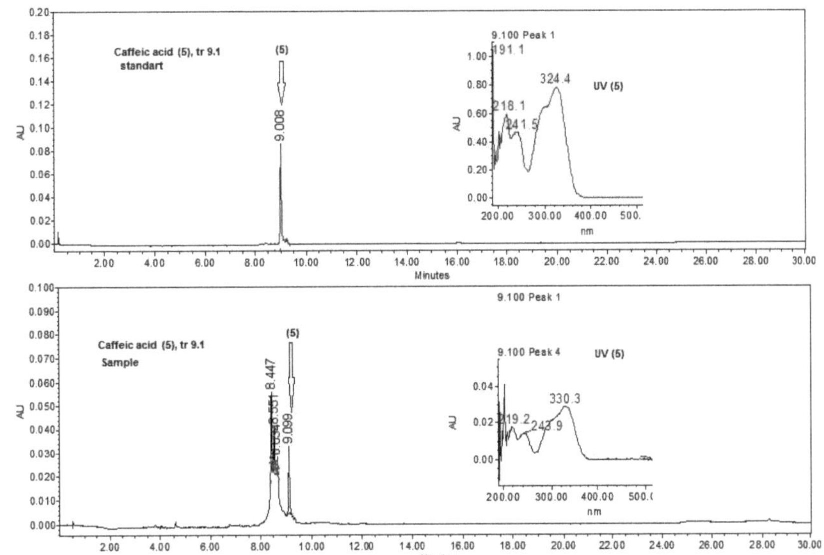

Figure 6. HPLC chromatograms of caffeic acid (**5**) compared to the standard and its UV spectrum.

(*S*,*E*)-1,3-Diphenylprop-2-en-1-ol (**7**): ^{1}H NMR (CD_3OD, 600 MHz) δ 5.31 (1H, d, *J* = 6.6 Hz, H-1), 6.39 (1H, dd, *J* = 6.6, 15.8 Hz, H-2), 6.64 (1H, d, *J* = 15.8 Hz, H-3), 7.42 (1H, dd, *J*= 1.3, 8.4 Hz, H-2′′ y H-6′′), 7.39 (dd, *J* = 1.3, 8.4 Hz, H-2′ y H-6′), 7.20 (dd, *J* = 7.4, 7.4 Hz, H-3′ y H-5′), 7.26 (dd, *J* = 7.4, 7,4 Hz, H-3′′ y H-5′′), 7.35 (dd, *J* = 7.5, 7,6 Hz, H-4′), 7.28 (dd, *J* = 7.6, 7.9 Hz, H-4′′); ^{13}C NMR (CD_3OD, 150 MHz) δ 75.8 (CH, C-1), 133.3 (CH, C-2), 131.1 (CH,

C-3), 127.4 (4CH, C-2′,C-2″, C-6′ y C-6″), 128.3 (2CH, C-3′y C-5′), 128.5 (2CH, C-3″y C-5″), 129.5 (CH, C-6″), 129.3 (CH, C-6′), 141.8 (C, C-1″), 138.2 (C, C-1′).

(1*R*,2*R*,3*R*)-5-(hydroxymethyl)cyclohex-4-ene-1,2,3-triol (**8**): ^{1}H NMR (CD_3OD, 600 MHz) δ 6.78 (1H, d, 3.3 Hz, H-4), 4.37 (1H, dd, 3.6, 4.2 Hz, H-3), 3.67 (1H, m, H-2), 3.98 (1H, dd, 5.4, 7.5 Hz), 2.72 (1H, dd, 4.9, 18.0, H-6a), 2.19 (1H, dd, 5.9, 18.0, H-6b), 3.74 (1H, dd, 2.7, 11.0 Hz, H-7a), 3.60 (1H, m, H-7b): ^{13}C NMR (CD_3OD, 150 MHz) δ 138.2 (CH, H-4), 67.37 (CH, C-3), 72.9 (CH, C-2), 68.3 (CH, C-1), 31.8 (CH_2, C-6), 131.1 (C, C-5), 64.6 (CH_2, C-7).

3. Discussion

Considering the current uses in traditional medicine and the antibacterial activity reports from the title plant, the aim of this work was to evaluate the antibacterial properties of *Acalypha arvensis* and the identification of molecules responsible for the activity.

The emergence and dissemination of methicillin-resistant *Staphylococcus* (MRS) strains is a worrying problem in public health. Therefore, new anti-MRS agents are urgently needed. Species of the genus *Acalypha* have shown antibacterial activity including *A. alnifolia, A. alopecuroidea, A. arvenis, A. fimbriata, A. gaumeri, A. hispida, A. indica, A. monostachya, A. platyphilla, A. racemose, A. wilkesiana,* and *A. torta* [28]. In this work, it was demonstrated that the ethanolic extract of *A. arvensis* exerts antibacterial effects against methicillin-sensitive and methicillin-resistant *Staphylococcus aureus, Klebsiella pneumoniae,* and *Pseudomonas aeruginosa* strains, which is consistent with what has been described in the literature, where its effect against strains of *Staphylococcus aureus, Salmonella typhi, Shigella flexneri, Mycobacterium intracellurare, Thychophyton mentagrophyte,* and *Saccharomyces cerevisiae* has been reported [18,19,29,30]. There are also reports of antimicrobial potential in other species of the same genus such as *A. diversifolia* where hexane, dichloromethane, and methanol extracts were tested against *Staphylococcus aureus* (ATCC 6538), *Bacillus subtilis* (ATCC 21556), *Klebsiella pneumonia* (ATCC 10031), and *Escherichia coli* (ATCC 9637) and were bioactive [31]. It has also been mentioned that methanolic extract of *Acalypha fruticosa* exhibit positive effects against *Staphylococcus aureus, Bacillus subtilis, Myotis flavus,* and *Staphylococcus epidermis,* while the aqueous extract shows activity against *Streptococcus pyogene, Staphylococcus epidermis, Proteus vulgaris,* and *Escherichia coli* [28].

The activity of AaEOH from *A. arvensis* could be due to phenolic acids and flavonoids that were identified in this extract. In general, tannins, flavonoids, coumarins, saponins, alkaloids, terpenes, coumarins, anthocyanins, and anthraquinones have been previously described in species of this genus [28]. In this study, corilagin (**1**), an ellagitannin, was isolated and identified; this compound has been isolated from different species as well, including those of the genus Euphorbiaceae, such as *Phyllantus niruri* [32]. Some pharmacological activities of corilagin have already been described, such as antiatherogenic [33], antioxidant [34], hepatoprotective [35], antitumor [36], and antibacterial [37]. Corilagin has been reported to exhibit antimicrobial activity due to its inhibitory effect against *S. aureus, E. coli, P. aeruginosa,* and *K. pneumoniae* with a MIC of 1024 pg/mL [38]. Tannins have antibacterial effects, inhibiting the growth of Gram-positive and Gram-negative bacteria, and most compounds have bacteriostatic properties. The MICs of several tannins range from 61.5 to 3200 μg/mL [39]. Some of the proposed mechanisms of how these compounds (gallotannins) act involve their iron chelating property. Iron is essential for optimal bacterial growth. Siderophores, low-molecular-weight organic compounds produced by bacteria, can solubilize iron in the external environment and make it available to bacteria. Gallotannins can chelate ferric iron from their environment, making iron unavailable to bacteria, leading to inhibition of bacterial growth due to iron deprivation. Furthermore, the iron chelation efficiency of gallotannins is correlated with the number of galloyl groups, with increasing degrees of galloylation reducing the iron-binding capacity due to steric effects. It has also been shown that tannins can inhibit bacterial cell-wall synthesis by inactivating the enzymes involved or by direct binding [40]. On the other hand, chlorogenic acid has been shown to have broad-spectrum antibacterial activity and some inhibition against *E. coli* and *S. aureus*. Reports indicate that the antibacterial mechanism of chlorogenic acid may be related to noncompetitive inhibition of arylamine acetyltransferase in bacteria, as well as changing the permeability of cell membranes by inhibiting a change in β-galactosidase. This reduces the concentration of sugar and acetone in the metabolic process of bacteria and hindering the metabolism and protein synthesis of the strain, resulting in insufficient energy and further affecting the bacteria's growth and reproduction [41].

It has been reported that plants rich in phenolic acids and flavonoids, two types of metabolites present in the ethanolic extract of *A. arvensis*, possess a broad spectrum of antimicrobial activity [42]. In particular, the amount and position of the hydroxyl groups in these types of compounds, such as rutin, quercetin glycoside, and caffeic acid present in *A. arvensis*, have been shown to be related to their antibacterial effect, mainly against *S. aureus*. The presence of these functional groups affects lipophilicity, damaging the phospholipids and proteins, resulting in an increase in cell permeability [43,44].

Since there are numerous toxic plants, it is important to evaluate their preclinical toxicological and pharmaceutical action before considering using them for a safe/beneficial treatment and therefore validating them as medicinal plants [45]. As such, it is in our interest to continue with further evaluation (in vivo) of the *Acalypha arvensis* extracts against biological models. Thus far, it has been reported that studies carried out with corilagin revealed that the applied therapeutic dose does not exert adverse effects on the liver [36]. Additionally, the compounds quercetin, gallic acid, corilagin, and ellagic acid protect against some cytotoxic effects of acetaminophen, microcystins, galactosamine, and lipopolysaccharides [46]. Therefore, isolation and chromatographic analysis of polyphenolic compounds such as flavonoids, hydroxycinnamic acid derivatives, and organic acids from plants aid in the understanding of their inherited bioactivity [47–49]. In the case of *A. arvensis* extracts, it was found that around (8–10 min) ellagitannin was found during chromatographic analysis. However, there were additional natural products that overlapped at that same retention time. This low resolution could be improved using micro-pillar matrix columns, which have proved to have better separation performance in HPLC [50].

4. Materials and Methods

4.1. Equipment and Reagents

NMR spectra were recorded on an Agilent DD2-600 at 600 MHz for 1H and 150 MHz for ^{13}C NMR, using CD_3OD as the solvent. Chemical shifts are reported in ppm relative to TMS. Thin-layer chromatography (TLC) was performed using TLC Silica gel 60, F254, and 20 × 20 cm aluminum sheets (Merck KGaA, Darmstadt, Germany). High-performance liquid chromatography (HPLC, Milford, MA, USA) analyses were performed on a Waters 2695 Separation module system, equipped with a photodiode array detector (Waters Co. 2996) and Empower 3 software (Waters Corporation, Milford, MA, USA).

4.2. Plant Material

The aerial parts of *Acalypha arvensis* were collected in Pechucalco 1st section of the municipality of Cunduacán, Tabasco, Mexico; during the month of April 2021. One specimen was deposited in the Herbarium of the Academic Division of Biological Sciences of the Universidad Juárez Autónoma de Tabasco for taxonomic identification and safekeeping (Voucher No. 036228). The fresh plant material of *Acalypha arvensis* was dried at room temperature, under shade for 72 h.

4.3. Extracts

Dried material (820 g) was milled in a grinder (Pulvex, particle size 4 mm). The extraction process was through a serial maceration with solvents of ascending polarity (*n*-hexane, ethyl acetate and ethanol). This to extract all the secondary metabolites according to their dissolution affinity. Initially, *n*-hexane (2.0 L, Merck) was added to the dry material and allowed to stand for 24 h at room temperature (25–30 °C). Subsequently, it was filtered (Whatman No. 4 paper) and concentrated in a rotary evaporator (Heidolph G3, Germany) under reduced pressure, to obtain the hexane extract (AaHex, 7.4 g). This process was repeated in triplicate. The same plant material dried after extraction with hexane was macerated with ethyl acetate (2.0 L, Merck, Mexico City, Mexico). The same procedure mentioned above was followed, to give the ethyl acetate extract (AaAcOEt, 19.2 g). Finally, the ethanolic (AaEtOH, 11.2 g) was obtained following the same protocol. All extracts were lyophilized (Heto Drywinner DW3) and tested by the biological model used for this study.

4.4. Isolation and Identification of Compounds (**1–8**)

Ethanolic extract (AaEtOH, 11.2 g) was adsorbed with silica gel (60 g, gel 60, Merck) and fractionated in a glass column (600 × 50 mm) packed with silica gel (100 g, 70–230 mesh, Merck) as stationary phase. Dichloromethane was used as the mobile phase with gradual increase of polarity using 10% *v*/*v* methanol, collecting 32 fractions of 200 mL. All samples were concentrated under reduce pressure using a rotary evaporator (Heidolph Laborota 4000) and lyophilized. Chromatographic analysis by CCF allowed the assemblies in five fractions: AaR1 (0.05 g), AaR2 (1.5 g), AaR3 (2.2), AaR4 (4.2 g), and AaR5 (1.8 g). Fraction AaR4 (4 g, column 2) was adsorbed on silica gel (7 g, Rp-18, Merck) as a stationary phase and water with a decrease in polarity with acetonitrile at 5% *v*/*v* was used as a mobile phase, collecting 50 fractions of 30 mL each. A dark yellow precipitate was obtained in fractions 3–5, which was identified as corilagin (**1**). In fractions 8–10, 13–20, and 23–27, chlorogenic acid (**2**), rutin (**3**), quercetin-3-*O*-glucoside (**4**) and caffeic acid (**5**) were identified using HPLC by comparison with commercial standards. In fraction 32–37, a mixture of a polyol called treitol (**6**), a dihydrochalcone known as (*S*,*E*)-1,3-diphenylprop-2-en-1-ol (**7**) and a shikimic acid derivative (1*R*,2*R*,3*R*)-5-(hydroxymethyl)cyclohex-4-ene-1,2,3-triol (**8**) were identified (see Scheme 1).

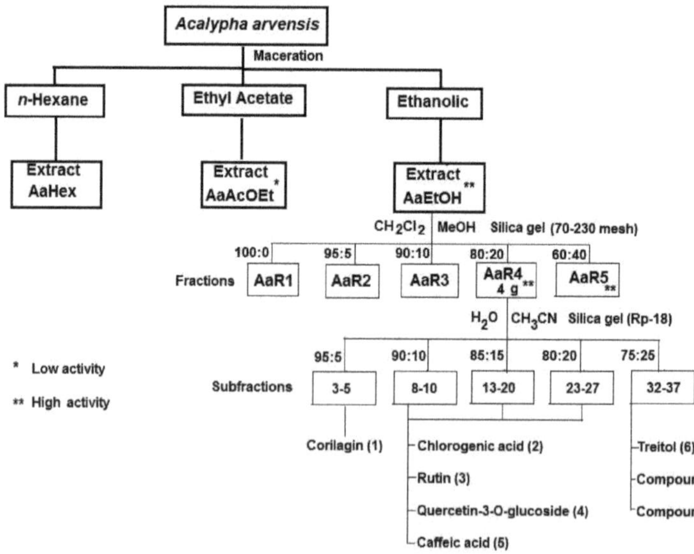

Scheme 1. Protocol used for the isolation of active compounds from *Acalypha arvensis*.

4.5. Antibacterial Activity

4.5.1. Strains Used

The 13 ATCC bacterial strains used were: Gram-positive; *Staphylococcus aureus* 29213(Sa), methicillin-resistant *Staphylococcus aureus* 43300(SaRM1) and 3359(SaRM2), *Staphylococcus epidermis* 1042(Se), *Staphylococcus haemolyticus* 1165(Sh), and *Enterococcus faecalis* 29212(Ef), Gram-negative; *Klebsiella pneumoniae* 13883(Kp1) and 700605(Kp2), *Pseudomonas aeruginosa* 27853(Pa), *Escherichia coli* 25922(Ec1), 1047(Ec2), and 4036(Ec3) and *Salmonella dublin* 9676(Sd). The strains were maintained on Trypticase Soy Agar (Merck) at 37 °C, 24 h.

4.5.2. In Vitro Evaluation of the Organic Extracts Using a Plate Dilution Method

The antibacterial activity was measured by determining the minimal inhibitory concentration (MIC) and it was carried out using the standard agar dilution method [51]. Briefly, the AaHex, AaAcOEt, and AaEtOH extracts and AaR2, AaR3, AaR4, and AaR5 fractions were dissolved in dimethyl sulfoxide (DMSO; 2% v/v) and sterile water (8% v/v); to obtain a concentration of 2, 1, 0.5, 0.25, 0.125 mg/mL. The inoculum for each organism was prepared from cultures containing 10^8 colony-forming units (CFU)/mL (MacFarland scale standard 0.5). The diluted (1:20) inoculum was applied as a drop, by means of a calibrated pipet that delivered 2 μL, resulting in a drop inoculum covering a circle of 5 mm diameter and containing 10^4 CFU. The plates were incubated for 24 h at 37 °C. Gentamicin (250 μg/mL; Sigma) was used as positive control (Control (+)). Observations were performed by duplicate, and results are expressed as the lowest concentration of extract or fraction able to produce a complete suppression of colony growth on agar (minimum inhibitory concentration).

5. Conclusions

The present study allowed us to conclude that the ethanolic extract from the plant species *Acalypha arvensis* presents antibacterial activity against a couple of strains of methicillin-sensitive and methicillin-resistant *Staphylococcus aureus*, *Klebsiella pneumoniae*, and *Pseudomonas aeruginosa*, which supports the known medicinal applications attributed to this plant. The presence of the ellagitannin called corilagin (**1**) was isolated and identified in this species for the first time. It is one of the main constituents present in both the extract and in the active fractions (AaR4 and AaR5). Therefore, it makes sense to state that corilagin (**1**) is one of the main natural products responsible for the observed biological activity. However, we cannot discard the other phenolic-type compounds (chlorogenic acid, quercetin glycoside, caffeic acid and rutin) isolated. Therefore, it can be concluded that the aerial parts of this plant could be a good alternative in the treatment of infections caused by methicillin-resistant bacteria.

Supplementary Materials: The following supporting information can be downloaded at: https://www.mdpi.com/article/10.3390/plants11030300/s1, Figure S1: Chemical structure, HPLC chromatogram and UV light spectrum of corilagin (**1**). Figure S2: ^{1}H-NMR (CD_3OD, 600 MHz) of corilagin (**1**). Figure S3: ^{13}C-NMR (CD_3OD, 150 MHz) of corilagin (**1**). Figure S4: ^{13}C-DEPT-NMR (CD_3OD, 150 MHz) of corilagin (**1**). Figure S5: ^{1}H-^{1}H(COSY)-NMR (CD_3OD, 600 MHz) of corilagin (**1**). Figure S6: ^{1}H-^{13}C(HSQC)-NMR (CD_3OD, 600 MHz for ^{1}H and 150 MHz for ^{13}C) of corilagin (**1**). Figure S7: ^{1}H-^{13}C(HMBC)-NMR (CD_3OD, 600 MHz for ^{1}H and 150 MHz for ^{13}C) of corilagin (**1**). Figure S8: ^{1}H-NMR (CD_3OD, 600 MHz) of the mixture of compounds (**6**–**8**). Figure S9: ^{13}C-NMR (CD_3OD, 150 MHz) of the mixture of compounds (**6**–**8**). Figure S10: ^{13}C-DEPT-NMR (CD_3OD, 150 MHz) of the mixture of compounds (**6**–**8**). Figure S11: ^{1}H-^{1}H(COSY)-NMR (CD_3OD, 600 MHz) of the mixture of compounds (**6**–**8**). Figure S12: ^{1}H-^{13}C(HSQC)-NMR (CD_3OD, 600 MHz) of the mixture of compounds (**6**–**8**). Figure S13: ^{1}H-^{13}C(HMBC)-NMR (CD_3OD, 600 MHz) of the mixture of compounds (**6**–**8**).

Author Contributions: E.A.B.-G., C.E.L.-G. and M.G.-C., conceived and designed the experiments; C.E.L.-G., M.D.P.-G., P.Á.-F. and A.Z. analyzed the chemical and biological data; E.A.B.-G., R.L.-R. and M.G.-C., performed the experiments; E.A.B.-G., M.G.-C., A.S.G.-R., M.D.P.-G. and A.G.-R. and A.B. wrote/edited the paper. All authors have read and agreed to the published version of the manuscript.

Funding: This research was funded by Universidad Juarez Autónoma de Tabasco Grant number: UJAT-PTC-286, PRODEP 2020.

Institutional Review Board Statement: Not applicable.

Informed Consent Statement: Not applicable.

Data Availability Statement: Data is contained within the article and supplementary material.

Acknowledgments: The authors wish to thank Arturo Pérez, Jonatan Orduño and Ixchel Palacios for technical assistance and analytical.

Conflicts of Interest: The authors declare no conflict of interest.

References

1. OMS. Datos Recientes Revelan los Altos Niveles de Resistencia a los Antibióticos en Todo el Mundo. Available online: https://www.who.int/es/news-room/detail/29-01-2018-high-levels-of-antibiotic-resistance-found-worldwide-new-data-shows (accessed on 2 December 2021).
2. Giono-Cerezo, S.; Santos-Preciado, J.I.; Rayo Morfín-Otero, M.D.; Torres-López, F.J.; Alcántar-Curiel, M.D. Resistencia antimicrobiana. Importancia y esfuerzos por contenerla. *Gac. Med. Mex.* **2020**, *156*, 172–180. [CrossRef]
3. Fauci, A.S. Emerging and reemerging infectious diseases: The perpetual challenge. *Acad. Med.* **2005**, *80*, 1079–1085. [CrossRef] [PubMed]
4. Watkins, K. Emerging infectious diseases: A review. *Curr. Emerg. Hosp. Med. Rep.* **2018**, *6*, 86–93. [CrossRef] [PubMed]
5. Romero-Cordero, S.; Kirwan, R.; Noguera-Julian, A.; Cardellach, F.; Fortuny, C.; Morén, C. A Mitocentric View of the Main Bacterial and Parasitic Infectious Diseases in the Pediatric Population. *Int. J. Mol. Sci.* **2021**, *22*, 3272. [CrossRef] [PubMed]
6. Bhatia, P.; Sharma, A.; George, A.J.; Anvitha, D.; Kumar, P.; Dwivedi, V.P.; Chandra, N.S. Antibacterial activity of medicinal plants against ESKAPE: An update. *Heliyon* **2021**, *7*, e06310. [CrossRef]
7. Velázquez Meza, M.E. Surgimiento y diseminación de *Staphylococcus aureus* meticilinorresistente. *Salud Pública México* **2005**, *47*, 381–387. [CrossRef] [PubMed]
8. León-Buitimea, A.; Garza-Cárdenas, C.R.; Garza-Cervantes, J.A.; Lerma-Escalera, J.A.; Morones-Ramírez, J.R. The Demand for New Antibiotics: Antimicrobial Peptides, Nanoparticles, and Combinatorial Therapies as Future Strategies in Antibacterial Agent Design. *Front. Microbiol.* **2020**, *11*, 1669. [CrossRef]
9. Cheesman, M.J.; Ilanko, A.; Blonk, B.; Cock, I.E. Developing New Antimicrobial Therapies: Are Synergistic Combinations of Plant Extracts/Compounds with Conventional Antibiotics the Solution? *Pharmacogn. Rev.* **2017**, *11*, 57–72.
10. Hemeg, H.A.; Moussa, I.M.; Ibrahim, S.; Dawoud, T.M.; Alhaji, J.H.; Mubarak, A.S.; Kabli, S.A.; Alsubki, R.A.; Tawfik, A.M.; Marouf, S.A. Antimicrobial effect of different herbal plant extracts against different microbial population. *Saudi J. Biol. Sci.* **2020**, *27*, 3221–3227. [CrossRef] [PubMed]
11. Farhadi, F.; Khameneh, B.; Iranshahi, M.; Iranshahy, M. Antibacterial activity of flavonoids and their structure-activity relationship: An update review. *Phytother. Res.* **2019**, *33*, 13–40. [CrossRef]
12. Górniak, I.; Bartoszewski, R.; Króliczewski, J. Comprehensive review of antimicrobial activities of plant flavonoids. *Phytochem. Rev.* **2019**, *18*, 241. [CrossRef]
13. Standley, P.C.; Steyermark, J.A. *Flora of Guatemala*; Chicago Natural History Museum: Chicago, IL, USA, 1949; Volume 24, pp. 25–170.
14. Méndez Gómez, E.; López Noverola, U.; López Naranjo, J.I.; Salaya Domínguez, J.M.; Díaz González, J.A. *Catálogo de Plantas Medicinales de Uso Actual en el Estado de Tabasco*, 1st ed.; Fundación Produce Tabasco-UJAT: Tabasco, Mexico, 2004; p. 39.
15. Duke, J.A. *Duke's Handbook of Medicinal Plants of Latin America*, 1st ed.; CRC Press: Boca Raton, FL, USA, 2008; pp. 4–5.
16. Argueta, A.L.; Cano, L.; Asselein, L.; Rodarte, M.E. *Atlas de las Plantas de la Medicina Tradicional Mexicana*, 1st ed.; Instituto Nacional Indigenista: Ciudad de México, Mexico, 1994; p. 236.
17. Vibrans, H. *Acalypha arvensis* Poepp. & Endl. (Hierba del Gusano). Available online: http://www.conabio.gob.mx/malezasdemexico/euphorbiaceae/acalypha-arvensis/fichas/ficha.htm (accessed on 2 December 2021).
18. Cáceres, A.; Cano, O.; Samayoa, B.; Aguilar, L. Plants used in Guatemala for the treatment of gastrointestinal disorders. 1. Screening of 84 plants against enterobacteria. *J. Ethnopharmacol.* **1990**, *30*, 55–73. [CrossRef]
19. Cáceres, A.; Alvarez, A.V.; Ovando, A.E.; Samayoa, B.E. Plants used in Guatemala for the treatment of respiratory diseases. 1. Screening of 68 plants against gram-positive bacteria. *J. Ethnopharmacol.* **1991**, *31*, 193–208. [CrossRef]
20. Jiménez, M.; Cruz, S.M.; Cáceres, A. Determinación de la actividad biocida de cinco especies del genero Acalypha (*A. guatemalensis, A. arvensis, A. polystaquia, A. hispida* y *A. pseudoalopecuroides*). *Rev. Científica (IIQB USAC)* **2005**, *3*, 35–38.
21. Yamada, H.; Nagao, K.; Dokei, K.; Kasai, Y.; Michihata, N. Total synthesis of (−)-Corilagin. *J. Am. Chem. Soc.* **2008**, *130*, 7566–7567. [CrossRef] [PubMed]
22. Kochumadhavan, A.; Mangal, P.; Kumar, L.S.; Meenakshi, B.M.; Venkanna, B.U.; Muguli, G. Corilagin: First time isolation from the whole plant of *Phyllanthus maderaspatensis* L. *Pharmacogn. Commun.* **2019**, *9*, 135–138. [CrossRef]
23. Sudjaroen, Y.; Hull, W.E.; Erben, G.; Würtele, G.; Changbumrung, S.; Ulrich, C.M.; Owen, R.W. Isolation and characterization of ellagitannins as the major polyphenolic components of Longan (*Dimocarpus longan* Lour) seeds. *Phytochemistry* **2012**, *77*, 226–237. [CrossRef] [PubMed]
24. Wang, J.C.; Hu, S.H.; Wang, J.T.; Chen, K.S.; Chia, Y.C. Hypoglycemic effect of extract of *Hericium erinaceus*. *J. Sci. Food Agric.* **2005**, *85*, 641–646. [CrossRef]
25. Walters, K.R.; Pan, Q.; Serianni, A.S.; Duman, J.G. Cryoprotectant biosynthesis and the selective accumulation of threitol in the freeze-tolerant Alaskan beetle, *Upis ceramboides*. *J. Biol. Chem.* **2009**, *284*, 16822–16831. [CrossRef]

26. Szymczyk, M. Unexpected course of Wittig reaction when using cinnamyl aldehyde as a substrate. *Phosphorus Sulfur Silicon Relat. Elem.* **2017**, *192*, 264–266. [CrossRef]
27. Pawlak, J.L.; Berchtold, G.A. Total synthesis of (−)-chorismic acid and (−)-shikimic acid. *J. Org. Chem.* **1987**, *52*, 1765–1771. [CrossRef]
28. Seebaluck, R.; Gurib-Fakim, A.; Mahomoodally, F. Medicinal plants from the genus *Acalypha* (Euphorbiaceae)—A review of their ethnopharmacology and phytochemistry. *J. Ethnopharmacol.* **2015**, *159*, 137–157. [CrossRef] [PubMed]
29. Cáceres, A.; Girón, L.M.; Alvarado, S.R.; Torres, M.F. Screening of antimicrobial activity of plants popularly used in Guatemala for the treatment of dermatomucosal diseases. *J. Ethnopharmacol.* **1987**, *20*, 223–237. [CrossRef]
30. Lentz, D.L.; Clark, A.M.; Hufford, C.D.; Meurer-Grimes, B.; Passreiter, C.M.; Cordero, J.; Ibrahimi, O.; Okunade, A.L. Antimicrobial properties of Honduran medicinal plants. *J. Ethnopharmacol.* **1998**, *63*, 253–263. [CrossRef]
31. Nino, J.; Mosquera, O.M.; Corea, Y.M. Antibacterial and antifungal activities of crude plant extracts from Colombian biodiversity. *Rev. Biol. Trop.* **2012**, *60*, 1535–1542. [CrossRef]
32. Moreira, J.; Klein-Júnior, L.C.; Filho, V.C.; Buzzi, F.C. Anti-hyperalgesic activity of corilagin, a tannin isolated from *Phyllanthus niruri* L. (Euphorbiaceae). *J. Ethnopharmacol.* **2013**, *146*, 318–323. [CrossRef]
33. Duhnam, N.W.; Miya, T.S. A note on a simple apparatus for detecting neurobiological deficit in rats and mice. *J. Am. Pharm. Assoc.* **1957**, *46*, 208–209.
34. Chen, Y.; Chen, C. Corilagin prevents tert-butyl hydroperoxide-induced oxidative stress injury in cultured N9 murine microglia cells. *Neurochem. Int.* **2011**, *59*, 290–296. [CrossRef]
35. Kinoshita, S.; Inoue, Y.; Nakama, S.; Ichiba, T.; Aniya, Y. Antioxidant and hepatoprotective actions of medicinal herb, *Terminalia catappa* L. from Okinawa Island and its tannin corilagin. *Phytomedicine* **2007**, *14*, 755–762. [CrossRef]
36. Hau, D.K.; Zhu, G.; Leung, A.K.; Wong, R.S.; Cheng, G.Y.; Lai, P.B.; Tong, S.; Lau, F.; Chan, K.; Wong, W.; et al. In vivo anti-tumor activity of corilagin on Hep3B hepatocellular carcinoma. *Phytomedicine* **2010**, *18*, 11–15. [CrossRef]
37. Burapadaja, S.; Bunchoo, A. Antimicrobial Activity of Tannins from *Terminalia citrina*. *Planta Med.* **1995**, *61*, 365–366. [CrossRef] [PubMed]
38. Kolodziej, H.; Kayser, O.; Latt'e, K.P.; Ferreira, D. Evaluation of the antimicrobial potency of tannins and related compounds using the microdilution broth method. *Planta Med.* **1999**, *65*, 444–446. [CrossRef] [PubMed]
39. Boakye, Y.D.; Agyare, C.; Hensel, A. Anti-infective properties and time-kill kinetics of *Phyllanthus muellerianus* and its major constituent, geraniin. *Med. Chem. Curr. Res.* **2016**, *6*, 95–104. [CrossRef]
40. Farha, A.K.; Yang, Q.-Q.; Kim, G.; Li, H.B.; Zhu, F.; Liu, H.Y.; Gan, R.Y.; Corke, H. Tannins as an alternative to antibiotics. *Food Biosci.* **2020**, *38*, 100751–100770. [CrossRef]
41. Miao, M.; Xiang, L. Pharmacological action and potential targets of chlorogenic acid. *Adv. Pharmacol.* **2020**, *87*, 71–88.
42. Kumar, S.; Pandey, A.K. Chemistry and biological activities of flavonoids: An overview. *Sci. World J.* **2013**, *2013*, 162750. [CrossRef]
43. Kępa, M.; Miklasińska-Majdanik, M.; Wojtyczka, R.D.; Idzik, D.; Korzeniowski, K.; Smoleń-Dzirba, J.; Wąsik, T.J. Antimicrobial Potential of Caffeic Acid against *Staphylococcus aureus* Clinical Strains. *Biomed. Res. Int.* **2018**, *2018*, 7413504. [CrossRef] [PubMed]
44. Adamczak, A.; Ożarowski, M.; Karpiński, T.M. Antibacterial Activity of Some Flavonoids and Organic Acids Widely Distributed in Plants. *J. Clin. Med.* **2019**, *9*, 109. [CrossRef] [PubMed]
45. Parra, A.L.; Yhebra, R.S.; Sardiñas, I.G.; Buela, L.I. Comparative study of the assay of *Artemia salina* L. and the estimate of the medium lethal dose (LD50 value) in mice, to determine oral acute toxicity of plant extracts. *Phytomedicine* **2001**, *8*, 395–400.
46. Thilakchand, K.R.; Mathai, R.T.; Simon, P.; Ravi, R.T.; Baliga-Rao, M.P.; Baliga, M.S. Hepatoprotective properties of the Indian gooseberry (*Emblica officinalis* Gaertn): A review. *Food Funct.* **2013**, *4*, 1431–1441. [CrossRef]
47. Rozing, G. Micropillar array columns for advancing nanoflow HPLC. *Microchem. J.* **2021**, *170*, 106629. [CrossRef]
48. Martínez-Hernández, G.B.; Jiménez-Ferrer, E.; Román-Ramos, R.; Zamilpa, A.; González-Cortazar, M.; León-Rivera, I.; Vargas-Villa, G.; Herrera-Ruiz, M. A mixture of quercetin 4′-O-rhamnoside and isoquercitrin from Tilia americana var. mexicana and its biotransformation products with antidepressant activity in mice. *J. Ethnopharmacol.* **2021**, *267*, 113619. [CrossRef]
49. Alegría-Herrera, E.; Herrera-Ruiz, M.; Román-Ramos, R.; Zamilpa, A.; Santillán-Urquiza, M.A.; Aguilar, M.I.; Aviles-Flores, M.; Fuentes-Mata, M.; Jiménez-Ferrer, E. Effect of Ocimum basilicum, Ocimum selloi, and rosmarinic acid on cerebral vascular damage in a chronic hypertension model. *Biol. Pharm. Bull.* **2019**, *42*, 201–211. [CrossRef] [PubMed]
50. García-Hernández, C.; Rojo-Rubio, R.; Olmedo-Juárez, A.; Zamilpa, A.; Mendoza-de Gives, P.; Antonio-Romo, I.A.; Aguilar-Marcelino, L.; Arece-García, J.; Tapia-Maruri, D.; González-Cortazar, M. Galloyl derivatives from Caesalpinia coriaria exhibit in vitro ovicidal activity against cattle gastrointestinal parasitic nematodes. *Exp. Parasitol.* **2019**, *200*, 16–23. [CrossRef] [PubMed]
51. Navarro García, V.M.; Rojas, G.; Gerardo Zepeda, L.; Aviles, M.; Fuentes, M.; Herrera, A.; Jiménez, E. Antifungal and Antibacterial Activity of Four Selected Mexican Medicinal Plants. *Pharm. Biol.* **2006**, *44*, 297–300. [CrossRef]

plants

MDPI

Article

An Insight into *Salvia haematodes* L. (Lamiaceae) Bioactive Extracts Obtained by Traditional and Green Extraction Procedures

Rosa Tundis [1,*], Nicodemo Giuseppe Passalacqua [2], Marco Bonesi [1] and Monica Rosa Loizzo [1]

1 Department of Pharmacy, Health and Nutritional Sciences, University of Calabria, 87036 Rende, Italy; marco.bonesi@unical.it (M.B.); monica_rosa.loizzo@unical.it (M.R.L.)

2 Museum of Natural History of Calabria and Botanic Garden, University of Calabria, 87036 Rende, Italy; nicodemo.passalacqua@unical.it

* Correspondence: rosa.tundis@unical.it; Tel.: +39-0984-493246

Abstract: Even though *Salvia* is one of the most known genera of the Lamiaceae family, some traditionally used *Salvia* species are still now less investigated. To that end, the present study aims to evaluate the chemical profile and the potential bioactivities of extracts and related fractions obtained from the endemic sage *Salvia haematodes* L. by applying a traditional extraction method such as Soxhlet apparatus (SHS) and the rapid solid–liquid dynamic extraction (RSLDE) by Naviglio extractor® (SHN), considered among the "green techniques" operating at room temperature and with minimum solvent employment and minimum energy. Acetylcholinesterase (AChE) and butyrylcholinesterase (BChE) inhibitory activity was measured by a modified Ellman's method. The antioxidant activity was investigated by using 2,2-diphenyl-1-picrylhydrazyl (DPPH), 2,2′-azino-bis (3-ethylbenzothiazoline-6-sulfonic acid) (ABTS), ferric reducing ability power (FRAP), and β-carotene bleaching tests. The SHN methanol fraction resulted the most active in all assays in particular in inhibiting lipid peroxidation with IC_{50} of 1.7 and 1.6 μg/mL, respectively, after 30 and 60 min of incubation. The SHN *n*-hexane fraction exhibited a selective activity against AChE with half-maximal inhibitory concentration (IC_{50}) of 22.9 μg/mL, while the SHS *n*-hexane extract was more active against BChE (IC_{50} of 30.9 μg/mL). Based on these results, these fractions were subjected to further bio-fractionation by Medium Pressure Liquid Chromatography (MPLC) and the relative obtained fractions were investigated for their AChE and BChE inhibitory activity. A comparative analysis with bio-activity and chemical profile was performed. The observed biological effects provided us with a good starting point for further studies on *S. haematodes* extracts and fractions such as agents beneficial for the treatment of AD.

Keywords: *Salvia haematodes*; rapid solid–liquid dynamic extraction (RSLDE); extractor Naviglio®; Soxhlet apparatus; medium pressure liquid chromatography (MPLC); Alzheimer's disease; antioxidant potential

Citation: Tundis, R.; Passalacqua, N.G.; Bonesi, M.; Loizzo, M.R. An Insight into *Salvia haematodes* L. (Lamiaceae) Bioactive Extracts Obtained by Traditional and Green Extraction Procedures. *Plants* **2022**, *11*, 781. https://doi.org/10.3390/plants11060781

Academic Editors: Juei-Tang Cheng, I-Min Liu, Szu-Chuan Shen, Jose Prieto Garcia and Antonella Smeriglio

Received: 4 February 2022
Accepted: 14 March 2022
Published: 15 March 2022

1. Introduction

Salvia haematodes L. (=*S. pratensis* L. subsp. *haematodes* (L.) Arcang.) belongs to *Salvia* subgen. *Sclarea* (Moench) Benth. (Lamiaceae) [1], and includes only herbaceous sages of the Old World having stamen with an elongated connective widening at the top. A phylogenetic analysis [2] revealed that this subgenus is included in a clade that is paraphyletic to subgen. *Salvia* L., showing a systematic distance between the typical sages (subgen. *Salvia*) and the sage under our investigation.

S. haematodes is a perennial herb with stems up to 100 cm, erect, branched, eglandular-pubescent below, glandular above. Basal leaves are simple, ovate, and long-petiolate (Figure 1). Cauline leaves are sessile and smaller. The inflorescences are from dense to lax, shortly branched with violet-blue flowers. *S. haematodes* is endemic to the Italian Peninsula, from the Marche to Calabria regions, where it grows as a hemicryptophyte scapose on Mediterranean hilly grasslands [3]. Few studies are present in the literature on this *Salvia* species. Its essential oil was predominantly characterized by the presence of monoterpenes, with sabinene as the most abundant constituent and significant amounts of borneol, 1,8-cineole, α-pinene, and (Z)-β-ocimene [4]. The main identified sesquiterpenes are γ-muurolene and β-caryophyllene. In another work, the ethanol extract of *S. haematodes* roots exhibited in vivo enhancement of wound healing, inhibition of carrageenin-induced pedal oedema, and induction of hypothermia [5]. Positive chronotropic and inotropic effects on isolated rabbit heart were also produced. *S. haematodes* ethanol extract demonstrated also in vivo analgesic properties, potentiation of pentobarbitone-induced narcosis, and antagonization of amphetamine-induced excitation [6].

Figure 1. *Salvia haematodes* L. from Calabria (Southern Italy): (**a**) full plant in flower, and (**b**) inflorescence.

Salvia species have been demonstrated to possess promising antioxidant, antimicrobial, anti-inflammatory, anti-mutagenic, and cytotoxic activities [7–12]. The protective effects in neurodegenerative diseases, including Alzheimer's diseases (AD), of *Salvia* species have been also extensively assayed. Among them, there are *S. officinalis*, *S. leriifolia*, *S. glutinosa*, *S. argentea*, *S. indica*, *S. bracteata*, *S. quezelii*, *S. cryptantha*, *S. caespitosa*, *S. viscosa*, *S. microstegia*, *S. lavandulaefolia*, *S. multicaulis*, *S. fruticosa*, *S. pinnata*, *S. tobeyi*, and *S. syriaca*, [9–13]. *S. officinalis* and *S. lavandulaefolia* are the two *Salvia* species that showed beneficial effects by enhancing cognitive performance both in patients affected by dementia or cognitive impairment and healthy subjects. Additionally, both species were demonstrated to be safe with no serious adverse effects compared with placebo.

Alzheimer's diseases (AD) is the most common cause of dementia in the ageing population, which leads to a progressive and irreversible decrease in mental function. The main degenerative condition is characterized by the formation of neurofibrillary tangles and amyloid plaques, and loss of neuron synapses.

Several studies shown early in the disease course a degeneration of cholinergic nuclei. Impairment of the cholinergic system is followed by alteration of attentional processes and cognitive decline. In fact, acetylcholine is a neurotransmitter found in many brain neurons that plays an important role in mental processes, such as memory and cognition. Moreover, it is implicated in smooth muscle contraction, dilatation of blood vessels, and increase in bodily secretions. Now, acetylcholinesterase (AChE) inhibitors represent the best treatment of AD [10]. In the brain, acetylcholine is hydrolysed by two cholinesterases, namely, acetylcholinesterase (AChE) and butyrylcholinesterase (BChE). It was found that in the brain of patients affected by AD, AChE is more abundant than BChE that contributes to the hydroxylation of acetylcholine in the cerebral cortex and hippocampus. However, it has been demonstrated that the activity of AChE is reduced by about 67% compared to the typical levels in the hippocampus and temporal lobe during the AD progression, whereas the activity of BChE was increased up to about 165% of the normal levels [13].

Initially, therapeutic approaches for improving cholinergic neurotransmission have focused on the use of AChE inhibitors, but successively several works demonstrated the relevance of both enzymes in the physio-pathology of AD, establishing the therapeutic importance to inhibit both AChE and BChE [13–15]. Cholinesterases inhibitors, preventing the degradation of acetylcholine, enhance the deficient brain cholinergic neurotransmission and are the first drugs authorized in Europe and the US for the symptomatic treatment of AD.

Numerous AChE inhibitors belong to the alkaloid class and include isoquinoline, steroidal, piperidine, indole, and quinolizidine alkaloids. Alkaloids are considered to be the most promising agents to treat AD due to their nitrogen containing structures. In fact, one of the AChE binding sites involves the interaction of the positively-charged nitrogen; although, another binding site has been found that allows the inhibition by non-alkaloid molecules. In fact, in recent years, different cholinesterases inhibitors with a non-alkaloidal structure have been isolated and tested. Among them, there are several flavonoids, phenolic compounds, coumarins, and several terpenes [16–18].

In this context and continuing our previous studies, the present work aimed to investigate, for the first time, the inhibitory capacity towards the enzymes playing a key role in neurodegenerative disorders, such as AD, of *S. haematodes* extracts in relation to their chemical profile. In this regard, the aerial parts of *S. haematodes* have been subjected to extraction by applying two different extraction procedures: Soxhlet apparatus and extractor Naviglio®. While the Soxhlet extraction is a classic exhaustive extraction technique widely applied to compounds that are sufficiently thermally stable, the rapid solid–liquid dynamic extraction (RSLDE) performed by extractor Naviglio® is considered among the "green techniques" operating at room temperature and with minimum solvents employment and minimum energy. To the best of our knowledge, this is the first work in which this extraction procedure was applied to a *Salvia* species.

2. Results and Discussion

2.1. Biological Activity of *S. haematodes* Extracts

S. haematodes aerial parts, collected in southern Italy, were subjected to extraction by using two different extraction procedures with methanol such as a solvent: (1) Soxhlet apparatus (SHS) and (2) extractor Naviglio® (SHN). Soxhlet extraction is a classic exhaustive extraction technique widely applied to compounds that are sufficiently thermally stable. The rapid solid–liquid dynamic extraction (RSLDE) performed by extractor Naviglio® is considered among the "green techniques" operating at room temperature and with minimum solvents employment and minimum energy.

The extraction by using the extractor Naviglio® is not carried out by osmosis or diffusion as occurs in most of the currently applied solid–liquid extraction techniques, but it is carried out by generating a negative pressure gradient between the internal and external sides of plant materials, followed by the restoration of the initial conditions of equilibrium. Solutions obtained by both extractive processes were combined and dried to obtain total extracts.

The extraction procedure that involves the use of the Soxhlet apparatus (SHS) has allowed obtaining a much higher extraction yield than that obtained with the extractor Naviglio® (SHN). Values of 4.0 and 15.8% for SHN and SHS, respectively, were in fact obtained.

Successively, to operate a separation of non-polar compounds, both total extracts were re-suspended in methanol and partitioned with *n*-hexane. The *n*-hexane solutions were combined and dried to obtain *n*-hexane extracts (yield 2.3% of and 7.2% for SHN and SHS, respectively).

Total extracts, and methanol and *n*-hexane fractions of SHN and SHS were assessed for their potential AChE and BChE inhibitory properties. Inhibitors of cholinesterase were demonstrated to be one of the most promising and used agents to treat AD. The inhibition of both AChE and BChE enzymes has complimentary implications in the treatment of AD. In fact, it was shown that as AD progresses the AChE activity in certain brain regions declines, whereas the BChE activity increases in part to compensate for the loss in the AChE activity. *S. haematodes* extracts inhibited both enzymes in a concentration-dependent manner. IC_{50} (half maximal inhibitory concentration) values and SI (Selectivity Index) are reported in Table 1.

Table 1. Cholinesterases (AChE and BChE) inhibitory activity (IC_{50} μg/mL) of *S. haematodes*.

S. haematodes	Extract/Fraction	AChE	BChE	SI #
SHN	Total	396.4 ± 3.4 [a]	559.9 ± 5.2 [a]	1.4
	Methanol	4.5% #	33.1% #	-
	n-Hexane	22.9 ± 1.0 [a]	408.4 ± 3.8 [a]	17.8
SHS	Total	249.8 ± 2.7 [a]	35.8 ± 0.9 [a]	0.1
	Methanol	458.2 ± 4.2 [a]	312.1 ± 3.4 [a]	0.7
	n-Hexane	214.2 ± 3.2 [a]	30.9 ± 1.1 [a]	0.1
Physostigmine		0.2 ± 0.02	2.4 ± 0.04	12

Data are expressed as mean ± S.D. (n = 3). SHN: *S. haematodes* aerial parts extracted by extractor Naviglio®; SHS: *S. haematodes* aerial parts extracted by Soxhlet apparatus. # SI (Selectivity Index): IC_{50} BChE/IC_{50} AChE. # percentage of inhibition at the concentration of 500 μg/mL. AChE and BChE tests: One-way ANOVA *** $p < 0.0001$ followed by a multicomparison Dunnett's test: [a] $p < 0.01$ compared with physostigmine.

Both total extracts of SHN and all SHS showed a weak inhibitory activity against AChE with IC_{50} values of 396.4 and 249.8 μg/mL, respectively. Interestingly, SHS total extract showed a good inhibitory activity against BChE with an IC_{50} value of 35.8 μg/mL. However, if we analyse literature data, *S. haematodes* exhibited a more potent cholinesterase inhibitory activity than other *Salvia* species including, for example, *S. verbenaca* and *S. aegyptiaca* methanol and decoction that showed a percentage of inhibition lower than 40% at 100 μg/mL, respectively [19]. A similar observation should be made by comparing our data with data obtained by Topçu et al. [20] that investigated fourteen extracts of *Salvia* species against AChE and BChE. Among them, *S. chrysophylla* methanol extract resulted in the most active against AChE with the percentage of inhibition of 64.65% at 200 μg/mL followed by *S. staminea* (55.17%), whereas only *S. staminea* (methanol extract) and *S. poculata* (ethanol extract) exerted a percentage of BChE inhibitory activity greater than 50% at the maximum concentration tested.

After partitioning total extracts, the most active sample against AChE was the *n*-hexane fraction of SHN with an IC_{50} value of 22.9 μg/mL. Conversely, a promising activity against BChE was found for the *n*-hexane fraction of SHS with an IC_{50} value of 30.9 μg/mL. The *n*-hexane fraction of SHN could potently selectively inhibit AChE activities with a Selectivity Index (SI) of 17.8.

The different and selective activity towards the two enzymes can be explained by evaluating the different composition of the two fractions. For this reason, both SHN and SHS *n*-hexane fractions were analysed by gas chromatography–mass spectrometry (GC-MS). In Table 2 the main identified compounds are listed based on their Retention Index (RI) on HP-5 column.

Table 2. The main constituents (%) of *S. haematodes n*-hexane fractions.

Compound	Class	RI [a]	SHN	SHS	I.M. [b]
Sabinene	MH	973	1.3 ± 0.04	tr	1, 2, 3
Eugenol	PH	1354	0.7 ± 0.02	-	1, 2
trans-Caryophyllene	SH	1415	7.7 ± 2.3	4.6 ± 0.7	1, 2, 3
α-Humulene	SH	1455	1.5 ± 0.2	tr	1, 2
γ-Muurolene	SH	1478	0.3 ± 0.01	-	1, 2
γ-Cadinene	SH	1515	0.4 ± 0.05	0.3 ± 0.04	1, 2
δ-Cadinene	SH	1526	1.4 ± 0.01	0.5 ± 0.06	1, 2
Caryophyllene oxide	OS	1580	tr	3.5 ± 1.0	1, 2
Neophytadiene	DI	1837	24.9 ± 2.3	17.8 ± 1.5	1, 2
Phytol	DI	2111	2.5 ± 1.2	1.4 ± 0.7	1, 2
Methyl myristate	FA	1726	0.9 ± 0.02	-	1, 2
Methyl palmitate	FA	1928	1.7 ± 0.3	6.7 ± 0.7	1, 2
Methyl linoleate	FA	1996	3.1 ± 0.4	2.4 ± 0.03	1, 2
Methyl heptadecanoate	FA	2030	1.2 ± 0.2	1.8 ± 0.1	1, 2
Methyl stearate	FA	2128	1.8 ± 0.1	3.7 ± 0.1	1, 2
Tetradecane	AL	1400	0.7 ± 0.02	2.9 ± 0.1	1, 2, 3
Heptadecane	AL	1700	0.5 ± 0.03	tr	1, 2, 3
Octadecane	AL	1800	0.7 ± 0.01	0.4 ± 0.05	1, 2, 3
Nonadecane	AL	1900	0.3 ± 0.01	3.5 ± 0.5	1, 2, 3
Docosane	AL	2200	1.1 ± 0.02	1.0 ± 0.06	1, 2, 3
Tetracosane	AL	2400	1.2 ± 0.01	tr	1, 2, 3
Pentacosane	AL	2500	1.4 ± 0.2	8.6 ± 0.7	1, 2, 3
Heptacosane	AL	2700	1.8 ± 0.04	1.0 ± 0.1	1, 2, 3
Octacosane	AL	2800	0.9 ± 0.01	1.3 ± 0.01	1, 2, 3
Stigmasterol	ST		17.2 ± 1.5	5.4 ± 0.9	1, 2
β-Sitosterol	ST		20.1 ± 2.1	2.3 ± 0.1	1, 2

Data are reported as the mean ± standard deviation (n = 3). tr: traces. -: not identified. [a] RI: Retention indices on the HP-5 column. [b] IM, identification method: 1: Cmparison of retention times; 2: Comparison of mass spectra with MS libraries, 3: Comparison with authentic compounds. Monoterpene Hydrocarbons: MH; Sesquiterpene Hydrocarbons: SH; Oxygenated Sesquiterpenes: OS; Phenolic compound: PH; Diterpenes: DI; Fatty acids derivatives: FA; Sterols: ST.

A total of twenty-seven constituents were identified. SHN *n*-hexane fraction showed as dominant compounds neophytadiene (24.9%), β-sitosterol (20.1%), and stigmasterol (17.2%).

However, different fatty acid methyl esters and alkanes were identified. Moreover, SHS *n*-hexane fraction was characterized by neophytadiene as the most abundant compound (17.8%). However, this fraction was richer in fatty acid derivatives. Sterols were not identified. The sesquiterpene hydrocarbon *trans*-caryophyllene characterized both fractions with percentage of composition of 7.7% and 4.6% for SHN and SHS, respectively. Eugenol, γ-muurolene, and methyl myristate were not found in SHS fraction.

A close correlation between oxidative stress and degenerative diseases has been demonstrated [18,21,22]. In fact, it was demonstrated as oxidative stress is involved in the development and/or progression of AD by promoting tau hyper-phosphorylation, β-amyloids plaques deposition, and the consequent loss of synapses and neurons.

This relationship between oxidative stress and AD suggests the important part of oxidative stress in the pathological process; consequently, antioxidant agents may be useful for the treatment of AD [20]. Taking into account these considerations, we have decided to assess also the antioxidant effects of *S. haematodes* extracts. Data are reported in Table 3.

Table 3. *In vitro* antioxidant activity of *S. haematodes*.

S. haematodes		DPPH Test (IC_{50} µg/mL)	ABTS Test (IC_{50} µg/mL)	FRAP Test * (µM Fe(II)/g)	β-Carotene Bleaching Test (IC_{50} µg/mL)	
	Extract/Fraction				30 min	60 min
SHN	Total	0.4 ± 0.07 [a]	28.1 ± 1.1 [a]	49.6 ± 1.7 [a]	5.3 ± 0.05 [a]	7.9 ± 0.08 [a]
	n-Hexane	17.9 ± 0.8 [a]	81.1 ± 2.8 [a]	10.3 ± 0.9 [a]	7.7 ± 0.02	14.0 ± 0.02
	Methanol	0.3 ± 0.03 [a]	15.6 ± 1.3 [a]	70.3 ± 2.8 [a]	1.7 ± 0.09 [a]	1.6 ± 1.5 [a]
SHS	Total	1.2 ± 0.02 [b]	42.1 ± 1.9 [a]	48.7 ± 1.7 [a]	2.9 ± 0.05 [a]	2.6 ± 0.05
	n-Hexane	41.9 ± 1.6 [a]	220.5 ± 2.5 [a]	1.2 ± 0.05 [a]	2.8 ± 0.03 [a]	2.9 ± 0.07 [c]
	Methanol	0.9 ± 0.04 [b]	18.0 ± 0.9 [a]	58.7 ± 1.2 [b]	2.7 ± 0.02 [a]	2.5 ± 0.04
Ascorbic acid		5.2 ± 0.8	1.2 ± 0.03	-		
BHT		-	-	63.4 ± 4.5		
Propyl gallate					1.3 ± 0.04	1.2 ± 0.03

Data are expressed as mean ± S.D. ($n = 3$). * Samples tested at the concentration of 2.5 mg/mL. DPPH test: One-way ANOVA *** $p < 0.0001$ followed by multicomparison Dunnett's test: [a] $p < 0.01$ compared with ascorbic acid, [b] $p < 0.05$ compared with ascorbic acid; ABTS test: One-way ANOVA *** $p < 0.0001$ followed by multicomparison Dunnett's test: [a] $p < 0.01$ compared with ascorbic acid; FRAP test: One-way ANOVA *** $p < 0.0001$ followed by multicomparison Dunnett's test: [a] $p < 0.01$ compared with ascorbic acid, [b] $p < 0.05$ compared with BHT; β-Carotene bleaching test (t = 30 min): One-way ANOVA *** $p < 0.0001$ followed by multicomparison Dunnett's test: [a] $p < 0.01$ compared with propyl gallate; β-Carotene bleaching test (t = 60 min): One-way ANOVA *** $p < 0.0001$ followed by multicomparison Dunnett's test: [a] $p < 0.01$ compared with ascorbic acid, [c] $p > 0.05$ compared with propyl gallate.

All samples exhibited antioxidant effects in a concentration-dependent manner. Generally, as might be expected for a higher concentration of polyphenolic compounds, the methanol fraction of both SHN and SHS, followed by the total extracts, evidenced the most promising activity. In particular, in the DPPH assay, all samples (IC_{50} values in the range 0.3–1.2 µg/mL), except the *n*-hexane fractions, were more active than the positive control ascorbic acid (IC_{50} value of 5.2 µg/mL).

The same trend was observed in the ABTS assay, but with IC_{50} values of 15.6 and 18.0 µg/mL for both SHN and SHS methanol fractions, respectively, and IC_{50} values of 28.1 and 42.1 µg/mL for the total extracts of SHN and SHS, respectively. However, all extracts are less active than the positive control ascorbic acid (IC_{50} value of 1.2 µg/mL). The ability of samples to induce a reduction in iron, assessed by applying FRAP test, revealed the following grade of potency in both SHN and SHS samples: MeOH fraction > Total extract > *n*-Hexane fraction. It is interesting to note that only SHN methanol fraction exhibited a ferric reducing ability power with a value of 70.3 µM Fe (II)/g greater than the positive control BHT (63.4 µM Fe (II)/g).

A promising ability to inhibit lipid peroxidation was evidenced by the β-carotene bleaching test for the methanol fraction of SHN with IC_{50} values of 1.7 and 1.6 µg/mL after 30 and 60 min of incubation, respectively. SHS samples were able to inhibit lipid peroxidation with IC_{50} values in the range 2.5–2.9 µg/mL. The greater antioxidant activity of SHN methanol fraction could be ascribed to the greater total phenol content (TPC) and total flavonoid content (TFC) that characterized this sample in comparison to the same extract obtained by Soxhlet extraction (Table 4).

Table 4. Total phenol content (TPC) and total flavonoid content (TFC) of *S. haematodes*.

S. haematodes Methanol Fraction	TPC [a]	TFC [b]	TFC/TPC
SHN	55.2 ± 1.3	36.5 ± 1.0	0.7
SHS	29.0 ± 1.1	7.2 ± 0.5	0.2

Data are expressed as mean ± S.D. ($n = 3$). [a] mg of chlorogenic acid equivalents (CA)/g of plant materials. [b] mg of quercetin equivalents (QE)/gram of plant materials.

In fact, methanol fraction obtained by using extractor Naviglio® showed a TPC and a TFC of about two times and five times, respectively, higher than SHS.

In inhibiting lipid peroxidation, promising results were obtained with the *n*-hexane fraction also obtained by the Soxhlet apparatus with IC_{50} values of 2.8 and 2.9 μg/mL after 30 and 60 min of incubation, respectively. In this case, a weak activity was obtained with the *n*-hexane fraction of SHN with IC_{50} values of 7.7 and 14.0 μg/mL after 30 and 60 min of incubation, respectively.

2.2. Medium Pressure Liquid Chromatography (MPLC) Fractionation

Both *n*-hexane fractions obtained from SHN and SHS exhibited an interesting anti-cholinesterase activity. For this reason, these were subjected to fractionation. Briefly, a portion of the *n*-hexane fraction was subjected to medium pressure liquid chromatography (MPLC) on silica gel. This procedure afforded nine fractions from SHN (N1–N9) and seven fractions from SHS (S1–S7). All fractions were tested and the data are reported in Table 5. However, these fractions were shown to be less potent than *S. haematodes n*-hexane fractions, despite some interesting data.

Table 5. AChE and BChE inhibitory activity (IC_{50}, μg/mL) of *S. haematodes* SHN and SHS fractions.

S. haematodes	Fraction	AChE	BChE	SI
SHN	N1	40.2 ± 1.5 [a]	61.5 ± 1.6 [a]	1.5
	N2	180.7 ± 4.2 [a]	166.3 ± 2.4 [a]	0.9
	N3	62.1 ± 2.6 [a]	73.9 ± 2.2 [a]	1.2
	N4	564.9 ± 3.4 [a]	129.8 ± 2.5 [a]	0.2
	N5	115.2 ± 1.8 [a]	52.7 ± 1.3 [a]	0.5
	N6	408.6 ± 3.5 [a]	53.9 ± 1.6 [a]	0.1
	N7	141.9 ± 2.8 [a]	55.9 ± 1.5 [a]	0.3
	N8	110.0 ± 2.0 [a]	31.5 ± 1.0 [a]	0.3
	N9	39.5 ± 1.2 [a]	43.4 ± 1.1 [a]	1.1
SHS	S1	40.4 ± 1.1 [a]	112.6 ± 2.4 [a]	2.8
	S2	114.4 ± 3.1 [a]	51.4 ± 1.1 [a]	0.4
	S3	90.9 ± 2.2 [a]	82.2 ± 1.4 [a]	0.9
	S4	418.7 ± 3.6 [a]	207.7 ± 4.4 [a]	0.5
	S5	444.8 ± 3.9 [a]	167.0 ± 3.5 [a]	0.4
	S6	458.7 ± 3.7 [a]	110.1 ± 1.2 [a]	0.2
	S7	307.8 ± 2.8 [a]	173.8 ± 4.0 [a]	0.6
Physostigmine		0.2 ± 0.02	2.4 ± 0.04	12

Data are expressed as mean ± S.D. ($n = 3$). SI: IC_{50} BChE/IC_{50} AChE. AChE and BChE tests: One-way ANOVA *** $p < 0.0001$ followed by multicomparison Dunnett's test: [a] $p < 0.01$ compared with physostigmine.

The most active fractions against AChE were N9 and N1, with IC_{50} value of 39.5 and 40.2 μg/mL, respectively, while fractions N2-N8 showed weak activity. In the same group of fractions, against BChE interesting results were obtained with fractions N8 and N9 with IC_{50} values of 31.5 and 43.4 μg/mL. Among SHS samples, fraction S1 exhibited an interesting IC_{50} value of 40.4 μg/mL against AChE while the other fractions S2–S7 showed IC_{50} values in the range 90.9–458.7 μg/mL. Against BChE of interest are S2 and S3 fractions (IC_{50} value of 51.4 and 82.2 μg/mL, respectively).

N1, N8, N9, and S1, such as the most promising biologically active fractions, were analysed by GC-MS to identify the components potentially responsible for the enzymes inhibitory activity. Fractions were found to contain different phytochemical classes.

The main components of fraction N1 were sabinene, tetradecane, heptadecane, octadecane, neophytadiene, methyl myristate, and methyl stearate, while eugenol, *trans*-caryophyllene, α-humulene, docosane, tetracosane, pentacosane, and methyl palmitate were identified in fraction N8.

Fraction N9 was characterized by the presence of *trans*-caryophyllene, α-humulene, γ-muurolene, γ-cadinene, δ-cadinene, phytol, stigmasterol, β-sitosterol, heptacosane, octacosane, methyl linoleate, and methyl heptadecanoate as dominant constituents.

Methyl linoleate, methyl heptadecanoate, *trans*-caryophyllene, caryophyllene oxide, nonadecane, neophytadiene, and phytol were recognised in fraction S1.

Some compounds identified in *S. haematodes* bioactive fractions were assayed as pure molecules for their potential activity such as AChE and BChE inhibitors [23–26] (Table S1, Supplementary Materials). Among them, the diterpene alcohol phytol exhibited IC_{50} values of 12.51 and 23.89 μg/mL against AChE and BChE, respectively [25]. A weak activity was found for eugenol with IC_{50} values of 42.4 and 63.5 μg/mL against AChE and BChE, respectively [27]. Analyzing the inhibitory activity of AChE, molecular docking revealed that eugenol forms hydrogen bonds with Trp84 and Glu199 in the catalytic domain of this enzyme as well as a hydrogen bond with His 440 in the peripheral anionic site [28]. However, also the presence of aliphatic hydrocarbons and fatty acids, as well as phytosterols, resulted in AChE and BChE inhibition observed in our studied fractions.

Kayamoto et al. [29] showed the ability of the fatty acid methyl linoleate to exert a selective activity against AChE with an IC_{50} value of 68.8 μM, whereas a weak inhibitory activity was reported against BChE (IC_{50} of 247.8 μM). Not active was methyl palmitate against both enzymes.

Recently, phytosterols have gained the attention of researchers due to their anti-atherogenic properties, cholesterol and lipid lowering effects, and immune-modulating activity. A report on the changes in cholesterol metabolism of patients affected by AD was also present in literature [30].

Phytosterols are able to cross the blood–brain barrier and to accumulate in the brain. Therefore, they might play an important role in modulating pathways linked to neurodegeneration. Several studies have described the antioxidant and neuroprotective properties of β-sitosterol and stigmasterol [31–35]. β-Sitosterol has been demonstrated to increase the action of antioxidant enzymes by the stimulation of the estrogenic receptor/PI3-kinase-reliant pathway and to increase the levels of antioxidant enzymes in colon carcinogenesis. Lipid peroxidation and glucose oxidase-mediated oxidative stress have been inhibited through the incorporation of the phytosterol into the cell membrane [32]. β-sitosterol exhibited in vitro an anti-cholinesterase (with IC_{50} values of 55 and 50 μg/mL against AChE and BChE, respectively) and antioxidant activity and in vivo inhibited enzymes involved in the metabolism of cholinesterases and exerted radicals scavenging effects [30].

3. Materials and Methods

3.1. Chemicals and Reagents

Solvents of analytical grade were obtained from VWR International s.r.l. (Milan, Italy). Ascorbic acid, propyl gallate, butylated hydroxytoluene (BHT), 2,2-azinobis(3-ethylbenzothiazoline-6-sulfonic) acid (ABTS) solution, 2,2-diphenyl-1-picrylhydrazyl (DPPH), tripyridyltriazine (TPTZ), β-carotene, Tween 20, linoleic acid, acetylcholinesterase (AChE) from *Electrophorus electricus* (EC 3.1.1.7, Type VI-S) and butyrylcholinesterase (BChE) from equine serum (EC 3.1.1.8), acetylthiocholine iodide (ATCI), butyrylthiocholine iodide (BTCI), 5,5′-dithio-bis(2-nitrobenzoic acid) (DTNB), and physostigmine were purchased from Sigma-Aldrich S.p.a. (Milan, Italy).

3.2. Plant Materials

The aerial parts of S. haematodes were collected in Calabria (Southern Italy) in May 2012, along the county road SP 241, near the crossing to SP 197 (39.610744° N, 16.275371° E, WGS84, 60 m a.s.l.).

Plant materials were examined for integrity and absence of dust and insect contamination. Aerial parts were harvested in order to obtain an adequate quantity for the analysis. A voucher specimen (n. CLU 23976) was retained at the Natural History Museum of Calabria and the Botanic Garden, University of Calabria (Italy).

3.3. Extraction and Fractionation by Medium Pressure Liquid Chromatography (MPLC)

The air-dried and powdered aerial parts of *S. haematodes* were extracted with methanol by using:

(a) Soxhlet apparatus (SHS) (conventional glass with an extraction chamber with a diameter of 8 cm and a height of 30 cm, accompanied by a flask of capacity of 1 L; 600 mL, 8 extractive cycles);
(b) extractor Naviglio® (SHN) (Nuova Estrazione S.a.s., Naples, Italy, 2 L capacity model; 30 extractive cycles each of which being 4 min). The combined extractive solutions were evaporated to dryness in vacuo using a rotary evaporator at 35–40 °C.

Yields of 2.3% and 15.8% were obtained for SHN and SHS, respectively. The total extracts were re-suspended in methanol (300 mL) and partitioned with *n*-hexane (7 × 260 mL). The combined solutions of *n*-hexane fraction were evaporated to dryness. A *n*-hexane fraction for SHN (yield of 4.0%) and a *n*-hexane fraction for SHS (yield of 7.2%) were obtained.

Both SHN and SHS *n*-hexane fractions, which exhibited an interesting anti-cholinesterase activity, were subjected to fractionation. Briefly, a portion of each extract was subjected to medium pressure liquid chromatography (MPLC) (Buchi Complete Flash System, column 920 mm x i.d. 26 mm, Specifications Pump Manager C-615, 2 Pump Modules C-605, Detector UV C-630, Fraction Collector Buchi C-660, Sepacore Record 1.0 Chromatography Software, adsorbent silica gel 20–45 mm, gradually increasing the eluent polarity from *n*-hexane/ethyl acetate 95:5 to ethyl acetate, flow rate 10 mL/min). This procedure afforded 9 fractions from SHN (N1–N9) and 7 fractions from SHS (S1–S7). All samples were stored at 4 °C for experimental use.

3.4. Chemical Analysis

The chemical composition of *S. haematodes* non-polar active fractions was assessed by using a Hewlett-Packard gas chromatograph (Agilent, Milan, Italy) equipped with a non-polar HP-5 capillary column (30 m × 0.25 mm, 0.25 μm), associated with a Hewlett-Packard mass spectrometer (Agilent, Milan, Italy) [36]. The ionization of the sample constituents was performed in electronic impact (EI, 70 electron volt). Helium was used as carrier gas (1.0 mL/min). The analyses were carried out as follows: isotherm at 50 °C for 5 min, temperature increase from 50 to 250 °C of 5 °C/min, and finally isotherm at 250 °C for 10 min. One microliter of diluted sample (1/10 *v*/*v*, in *n*-hexane) was injected. The identification of compounds was based on the comparison of their retention index (RI), either with those in literature or with those of available authentic standards (Sigma-Aldrich, Milan, Italy), on the comparison of the mass spectral data with the Wiley 138 library, and referring to the spectral data of pure compounds.

The methanol fractions were examined for their Total Phenol Content (TPC) and Total Flavonoid Content (TFC) as previously reported [36]. In brief, TPC was spectrophotometrically assessed by using Folin–Ciocalteu reagent. Absorbance was read at 765 nm using a UV–Vis Jenway 6003 spectrophotometer (Milan, Italy). In the TFC test, a method based on the formation of a flavonoid–aluminium complex was applied. Absorbance was read at 510 nm.

3.5. Analysis of Achetylcholinesterase (AChE) and Butyrylcholinesterase (BChE) Inhibitory Activity

The inhibition of AChE and BChE enzymes was measured by using a modified colorimetric Ellman's method [36] based on the reaction of released thiocholine to give a coloured product with a chromogenic reagent such as AChE from *E. electricus* (EC 3.1.1.7, Type VI-S) and BChE equine serum (EC 3.1.1.8). Acetylthiocholine (ATCI) iodide and butyrylthiocholine iodide (BTCI) were employed as the substrates of the reaction. In brief, enzyme, essential oils, and phosphate buffer were mixed in microplates and incubated in an ice bath at 4 °C. After 30 min, physostigmine was added. The reaction started by adding 5,5′-dithiobis(2-nitrobenzoic-acid) (DTNB) solution and substrate. The microplate was placed in a thermostatic water bath (Branson model 3800-CPXH, Milan, Italy) for

20 min at 37 °C. The reaction was stopped by placing the microplate in an ice bath and adding physostigmine.

The absorbance was measured at 405 nm. Results are calculated as IC_{50} values (μg/mL).

3.6. In Vitro Antioxidant Tests

Four antioxidant assays were herein applied to investigate the antioxidant effects of *S. haematodes* extracts. The radicals scavenging activity was evaluated using two spectrophotometric methods, such as DPPH and ABTS tests according to the procedure previously reported [37]. DPPH is a radical characterized by an intense purple colour, known for its stability due to the delocalization of the radical in aromatic rings. In this assay, the radical is neutralized by accepting either a hydrogen atom or an electron from an antioxidant agent or a reducing agent. When an odd electron pairs up with another electron, the initial colour gradually decolorizes into pale yellow. In the test, in brief, the DPPH solution (1.0×10^{-4} M) and *S. haematodes* extracts at different concentrations (in the range from 62.5 to 1000 μg/mL) were mixed. After 30 min, the absorbance was read at 517 nm. Compared to DPPH, which is stable by nature, the $ABTS^+$ radical is a radical that should be generated by chemical reactions. When this radical (unstable form) accepts an electron from an antioxidant agent, the blue-green colour changes into a pale blue colour, which is the regeneration of the stable form of ABTS. To obtain ABTS radical cation solution ($ABTS^+$), ABTS solution (7 mM) and potassium persulphate (2.45 mM) were mixed. After 12 h, $ABTS^+$ was diluted with ethanol to final absorbance of 0.70 at 734 nm. Successively, 2 mL of diluted $ABTS^+$ solution was added to extracts (25 μL) at concentrations in the range 1–400 μg/mL. After 6 min, the absorbance was read at 734 nm. In both DPPH and ABTS tests, ascorbic acid was used as a positive control.

The ability to reduce iron ions was assessed using FRAP test [37]. A solution of tripyridyltriazine (TPTZ), $FeCl_3$, HCl, and acetate buffer at pH 3.6 was prepared in order to obtain the FRAP reagent.

The FRAP reagent (2.0 mL), water (900 μL), and *S. haematodes* samples (100 μL at the concentration of 2.5 mg/mL) were mixed. After 30 min of incubation, the absorbance was measured at 595 nm. Butylated hydroxytoluene (BHT) was used as a positive control.

The capacity of *S. haematodes* aerial parts to protect lipid peroxidation was analysed by applying the β-carotene bleaching test, as previously reported [37]. Briefly, a mixture of β-carotene, linoleic acid, and 100% Tween 20 was prepared and the obtained emulsion was added to a 96-well microplate containing samples at concentrations in the range 2.5–100 μg/mL. The absorbance was measured at 470 nm against a blank at $t = 0$ and after 30 and 60 min of incubation. Propyl gallate was used as a positive control.

3.7. Statistical Analysis

The concentration giving 50% inhibition (IC_{50}) was obtained by nonlinear regression by using Prism GraphPad Prism version 4.0 for Windows (San Diego, CA, USA). The concentration–response curve was obtained by plotting the percentage inhibition vs. concentration. Differences within and between groups were evaluated by One-way analysis of variance test (ANOVA) followed by a multi-comparison Dunnett's test, used to compare each group with the positive control.

4. Conclusions

Salvia species have been traditionally used for the treatment of several ailments, with particular reference to cognitive and neurological conditions [38].

The analysis of literature confirms that many *Salvia* species and their main bioactive constituents influence some biological processes that may have an important impact on cognitive and neurological functions. Currently, *S. officinalis* and *S. lavandulaefolia* are the two *Salvia* species investigated in human studies, so the potency and the efficacy of other *Salvia* species are uncertain and deserve to be better investigated. Among these is *S. haematodes*.

To the best of our knowledge, this work is the first report on the AChE and BChE inhibitory activity of this *Salvia* species. In our study, the aerial parts of this *Salvia* species were extracted by using two different extraction procedures with methanol as a solvent, such as Soxhlet apparatus and extractor Naviglio® in order to investigate the impact on the extract composition and, consequently, on their biological effects. Despite the use of the traditional extraction technique of the Soxhlet apparatus, we obtained a better extraction yield, by analysing the antioxidant activity of samples it is evident that the methanol fraction obtained by extractor Naviglio® has a better antioxidant activity in all the carried-out tests, data that are supported by the higher content of TPC and TFC. This is not valid when we examine data relating to the cholinesterase inhibitory activity of the *n*-hexane extracts. In this case, in fact, the SHN *n*-hexane extract was more active against AChE, while the SHS *n*-hexane extract was more active on BChE. The bio-fractionation of these *n*-hexane fractions leads to less active samples. This anti-cholinesterase activity may be due to synergistic effects shown by the components of the mixtures in the test system used in this investigation.

Results obtained in this work provide the basis for further studies, necessary to confirm the activity of *S. haematodes* extracts and related fractions and the identification of some biologically active compounds. Several species of *Salvia* are commonly ingested across numerous cultures, which increases confidence about its safety. Moreover, in this case, further confirmation about the safety of *S. haematodes* is necessary.

Supplementary Materials: The following supporting information can be downloaded at: https://www.mdpi.com/article/10.3390/plants11060781/s1, Table S1. AChE and BChE inhibitory activity (IC50, μg/mL and/or μM) of pure compounds from literature.

Author Contributions: Conceptualization, R.T.; chemical analysis, M.B. and R.T.; biological investigation, M.R.L.; writing—original draft preparation, R.T., M.R.L. and N.G.P.; writing—review and editing, R.T., M.R.L. and N.G.P.; supervision, R.T. All authors have read and agreed to the published version of the manuscript.

Funding: This research received no external funding.

Acknowledgments: Authors wish to thank Francesca Cozza for her technical support.

Conflicts of Interest: The authors declare not conflict of interest.

References

1. Bentham, G. Labiatae. In *Genera Plantarum*; Bentham, G., Hooker, J.D., Eds.; Reeve and Co.: London, UK, 1876; Volume 2, pp. 1160–1196.
2. Walker, J.B.; Sytsma, K.J.; Treutlein, J.; Wink, M. *Salvia* (Lamiaceae) is not monophyletic: Implications for the systematics, radiation, and ecological specializations of *Salvia* and tribe Mentheae. *Am. J. Bot.* **2004**, *91*, 1115–1125. [CrossRef] [PubMed]
3. Pignatti, S.; Guarino, R.; La Rosa, M. *Flora d'Italia*; Edagricole: Milan, Italy, 2017; Volume 2.
4. Senatore, F.; De Feo, V. Essential oils from *Salvia* spp. (Lamiaceae). II. Chemical composition of the essential oil from *Salvia pratensis* L. su bsp. *haematodes* (L.) Briq. inflorescences. *J. Essent. Oil Res.* **1998**, *10*, 135–137. [CrossRef]
5. Akbar, S. Pharmacological investigations on the ethanolic extract of *Salvia haematodes*. *Fitoterapia* **1989**, *60*, 270–272.
6. Akbar, S.; Tariq, M.; Nisa, M. A study on CNS depressant activity of Salvia haematodes Wall. *Int. J. Crude Drug Res.* **1985**, *22*, 41–44. [CrossRef]
7. Loizzo, M.R.; Abouali, M.; Salehi, P.; Sonboli, A.; Kanani, M.; Menichini, F.; Tundis, R. *In vitro* antioxidant and antiproliferative activities of nine *Salvia* species. *Nat. Prod. Res.* **2014**, *28*, 2278–2285. [CrossRef]
8. Tundis, R.; Loizzo, M.R.; Menichini, F.; Bonesi, M.; Colica, C.; Menichini, F. *In vitro* cytotoxic activity of *Salvia leriifolia* Benth. extract and isolated constituents against a panel of human cancer cell lines. *Chem. Biodiv.* **2011**, *8*, 1152–1162. [CrossRef] [PubMed]
9. Loizzo, M.R.; Tundis, R.; Conforti, F.; Menichini, F.; Bonesi, M.; Nadjafi, F.; Frega, N.G.; Menichini, F. *Salvia leriifolia* Benth. (Lamiaceae) extract demonstrates *in vitro* antioxidant properties and cholinesterase inhibitory activity. *Nutr. Res.* **2010**, *30*, 823–830. [CrossRef] [PubMed]
10. Loizzo, M.R.; Menichini, F.; Tundis, R.; Bonesi, M.; Conforti, F.; Nadjafi, F.; Statti, G.A.; Frega, N.G.; Menichini, F. *In vitro* biological activity of *Salvia leriifolia* Benth essential oil relevant to the treatment of Alzheimer's disease. *J. Oleo Sci.* **2009**, *58*, 443–446. [CrossRef] [PubMed]
11. Tundis, R.; Leporini, M.; Bonesi, M.; Rovito, S.; Passalacqua, N.G. *Salvia officinalis* L. from Italy: A comparative chemical and biological study of its essential oil in the Mediterranean context. *Molecules* **2020**, *25*, 5826. [CrossRef] [PubMed]

12. Orhan, I.E.; Senol, F.S.; Ercetin, T.; Kahraman, A.; Celep, F.; Akaydin, G.; Sener, B.; Dogan, M. Assessment of anticholinesterase and antioxidant properties of selected sage (*Salvia*) species with their total phenol and flavonoid contents. *Ind. Crops Prod.* **2013**, *41*, 21–30. [CrossRef]
13. Perry, N.S.L.; Houghton, P.J.; Jenner, P.; Keith, A.; Perry, E.K. *Salvia lavandulaefolia* essential oil inhibits cholinesterase *in vivo*. *Phytomedicine* **2002**, *9*, 48–51. [CrossRef]
14. Ibach, B.; Haen, E. Acetylcholinesterase inhibition in Alzheimer's Disease. *Curr. Pharm. Des.* **2004**, *10*, 231–251. [CrossRef] [PubMed]
15. Marucci, G.; Buccioni, M.; Dal Ben, D.; Lambertucci, C.; Volpini, R.; Amenta, F. Efficacy of acetylcholinesterase inhibitors in Alzheimer's disease. *Neuropharmacology* **2021**, *190*, 108352. [CrossRef] [PubMed]
16. Perry, E.; McKeith, I.; Ballard, C. Butyrylcholinesterase and progression of cognitive deficits in dementia with Lewy bodies. *Neurology* **2003**, *60*, 1852–1853. [CrossRef] [PubMed]
17. Giacobini, E. Cholinesterase inhibitors for Alzheimer's disease: From tacrine to future applications. *Neurochem. Int.* **1998**, *32*, 413–419. [CrossRef]
18. Houghton, P.J.; Rena, Y.; Howes, M.-J. Acetylcholinesterase inhibitors from plants and fungi. *Nat. Prod. Rep.* **2006**, *23*, 181–199. [CrossRef] [PubMed]
19. Mamache, W.; Amira, S.; Ben Souici, C.; Laouer, H.; Benchikh, F. *In vitro* antioxidant, anticholinesterases, anti-α-amylase, and anti-α-glucosidase effects of Algerian *Salvia aegyptiaca* and *Salvia verbenaca*. *J. Food Biochem.* **2020**, *44*, e13472. [CrossRef] [PubMed]
20. Topcu, G.; Kolak, U.; Ozturk, M.; Boga, M.; Hatipoglu, S.D.; Bahadori, F.; Culhaoglu, B.; Dirmenci, T. Investigation of Anti-cholinesterase activity of a series of *Salvia* extracts and the constituents of *Salvia staminea*. *Nat. Prod. J.* **2013**, *3*, 3–9. [CrossRef]
21. Perry, G.; Cash, A.D.; Smith, M.A. Alzheimer disease and oxidative stress. *J. Biomed. Biotechnol.* **2002**, *2*, 120–123. [CrossRef]
22. Chen, Z.; Zhong, C. Oxidative stress in Alzheimer's disease. *Neurosci. Bull.* **2014**, *30*, 271–281. [CrossRef]
23. Miyazawa, M.; Yamafuji, C. Inhibition of acetylcholinesterase activity by tea tree oil and constituent terpenoids. *Flavour Fragr. J.* **2006**, *21*, 198–201. [CrossRef]
24. Bonesi, M.; Menichini, F.; Tundis, R.; Loizzo, M.R.; Conforti, F.; Passalacqua, N.G.; Statti, G.A.; Menichini, F. Acetylcholinesterase and butyrylcholinesterase inhibitory activity of *Pinus* species essential oils and their constituents. *J. Enz. Inhib. Med. Chem.* **2010**, *25*, 622–628. [CrossRef] [PubMed]
25. Elufioye, T.O.; Obuotor, E.M.; Agbedahunsi, J.M.; Adesanya, S.A. Anticholinesterase constituents from the leaves of *Spondias mombin* L. (Anacardiaceae). *Biol. Targets Ther.* **2017**, *11*, 107–114. [CrossRef] [PubMed]
26. Karakaya, S.; Yilmaz, S.V.; Özdemir, Ö.; Koca, M.; Pınar, N.M.; Demirci, B.; Yıldırım, K.; Sytar, O.; Hasan Turkez, H.; Baser, K.H.C. A caryophyllene oxide and other potential anticholinesterase and anticancer agent in *Salvia verticillata* subsp. *amasiaca* (Freyn & Bornm.) Bornm. (Lamiaceae). *J. Essent. Oil Res.* **2020**, *32*, 512–525.
27. Dalai, M.K.; Bhadra, S.; Chaudhary, S.K.; Bandyopadhyay, A.; Mukherjee, P.K. Anti-cholinesterase activity of the standardized extract of *Syzygium aromaticum* L. *Phcog. Mag.* **2014**, *10*, S276–S282. [PubMed]
28. Farag, M.A.; Ezzat, S.M.; Salama, M.M.; Tadros, M.G.; Serya, R.A.T. Anti-acetylcholinesterase activity of essential oils and their major constituents from four *Ocimum* species. *Z. Naturforsch.* **2016**, *71*, 393–402. [CrossRef] [PubMed]
29. Kawamoto, H.; Takeshita, F.; Murata, K. Inhibitory effects of essential oil extracts from *Panax* plants against β-secretase, cholinesterase, and amyloid aggregation. *Nat. Prod. Commun.* **2019**, *14*, 1–7. [CrossRef]
30. Sharma, N.; Tan, M.A.; An, S.S.A. Phytosterols: Potential metabolic modulators in neurodegenerative diseases. *Int. J. Mol. Sci.* **2021**, *22*, 12255. [CrossRef]
31. Ayaz, M.; Junaid, M.; Ullah, F.; Subhan, F.; Sadiq, A.; Ali, G.; Ovais, M.; Shahid, M.; Ahmad, A.; Wadood, A.; et al. Anti-Alzheimer's studies on β-sitosterol isolated from *Polygonum hydropiper* L. *Front. Pharmacol.* **2017**, *8*, 697. [CrossRef] [PubMed]
32. Baskar, A.A.; Al Numair, K.S.; Gabriel Paulraj, M.; Alsaif, M.A.; Muamar, M.A.; Ignacimuthu, S. β-Sitosterol prevents lipid peroxidation and improves antioxidant status and histoarchitecture in rats with 1,2-dimethylhydrazine-induced colon cancer. *J. Med. Food* **2012**, *15*, 335–343. [CrossRef]
33. Shi, C.; Wu, F.; Zhu, X.; Xu, J. Incorporation of β-sitosterol into the membrane increases resistance to oxidative stress and lipid peroxidation *via* estrogen receptor-mediated PI3K/GSK3b signaling. *Biochim. Biophys. Acta* **2013**, *1830*, 2538–2544. [CrossRef] [PubMed]
34. Vivancos, M.; Moreno, J.J. β-Sitosterol modulates antioxidant enzyme response in RAW 264.7 macrophages. *Free Radic. Biol. Med.* **2005**, *39*, 91–97. [CrossRef] [PubMed]
35. Khan, I.; Muhammad Zahoor, M.; Zeb, A.; Sahibzada, M.U.K.; Ul Bari, W.; Naz, S. Isolation, characterization, pharmacological evaluation and in silico modeling of bioactive secondary metabolites from *Ziziphus oxyphylla* a member of Rhamnaceae family. *Trop. J. Pharm. Res.* **2020**, *19*, 351–359. [CrossRef]
36. Tenuta, M.C.; Deguin, B.; Loizzo, M.R.; Dugay, A.; Acquaviva, R.; Malfa, G.A.; Bonesi, M.; Bouzidi, C.; Tundis, R. Contribution of flavonoids and iridoids to the hypoglycaemic, antioxidant, and nitric oxide (NO) inhibitory activities of *Arbutus unedo* L. *Antioxidants* **2020**, *9*, 184. [CrossRef] [PubMed]

37. Leporini, M.; Bonesi, M.; Loizzo, M.R.; Passalacqua, N.G.; Tundis, R. The essential oil of *Salvia rosmarinus* Spenn. from Italy as a source of health-promoting compounds: Chemical profile and antioxidant and cholinesterase inhibitory activity. *Plants* **2020**, *9*, 798. [CrossRef] [PubMed]
38. Lopresti, A.L. *Salvia* (sage): A review of its potential cognitive-enhancing and protective effects. *Drugs R D* **2017**, *17*, 53–64. [CrossRef]

plants

MDPI

Article

By-Products of the Black Soybean Sauce Manufacturing Process as Potential Antioxidant and Anti-Inflammatory Materials for Use as Functional Foods

Shu-Ling Hsieh [1], Yi-Wen Shih [2], Ying-Ming Chiu [3], Shao-Feng Tseng [4], Chien-Chun Li [5] and Chih-Chung Wu [2,*]

1 Department of Seafood Science, National Kaohsiung University of Science and Technology, Kaohsiung 81157, Taiwan; slhsieh@nkust.edu.tw
2 Department of Food and Nutrition, Providence University, Taichung 43301, Taiwan; a0922287120@gmail.com
3 Department of Allergy, Immunology, and Rheumatology, Tungs' Taichung Metro Harbor Hospital, Taichung 43503, Taiwan; ymcgreen@yahoo.com.tw
4 Department of Quality Control and Research, Ta-Tung Soya Sauce Co. Ltd., Yunlin 64069, Taiwan; blackbeanbeauty@gmail.com
5 Department of Nutrition, Chung Shan Medical University, Taichung 40201, Taiwan; chienchien367@gmail.com
* Correspondence: ccwumail@pu.edu.tw; Tel.: +886-4-26328001 (ext. 15318)

Abstract: To assess the potential of by-products of the black bean fermented soybean sauce manufacturing process as new functional food materials, we prepared black bean steamed liquid lyophilized product (BBSLP) and analysed its antioxidant effects in vitro. RAW264.7 macrophages were cultured and treated with BBSLP for 24 h, and 1 μg/mL lipopolysaccharide (LPS) was then used for another 24 h to induce inflammation. The cellular antioxidant capacity and inflammatory response were then analysed. Activation of nuclear factor kappa B (NF-κB) signaling in RAW264.7 macrophages was also analysed. Results showed BBSLP had 2,2-diphenyl-1-(2,4,6-trinitrophenyl)hydrazyl (DPPH) and 2,2′-azino-bis(3-ethylbenzothiazoline-6-sulfonic acid) diammonium ($ABTS^+$) radical-scavenging abilities and reducing power in vitro. The levels of both reactive oxygen species (ROS) and thiobarbituric acid reactive substances (TBARS) were reduced after RAW264.7 macrophages were treated with BBSLP after LPS induction. After RAW264.7 macrophage treatment with BBSLP and induction by LPS, the levels of inflammatory molecules, including nitric oxide (NO), prostaglandin E_2 (PGE_2), IL-1α, IL-6 and TNF-α, decreased. NF-κB signaling activity was inhibited by reductions in IκB phosphorylation and NF-κB DNA-binding activity after RAW264.7 macrophages were treated with BBSLP after LPS induction. In conclusion, BBSLP has antioxidant and anti-inflammatory capabilities and can be a supplement material for functional food.

Keywords: black bean soybean sauce; by-product; antioxidant; inflammation; functional supplements

Citation: Hsieh, S.-L.; Shih, Y.-W.; Chiu, Y.-M.; Tseng, S.-F.; Li, C.-C.; Wu, C.-C. By-Products of the Black Soybean Sauce Manufacturing Process as Potential Antioxidant and Anti-Inflammatory Materials for Use as Functional Foods. *Plants* **2021**, *10*, 2579. https://doi.org/10.3390/plants10122579

Academic Editors: Juei-Tang Cheng, I-Min Liu and Szu-Chuan Shen

Received: 27 October 2021
Accepted: 23 November 2021
Published: 25 November 2021

1. Introduction

Soybean sauces are commonly used as seasonings and sauces in Asia. Among them, black bean-fermented soybean sauce which uses black beans (*Glycine max* (L.) Merr.) as the major raw material is one of the most favored. During the soybean sauce manufacturing process, raw soybean steaming is an important step before yeast inoculation [1]. However, steamed soybean liquid waste, particularly by-products, may contain nutrients and other phytochemicals. These by-products may not only provide nutrient and active compounds for supplements but also possess sustainable and circular economic characteristics if they can be developed into a healthy food material. Considering global sustainability, the circular economy is a new alternative approach to the traditional economy [2]. Therefore, recycling certain by-products from food manufacturing processes as new food and pharmaceutical materials is not only a solution for food waste and supply issues, but also a new preferred resource for human health [2].

The black soybean cultivar has abundant polyphenols in its seed coating [3]. During the black soybean sauce manufacturing process, the beans are steamed at a high temperature and pressure [4]. This process is similar to soybean extraction with high-temperature steam under high pressure. These steamed soybean extracts contain polyphenols and some of the active components of black soybeans [5]. Previous studies have shown that black soybeans can reduce cardiovascular disease [6], regulate blood sugar [7] and have anticancer effects, [8,9], improve bone resorption in menopause [10], and possess antioxidative [11] and anti-inflammatory [12] properties. Kim et al. [13] showed that raw black soybeans have abundant total free polyphenols, flavonoids and phenolic acids. Anthocyanidin is a type of isoflavone that is present in black soybean coats [9]. Recycling this black soybean steamed liquid (BBSL) has the potential to providing an effective physiological material for functional food development.

The inflammatory response is a physiological protective mechanism of the body that occurs in response to infection; however, the occurrence of chronic inflammation with chronic inflammatory cells, including macrophages, lymphocytes and plasma cells, among others, is increasing [14]. During the inflammatory response process, reactive oxygen species (ROS), nitric oxide (NO) and prostaglandin E_2 (PGE_2) act as messengers for different physiological functions and pathological processes [15]. Conversely, intracellular cytokines, such as IL-1β, IL-6, IL-10 and TNF-α, are secreted from macrophages and lymphocytes [16]. These cytokines mediate the immune response and influence the macrophage microenvironment [17]. Excess or long-term chronic inflammation will lead to chronic diseases, such as cardiovascular disease (CVD), sepsis, diabetes mellitus (DM), and chronic kidney disease (CKD), increasing human health risks [18].

Some inflammatory response factors, such as IL-1β, IL-6, and TNF-α, and inflammation-mediated molecular enzymes, including inducible nitroxide synthase (iNOS) and cyclooxygenase-2 (COX-2), are primarily controlled by nuclear factor kappa B (NF-κB), which is well recognized to play an important role in inflammation [15,19–22].

Currently, the food supply chain has many challenges due to decreased natural resources and increased food waste [23]. Thus, recycling reusable by-products of plant foods manufacturing process as new foods or components with active physiological effects is a forward-looking issue. For further more study and assessment the potential of this by-product of the soybean sauce manufacturing process on functional food material. The present study aims to investigate the capability of antioxidation and anti-inflammation of black bean steamed liquid lyophilized product (BBSLP).

2. Results

2.1. *pH Value, Flavonoids, Total Phenol, and Protein Contents in BBSL and BBSLP*

To know the application of fresh BBSL, the pH was measured in this study. As shown in Table 1, the pH of fresh BBSL was 5.84 ± 0.12. The flavonoids, total phenol, and protein contents of BBSL were 11.5 ± 1.5 mg rutin equivalents (RUE)/mL, 3.83 ± 0.3 mg gallic acid equivalents (GAE)/mL and 3.5 ± 0.8 mg/mL, respectively (Table 1). In addition, after fresh BBSL was concentrated by a rotary vacuum dryer and frozen dry by a frozen dryer, the flavonoids, total phenol, and protein contents of BBSLP were 0.1 ± 0.01 mg RUE/mg, 0.03 ± 0.01 mg GAE/mg and 0.31 ± 0.04 mg/mg, respectively (Table 1).

Table 1. pH, Flavonoids, total polyphenols and crude protein levels of BBSL and BBSLP.

	pH Value	Flavonoids	Total Polyphenols	Crude Protein
BBSL *	5.84 ± 0.12	11.5 ± 1.5 mg RUE mL^{-1}	3.83 ± 0.3 mg GAE mL^{-1}	3.5 ± 0.8 mg mL^{-1}
BBSLP	-	0.1 ± 0.01 mg RUE mg^{-1}	0.03 ± 0.01 mg GAE mg^{-1}	0.31 ± 0.04 mg mg^{-1}

* BBSL: black bean steamed liquid; BBSLP: black bean steamed liquid lyophilized product; GAE: gallic acid equivalents; RUE: rutin equivalents. Values are presented as means ± SD (*n* = 3–5).

2.2. BBSLP Showed Antioxidant Effects In Vitro Conditions

In the 2,2-diphenyl-1-(2,4,6-trinitrophenyl)hydrazyl (DPPH) radical-reducing ability test the scavenging abilities of 0.5, 1 and 2 mg/mL BBSLP were 42.5 ± 8.7%, 54.4 ± 0.4% and 79.6 ± 0.2%, respectively. The observed scavenging ability in the vitamin C-treated group was 88.0 ± 5.8% (Figure 1A). Although the DPPH radical-reducing ability in the BBSLP groups was lower than that in the vitamin C group, it increased in a dose-dependent manner.

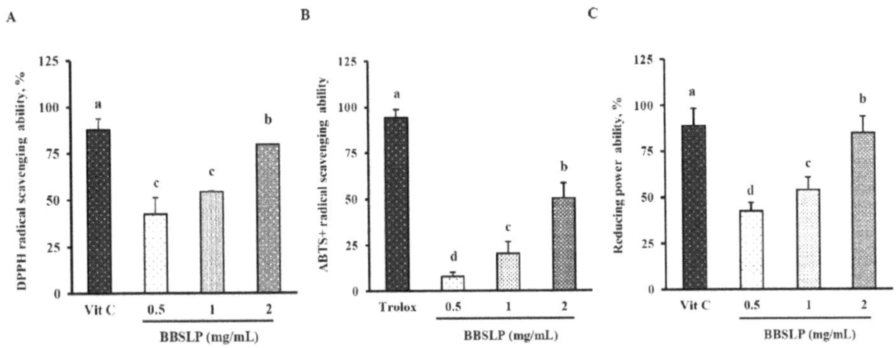

Figure 1. In vitro antioxidative ability of the BBSLP. DPPH radical scavenging activity (**A**), $ABTS^+$ radical scavenging activity (**B**) and reducing power (**C**). Vitamin C was used as the positive control in the DPPH radical scavenging assay and reducing power ability assay, Trolox was used as the positive control in the $ABTS^+$ radical scavenging. Values are presented as means ± SD (n = 3–5). ^{abc} Values are significantly different from the other groups as determined by Duncan's test ($p < 0.05$). Black bean steamed liquid lyophilized product (BBSLP).

The EC_{50} value of BBSLP for DPPH radical-scavenging ability was 0.81 mg/mL. Figure 1B shows that the Trolox-treated group demonstrated 95.5 ± 4.3% 2,2′-azino-bis(3-ethylbenzothiazoline-6-sulfonic acid) diammonium ($ABTS^+$) radical scavenging. The $ABTS^+$ radical scavenging abilities in the 0.5, 1 and 2 mg/mL BBSL groups were 7.8 ± 2.3%, 20.5 ± 6.3% and 50.2 ± 8.3%, respectively, and showed a dose-dependent increase ($p < 0.05$). The effect concentration (EC_{50}) value of BBSLP for $ABTS^+$ radical scavenging ability was 1.50 mg/mL. In addition, the reducing power of 0.5, 1 and 2 mg/mL BBSLP reached 42.5 ± 84.6% in a dose-dependent manner ($p < 0.05$). The vitamin C-treated group showed 88.7 ± 9.6% reducing power (Figure 1C). The EC_{50} value of BBSLP for reducing power was 0.80 mg/mL.

2.3. BBSLP Maintained the Viability of RAW264.7 Macrophages after Lipopolysaccharide (LPS) Induction

Our preliminary experimental results showed that the viability of RAW264.7 macrophages treated with 0.1 to 5 μg/mL BBSLP was not significantly different compared with that of the control cells (data not shown). For reasons related to BBSLP yield and solubility, we used 0.1, 0.5 and 1 μg/mL BBSLP as the experimental doses in the following study.

In the present study, the cell viability of RAW264.7 macrophages did not significantly differ between the 0.1, 0.5 or 1 μg/mL BBSLP (approximately 97–101%) under LPS induction or the control group (100%) (Figure 2A). Based on morphological examination results using inverted microscopy (Figure 2B), the cell number and morphology did not significantly differ between any BBSLP-treated group under LPS induction and the control group. Under LPS induction, RAW264.7 macrophage treatment with 0.1, 0.5, or 1 μg/mL BBSLP did not affect viability.

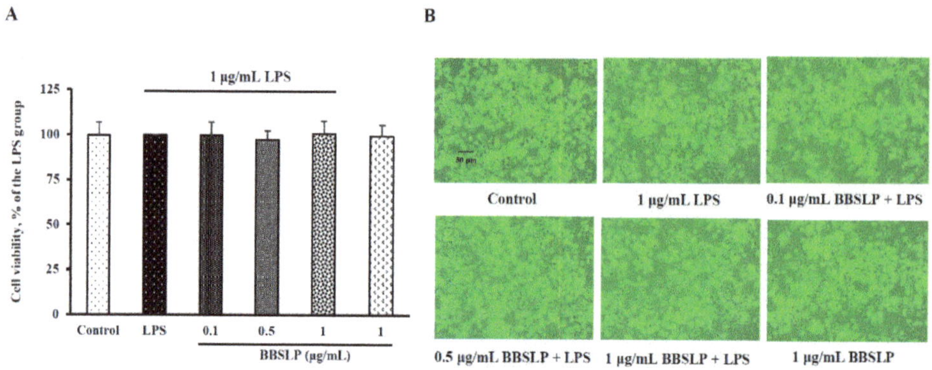

Figure 2. Effects of BBSLP on the viability of LPS-induced RAW264.7 cells. RAW264.7 cells (1×10^5 cells/30-mm plate) were seeded and cultured overnight, treated with 0.1, 0.5 or 1 μg/mL BBSLP for 24 h and then induced or not with 1 μg/mL LPS for another 24 h. The group treated with 1 μg/mL LPS alone served as an induced control group. BBSLP was diluted in sterilized H_2O, and cells treated with sterilized H_2O alone served as the control group. Cells treated with 1 μg/mL BBSLP without LPS treatment for 48 h were used as another control group. Cell viability (**A**) and morphological changes (**B**) were examined. Values are presented as means ± SD (n = 3–5).

2.4. BBSLP Reduced Oxidative Stress in RAW264.7 Macrophages after LPS Induction

When RAW264.7 macrophages were treated with LPS alone, the thiobarbituric acid reactive substances (TBARS) level was significantly increased by 229% ($p < 0.05$) (Figure 3A). However, when RAW264.7 macrophages were treated with 0.1, 0.5 or 1 μg/mL BBSLP and then stimulated with LPS, the TBARS levels were significantly decreased by 38 to 70% compared with those in the LPS alone-treated group ($p < 0.05$) (Figure 3A). The TBARS levels in the group treated with only 5 μg/mL BBSLP did not differ from those in the control group. The ROS levels (100%) of RAW264.7 macrophages treated with only LPS were significantly higher than those in control cells (23.1 ± 5.3%) ($p < 0.05$) (Figure 3B). However, the ROS levels in RAW264.7 macrophages did significantly decrease in the groups treated with 0.1, 0.5 or 1 μg/mL BBSLP after stimulation with LPS (approximately 21.2–66.5%) ($p < 0.05$).

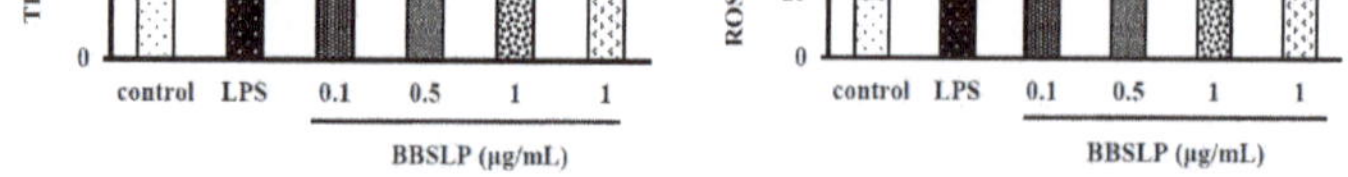

Figure 3. Effects of BBSLP on LPS-induced RAW264.7 cell oxidative stress. RAW264.7 cells (1×10^5 cells/30-mm plate) were seeded and cultured overnight, treated with 0.1, 0.5 or 1 μg/mL BBSLP for 24 h and then induced or not with 1 μg/mL LPS for another 24 h. The group treated with 1 μg/mL LPS alone served as an induced control group. BBSLP was diluted in sterilized H_2O, and cells treated with sterilized H_2O alone served as the control group. Cells treated with 1 μg/mL BBSLP without LPS treatment for 48 h were used as another control group. TBARS levels (**A**) and ROS levels (**B**) were examined. Values are presented as means ± SD (n = 3–5). [abc] Values are significantly different from the other groups as determined by Tukey's test ($p < 0.05$).

2.5. BBSLP Reduced NO and PGE_2 Production in RAW264.7 Macrophages after LPS Induction

NO and PGE_2 production was significantly increased after RAW264.7 macrophages were induced with LPS compared with the control group ($p < 0.05$, Figure 4A,B). However, when RAW264.7 macrophages were cotreated with 1 μg/mL BBSLP and LPS, NO levels were decreased by 15% compared with those in the LPS group ($p < 0.05$). A 60–68% decrease in PGE_2 was noted in cells cotreated with 0.1 to 1 μg/mL BBSLP and LPS ($p < 0.05$). Immunoblot analysis showed that the iNOS levels in RAW264.7 macrophages treated with 0.1, 0.5 and 1 μg/mL BBSLP were 82%, 77% and 69% that of cells treated with LPS alone, respectively ($p < 0.05$, Figure 4C,D). When RAW264.7 macrophages were treated with 0.1, 0.5 and 1 μg/mL BBSLP, the COX-2 protein levels were significantly reduced by 84%, 84% and 82%, respectively, compared with those in the LPS-treated group ($p < 0.05$, Figure 4C,D).

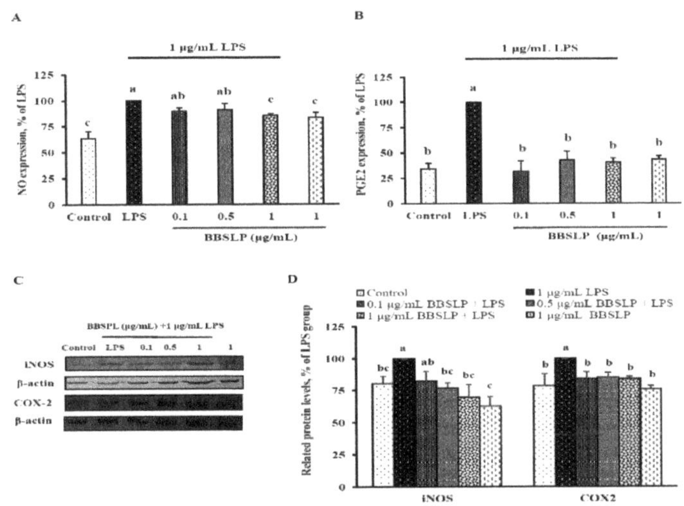

Figure 4. Effects of BBSLP on NO, PGE_2, iNOS and COX-2 levels in RAW264.7 cells induced by LPS. RAW264.7 cells (1×10^5 cells/30-mm plate) were seeded and cultured overnight, treated with 0.1, 0.5 or 1 μg/mL BBSLP for 24 h and then induced or not with 1 μg/mL LPS for another 24 h. The group treated with 1 μg/mL LPS alone served as an induced control group. BBSLP was diluted in sterilized H_2O, and cells treated with sterilized H_2O alone served as the control group. Cells treated with 1 μg/mL BBSLP without LPS treatment for 48 h were used as another control group. NO (**A**), PGE_2 (**B**), iNOS and COX-2 (**C**) protein expression and quantified iNOS and COX-2 levels (**D**) were examined. Values are presented as means ± SD (n = 3–5). ^{abc} Values are significantly different from the other groups as determined by Tukey's test ($p < 0.05$).

2.6. BBSLP Decreased IL-1β, IL-6 and TNF-α Levels in RAW264.7 Macrophages after LPS Induction

Figure 5A shows that the IL-1β level was significantly increased after an inflammatory response was induced in RAW264.7 macrophages by LPS compared with the control group ($p < 0.05$); however, when RAW264.7 macrophages were treated with 0.5 or 1 μg/mL BBSLP combined with LPS, the IL-1β levels were 59.9 ± 8.7% and 32.0 ± 8.7%, they were significantly lower than the LPS induction alone group (100%, $p < 0.05$, Figure 5A). The IL-6 levels in RAW264.7 cells decreased significantly by 7% after treatment with 1 μg/mL BBSLP compared with those after LPS induction alone ($p < 0.05$, Figure 5B). Figure 5C also shows that the TNF-α level in the group treated with only 1 μg/mL BBSLP decreased significantly compared with the LPS alone group ($p < 0.05$, Figure 5B). Notably, IL-10 production did not differ among the control group, LPS-treated group, the group treated with various concentrations of BBSLP combined with LPS, and the group treated with BBSLP alone (Figure 5D).

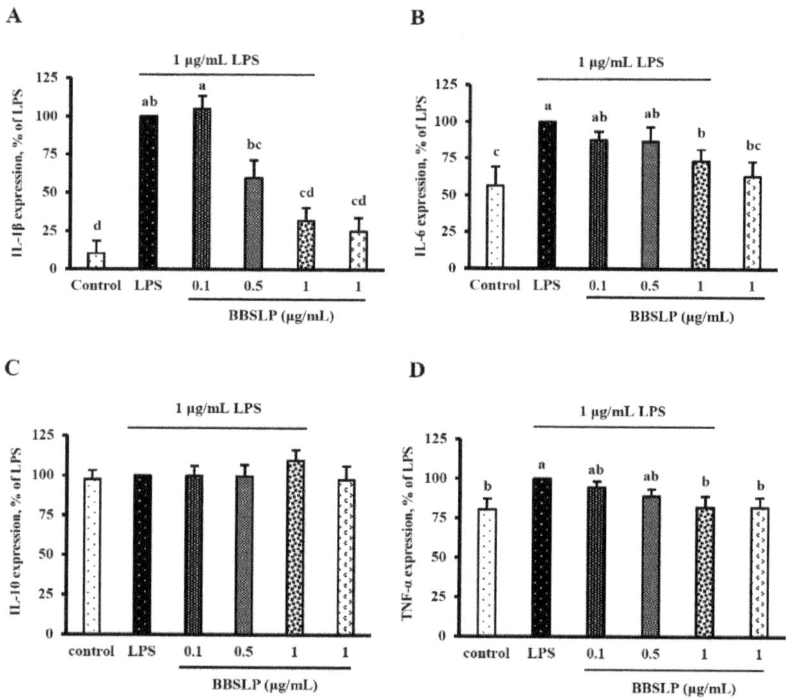

Figure 5. Effects of BBSLP on the inflammatory response in RAW264.7 cells induced by LPS. RAW264.7 cells (1×10^5 cells/30-mm plate) were seeded and cultured overnight, treated with 0.1, 0.5 or 1 μg/mL BBSLP for 24 h and then induced or not with 1 μg/mL LPS for another 24 h. The group treated with 1 μg/mL LPS alone served as an induced control group. BBSLP was diluted in sterilized H_2O, and cells treated with sterilized H_2O alone served as the control group. Cells treated with 1 μg/mL BBSLP without LPS treatment for 48 h were used as another control group. Levels of NOIL-1β (**A**), IL-6 (**B**), IL-10 (**C**), and TNF-α (**D**) were examined. Values are presented as means ± SD (n = 3–5). ^{abc} Values are significantly different from the other groups as determined by Tukey's test ($p < 0.05$).

2.7. BBSLP Reduced the Activation of NF-kB Signalling in RAW264.7 Cells after LPS Induction

Figure 6A,B show that IκB phosphorylation was significantly reduced by 20–50% after 0, 1, 0.5 or 1 μg/mL BBSLP treatment in RAW264.7 cells induced with LPS ($p < 0.05$). However, BBSLP did not affect the protein contents of cytosolic I-κB in EA.hy926 cells (Figure 6A,B). The nuclear NF-κB levels were significantly decreased by 24% and 16%, respectively, after 0.5 or 1 μg/mL BBELP treatment in RAW264.7 cells induced by LPS ($p < 0.05$, Figure 6A,B). The DNA-binding activity of nuclear NF-κB was significantly suppressed by 49% in cells treated with 100 μg/mL BBSLP (Figure 6C).

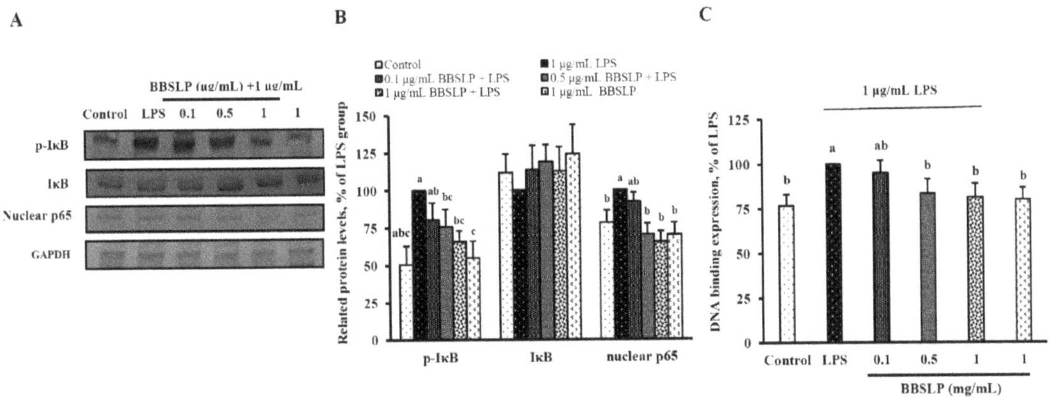

Figure 6. Effects of BBSLP on NF-κB signalling activation in LPS-induced RAW264.7 cells. RAW264.7 cells (1×10^5 cells/30-mm plate) were seeded and cultured overnight, treated with 0.1, 0.5 or 1 μg/mL BBSLP for 24 h and then induced or not with 1 μg/mL LPS for another 24 h. The group treated with 1 μg/mL LPS alone served as an induced control group. BBSLP was diluted in sterilized H_2O, and cells treated with sterilized H_2O alone served as the control group. Cells treated with 1 μg/mL BBSLP without LPS treatment for 48 h were used as another control group. Phosphorylated IκB (p-IκB), IκB and nuclear p65 expression (**A**), quantified p-IκB, IκB and nuclear p65 levels (**B**) and NF-κB-DNA binding activity (**C**) were examined. Values are presented as means ± SD (n = 3–5). [abc] Values are significantly different from the other groups as determined by Tukey's test ($p < 0.05$).

3. Discussion

The present study showed that BBSLP has the ability of free radical scavage and reducing power enhance in vitro. And, the potential antioxidant and anti-inflammatory effects of BBSLP in LPS-induced RAW264.7 macrophages. Because BBSLP significantly reduced oxidative stress, including an ability to decrease levels of free radicals and lipid peroxidation, and reduced pro-inflammation molecules, including IL-1β, IL-6 and TNF-α levels, in LPS-induced RAW264.7 cells.

Our results found that BBSLP contained flavonoids and polyphenols, which were may be involved in its high antioxidative and anti-inflammatory properties. Glevitzky et al. [24] showed that there are a high intercorrelation between the number of phenolic groups within the basic structure of flavonoids and their antioxidant activity and between the antioxidant activity and the number of -OH phenolic groups also with a high correlation. However, polyphenols or the other components of BBSL and BBSLP whether playing a major or important role in antioxidation or not, are still need furthermore composition analysis and investigation. Inflammatory response mediators, including PGE_2 and NO, and inflammatory cytokines, including IL-1β, IL-6, IL-10 and TNF-α, were all regulated, reducing inflammation in LPS-induced RAW264.7 cells. Furthermore, BBSLP could down-regulate NF-κB signalling activation, which led to reduced iNOS, COX-2, IL-1β, IL-6, IL-10 and TNF-α transcription.

Oxidative stress usually triggers the inflammatory response in various cells and tissues [25]. Macrophages then use ROS production to scavenge xenobiotics, including bacteria and oxidized low-density lipoprotein (ox-LDL) [26]. Long-term inflammation, chronic inflammation and oxidative stress lead to chronic diseases, such as CKD, CVD, DM and sepsis with a poor prognosis. In the present study, BBSLP displayed excellent free radical-scavenging ability, reducing power and SOD activity in an in vitro model and significantly reduced ROS production and TBARS levels in LPS-induced RAW264.7 cells. The above excellent antioxidative effects were from the polyphenols and isoflavones contained in BBSLP. Takahashi et al. [27] showed that the seed coats of black soybeans have a higher total polyphenol content than those of yellow soybeans. Black soybeans may more effectively inhibit LDL oxidation than yellow soybeans because of the higher total polyphenol contents in their seed coat. Additionally, cyanidin-3-glucoside, petunidin-3-glucoside and peonidin-3-glucoside, three major anthocyanins, have been detected in black soybean seed coats [3]. DPPH radical scavenging, $ABTS^+$ radical scavenging and ferric reducing antioxidant power (FRAP) analysis results have also shown that the black soybean seed coat is a more efficient reducing agent than dehulled black soybeans and yellow soybean coats [3]. In addition, black soybeans are rich in polyphenols, including isoflavones, anthocyanidins and flavan-3-ols. Moreover, black soybeans can prevent CVD risks by increasing polyphenol concentrations and decreasing oxidative stress in healthy women [28]. Previous studies in different experimental models have shown that polyphenols and isoflavones also have anti-inflammatory characteristics. Takekawa et al. [29] showed that genistein, a soybean polyphenol, can significantly suppress water immersion restraint (WIR) stress-induced gastric mucosal injury. The underlying mechanism involves a significant elevation of SOD activity and significant suppression of both TBARS levels and the production of TNF-α to protect against gastric mucosal injury [29]. Additionally, puerarin, an isoflavonoid extracted from Kudzu roots, reduces malondialdehyde levels, increases SOD activity and alleviates TNF-α, IL-1β and IL-6 protein levels in the hippocampus. Antioxidation and anti-inflammation are induced by the streptozotocin (STZ) group to protect DM rats from cognitive deficits [30].

In addition, resveratrol, a polyphenol constituent of grapes, acts as a COX suppressor, reducing the inflammatory response similarly to a nonsteroidal anti-inflammatory drug (NSAID). A molecular basis for the mutually beneficial relationship between plants and humans has been speculated [31]. Hussain et al. [32] reported the anti-inflammatory and antioxidative properties of polyphenols, the mechanisms by which polyphenols inhibit molecular signalling pathways that have been activated by oxidative stress, and the roles of polyphenols in inflammation-mediated chronic disorders. The above data and previous reports indicate that BBSLP, an extract rich in polyphenols and isoflavones, can significantly reduce oxidative stress, inhibit IL-1β, IL-6 and TNF-α and increase IL-10 as an anti-inflammatory material in RAW264.7 macrophages. BBSLP may be helpful for the development of future antioxidant therapeutics and new anti-inflammatory drugs [33].

In the present study, BBSLP significantly inhibited NF-κB signalling activity by reducing I-κB phosphorylation and NF-κB-DNA binding activity. NF-κB signalling plays an important role in iNOS, COX-2, IL-1β, IL-6 and TNF-α transcription [15,19–22]. Previous studies have shown that one of the major mechanisms of reducing inflammation is to reduce NF-κB signalling activity and the expression of proinflammatory molecules [23]. Bao et al. [34] showed that chlorogenic acid, a major polyphenol compound from coffee, can prevent diabetic nephropathy by inhibiting oxidative stress and inflammation through the reduction of NF-κB signalling activity. Singh et al. [35] also reported that polyphenols have antioxidative and anti-neuroinflammatory properties by regulating NF-κB activation in neurodegenerative diseases.

Currently, the food supply chain is facing substantial pressures, including the availability of fewer natural resources and increased food waste [23]. One important way to increase the food supply and decrease the environmental consequences of current food production is to reduce food waste levels and their economic, environmental and social implications [36]. Previous studies have shown that various by-products from the manufacturing of animals and plants for food contain various fatty acids [37], phytochemicals [38,39], and amino acids [40], all of which are beneficial to the food supply chain and/or can act as health promoters for food sustainability. In the present study, BBSLP showed potential antioxidant and anti-inflammatory effects. In the present study, BBSLP showed preliminary potential on antioxidant and anti-inflammatory effects in vitro. However, the molecular mechanisms of the physiological effects require further study in animal models or human clinical trials. On the other hand, how does BBSLP apply in function foods? Where BBSLP to be developed into a functional food material, a functional assessment of this product is needed. The above questions are an important issue for BBSLP application.

4. Materials and Methods

4.1. Materials

BBSL was a gift from Ta-Tung Soya Sauce Co., Ltd., located in Siluo Town (Yunlin, Taiwan). BBSL is collected from the soybean sauce manufacturing process. After fresh black beans are washed and steamed by 120 °C streams in a closed steam tank for 1 h. BBSL were directly collected as experimental materials. LPS was purchased from Sigma-Aldrich Co. (St. Louis, MO, USA).

4.2. Preparation of BBSLP

According to our previous methods [41], BBSLP was prepared by our laboratory. Fresh BBSL was concentrated in a rotary evaporator (N-1110, Tokyo Rikakikai Co., Ltd., Tokyo, Japan) and then dried in a freeze dryer (Freezone 4.5, Labconco, Kansas City, MO, USA) at −43 °C. The BBSLP was stored at −20 °C until use. The percent yield of BBSLP was 1.16% (*w*/*v*).

4.3. Determination of the pH Value and Total Flavonoids, Phenols and Protein in BBSL and BBSLP

The pH values of fresh BBSL were measured using a pH metre (MP220 pH meter, Mettler Toledo, Greifensee, Switzerland). The total phenol contents were analysed using a colorimetric method according to Padmavati et al. [42]. One hundred microlitres of 1 N Folin-Ciocalteu reagent (Sigma-Aldrich Co.) was added to 100 μL of diluted BBSL or BBSLP (dissolved in reverse-osmosis (RO) H_2O). Then, 500 μL of 7.5% Na_2CO_3 solution was added to react for 30 min, and the absorbance of each sample was measured at an optical density (OD) of 760 nm in a Biokinetics microplate reader (Bio-Tek Instruments, Winooski, VT, USA). Calibration curves were constructed using 0, 0.125, 0.25 and 0.5 mg/mL gallic acid (GA). The total phenolic content is represented as mg GAE/mL BBSL or mg GAE/g BBSLP.

The flavonoid content was analysed using the colorimetric method according to Jia et al. [43]. One hundred microlitres of 5% $NaNO_3$ solution was added to 100 μL of the BBSL or BBSLP solution to react for 5 min. Then, 50 μL of 10% $AlCl_3$ solution was added. Finally, 600 μL of 4% NaOH solution was added to the mixture for 30 min. The absorbance of the mixture was measured at OD 510 nm on the Biokinetics microplate reader. Calibration curves were constructed with 0, 0.02, 0.06, 0.08 and 0.1 mg/mL rutin (RU) as the standard. The total flavonoid content is represented as mg RUE/mL BBSL or mg RUE/g BBSLP.

Crude protein contents were analysed according to Lowry et al. [44]. Fifty microlitres of each standard of BBSL or BBSLP was added to 50 μL of trichloroacetic acid and standing for reacted for 30 min at room temperature. The mixture was centrifuged at 15,000× *g* for 20 min at 4 °C. Then, the supernatant was discarded. The precipitate was dissolved in 1 mL of NaOH and standing for reacted for 30 min at room temperature. Then, 1.0 mL of modified Lowry Reagent (Sigma-Aldrich Co.) was added, then mixing and incubation at room temperature for 10 min. Five hundred μL Prepared 1X Folin-Ciocalteu's phenol reagent (Sigma-Aldrich Co.) was added, and the mixture was standing in a water bath at 37 °C for 30 min. After 30 min, the absorbance was measured at 660 nm on the Biokinetics microplate reader.

4.4. In Vitro Antioxidant Ability of BBSLP

In this study, DPPH (Sigma-Aldrich Co.) radical scavenging activity by BBSLP was analysed according to the method of Shimada et al. [45]. For the measure of the DPPH radical-scavenging activity, 1.5 mL of the sample solution with varying BBSLP concentrations (0.5, 1, 2 and 4 mg/mL) were added 1.5 mL of 0.15 mM DPPH in 50% ethanol. The mixture was mixed and incubated at room temperature in the dark for 30 min. The optical density at 517 nm was measured using the Biokinetics microplate reader. In this test, 1 mg/mL vitamin C was used as a control. The scavenging activity was calculated as $(1 - A_{BBSLP}$ or $A_{vitamin\ C}/A_{blank}) \times 100$.

According to the method of Shimada et al. [45] to analyse the reducing power activity of BBSLP in vitro. Here, 0.5 mL of the 0.5, 1, 2 and 4 mg/mL BBSLP, respectively were mixed with 2.5 mL of 0.2 M phosphate buffer (pH 6.6) and 2.5 mL of 1% potassium ferricyanide, then the mixture was incubated at 50 °C for 20 min. A 2.5 mL aliquot of 10% trichloroacetic acid was added to the mixture, and the mixture was then centrifuged at 3000× *g* for 10 min. The supernatant (2.5 mL) was mixed with 2.5 mL of distilled water and 2.5 mL of 0.1% ferric chloride, and the absorbance at 700 nm was read using the microplate reader. The reducing power was calculated as (A_{BBSLP} or $A_{vitamin\ C} - A_{blank}$)/$A_{vitamin\ C} \times 100$. A vitamin C (1 mg/mL) was used as a control.

The ABTS$^+$ radical scavenging ability of 0.5, 1, 2 or 4 mg/mL BBSLP was analysed according to the method described by Re et al. [46]. Use a 10 µL of the sample solution with varying BBSLP concentrations (0.5, 1, 2 and 4 mg/mL) were added 990 µL of 2 mM 2,2′-azino-bis(3-ethylbenzothiazoline-6-sulfonate) radical cation (ABTS$^{\bullet+}$) solution. The mixture was mixed and incubated at room temperature in the dark for 10 min. The optical density at 737 nm was measured using the microplate reader. The ABTS$^+$ radical-scavenging ability was calculated as (A_{blank} or $A_{BBSLP} - A_{Trolox}$)/$A_{blank} \times 100$. In this test, 1 mg/mL Trolox was used as a control.

4.5. Cell Culture and Treatment

RAW264.7 macrophages and mouse monocyte macrophages were purchased from the Bioresource Collection and Research Center (Hsinchu, Taiwan). Dulbecco's modified Eagle's medium containing 42 mM L-glutamine, 100 units/mL penicillin, 100 µg/mL streptomycin and 10% (*v/v*) heat-inactivated fetal bovine serum (FBS; Gibco, Thermo Fisher Scientific, Inc., Waltham, MA, USA) was used as the culture medium. All cultured cells were incubated in an atmosphere of 5% CO_2/95% air at 37 °C.

In this study, 1×10^5 RAW264.7 macrophages per 30 mm plate or 1×10^6 per 60 mm plate were cultured for various biochemical tests. RAW264.7 macrophages were incubated with 0.5, 1 or 5 µg/mL BBSLP for 24 h and then induced with 1 µg/mL LPS (Sigma-Aldrich Co.) for another 24 h. LPS was used to induce inflammation [47]. The induced control group was treated with 1 µg/mL LPS alone. BBSLP was soluble in sterilized H_2O, and cells treated with sterilized H_2O alone served as the control group. BBSLP (1 µg/mL) without LPS treatment for 48 h made up another control group.

4.6. Cell Viability Analysis

To determine the optimum test concentration of BBSLP for use in this study, the cell viability of RAW264.7 macrophages was analysed according to the method of Denizot and Lang [48]. After RAW264.7 macrophages were incubated in DMEM containing 0.5 mg/mL thiazolyl blue formazan (MTT; Sigma-Aldrich Co.) for an additional 3 h, the medium was removed and extracted with isopropanol for 15 min. The isopropanol fraction was measured with the Biokinetics microplate reader at OD 570 nm. To evaluate morphological changes, a phase-contrast inverted fluorescence microscope (Olympus IX51, Olympus, Tokyo, Japan) was used.

4.7. Measurement of Lipid Peroxidation and ROS Levels

The effect of BBSLP on lipid peroxidation in RAW264.7 macrophages induced by LPS was determined according to the method of Fraga et al. [49]. The lipid peroxidation indicator thiobarbituric acid reactive substances (TBARS) was extracted and measured with a fluorescence microplate reader (excitation wavelength 515 nm and emission wavelength 555 nm, Bio-Tek Instruments, Winooski, VT, USA). The protein levels were determined according to the method described by Lowry et al. [44]. The TBARS level is shown in nmol TBARs/mg protein. The levels of ROS in RAW364.7 cells were determined using a Cellular ROS Assay Kit (ab113851, Abcam Inc., Cambridge, MA, USA).

4.8. Determination of Nitrite (NO) and Prostaglandin E_2 (PGE_2)

To determine the inflammation level in RAW 264.7 cells, the Griess assay was used [50]. Griess reagent (1% sulfanilamide/0.1% N-(1-naphtyl)ethylene diamine dihydrochloride in 2.5% H_3PO_4) was mixed with an equal part of the cell culture medium of control or various experimental groups RAW 264.7 cells. In this test, the NO_2 content was used as an indicator of NO content in RAW364.7 macrophages. The OD 550 nm was determined and calibrated using a standard curve of $NaNO_2$ prepared in culture medium.

The PGE_2 levels in RAW264.7 macrophages were determined using an enzyme immunoassay (EIA) kit (ADI-900-001, Cayman Chemical, Ann Arbor, MI, USA). The cell culture supernatants were collected after experimental treatment and centrifuged at $1000\times g$ for 15 min to remove the particulate matter. The medium and PGE2 EIA conjugate was added to a 96-well plate pre-coated with goat anti-mouse IgG and left to react for 1 h, followed by a final wash to remove any unbound antibody-enzyme reagent. A substrate solution was added and the intensity of the color produced was measured at 412 nm. The concentration of PGE_2 in each sample was calculated according to PGE_2 standards.

4.9. Measurement of IL-1β, IL-6 and TNF-α

The levels of IL-1β, IL-6 and TNF-α in RAW264.7 macrophages were analysed using rat IL-1β/IL-1F2 DuoSet ELISA (R&D, DY501-05), rat IL-6 DuoSet ELISA (R&D, DY506-05), and rat TNF-α DuoSet ELISA (R&D, DY510-05) kits (R&D Systems, Inc., Minneapolis, MN, USA), respectively, according to the manufacturer's instructions. In brief, capture antibodies, cultured medium supernatants, detection antibodies, streptavidin-conjugated horseradish-peroxidase were processed on the plate in order, and the color subtract tetramethylbenzidine was used. The absorbance was measured and the concentration was calculated according to the standard.

4.10. Immunoblot Analyses of iNOS, COX-2 and NF-κB Signalling Molecule Expression

The iNOS, COX-2 and NF-κB signalling molecule expression was analysed using the method described by Hsieh et al. [51]. At the end of the treatment, the cells were collected in 200 μL of lysis buffer (10 mM Tris-HCl, 5 mM EDTA, 0.2 mM phenylmethylsulfonyl fluoride and 20 μg/mL aprotinin, pH 7.4), and the protein content was determined according to the method of Lowry et al. [44].

Equal amounts (approximately 10–20 μg per sample) of cellular protein were separated by 10% sodium dodecyl sulfate (SDS) polyacrylamide gel electrophoresis (PAGE) [52], after which the samples were transferred to polyvinylidene difluoride (PVDF) membranes [53]. The PVDF membranes were then incubated with anti-iNOS, anti-COX-2, anti-p-IκB, anti-IκB, anti-NF-κB (p65) or anti-GAPDH antibodies at 4 °C overnight, followed by incubation with a peroxidase-conjugated secondary antibody. For density analysis, blots were treated with enhanced chemiluminescence substrate solutions and exposed using a ChemiDoc XRSt System (Bio-Rad Laboratories, Hercules, CA, USA). An NF-κB (p65) transcription factor activity assay kit (Cayman Chemical Co.) was used to analyse the NF-κB DNA binding activity of the nuclear fraction.

4.11. Statistical Analysis

The SPSS Statistical Analysis Software for Windows, version 20.0 (SPSS Inc., Chicago, IL, USA) was used to analyse the experimental data in the present study. One-way analysis of variance (ANOVA) and Duncan's or Tukey's multiple-range test were used to evaluate the significance of differences between each mean value. A p-value less than 0.05 was used to indicate a statistically significant result.

5. Conclusions

In conclusion, the presented data in this study demonstrate that BBSLP could reduce oxidative stress and pro-inflammation factors in LPS-induced RAW264.7 cells through the inhibition of the NF-κB signaling pathway, indicating that it may have potent antioxidant and anti-inflammatory capabilities, suggesting that BBSLP could be developed as a supplement material for functional foods.

Author Contributions: Conceptualization, S.-L.H., S.-F.T. and C.-C.W.; Methodology, S.-L.H. and Y.-M.C.; Validation, Y.-W.S., C.-C.L. and C.-C.W.; Formal Analysis, Y.-M.C.; Investigation, Y.-W.S. and S.-L.H.; Resources, Y.-M.C. and S.-F.T.; Data Curation, Y.-W.S. and C.-C.W.; Writing—Original Draft Preparation, C.-C.W.; Writing—Review & Editing, S.-L.H. and C.-C.W.; Supervision, C.-C.W.; Project Administration, C.-C.W. All Authors read and approved the article and agree to be accountable for all aspects of the research and ensure that the accuracy or integrity of any part of the work is appropriately investigated and resolved. All authors have read and agreed to the published version of the manuscript.

Funding: This work was supported by Ta-Tung Soya Sauce Co. Ltd. through a grant (grant no. PU108-11150-A002).

Institutional Review Board Statement: Not applicable.

Informed Consent Statement: Not applicable.

Data Availability Statement: The datasets used and/or analyzed during the present study are available from the corresponding author on reasonable request.

Conflicts of Interest: The authors declare no conflict of interest.

References

1. O'Toole, D.K. The role of microorganisms in soy sauce production. *Adv. Appl. Microbiol.* **2019**, *108*, 45–113. [CrossRef] [PubMed]
2. Teigiserova, D.; Hamelin, L.; Thomsen, M. Towards transparent valorization of food surplus, waste and loss: Clarifying definitions, food waste hierarchy, and role in the circular economy. *Sci. Total. Environ.* **2019**, *706*, 136033. [CrossRef] [PubMed]
3. Domae, C.; Nanba, F.; Maruo, T.; Suzuki, T.; Ashida, H.; Yamashita, Y. Black soybean seed coat polyphenols promote nitric oxide production in the aorta through glucagon-like peptide-1 secretion from the intestinal cells. *Food Funct.* **2019**, *10*, 7875–7882. [CrossRef] [PubMed]
4. Diez-Simon, C.; Eichelsheim, C.; Mumm, R.; Hall, R.D. Chemical and sensory characteristics of soy sauce: A Review. *J. Agric. Food Chem.* **2020**, *68*, 11612–11630. [CrossRef]
5. Choi, Y.-M.; Yoon, H.; Lee, S.; Ko, H.-C.; Shin, M.-J.; Lee, M.C.; Hur, O.S.; Ro, N.Y.; Desta, K.T. Isoflavones, anthocyanins, phenolic content, and antioxidant activities of black soybeans (*Glycine max* (L.) Merrill) as affected by seed weight. *Sci. Rep.* **2020**, *10*, 19960. [CrossRef]
6. Yamashita, Y.; Nakamura, A.; Nanba, F.; Saito, S.; Toda, T.; Nakagawa, J.; Ashida, H. Black soybean improves vascular function and blood pressure: A randomized, placebo controlled, crossover trial in humans. *Nutrients* **2020**, *12*, 2755. [CrossRef]
7. Chen, Z.; Wang, C.; Pan, Y.; Gao, X.; Chen, H. Hypoglycemic and hypolipidemic effects of anthocyanins extract from black soybean seed coat in high fat diet and streptozotocin-induced diabetic mice. *Food Funct.* **2017**, *9*, 426–439. [CrossRef] [PubMed]
8. Takagi, A.; Kano, M.; Kaga, C. Possibility of breast cancer prevention: Use ofsoy isoflavones and fermented soy beverage produced using probiotics. *Int. J. Mol. Sci.* **2015**, *16*, 10907–10920. [CrossRef]
9. Ganesan, K.; Xu, B. A Critical review on polyphenols and health benefits of black soybeans. *Nutrients* **2017**, *9*, 455. [CrossRef]
10. Tit, D.M.; Bungau, S.; Iovan, C.; Cseppento, D.C.N.; Endres, L.; Sava, C.; Sabau, A.M.; Furau, G.; Furau, C. Effects of the hormone replacement therapy and of soy isoflavones on bone resorption in postmenopause. *J. Clin. Med.* **2018**, *7*, 297. [CrossRef]
11. Rao, D.E.; Chaitanya, K.V. Changes in the antioxidant intensities of seven different soybean (*Glycine max* (L.) Merr.) cultivars during drought. *J. Food Biochem.* **2020**, *44*, e13118.
12. Kim, J.N.; Han, S.N.; Ha, T.J.; Kim, H.-K. Black soybean anthocyanins attenuate inflammatory responses by suppressing reactive oxygen species production and mitogen activated protein kinases signaling in lipopolysaccharide-stimulated macrophages. *Nutr. Res. Pract.* **2017**, *11*, 357–364. [CrossRef]
13. Kim, M.Y.; Jang, G.Y.; Lee, Y.; Li, M.; Ji, Y.M.; Yoon, N.; Lee, S.H.; Kim, K.M.; Lee, J.; Jeong, H.S. Free and bound form bioactive compound profiles in germinated black soybean (*Glycine max* L.). *Food Sci. Biotechnol.* **2016**, *25*, 1551–1559. [CrossRef]
14. Rajendran, P.; Chen, Y.; Chen, Y.; Chung, L.; Tamilselvi, S.; Shen, C.; Day, C.H.; Chen, R.; Viswanadha, V.P.; Kuo, W.; et al. The multifaceted link between inflammation and human diseases. *J. Cell. Physiol.* **2018**, *233*, 6458–6471. [CrossRef]
15. Infantino, V.; Pierri, C.L.; Iacobazzi, V. Metabolic routes in inflammation: The citrate pathway and its potential as therapeutic target. *Curr. Med. Chem.* **2020**, *26*, 7104–7116. [CrossRef] [PubMed]
16. Kondo, N.; Kuroda, T.; Kobayashi, D. Cytokine networks in the pathogenesis of rheumatoid arthritis. *Int. J. Mol. Sci.* **2021**, *22*, 10922. [CrossRef]
17. Li, L.; Yu, R.; Cai, T.; Chen, Z.; Lan, M.; Zou, T.; Wang, B.; Wang, Q.; Zhao, Y.; Cai, Y. Effects of immune cells and cytokines on inflammation and immunosuppression in the tumor microenvironment. *Int. Immunopharmacol.* **2020**, *88*, 106939. [CrossRef] [PubMed]

18. Donate-Correa, J.; Ferri, C.M.; Sánchez-Quintana, F.; Pérez-Castro, A.; González-Luis, A.; Martín-Núñez, E.; Mora-Fernández, C.; Navarro-González, J.F. Inflammatory cytokines in diabetic kidney disease: Pathophysiologic and therapeutic implications. *Front. Med.* **2021**, *7*, 1137. [CrossRef]
19. Mitchell, J.P.; Carmody, R.J. NF-κB and the transcriptional control of inflammation. *Int. Rev. Cell Mol. Biol.* **2018**, *335*, 41–84.
20. Zhao, H.; Wu, L.; Yan, G.; Chen, Y.; Zhou, M.; Wu, Y.; Li, Y. Inflammation and tumor progression: Signaling pathways and targeted intervention. *Signal. Transduct. Target. Ther.* **2021**, *6*, 1–46. [CrossRef] [PubMed]
21. Toker, H.; Görgün, E.P.; Korkmaz, E.M. Analysis of IL-6, IL-10 and NF-kappaB gene polymorphisms in aggressive and chronic periodontitis. *Cent. Eur J. Public Health.* **2017**, *25*, 157–162. [CrossRef]
22. Kunnumakkara, A.B.; Shabnam, B.; Girisa, S.; Harsha, C.; Banik, K.; Devi, T.B.; Choudhury, R.; Sahu, H.; Parama, D.; Sailo, B.L.; et al. Inflammation, NF-kappaB, and chronic diseases: How are they linked? *Crit. Rev. Immunol.* **2020**, *40*, 1–39. [CrossRef] [PubMed]
23. Dutta, S.; He, M.; Xiong, X.; Tsang, D.C. Sustainable management and recycling of food waste anaerobic digestate: A review. *Bioresour. Technol.* **2021**, *341*, 125915. [CrossRef] [PubMed]
24. Glevitzky, I.; Dumitrel, G.A.; Glevitzky, M.; Pasca, B.; Otřísal, P.; Bungau, S.; Cioca, G.; Pantis, C.; Popa, M. Statistical analysis of the relationship between antioxidant activity and the structure of flavonoid compounds. *Rev. Chim.* **2019**, *70*, 3103–3107. [CrossRef]
25. de Bhailís, B.; Chrysochou, C.; Kalra, P. Inflammation and oxidative damage in ischaemic renal disease. *Antioxidants* **2021**, *10*, 845. [CrossRef]
26. Marchio, P.; Guerra-Ojeda, S.; Vila, J.M.; Aldasoro, M.; Victor, V.M.; Mauricio, M.D. Targeting early atherosclerosis: A focus on oxidative stress and inflammation. *Oxid. Med. Cell. Longev.* **2019**, *2019*, 8563845. [CrossRef] [PubMed]
27. Takahashi, R.; Ohmori, R.; Kiyose, C.; Momiyama, Y.; Ohsuzu, A.F.; Kondo, K. Antioxidant activities of black and yellow soybeans against low density lipoprotein Oxidation. *J. Agric. Food Chem.* **2005**, *53*, 4578–4582. [CrossRef]
28. Yamashita, Y.; Wang, L.; Nakamura, A.; Nanba, F.; Saito, S.; Toda, T.; Nakagawa, J.; Ashida, H. Black soybean improves the vascular function through an increase in nitric oxide and a decrease in oxidative stress in healthy women. *Arch. Biochem. Biophys.* **2020**, *688*, 108408. [CrossRef]
29. Takekawa, S.; Matsui, T.; Arakawa, Y. The protective effect of the soybean polyphenol genistein against stress-induced gastric mucosal lesions in rats, and its hormonal mechanisms. *J. Nutr. Sci. Vitaminol.* **2006**, *52*, 274–280. [CrossRef]
30. Liu, X.; Mo, Y.; Gong, J.; Li, Z.; Peng, H.; Chen, J.; Wang, Q.; Ke, Z.; Xie, J. Puerarin ameliorates cognitive deficits in streptozotocin-induced diabetic rats. *Metab. Brain Dis.* **2015**, *31*, 417–423. [CrossRef]
31. Inoue, H.; Nakata, R. Resveratrol targets in inflammation. *Endocr. Metab. Immune Disord. Drug Targets.* **2015**, *15*, 186–195. [CrossRef]
32. Hussain, T.; Tan, B.; Yin, Y.; Blachier, F.; Tossou, M.C.; Rahu, N. Oxidative stress and inflammation: What polyphenols can do for us? *Oxid. Med. Cell Longev.* **2016**, *2016*, 7432797. [CrossRef]
33. Thawkar, B.S.; Kaur, G.J. Inhibitors of NF-kappaB and P2X7/NLRP3/Caspase 1 pathway in microglia: Novel therapeutic opportunities in neuroinflammation induced early-stage Alzheimer's disease. *J. Neuroimmunol.* **2019**, *326*, 62–74. [CrossRef] [PubMed]
34. Bao, L.; Li, J.; Zha, D.; Zhang, L.; Gao, P.; Yao, T.; Wu, X. Chlorogenic acid prevents diabetic nephropathy by inhibiting oxidative stress and inflammation through modulation of the Nrf2/HO-1 and NF-qB pathways. *Int. Immunopharmacol.* **2018**, *54*, 245–253. [CrossRef] [PubMed]
35. Singh, S.S.; Rai, S.N.; Birla, H.; Zahra, W.; Rathore, A.S.; Singh, S.P. NF-kappaB-mediated neuroinflammation in Parkinson's disease and potential therapeutic effect of polyphenols. *Neurotox. Res.* **2020**, *37*, 491–507. [CrossRef] [PubMed]
36. Godfray, H.C.; Crute, I.R.; Haddad, L.; Lawrence, D.; Muir, J.F.; Nisbett, N.; Pretty, J.; Robinson, S.; Toulmin, C.; Whiteley, R. The future of global food system. *Philos. Trans. R Soc. Lond B Biol. Sci.* **2010**, *365*, 2769–2777. [CrossRef]
37. Benítez, J.J.; Castillo, P.M.; Del Río, J.C.; León-Camacho, M.; Domínguez, E.; Heredia, A.; Guzmán-Puyol, S.; Athanassiou, A.; Heredia-Guerrero, J.A. Valorization of tomato processing by-products: Fatty acid extraction and production of bio-based materials. *Materials* **2018**, *11*, 2211. [CrossRef]
38. Li, Q.; Sun, J.; Mohammadtursun, N.; Wu, J.; Dong, J.; Li, L. Curcumin inhibits cigarette smoke-induced inflammation via modulating the PPARγ-NF-κB signaling pathway. *Food Funct.* **2019**, *10*, 7983–7994. [CrossRef]
39. Lee, K.J.; Baek, D.-Y.; Lee, G.-A.; Cho, G.-T.; So, Y.-S.; Lee, J.-R.; Ma, K.-H.; Chung, J.-W.; Hyun, D.Y. Phytochemicals and antioxidant activity of Korean black soybean (*Glycine max* L.) Landraces. *Antioxidants* **2020**, *9*, 213. [CrossRef]
40. Kerr, B.J.; Urriola, P.E.; Jha, R.; E Thomson, J.; Curry, S.M.; Shurson, G.C. Amino acid composition and digestible amino acid content in animal protein by-product meals fed to growing pigs1. *J. Anim. Sci.* **2019**, *97*, 4540–4547. [CrossRef]
41. Hsieh, S.-L.; Tsai, P.-J.; Liu, Y.-C.; Wu, C.-C. Potential effects of antioxidant and serum cholesterol-lowering effects of *Gynura bicolor* water extracts in Syrian Hamster. *Evid. Based Complement. Altern. Med.* **2020**, *2020*, 2907610. [CrossRef]
42. Azeem, S.M.A.; Al Mohesen, I.A.; Ibrahim, A.M. Analysis of total phenolic compounds in tea and fruits using diazotized aminobenzenes colorimetric spots. *Food Chem.* **2020**, *332*, 127392. [CrossRef]
43. Paiva, L.; Lima, E.; Motta, M.; Marcone, M.; Baptista, J. Variability of antioxidant properties, catechins, caffeine, L-theanine and other amino acids in different plant parts of *Azorean Camellia* sinensis. *Curr. Res. Food Sci.* **2020**, 3, 227–234. [CrossRef] [PubMed]

44. Hsieh, S.L.; Wang, J.C.; Huang, Y.S.; Wu, C.C. Ethanol extract of *Gynura bicolor* reduces atherosclerosis risk through enhancing antioxidant capacity and reducing adhesion molecule levels. *Pharm. Biol.* **2021**, *59*, 504–512. [CrossRef] [PubMed]
45. Shimada, K.; Fujikawa, K.; Yahara, K.; Nakamura, T. Antioxidative properties of xanthan on the autoxidation of soybean oil in cyclodextrin emulsion. *J. Agric. Food Chem.* **1992**, *40*, 945–948. [CrossRef]
46. Re, R.; Pellegrini, N.; Proteggente, A.; Pannala, A.; Yang, M.; Rice-Evans, C. Antioxidant activity applying an improved ABTS radical cation decolorization assay. *Free Radic. Biol. Med.* **1999**, *26*, 1231–1237. [CrossRef]
47. Han, Y.; Yuan, C.; Zhou, X.; Han, Y.; He, Y.; Ouyang, J.; Zhou, W.; Wang, Z.; Wang, H.; Li, G. Anti-inflammatory activity of three triterpene from *Hippophae rhamnoides* L. in lipopolysaccharide-stimulated RAW264.7 Cells. *Int. J. Mol. Sci.* **2021**, *22*, 12009. [CrossRef]
48. Denizot, F.; Lang, R. Rapid colorimetric assay for cell growth and survival. Modifications to the tetrazolium dye procedure giving improved sensitivity and reliability. *J. Immunol. Methods* **1986**, *89*, 271–277. [CrossRef]
49. Fraga, C.G.; Leibovitz, B.E.; Tappel, A.L. Lipid peroxidation measured as thiobarbituric acid-reactive substances in tissue slices: Characterization and comparison with homogenates and microsomes. *Free. Radic. Biol. Med.* **1988**, *4*, 155–161. [CrossRef]
50. Green, L.C.; Wagner, D.A.; Glogowski, J.; Skipper, P.L.; Wishnok, J.S.; Tannenbaum, S.R. Analysis of nitrate, nitrite, and [15N]-nitrate in biological fluids. *Anal. Biochem.* **1992**, *126*, 131–138. [CrossRef]
51. Hsieh, S.-L.; Wang, J.-J.; Su, K.-H.; Kuo, Y.-L.; Hsieh, S.; Wu, C.-C. Suppressive Effects of the *Gynura bicolor* Ether Extract on endothelial permeability and leukocyte transmigration in human endothelial cells induced by TNF-α. *Evid.-Based Complement. Altern. Med.* **2020**, *2020*, 1–13. [CrossRef] [PubMed]
52. Gallagher, S.R. One-dimensional SDS gel electrophoresis of proteins. *Curr. Protoc. Toxicol.* **2001**. [CrossRef]
53. Towbin, H.; Staehelin, T.; Gordon, J. Electrophoretic transfer of proteins from polyacrylamide gels to nitrocellulose sheets: Procedure and some applications. *Proc. Natl. Acad. Sci. USA* **1979**, *76*, 4350–4354. [CrossRef] [PubMed]

plants

MDPI

Article

Formulation of Fast Dissolving β-Glucan/Bilberry Juice Films for Packaging Dry Powdered Pharmaceuticals for Diabetes

Ionut Avramia * and Sonia Amariei

Faculty of Food Engineering, Stefan cel Mare University of Suceava, 720229 Suceava, Romania
* Correspondence: ionut.avramia@usm.ro

Abstract: The aim of this study was to develop fast dissolving films based on β-glucan and bilberry juice due to the bioactive potential of β-glucan and antidiabetic effect of bilberry juice. The benefit of incorporation of bioactive compounds into the films is due to the removal of unnecessary excipients and to confer protection as well as increase stability and shelf life to the packaged product. Due to the fast dissolving requirements of the European Pharmacopeia, which reduced the dissolution time from 180 to 60 s, indicating less than a minute, hygroscopic materials, such as sodium alginate and a suitable plasticizer, such as glycerin were incorporated. Moreover, the influence of ingredients and surfactants, such as soybean oil was studied in the design of fast dissolving films. Additionally, the steady state rate water vapor transmission rate (WVTR), water vapor permeability (WVP), and FT-IR spectroscopy tests were performed at high resolution to ensure the reliability of the films and composition as well as to validate the results. Our data suggest that the addition of surfactants contributed to the development of fast dissolving films without influencing the diffusion of water vapor. Low levels of WVTR and short dissolution time made from β-glucan and bilberry juice are a convenient candidate for packaging dry powdered pharmaceuticals for diabetes.

Keywords: fast dissolving films; yeast β-glucan; bilberry juice; diabetes

Citation: Avramia, I.; Amariei, S. Formulation of Fast Dissolving β-Glucan/Bilberry Juice Films for Packaging Dry Powdered Pharmaceuticals for Diabetes. *Plants* **2022**, *11*, 2040. https://doi.org/10.3390/plants11152040

Academic Editors: Juei-Tang Cheng, I-Min Liu and Szu-Chuan Shen

Received: 12 July 2022
Accepted: 3 August 2022
Published: 4 August 2022

1. Introduction

Films based on different bioactive compounds are defined as a new generation of edible films due to the active agents, which act as antimicrobials, anti-inflammatory, antioxidants, immunostimulants, anti-cancer, etc. [1–3]. There are plenty of natural compounds that can be incorporated into the film-forming solution for film making, depending on their role in therapeutics and the related medical condition. Films with strong antioxidant properties were successfully developed using natural compounds, such as plants from *Moringaceae* family [4], chitosan/ellagic acid films were found to have strong antimicrobial activity against *S. aureus* and *P. aeruginosa* [5], while β-glucan/pomegranate juice films had potential in the management of diabetes through the variety of tannins and bioactivities [6].

Diabetes is the most common noncontagious disease in the world [7]. A study conducted in 2019 estimates that this serious, long-term condition with a major impact on the lives and well-being of individuals around the world affects about 463 million people. It was estimated that, by 2030 this number will increase to 578 million people and in 2045 it will increase to 700 million people (equivalent to 10.9% of the population), with over 200 million more than 2019 [8].

Functional ingredients, such as bilberry (*Vaccinium myrtillus* L.) are recognized since centuries to have antidiabetic potential and have been used to control blood sugar levels [9]. A comprehensive review in 2022 of the phytochemical and pharmacological antidiabetic properties of bilberries by Chehri et al. (2022) highlighted the beneficial properties of the most significant components from the fruit. In addition to the antidiabetic effects, the authors analyzed the cardioprotective effects, anti-obesity, anti-inflammatory or ocular disorder effects, with all of these found in diabetes-related complications [10].

Yeast β-glucan is a complex polysaccharide of β-1,3, which is linked with β-1,6 glucose polymers that are found in the cell wall of yeast [11]. The biological effects of the β-glucans are related to the immunity, anti-cancer, and anti-inflammatory properties of the body with a mechanism of action not yet fully understood [12]. It is known to act by binding to Pattern Recognition Receptors, such as Dectin-1, LacCer, CR3, and SR3 [13]. By stimulating the Dectin-1 signaling pathway, β-glucans might confer protection of β pancreatic cells against the T cells in T1D [14]. Several authors have studied the oral dispersible polysaccharides film with pullulan (a linear structure of glucan) as a drug delivery system for treatment of diabetes [15], while other researchers used immediate-release layers of coatings to prepare fixed-dose combination tablets for diabetes [16].

Taken together, a fast dissolving film that allows the incorporation of these bioactive compounds is the aim of this study. The composition of film-forming solution (FFS) has been chosen to examine the rapid dissolving forms of films. Therefore, the addition of hygroscopic materials, plasticizers, and surfactants was evaluated to observe significant changes between samples. Moreover, for an adequate packaging film, additional tests including thickness, WVTR, WVP, film opacity, water activity, moisture content, and color profile were performed. Finally, further statistical correlation between dependent and independent variables, such as chemical film composition was determined.

2. Results

2.1. *Optimization of the Film Composition Considering Dietary Intake and Total Solids*

In order to find an optimal composition for a fast dissolving film with a high content of bioactive compounds, seven samples with different amounts of β-glucan and bilberry juice were cast onto a plastic petri dish with or without soybean oil as surfactant. This allows for the identification of any changes in the surface tension reduction as well as improvements on wettability and adhesion of the film. While the bilberry juice rich in anthocyanins, quercetin derivatives, proanthocyanidins or chlorogenic acid phytochemicals has no dietary intake recommendation, β-glucans are limited by the European Food Safety Authority up to 1.275 g/day of dietary use for the general adult population [17].

From the data presented in Table 1, it can be seen that an equal amount of sodium alginate was introduced in all seven samples, which in relation to the total volume has a concentration of 0.53 % (*w*/*v*). This falls within the limits of 0.125–1.5% and exhibits a pseudoplastic shear flow behavior [18]. Additionally, film sizing was performed to a corresponding area density between 7.82 and 12.19 mg × cm^{-2}. The plasticizer (glycerin) with the best compatibility related to β-glucan and most popular for anthocyanin-based films [19,20] was added to the all polymer samples, while the soybean oil was included in three different samples as 2% (*w*/*w*) of the total solids (β-glucan, sodium alginate, and bilberry juice).

Table 1. Composition of β-glucan/bilberry juice blends.

Sample	β-Glucan (BG), g	Sodium Alginate (SA), g	Bilberry Juice (BJ), g	Dry Weight of Bilberry Juice (Determined), g	Total Solids (BG+SA+BJ), g	Glycerin, 25% (*w*/*w*) of the Total Solid Weight, g	Soybean Oil, 2% (*w*/*w*) of the Total Solid Weight, %	Water, Up to Total Volume, mL
1	1	0.8	10	1.15	2.95	0.7375	0	150
2	1	0.8	10	1.15	2.95	0.7375	2	150
3	1	0.8	20	2.3	4.10	1.0250	0	150
4	1	0.8	20	2.3	4.10	1.0250	2	150
5	1.5	0.8	10	1.15	3.45	0.8625	0	150
6	1.5	0.8	20	2.3	4.60	1.1500	0	150
7	1.5	0.8	20	2.3	4.60	1.1500	2	150

2.2. Film Thickness

The film thickness expressed in µm is an important characteristic in packaging materials. Different thicknesses are essential to the other properties, such as water barrier, transparency or color attributes [21,22]. The film thickness of β-glucan/bilberry juice films ranges between 66.43 and 119.7 µm and was significant ($p < 0.001$) depending on the total solids in the film-forming solution (2.95 and 4.6 g, respectively). This behavior is in accordance with the observations of Arham et al. (2016) who observed an interaction between film thickness and based materials [23]. Similar results have been obtained by Peltzer et al. (2018) and Zhao et al. (2022), in which thicknesses up to 200 µm contributed to the stability and uniformity of the film making suitable for packaging applications [24,25].

2.3. Water Vapor Transmission Rate (WVTR)

WVTR of the β-glucan/bilberry juice films generally has low values between 3.2562 and 7.1111 $g \times h^{-1} \times m^{-2}$ (Table 2). It was observed that the increasing trend of WVTR was determined by the increase in the total content of substances introduced in the film-forming solution. Additionally, studies conducted by Rahmawati et al. (2020) showed that the amount of plasticizer strongly influenced the water absorption rates [26]. Due to the hydrophilic nature of glycerin, which has three hydroxyl groups, the influence of the plasticizer on the water vapor that permeates the film was investigated. The results are shown in Table 3.

Table 2. Physicochemical characteristics of β-glucan/bilberry juice film.

Physichochemical Parameters	β-Glucan/Bilberry Juice Film							F-Value
	Sample 1	Sample 2	Sample 3	Sample 4	Sample 5	Sample 6	Sample 7	
Thickness (µm)	66.43 (8.53) ^{d}	73.03 (6.03) ^{cd}	112.06 (6.66) ^{a}	68.96 (9.51) ^{d}	83.73 (7.57) ^{c}	119.7 (9.39) ^{a}	97.8 (6.48) ^{b}	22.24 ***
WVTR ($g \times h^{-1} \times m^{-2}$)	3.7956 (0.2) ^{c}	3.6140 (0.16) ^{c}	3.2562 (0.35) ^{c}	5.9213 (0.48) ^{b}	6.8927 (0.68) ^{a}	7.1111 (0.62) ^{a}	5.9036 (0.56) ^{b}	34.4 ***
WVP ($g \times mm \times kPa^{-1} \times h^{-1} \times m^{-2}$)	0.1057 (0.009) ^{e}	0.1107 (0.004) ^{de}	0.1542 (0.02) ^{cd}	0.1710 (0.01) ^{c}	0.2442 (0.04) ^{b}	0.3568 (0.01) ^{a}	0.2432 (0.03) ^{b}	38.82 ***
L	31.41 (0.34) ^{bcd}	30.33 (1.37) ^{cd}	29.63 (1.09) ^{d}	32.21 (0.80) ^{bc}	35.69 (0.58) ^{a}	35.5 (1.67) ^{a}	32.38 (1.53) ^{b}	11.55 ***
a*	1.19 (0.06) ^{de}	1.35 (0.07) ^{d}	2.82 (0.21) ^{a}	2.00 (0.1) ^{b}	1.16 (0.02) ^{e}	1.75 (0.1) ^{c}	2.03 (0.03) ^{b}	94.41 ***
b*	1.25 (0.02) ^{de}	1.69 (0.11) ^{ab}	1.40 (0.18) ^{cd}	1.11 (0.07) ^{e}	1.53 (0.09) ^{bc}	1.53 (0.18) ^{bc}	1.79 (0.2) ^{a}	8.38 ***
Opacity (Abs × mm^{-1})	0.39 (0.09) ^{d}	0.64 (0.007) ^{c}	0.85 (0.12) ^{bc}	1.11 (0.19) ^{a}	1.21 (0.02) ^{a}	1.05 (0.04) ^{ab}	1.14 (0.21) ^{a}	17.38 ***
Dissolution Time (s)	36.66 (1.52) ^{e}	22.33 (0.57) ^{f}	94 (3.6) ^{b}	46.33 (1.52) ^{d}	35.33 (2.51) ^{e}	105.66 (2.08) ^{a}	50.33 (1.52) ^{c}	674.08 ***
a_w	0.3562 (0.004) ^{d}	0.3623 (0.003) ^{cd}	0.3584 (0.002) ^{d}	0.3606 (0.006) ^{d}	0.3915 (0.0006) ^{a}	0.3690 (0.003) ^{c}	0.3765 (0.003) ^{b}	28.22 ***
MC (%)	11.14 (0.13) ^{d}	10.19 (0.24) ^{e}	13.69 (0.29) ^{b}	11.16 (0.22) ^{d}	14.32 (0.16) ^{a}	11.89 (0.12) ^{c}	14.39 (0.29) ^{a}	181.03 ***

Apart from thickness which is determined in 10 data points ($n = 10$), each value is the mean of three replicates ± standard deviation ($n = 3$). *** Statistically significant at $p < 0.001$. $^{a–f}$ Different letters in the same rows indicate significant differences between samples. WVTR: Water Vapor Transmission Rate; WVP: Water Vapor Permeability; L, a*, b*: Color profile; a_w: Water activity; MC: Moisture content.

Table 3. The effect of glycerin content on WVTR values.

Sample	Amount of Glycerin, g	WVTR, ($g \times h^{-1} \times m^{-2}$)
S1, S2	0.7375	3.7048 (0.19) ^{b}
S5	0.8625	6.8927 (0.68) ^{a}
S3, S4	1.025	4.5887 (1.5) ^{b}
S6, S7	1.15	6.5073 (0.84) ^{a}
		F-value 12.14 ***

*** Statistically significant at $p < 0.001$. ^{a,b} Different letters in the same column indicate significant differences.

Indeed, WVTR has a tendency to increase with the glycerin content. The maximum value of 7.1111 $g \times h^{-1} \times m^{-2}$ (Table 2) has the highest value of glycerin content of 1.15 g in the film-forming solution. Interestingly, Sample 7 with the same amount of glycerin has a low value of WVTR of 5.9036 $g \times h^{-1} \times m^{-2}$. One of the reasons for the difference is due to the fact that in the sample with the highest WVTR value the film does not contain soybean oil, which is known to have hydrophobic nature. Therefore, by decreasing the intermolecular interaction, the mobility of the molecule promotes the migration of water vapor through membranes. Compared with other bilberry-based films, which have values between 52.91 and 61.87 g $\times$ h^{-1} $\times$ m^{-2} depending on the bilberry concentration [27], WVTR showed low values with a minimum of 3.2562 g $\times$ h^{-1} $\times$ m^{-2}. The addition of β-glucan or pullulan to the films has also been shown to increase moisture barrier properties [28]. Ultimately, of course, our data are 32 higher than the value of pure low density polyethylene (LDPE) film values of 0.1012 g $\times$ h^{-1} $\times$ m^{-2} found by Reesha et al. (2015) [29]. To date, a bioactive film with potential packaging product applications showed the best barrier properties.

2.4. Water Vapor Permeability (WVP)

The results of the water vapor flux through the film (WVP) determined by dividing WVTR value to the differential water vapor partial pressure across the film and multiplied by the thickness of the film (in mm) are presented in Table 2. It was observed that WVP values ranged between 0.1057 and 0.3568 g $\times$ mm $\times$ kPa^{-1} $\times$ h^{-1} $\times$ m^{-2}, particularly with the increase in bilberry juice and β-glucan content. An increased value of the WVP parameter indicates that the film is more susceptible to water vapor flux penetration [24]. With an investigation of the physicochemical properties of films, and with the knowledge that high polar polymers and the addition of plasticizers resulted in an increased WVP values, Henrique et al. (2007) concluded that vapor permeability can be related to the quantity of –OH groups in the molecule [30]. On the other hand, Garcia et al. (1999) mentioned that coatings without plasticizing agents led to significantly ($p < 0.05$) higher values of WVP than those with plasticizer due to the formation of pores and cracks [31]. Undoubtedly, we can conclude that our data are lower than those found in films made only from β-glucan without the addition of plasticizer. This is the case of research conducted by Sarossy et al. (2013), which reported WVP values of 0.4625 g $\times$ mm $\times$ kPa^{-1} $\times$ h^{-1} $\times$ m^{-2} (or 11.1 g $\times$ mm $\times$ kPa^{-1} $\times$ m^{-2} $\times$ d^{-1}) [32] or by those determined by Peltzer et al. (2018) with an amount of 2.8×10^{-10} g $\times$ s^{-1} $\times$ m^{-1} $\times$ Pa^{-1} [24].

2.5. Dissolution Time

One of the most important aspects in developing a fast dissolving film is the time that should not exceed 1 min [33]. Significant differences between samples ($p < 0.001$) are observed in Table 2 with a higher variation between sample means relative to the variation within the samples (an F-value of 674.08). The values of the dissolution time range from 22.33 to 105.66 s. This is mainly due to the composition of the film-forming solution. While soybean oil behaves as a good surfactant [34], Rodriguez et al. (2006) investigated the combined effect of plasticizers and surfactants on the physical properties of films. The authors concluded that surfactants improved the wettability properties of the film solutions by decreasing the surface tension and in combination with glycerin allowed a higher molecular mobility [35]. To verify whether the composition significantly influenced the dissolution time, Table 4 investigated the film composition and the addition of surfactant on dissolution time.

In Table 4, we can observe that the incorporation of β-glucan between 1 and 1.5 g in films does not have a significant influence on dissolution time ($p > 0.05$; $p = 0.3$). On the other hand, data analysis showed that soybean oil and bilberry juice content have significant influence on the dissolution time of the films ($p < 0.05$ and $p < 0.001$, respectively). The obtained F-value of 20.06, which is significant at 0.1%, indicates that there is a significant influence of bilberry juice on dissolution time. High content in the amount of bilberry juice had a negative effect in terms of dissolution time of the films with an average of 74.08 s at 20 g compared with 31.44 s at 10 g of added bilberry juice. At the same time, the addition of 2% soybean oil to the film-forming solution showed a significant decrease (5% level of significance) in the dissolution time.

Table 4. The influence of β-glucan, soybean oil, and bilberry juice on dissolution time.

Physichochemical Parameter	Composition		
	β-glucan		
	1 g	1.5 g	F-value
Dissolution Time (s)	49.83 (28.14) ^{a}	63.77 (32.13) ^{a}	1.12 ^{ns}
	Soybean oil		
	0%	2%	F-value
Dissolution Time (s)	67.91 (33.68) ^{b}	39.66 (13.16) ^{a}	5.62 *
	Bilberry juice		
	10 g	20 g	F-value
Dissolution Time (s)	31.44 (7.01) ^{b}	74.08 (27.35) ^{a}	20.06 ***

^{ns}: Not significant; * $p < 0.05$; *** $p < 0.001$. ^{a,b} Different letters in the same rows indicate significant differences.

By correlation, for example, we can observe that in Table 1 the composition of Sample 7 differs from Sample 6 only by the addition of soybean oil, with other compounds remaining in the same proportion. On the other hand, in Table 2, we can observe a major difference in the dissolution time of 50.33 s in the sample with the addition of surfactant and 105.66 s without the soybean oil. Finally, we can summarize that the addition of surfactant positively influences the dissolution time and acts as a solubilizing agent.

2.6. Water Activity Tests (a_w) and Moisture Content (MC)

While the statistical analysis of the water activity tests showed significant differences between samples ($p < 0.001$), the data indicate narrow values between 0.3562 and 0.3915. However, our results are below the critical values where microbiological spoilage can occur. Majumdar et al. (2018) summarized in a review that a minimum water activity value of 0.6 is required to initiate the growth of microorganisms [36]. These data are also consistent with more in-depth research by Beuchat (1983), which concluded that below 0.61 a_w there can be no microbial growth, between 0.61 and 0.85 a_w food spoilage starts with mold and yeast formation, and above 0.85 a_w bacteria start to grow [37].

Another important parameter in food packaging applications is that the moisture content varied between 10.19 and 14.39%. Moisture content of films is closely related to the total amount of water molecules in the network microstructure of the composite films [38]. Abdalrazeq et al. (2019) stated that a high MC considerably limits the use of coatings for packaging materials and found the highest moisture value of 33.27% in film samples with 50% of glycerin prepared at pH 7 [39].

2.7. Color and Opacity

Consumer acceptance is influenced by two parameters of the film appearance: Color and opacity. The opacity and L, a*, b* values presented in Table 2 were significantly different. The two main factors that varied in the film-forming solution, β-glucan and bilberry juice, have been evaluated in Table 5 for their influence on the opacity, brightness characteristics of L value (between 0 and 100), redness/greenness (a* value), and yellowness/blueness (b* value).

Table 5. The influence of the β-glucan and bilberry juice on opacity and color of the films.

Physichochemical Parameters	Composition		
	β-glucan		
	1 g	1.5 g	F-value
L	30.89 (1.32) ^{b}	34.37 (1.91) ^{a}	24.25 ***
a*	1.84 (0.67) ^{a}	1.64 (0.38) ^{a}	0.59 ^{ns}
b*	1.36 (0.24) ^{b}	1.61 (0.19) ^{a}	6.48 **
Opacity (Abs × mm^{-1})	0.74 (0.29) ^{b}	1.13 (0.12) ^{a}	13.25 ***

Table 5. *Cont.*

Physichochemical Parameters	Composition		
	Bilberry juice		
	10 g	20 g	F-value
L	32.48 (2.57) [a]	32.32 (2.29) [a]	0.02 [ns]
a*	1.23 (0.10) [b]	2.15 (0.43) [a]	37.99 ***
b*	1.49 (0.2) [a]	1.46 (0.29) [a]	0.08 [ns]
Opacity (Abs × mm^{-1})	0.74 (0.36) [b]	1.04 (0.17) [a]	5.77 *

[ns:] Not significant; * $p < 0.05$; ** $p < 0.01$; *** $p < 0.001$. [a,b] Different letters in the same rows indicate significant differences.

The results in Table 5 showed that a high amount of β-glucan is reflected by an increased brightness, a tendency to blueness, and an opaquer film, while the a* parameter does not show a significant variation ($p > 0.05$). Regarding bilberry juice, a significant variation ($p < 0.001$) can be observed between the addition of 10 and 20 g to the red color of the samples (a* value). L and b* parameters for bilberry juice are not statistically significant ($p > 0.05$), while the opacity increased significantly ($p < 0.05$).

2.8. Scanning Electron Microscopy (SEM)

Figure 1 shows the microstructure of the β-glucan/bilberry juice films in the cross section at 1 kx to analyze the differences between surfactant and non-surfactant samples and to observe potential microcracks. As can be seen, on the transverse section, the samples with 2% surfactant (Figure 1B,D,G) showed a more compact surface, while the remaining samples present porous surfaces with micropores. The presence of pores makes the film less efficient in the water vapor barrier performance.

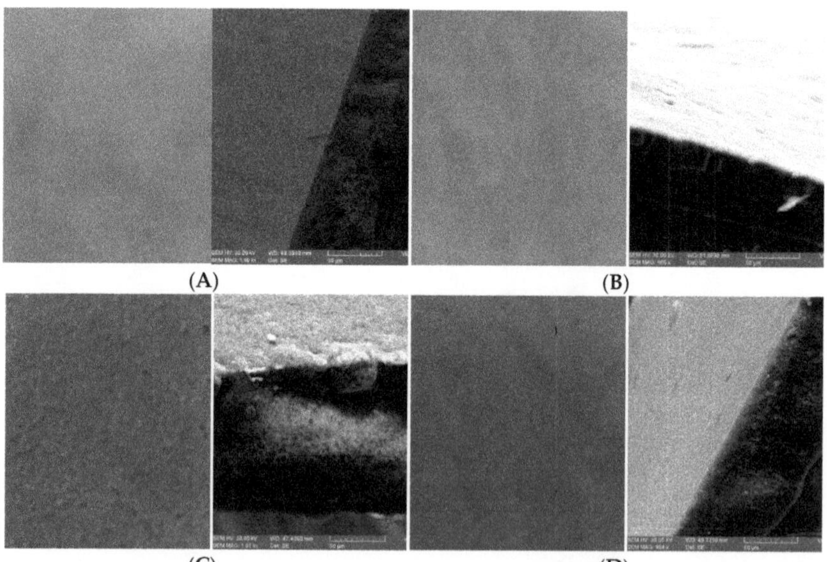

Figure 1. *Cont.*

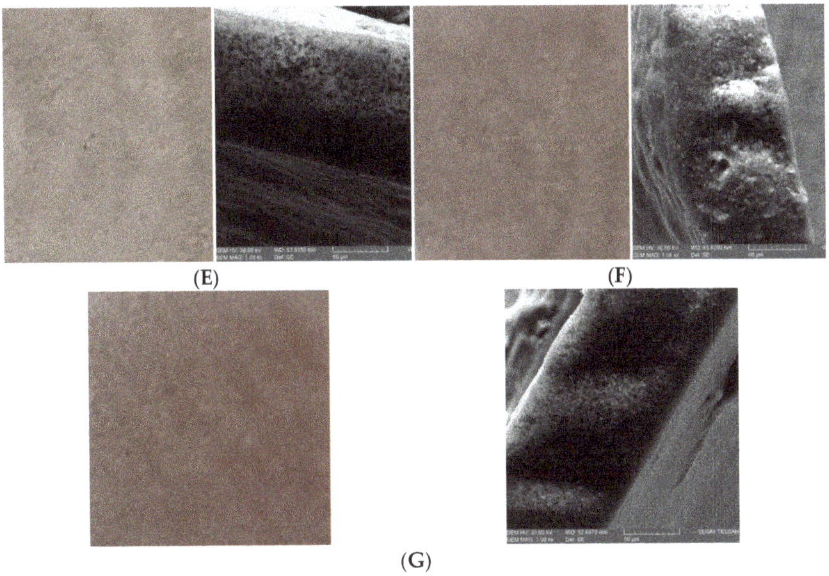

Figure 1. β-glucan/bilberry juice images and SEM micrographs of the film samples. (**A–G**) indicate the film samples from 1–7.

2.9. FT-IR Spectroscopy

The ATR-FTIR spectra (Figure 2A) of the most abundant group of polyphenols from bilberry juice (rich in anthocyanins) are identified by peaks found near the wavenumbers of ~1716.01 cm^{-1} (C=O stretching for aromatic nucleus) [40], ~1652.15 cm^{-1} characteristic for benzene skeleton vibration in anthocyanins, while the 3317.61 and 2924.44 cm^{-1} are assigned to O–H stretching vibration of water and CH, CH_2, and CH_3 groups, respectively [41], and ~1417.07 cm^{-1} corresponds to the C–H deformation [42]. Figure 2B showed that the incorporation of bilberry juice in films preserves characteristic peaks in the fingerprint region of the BJ, which indicates that the heat treatment of 15 min from the addition of bilberry juice to obtain the film-forming solution did not affect the bioactive compounds in the obtained β-glucan/bilberry film. Moreover, peaks near the wavenumbers of ~1149.20, 1024.65, and 920.90 cm^{-1} are characteristic for yeast β-glycosidic configuration, sodium alginate (carboxyl stretching bands), and glycerin, respectively [43–45].

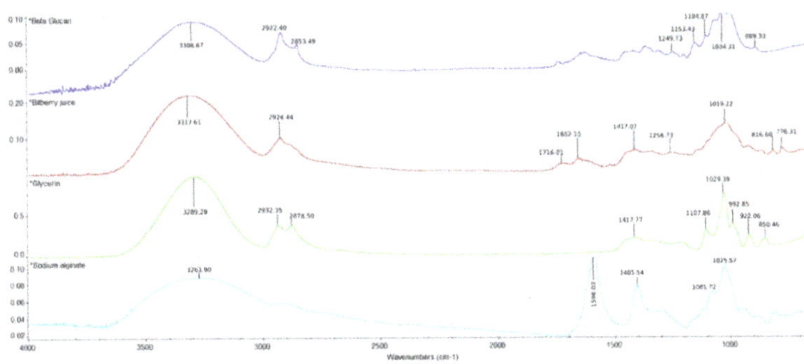

(**A**)

Figure 2. *Cont.*

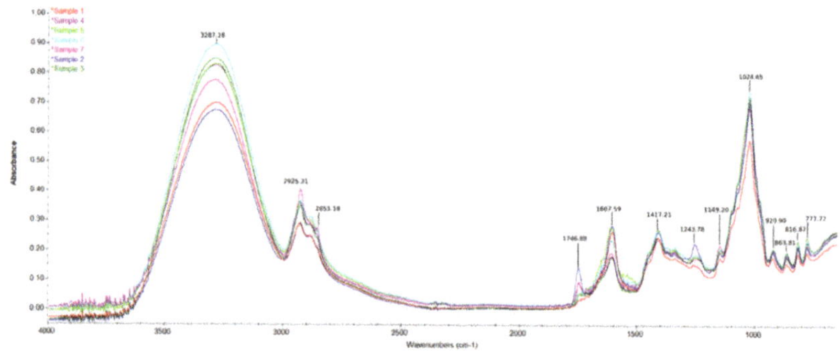

(B)

Figure 2. ATR−FTIR spectra: (**A**) Individual FT−IR spectra for pure samples of BG, BJ, GLY, SA; (**B**) FT−IR spectra of the β-glucan/bilberry juice film samples (4000−650 cm^{-1}).

3. Discussion

In this study, fast dissolving films from β-glucans and bilberry juice were successfully prepared. In addition to the role of plasticizer and sodium alginate, which confers vital primary film structure characteristics for a rapid dissolution, the addition of surfactant significantly reduces the dissolution time by improving the solubility of the β-glucans/bilberry juice films. In the case of fast dissolving films, this becomes very important, particularly when dispersing compounds, such as β-glucans or bioactive compounds from bilberry juice with antidiabetic properties, such as anthocyanins that attenuate the glycemic response. Our experimental data showed that the dissolution time of the films was halved by adding 2% surfactant. The film with the best dissolution properties compared with an increased content of 20 g of bilberry juice and 1.5 g of β-glucan is represented by Sample 7, which dissolves in 50.33 ± 1.52 s.

Moreover, we can conclude that in all of the samples, the water vapor barrier properties are remarkably low. Furthermore, the low water absorption rate (WVTR) between 3.2562 and 7.1111 g × h^{-1} × m^{-2} makes β-glucans/bilberry juice films suitable for packaging dry powdered pharmaceuticals. Ultimately, of course, these values are up to 32 times higher than a plastic film (e.g., LDPE). However, for a fast dissolving bioactive film that is intended for edible packaging, the values are outstanding. Water activity tests and moisture content proved the film stability by values below 0.6 a_w and MC up to 14.39%, which will not affect the quality of the packaged products. Since the film-forming solution contains a high amount of β-glucan and bilberry juice (rich in anthocyanidins), a considerably more opaque and reddish film will form after drying. Images of β-glucan/bilberry juice films and laboratory tests had detected the influence of these changes on different levels of significance, while SEM micrographs identified porous and compact structures on the cross section depending on the analyzed samples. FT-IR spectra revealed all of the compounds in film samples. Therefore, no degradation was found for β-glucans and bilberry juice.

The bioactive films developed in the present work could be used for packaging materials along with the bioactive delivering system of releasing compounds in aqueous solutions, which is necessary for people with a special medical condition, such as diabetes. Dispersion of the packaged product will be carried out simultaneously with the dissolution of bioactive film.

4. Materials and Methods

4.1. Chemicals

Bilberry juice (*Vaccinium myrtillus* L.) was purchased from a local market (distributed by SC. Deco Italia SRL, Suceagu, Cluj, Romania). According to the manufacturer, it is a 100% natural juice from bilberry fruit. The determined dry weight of the clear juice was

11.5% (w/w) with a pH value of 3.3. β-glucan was extracted from spent brewer's yeast provided by the SC. Bermas SA. brewery (Suceava, Romania). Other film components were: Sodium alginate, Product No. 9180.1 (Carl Roth, Karlsruhe, Germany), Glycerin, Product No. G7893 (Sigma-Aldrich, ACS reagent $\geq$ 99.5%, St. Louis, MO, USA), and Soybean oil (oil of genetically unmodified soybeans, Dachim SRL, Cluj, Romania).

4.1.1. β-Glucan Isolation

β-glucan, especially from spent brewer's yeast, is known to have particular potential in the inducement of innate immune response due to the triple helix structure of the insoluble polysaccharide conformation [46]. The most reliable method used for yeast β-glucan isolation is the alkaline-acid process [47]. Briefly, yeast slurry was purified and debittered at 50 °C with NaOH 2 N (up to pH 10) for 10 min according to [48]. Yeast cells were autolyzed at 55 °C/24 h and then were subjected to an alkaline extraction with NaOH 1.5 N at 90 °C/2 h in a ratio of 1.5 (w/v) according to [49], followed by an acid treatment with HCl solution for 2 h at 75 °C. The wet extract was washed three times and labeled as yeast β-glucan since it contains this polysaccharide as the principal component [19]. The FT-IR spectra showed characteristic bands for β-1,3 configuration at the wavenumbers of 1153.43 and 1104.87 cm^{-1}, while the β-1,6 glucan specific for yeast glucan was found at 889.33 cm^{-1} [43,50].

4.1.2. Film Preparation and Casting

The film-forming solution was prepared from the isolated β-glucan in different proportions and the addition of sodium alginate into each beaker. Therefore, bilberry juice with a dry weight of 11.5% (w/w) has been measured and prepared for each sample. Glycerin was added as plasticizer as 25% related to the dry weight of the solids of β-glucan, sodium alginate, and bilberry juice. Two percent (w/w) soybean oil of the total solid weight was added as a surfactant to samples 2, 4, and 7 in order to observe whether there are significant changes between the physical chemical parameters analyzed. Distilled water was added up to a total volume of 150 mL. The mixture was subjected to heating at 80 °C under continuous stirring (900 rpm) for 15 min. After 15 min of stirring in the homogeneous solution, the measured bilberry juice was added in each sample and stirred continuously for another 15 min for incorporation and for slight sterilization (30 min in total).

β-glucan/bilberry juice films were prepared by the casting technique as reported by [51]. Accordingly, equal suspensions from the film-forming solution were poured onto plastic petri dishes and dried at 40 °C for 48 h. The dried films were stored at room temperature prior to the analysis.

4.2. Methods

4.2.1. Determination of Thickness

The film thickness was determined with the thickness gauge PosiTector 6000 (DeFelsko, Ogdensburg, NY, USA) with an accuracy of 0.1 µm. Measurements were taken at 10 different points, and the average was used to calculate the film properties. Thickness of the films was expressed in µm.

4.2.2. Determination of Water Vapor Transmission Rate (WVTR)

Water vapor transmission rates were measured using the described standard ASTM E96/96M method [52]. The dry cup method involved sealing films horizontally on a petri dish containing about 10 g $CaCl_2$ as desiccant to create 0% RH inside the cups. Samples with desiccant were placed in an environmental chamber with a NaCl solution, which provides 75% RH. The WVTR of the films was calculated by dividing the slope to the area of exposed film using the following equation:

$$WVTR = \frac{\Delta W}{\Delta t \times A} \left(\mathrm{g} \times \mathrm{h}^{-1} \times \mathrm{m}^{-2}\right) \quad (1)$$

where $\Delta W/\Delta t$ is the amount of water gained in the unit of time (g/h) and A is the area exposed to the water vapor diffusion (m^2).

4.2.3. Determination of Water Vapor Permeability (WVP)

The permeation characteristic of the β-glucan/bilberry juice films was investigated by dividing the *WVTR* values to the water vapor partial pressure across the film and multiplying by the film thickness (in mm) as described by [53]. The *WVP* was expressed by the following equation:

$$WVP = \frac{WVTR \times L}{\Delta p} \left(g \times mm \times kPa^{-1} \times h^{-1} \times m^{-2}\right) \quad (2)$$

where *WVTR* is the water vapor transmission rate ($g \times h^{-1} \times m^{-2}$), L is the thickness of the film (mm), and Δp is the water vapor partial pressure across the film (kPa) calculated according to the formula:

$$\Delta p = S \times (R_1 - R_2)\ (kPa) \quad (3)$$

where S is the saturated vapor pressure of water (3.1687 kPa at 25 °C [54]) and the moisture gradients R_1 and R_2 are 0.75 and 0, respectively.

4.2.4. Dissolution Time

The film samples were cut into squares of 2 × 2 cm, immersed in 50 mL of distilled water, and then vigorously shaken until dissolution. The dissolution time (s) was recorded using a chronometer.

4.2.5. Determination of Color and Opacity

Film opacity was determined by measuring the absorbance at 600 nm and dividing by the film thickness [55]. The absorbance was acquired in UV-VIS-NIR Shimadzu 3600 spectrophotometer (Tokyo, Japan). The following equation for *opacity* is shown below:

$$Opacity = Abs_{600nm}/L \quad (4)$$

where Abs_{600nm} is the absorbance (600 nm) and L is the film thickness (mm).

The color was quantified according to the CIELab color space (lightness (L), redness (a*), and yellowness (b*)) using a portable chromameter CR-400 (Konica Minolta, Tokyo, Japan).

4.2.6. Water Activity Tests and Moisture Content

The water activity was measured with a water activity analyzer AquaLab 4TE (Meter Group, Inc., Pullman, WA, USA). With the use of chilled-mirror dew point technology, the instrument was able to determine a_w values with a resolution of 10^{-4} between 0.03 and 1.

Moisture content was determined gravimetrically according to [56], samples were weighed before and after drying at 105 °C for 24 h, and the difference in weight loss was expressed as the moisture content in films, according to the following equation:

$$MC = \frac{W_0 - W_1}{W_0} \times 100\ (\%) \quad (5)$$

where W_0 is the initial weight of the sample and W_1 is the final weight of the dried film.

4.2.7. FT-IR Spectroscopy

FT-IR spectra with attenuated total reflectance unit (ATR) analysis were achieved using Nicolet iS-20 FT-IR spectrometer (Thermo Scientific™, Karlsruhe, Dieselstraße, Germany). Measurements were conducted by placing samples directly on the ZnSe crystal plate. The spectra were collected in the region of 4000–650 cm^{-1} by 32 scans per spectrum at a resolution of 4 cm^{-1}.

4.2.8. Scanning Electron Microscopy (SEM)

The cross-section morphology of the films was observed using a scanning electron microscope VEGA II LMU (Tescan, Brno, Czech Republic) under HighVac conditions using a secondary electron (SE) detector operated at an accelerating voltage of 30 kV and a magnification of 1 kx without preliminary coating on the investigated surface.

4.2.9. Dry Weight Determination (w/w)

Dry weight was determined by weighing about 10 g of juice, solids and dry in an oven at 105 °C to constant weight. Samples were transferred into a desiccator to prevent moisture uptake. The measurements for dry weight were made according to the following equation:

$$DW = \frac{w_2 - w_3}{w_2 - w_1} \times 100\ (\%\ w/w) \tag{6}$$

where w_1 is the weight of crucible; w_2 is the initial weight of crucible with sample, g; and w_3 is the final weight of crucible with sample after drying, g.

4.2.10. Statistical Analysis of the Results

A one-way ANOVA test was used to determine whether there was a statistically significant difference between the means of the independent groups. Table 2 summarized the mean values and standard deviation of the physicochemical test results on a 95% confidence level. All of the tests, except for thickness, which require a mean of minimum 10 data points were expressed as the average and standard deviation of triplicates.

5. Conclusions

This study demonstrated good compatibility between yeast β-glucans and bilberry juice in the development of fast dissolving films. The addition of bilberry juice in films is of particular importance in the management of metabolic disorders, such as diabetes. Considering the results, films based on β-glucans/bilberry juice with improved fast dissolving time and good water vapor barrier properties can be a potential novel film for packaging dry powdered pharmaceuticals.

Author Contributions: Conceptualization, S.A. and I.A.; methodology, I.A. and S.A.; software, I.A.; validation, S.A. and I.A.; formal analysis, I.A.; investigation, I.A.; resources, S.A.; writing—original draft preparation, I.A.; writing—review and editing, I.A.; visualization, I.A.; supervision, S.A.; project administration, S.A.; funding acquisition, S.A. All authors have read and agreed to the published version of the manuscript.

Funding: This work was funded by the Ministry of Research, Innovation and Digitalization within Program 1—Development of national research and development system, Subprogram 1.2—Institutional Performance–RDI excellence funding projects, under contract no. 10PFE/2021.

Institutional Review Board Statement: Not applicable.

Informed Consent Statement: Not applicable.

Data Availability Statement: Not applicable.

Conflicts of Interest: The authors declare no conflict of interest.

References

1. Oliveira Filho, J.G.; Egea, M.B. Edible Bioactive Film with Curcumin: A Potential "Functional" Packaging? *Int. J. Mol. Sci.* **2022**, *23*, 5638. [CrossRef] [PubMed]
2. Benbettaïeb, N.; Karbowiak, T.; Debeaufort, F. Bioactive edible films for food applications: Influence of the bioactive compounds on film structure and properties. *Crit. Rev. Food Sci. Nutr.* **2019**, *59*, 1137–1153. [CrossRef] [PubMed]
3. Zhang, L.; Yu, D.; Regenstein, J.M.; Xia, W.; Dong, J. A comprehensive review on natural bioactive films with controlled release characteristics and their applications in foods and pharmaceuticals. *Trends Food Sci. Technol.* **2021**, *112*, 690–707. [CrossRef]
4. Chan, D.I.; Víctor, M.; López, M.T.; Lourdes, M. De Preparation and characterization of chitosan-based bioactive films incorporating *Moringa oleifera* leaves extract. *J. Food Meas. Charact.* **2021**, *15*, 4813–4824. [CrossRef]

5. Vilela, C.; Pinto, R.J.B.; Coelho, J.; Domingues, M.R.M.; Daina, S.; Sadocco, P.; Santos, S.A.O.; Freire, C.S.R. Bioactive chitosan/ellagic acid films with UV-light protection for active food packaging. *Food Hydrocoll.* **2017**, *73*, 120–128. [CrossRef]
6. Avramia, I.; Amariei, S. Formulation, Characterization and Optimization of β–Glucan and Pomegranate Juice Based Films for Its Potential in Diabetes. *Nutrients* **2022**, *14*, 2142. [CrossRef]
7. Kermansaravi, F.; Metanat, M.; Sharifi-Mood, B. Evaluation of Active Pulmonary Tuberculosis Among Patients With Diabetes. *Int. J. Infect.* **2014**, *1*, 56–63. [CrossRef]
8. Saeedi, P.; Petersohn, I.; Salpea, P.; Malanda, B.; Karuranga, S.; Unwin, N.; Colagiuri, S.; Guariguata, L.; Motala, A.A.; Ogurtsova, K. Global and regional diabetes prevalence estimates for 2019 and projections for 2030 and 2045: Results from the International Diabetes Federation Diabetes Atlas. *Diabetes Res. Clin. Pract.* **2019**, *157*, 107843. [CrossRef]
9. Mizuno, C.S.; Rimando, A.M. Blueberries and Metabolic Syndrome. *Silpakorn Univ. Sci. Tech. J.* **2009**, *3*, 7–17.
10. Chehri, A.; Yarani, R.; Yousefi, Z.; Shakouri, S.K.; Ostadrahimi, A.; Mobasseri, M.; Araj-Khodaei, M. Phytochemical and pharmacological anti-diabetic properties of bilberries (*Vaccinium myrtillus*), recommendations for future studies. *Prim. Care Diabetes* **2022**, *16*, 27–33. [CrossRef]
11. Lowman, D.W.; West, L.J.; Bearden, D.W.; Wempe, M.F.; Power, T.D.; Ensley, H.E.; Haynes, K.; Williams, D.L.; Kruppa, M.D. New insights into the structure of (1→3,1→6)-β-D-glucan side chains in the Candida glabrata cell wall. *PLoS ONE* **2011**, *6*, e27614. [CrossRef]
12. Petravić-Tominac, V.; Panjkota-Krbavčić, I. Biological effects of yeast β-glucans. *Agric. Conspec. Sci.* **2010**, *75*, 149–158.
13. Avramia, I.; Amariei, S.; Ursachi, F.-V. Bioactive Î²-Glucans, Adjuvants in SARS-CoV-2 Antiviral Therapy. *Biomed. J. Sci. Tech. Res.* **2021**, *33*, 25977–25980.
14. Karumuthil-Melethil, S.; Gudi, R.; Johnson, B.M.; Perez, N.; Vasu, C. Fungal β-glucan, a Dectin-1 ligand, promotes protection from type 1 diabetes by inducing regulatory innate immune response. *J. Immunol.* **2014**, *193*, 3308–3321. [CrossRef]
15. Malviya, V.R.; Tawar, M.G. Preparation and Evaluation of Oral Dispersible Strips of Teneligliptin Hydrobromide for Treatment of Diabetes Mellitus. *Int. J. Pharm. Sci. Nanotechnol.* **2020**, *13*, 4745–4752.
16. Kim, J.Y.; Kim, D.W.; Kuk, Y.M.; Park, C.W.; Rhee, Y.S.; Oh, T.O.; Weon, K.Y.; Park, E.S. Investigation of an active film coating to prepare new fixed-dose combination tablets for treatment of diabetes. *Int. J. Pharm.* **2012**, *427*, 201–208. [CrossRef]
17. European Commission. Commission Implementing Decision (EU) 2017/2078—Of 10 November 2017—Authorising an extension of use of yeast beta-glucans as a novel food ingredient under Regulation (EC) No 258/97 of the European Parliament and of the Council. *Off. J. Eur. Union* **2017**, *2009*, 12–15. Available online: https://eur-lex.europa.eu/legal-content/EN/TXT/PDF/?uri=CELEX:32017D2078&from=EN (accessed on 5 July 2022).
18. Mancini, M.; Moresi, M.; Sappino, F. Rheological behaviour of aqueous dispersions of algal sodium alginates. *J. Food Eng.* **1996**, *28*, 283–295. [CrossRef]
19. Novák, M.; Synytsyaa, A.; Gedeonb, O.; Slepičkac, P.; Procházkab, V.; Synytsyad, A.; Blahovece, J.; Hejlováe, A.; Čopíkováa, J. Yeast β(1-3),(1-6)-d-glucan films: Preparation and characterization of some structural and physical properties. *Carbohydr. Polym.* **2012**, *87*, 2496–2504. [CrossRef]
20. Neves, D.; Andrade, P.B.; Videira, R.A.; de Freitas, V.; Cruz, L. Berry anthocyanin-based films in smart food packaging: A mini-review. *Food Hydrocoll.* **2022**, *133*, 107885. [CrossRef]
21. Lim, L.I.; Tan, H.L.; Pui, L.P. Development and characterization of alginate-based edible film incorporated with hawthorn berry (*Crataegus pinnatifida*) extract. *J. Food Meas. Charact.* **2021**, *15*, 2540–2548. [CrossRef]
22. Alizadeh-Sani, M.; Tavassoli, M.; McClements, D.J.; Hamishehkar, H. Multifunctional halochromic packaging materials: Saffron petal anthocyanin loaded-chitosan nanofiber/methyl cellulose matrices. *Food Hydrocoll.* **2021**, *111*, 106237. [CrossRef]
23. Arham, R.; Mulyati, M.T.; Metusalach, M.; Salengke, S. Physical and mechanical properties of agar based edible film with glycerol plasticizer. *Int. Food Res. J.* **2016**, *23*, 1669–1675. [CrossRef]
24. Peltzer, M.A.; Salvay, A.G.; Delgado, J.F.; de la Osa, O.; Wagner, J.R. Use of Residual Yeast Cell Wall for New Biobased Materials Production: Effect of Plasticization on Film Properties. *Food Bioprocess Technol.* **2018**, *11*, 1995–2007. [CrossRef]
25. Zhao, Y.; Du, J.; Zhou, H.; Zhou, S.; Lv, Y.; Cheng, Y.; Tao, Y.; Lu, J.; Wang, H. Biodegradable intelligent film for food preservation and real-time visual detection of food freshness. *Food Hydrocoll.* **2022**, *129*, 107665. [CrossRef]
26. Rahmawati, H.; Mutmainna, I.; Ilyas, S.; Fahri, A.N.; Wahyuni, A.S.H.; Afrianti, E.; Setiawan, I.; Tahir, D. Effect of carbon for enhancing degradation and mechanical properties of bioplastics composite cassava starch/glycerin/carbon. *AIP Conf. Proc.* **2020**, *2219*, 100004.
27. Dordevic, S.; Dordevic, D.; Sedlacek, P.; Kalina, M.; Tesikova, K.; Antonic, B.; Tremlova, B.; Treml, J.; Nejezchlebova, M.; Vapenka, L.; et al. Incorporation of natural blueberry, red grapes and parsley extract by-products into the production of chitosan edible films. *Polymers* **2021**, *13*, 3388. [CrossRef]
28. Chang, J.; Li, W.; Liu, Q.; Zhou, Y.; Chen, X.; Lyu, Q.; Liu, G. Preparation, properties, and structural characterization of β-glucan/pullulan blend films. *Int. J. Biol. Macromol.* **2019**, *140*, 1269–1276. [CrossRef]
29. Reesha, K.V.; Satyen Kumar, P.; Bindu, J.; Varghese, T.O. Development and characterization of an LDPE/chitosan composite antimicrobial film for chilled fish storage. *Int. J. Biol. Macromol.* **2015**, *79*, 934–942. [CrossRef]
30. Henrique, C.M.; Teófilo, R.F.; Sabino, L.; Ferreira, M.M.C.; Cereda, D.M.P. Classification of cassava starch films by physicochemical properties and water vapor permeability quantification by FTIR and PLS. *J. Food Sci.* **2007**, *72*, E184–E189. [CrossRef]

31. García, M.A.; Martino, M.N.; Zaritzky, N.E. Edible starch films and coatings characterization: Scanning electron microscopy, water vapor, and gas permeabilities. *Scanning* **1999**, *21*, 348–353. [CrossRef]
32. Sárossy, Z.; Tenkanen, M.; Pitkänen, L.; Bjerre, A.-B.; Plackett, D. Extraction and chemical characterization of rye arabinoxylan and the effect of β-glucan on the mechanical and barrier properties of cast arabinoxylan films. *Food Hydrocoll.* **2013**, *30*, 206–216. [CrossRef]
33. Cilurzo, F.; Cupone, I.E.; Minghetti, P.; Selmin, F.; Montanari, L. Fast dissolving films made of maltodextrins. *Eur. J. Pharm. Biopharm.* **2008**, *70*, 895–900. [CrossRef] [PubMed]
34. Xu, Q.; Nakajima, M.; Liu, Z.; Shiina, T. Soybean-based surfactants and their applications. In *Soybean-Applications Technology*; IntechOpen: London, UK, 2011; pp. 341–364. Available online: https://www.intechopen.com/chapters/15794 (accessed on 5 July 2022).
35. Rodríguez, M.; Osés, J.; Ziani, K.; Maté, J.I. Combined effect of plasticizers and surfactants on the physical properties of starch based edible films. *Food Res. Int.* **2006**, *39*, 840–846. [CrossRef]
36. Majumdar, A.; Pradhan, N.; Sadasivan, J.; Acharya, A.; Ojha, N.; Babu, S.; Bose, S. Food degradation and foodborne diseases: A microbial approach. In *Microbial Contamination and Food Degradation*; Elsevier: Amsterdam, The Netherlands, 2018; pp. 109–148.
37. Beuchat, L.R. Influence of Water Activity on Growth. *Metab. Act. Surviv. Yeasts Molds* **1983**, *46*, 135–141.
38. Kanmani, P.; Rhim, J.W. Antimicrobial and physical-mechanical properties of agar-based films incorporated with grapefruit seed extract. *Carbohydr. Polym.* **2014**, *102*, 708–716. [CrossRef]
39. Abdalrazeq, M.; Giosafatto, C.V.L.; Esposito, M.; Fenderico, M.; Di Pierro, P.; Porta, R. Glycerol-plasticized films obtained from whey proteins denatured at alkaline pH. *Coatings* **2019**, *9*, 322. [CrossRef]
40. Espinosa-Acosta, G.; Ramos-Jacques, A.L.; Molina, G.A.; Maya-Cornejo, J.; Esparza, R.; Hernandez-Martinez, A.R.; Sánchez-González, I.; Estevez, M. Stability analysis of anthocyanins using alcoholic extracts from black carrot (*Daucus carota* ssp. Sativus var. Atrorubens alef.). *Molecules* **2018**, *23*, 2744. [CrossRef]
41. Fei, P.; Zeng, F.; Zheng, S.; Chen, Q.; Hu, Y.; Cai, J. Acylation of blueberry anthocyanins with maleic acid: Improvement of the stability and its application potential in intelligent color indicator packing materials. *Dye. Pigment.* **2021**, *184*, 108852. [CrossRef]
42. Mosoarca, G.; Vancea, C.; Popa, S.; Dan, M.; Boran, S. The Use of Bilberry Leaves (*Vaccinium myrtillus* L.) as an Efficient Adsorbent for Cationic Dye Removal from Aqueous Solutions. *Polymers* **2022**, *14*, 978. [CrossRef]
43. Galichet, A.; Sockalingum, G.D.; Belarbi, A.; Manfait, M. FTIR spectroscopic analysis of Saccharomyces cerevisiae cell walls: Study of an anomalous strain exhibiting a pink-colored cell phenotype. *FEMS Microbiol. Lett.* **2001**, *197*, 179–186. [CrossRef]
44. Deepa, B.; Abraham, E.; Pothan, L.A.; Cordeiro, N.; Faria, M.; Thomas, S. Biodegradable nanocomposite films based on sodium alginate and cellulose nanofibrils. *Materials* **2016**, *9*, 50. [CrossRef]
45. Tee, Y.B.; Tee, L.T.; Daengprok, W.; Talib, R.A. Chemical, physical, and barrier properties of edible film from flaxseed mucilage. *BioResources* **2017**, *12*, 6656–6664. [CrossRef]
46. Bastos, R.; Oliveira, P.G.; Gaspar, V.M.; Mano, J.F.; Coimbra, M.A.; Coelho, E. Brewer's yeast polysaccharides—A review of their exquisite structural features and biomedical applications. *Carbohydr. Polym.* **2021**, *277*, 118826. [CrossRef]
47. Ruphuy, G.; Saloň, I.; Tomas, J.; Šalamúnová, P.; Hanuš, J.; Štěpánek, F. Encapsulation of poorly soluble drugs in yeast glucan particles by spray drying improves dispersion and dissolution properties. *Int. J. Pharm.* **2020**, *576*, 18990. [CrossRef]
48. Zechner-Krpan, V.; Petravić-Tominac, V.; Gospodarić, I.; Sajli, L.; Daković, S.; Filipović-Grčić, J. Characterization of β-glucans isolated from brewer's yeast and dried by different methods. *Food Technol. Biotechnol.* **2010**, *48*, 189–197.
49. Petravić-Tominac, V.; Zechner-Krpan, V.; Berković, K.; Galović, P.; Herceg, Z.; Srečec, S.; Špoljarić, I. Rheological properties, water-holding and oil-binding capacities of particulate β-glucans isolated from spent Brewer's yeast by three different procedures. *Food Technol. Biotechnol.* **2011**, *49*, 56–64.
50. Abdel Naser, A.; Zohri, H.M. Biotechnological β-glucan Production from Returned Baker's Yeast and Yeast Remaining after Ethanol Fermentation. *Egypt. Sugar J.* **2019**, *13*, 29–43.
51. Velickova, E.; Winkelhausen, E.; Kuzmanova, S.; Moldao-Martins, M.; Alves, V.D. Characterization of multilayered and composite edible films from chitosan and beeswax. *Food Sci. Technol. Int.* **2015**, *21*, 83–93. [CrossRef]
52. *ASTM D1653*; ASTM Standard Test Methods for Water Vapor Transmission of Materials. ASTM International: West Conshohocken, PA, USA, 2018.
53. Ghanbarzadeh, B.; Musavi, M.; Oromiehie, A.R.; Rezayi, K.; Razmi, E.; Milani, J. Effect of plasticizing sugars on water vapor permeability, surface energy and microstructure properties of zein films. *LWT-Food Sci. Tech.* **2007**, *40*, 1191–1197. [CrossRef]
54. Wexler, A. Vapor pressure formulation for water in range 0 to 100 °C. A revision. *J. Res. Natl. Bur. Stand. A Phys. Chem.* **1976**, *80*, 775–785. [CrossRef]
55. Riaz, A.; Lei, S.; Akhtar, H.M.S.; Wan, P.; Chen, D.; Jabbar, S.; Abid, M.; Hashim, M.M.; Zeng, X. Preparation and characterization of chitosan-based antimicrobial active food packaging film incorporated with apple peel polyphenols. *Int. J. Biol. Macromol.* **2018**, *114*, 547–555. [CrossRef]
56. Nabi, M.; Jafar, D. Preparation and characterization of a novel biodegradable film based on sulfated polysaccharide extracted from seaweed *Ulva intestinalis*. *Food Sci. Nutr.* **2021**, *9*, 4108–4116. [CrossRef]

plants

MDPI

Review

Prospects for Protective Potential of *Moringa oleifera* against Kidney Diseases

Tanzina Akter [1,†], Md Atikur Rahman [1,†], Akhi Moni [1], Md. Aminul Islam Apu [1], Atqiya Fariha [1], Md. Abdul Hannan [1,2] and Md Jamal Uddin [1,3,*]

1 ABEx Bio-Research Center, East Azampur, Dhaka 1230, Bangladesh; tanzinatuli842@gmail.com (T.A.); atik.rahmanbt@gmail.com (M.A.R.); akhimoni840818@gmail.com (A.M.); aminul.btge@gmail.com (M.A.I.A.); atqiyafariha047@gmail.com (A.F.); hannanbmb@bau.edu.bd (M.A.H.)

2 Department of Biochemistry and Molecular Biology, Bangladesh Agricultural University, Mymensingh 2202, Bangladesh

3 Graduate School of Pharmaceutical Sciences, College of Pharmacy, Ewha Womans University, Seoul 03760, Korea

* Correspondence: hasan800920@gmail.com

† These authors are contributed equally to this work.

Abstract: Kidney diseases are regarded as one of the major public health issues in the world. The objectives of this study were: (i) to investigate the causative factors involved in kidney disease and the therapeutic aspects of *Moringa oleifera*, as well as (ii) the effectiveness of *M. oleifera* in the anti-inflammation and antioxidant processes of the kidney while minimizing all potential side effects. In addition, we proposed a hypothesis to improve *M. oleifera* based drug development. This study was updated by searching the key words *M. oleifera* on kidney diseases and *M. oleifera* on oxidative stress, inflammation, and fibrosis in online research databases such as PubMed and Google Scholar. The following validation checking and scrutiny analysis of the recently published articles were used to explore this study. The recent existing research has found that *M. oleifera* has a plethora of health benefits. Individual medicinal properties of *M. oleifera* leaf extract, seed powder, stem extract, and the whole extract (ethanol/methanol) can up-increase the activity of antioxidant enzymes like superoxide dismutase (SOD), catalase (CAT), and glutathione (GSH), while decreasing the activity of inflammatory cytokines such as TNF-α, IL-1β, IL-6, and COX-2. In our study, we have investigated the properties of this plant against kidney diseases based on existing knowledge with an updated review of literature. Considering the effectiveness of *M. oleifera*, this study would be useful for further research into the pharmacological potential and therapeutic insights of *M. oleifera*, as well as prospects of *Moringa*-based effective medicine development for human benefits.

Keywords: *Moringa oleifera*; antioxidant; anti-aging; fibrosis; inflammation; kidney disease

Citation: Akter, T.; Rahman, M.A.; Moni, A.; Apu, M.A.I.; Fariha, A.; Hannan, M.A.; Uddin, M.J. Prospects for Protective Potential of *Moringa oleifera* against Kidney Diseases. *Plants* **2021**, *10*, 2818. https://doi.org/10.3390/plants10122818

Academic Editors: Juei-Tang Cheng, I-Min Liu and Szu-Chuan Shen

Received: 27 November 2021
Accepted: 16 December 2021
Published: 20 December 2021

1. Introduction

Kidney diseases are considered among the major health problems worldwide. Acute kidney injury (AKI) is closely connected with chronic kidney diseases (CKD). Since 1990, CKD has been included in the list of non-communicable conditions investigated by the global burden of disease study. As the disease's growth rate accelerates, it has become a global concern. The majority of incidents occur in low and lower-middle income countries [1–3]. The kidneys gradually lose their ability to function in CKD patients, and the glomerular filtration rate (GFR) falls below 60 mL/min per 1.73 m^2 [1,2]. Mainly people who have been already suffering from diabetes, heart disease, or high blood pressure are at a high risk of developing CKD. Few drugs, such as prolyl hydroxylase domain inhibitors against anemia in CKD [3], can be used to treat CKD complications. The main pathologies involved in kidney complications are inflammation, oxidative stress, apoptosis, and fibrosis [4]. Unfortunately, no potential drug for treating kidney diseases exists at this time. Therefore, the search for a potential drug with fewer side effects to combat this disease is becoming increasingly important. *M. oleifera* Lam., also known as drumstick tree, is a *Moringaceae* family member that grows in the Indian subcontinent. This plant's various parts have medicinal applications, such as antifungal, antiviral, anti-inflammatory, etc. [5–8]. *Moringa* leaves also have a low calorific value and can be included in the diet of obese individuals [9]. Furthermore, it contains numerous bioactive phytochemicals such as flavonoids, saponin, vanillin, omega fatty acids, carotenoids, ascorbates, tocopherols, beta-sitosterol, moringine, kaempferol, and quercetin that have been reported in its flowers, roots, fruits, and seeds, and can play a variety of roles in medicine [10–13]. In general, the choice of the most suitable bioactive substance for therapeutic purposes necessarily depends on the chemical formula of that specific compound, its structure giving its unique properties, and implicitly its mode of action [14]. Kaempferol has been shown to promote cancer cell apoptosis, such as MCF-7 and A549 cells [15]. Due to its anti-inflammatory and antioxidant properties, quercetin has the potential to be hepatoprotective, hypocholesterolemic, hypolipidemic, and anti-atherosclerotic [16]. *Moringa* has an anti-hyperglycemic effect, according to researchers who studied it in vivo on mice models [17].

Previous studies indicate that the juice of the super food *M. oleifera* enhances antimicrobial defense [18] and regulates insulin level, as well as glucose uptake in muscles [19,20]. Interestingly, *M. oleifera* showed a significant reduction of hyperglycemia, low-density lipoprotein (LDL) cholesterol, total cholesterol, fatty substances, FPG, and VLDL-cholesterol [21]. *M. oleifera* is also beneficial for skin, hair, liver, eye, blood pressure, treating anemia, kidney disease, and diabetes [22]. Several recent studies have documented the beneficial impacts of *M. oleifera* in alleviating renal diseases in animal model. Nafiu et al. [23] marked that gentamicin-induced impairment and oxidative stress significantly reduced by ethanolic extract of *Moringa oleifera* seeds in plasma, urine and kidney homogenate of rats. Akinrinde et al. [24] observed that *M. oleifera* extract attenuates the deleterious effects of renal ischemia-reperfusion through alleviation of oxidative stress. Soliman et al. [25] explored the ameliorative effects of *M. oleifera* against oxidative stress and methotrexate-induced hepato–renal dysfunction. Recently, Abu-Zeid et al. [26] discovered that the ecofriendly selenium nanoparticle using *M. oleifera* and/or *M. oleifera* ethanolic leaf extract reduces melamine-induced nephrotoxicity by alleviating of renal function impairments, oxidative stress, and apoptosis in rat kidney. Despite the great progress of *M. oleifera* in this field in recent years, less attention has been given to the effectiveness of *M. oleifera*, particularly against kidney related diseases. Therefore, there are still some issues which need further exploration, such as the protective effects of *M. oleifera* in kidney related disease difficulties and its prospects in drug development for human benefits.

This review updates the existing knowledge concerning the causative factors involved in kidney disease, as well as the therapeutic aspects of *M. oleifera*. Furthermore, this study provides a hypothesis on how *M. oleifera* would be effective in the anti-inflammation and antioxidant processes of the kidney, with the least amount of side effects.

2. Methods

This systematic review was carried out following the Preferred Reporting Items for Systematic Reviews and Meta-Analyses (PRISMA) guidelines [27]. Databases such as Scopus, PubMed, and Google Scholar were accessed to retrieve information using the keywords 'MeSH terms', on 'kidney diseases' and 'oxidative stress' and 'inflammation', and 'fibrosis' and '*Moringa oleifera*'. The information was retrieved from 2011 to 15 June 2021. Automatic search tools were used to exclude some of the articles, while others were screened manually. Articles published in languages other than English were excluded. Reviews, book chapters, expert opinions, conference papers, and letters to editors were also excluded from this review. A total of 151 research articles were retrieved from the databases and discussed in this study (Figure 1). All information compiled in the table was obtained from these research articles.

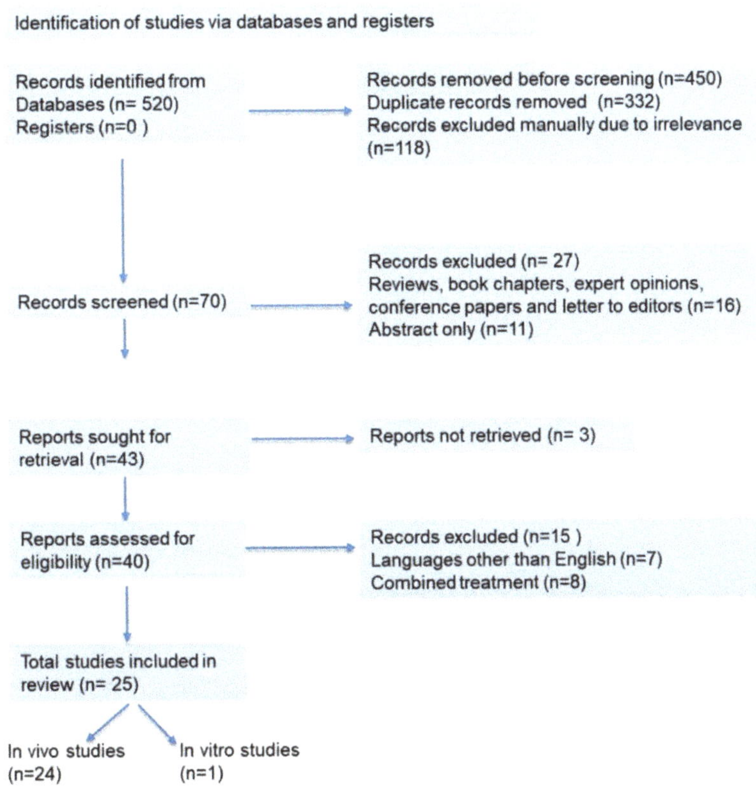

Figure 1. PRISMA 2020 flow diagram for the systematic review.

3. Phytochemical Content and Pharmacological Potential of *M. oleifera* on Kidney Diseases

M. oleifera contains several bioactive phytochemicals including flavonoids and isothiocyanates [10]; polyphenols, carotenoids, alkaloids, and terpenoids [11]; and triterpenoids, moringyne, monopalmitic, di-oleic triglyceride, campesterol, stigmasterol, β-sitosterol, avenasterol, and vitamin A [12]. These bioactive phytochemicals are found in *M. oleifera* roots, fruits, and seeds. These phytochemicals have medicinal properties which have been shown to be effective antioxidant, antimicrobial, inflammatory, and anti-carcinogenic agents [28]. More studies are required to explore the role of bioactive phytochemicals specially in kidney diseases.

M. oleifera also possesses a variety of pharmacological properties, which are closely associated with the presence of its bioactive compounds. Therefore, in the following section we highlighted the pharmacological potential of *M. oleifera*. *M. oleifera* showed pharmacological potential against some plausible factors such as oxidative stress, inflammation, fibrosis, and other pathologies responsible for kidney diseases. The potential effects of *M. oleifera* against risk factors associated with kidney disease in the following sections as shown in Figures 2 and 3.

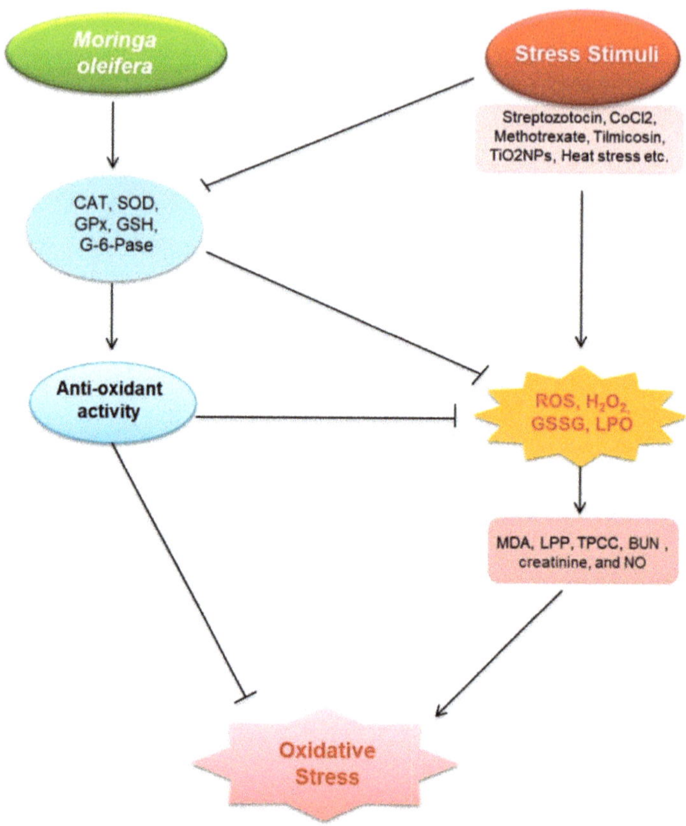

Figure 2. Renoprotective effects of *M. oleifera* against oxidative stress. Stress stimuli (streptozotocin, $CoCl_2$, methotrexate, tilmicosin, TiO_2NPs, acetaminophen (APAP), glycerol, and *Salmonella*) increased malondialdehyde (MDA), lipid peroxidation products (LPP), total protein carbonyl content (TPCC), blood urea nitrogen (BUN), creatinine, and nitric oxide (NO) production via triggering reactive oxygen species (ROS), H_2O_2, glutathione disulfide (GSSG), and lactoperoxidase (LPO). Oxidative stress emerged as a result of these events. MO—induced models, on the other hand, increased the expression of catalase (CAT); superoxide dismutase (SOD); glutathione peroxidase (GPx); glutathione (GSH), total antioxidant capacity (TAC); delta-amino levulinic acid dehydratase (ALAD), and G-6-Pase, which then activates glutathione (GSH). These stressors inhibit the expression of oxidative stress suppressive factors. ROS, H_2O_2, GSSG, and LPO, all related to oxidative stress, were decreased by GSH. GSH is also capable of reducing oxidative stress.

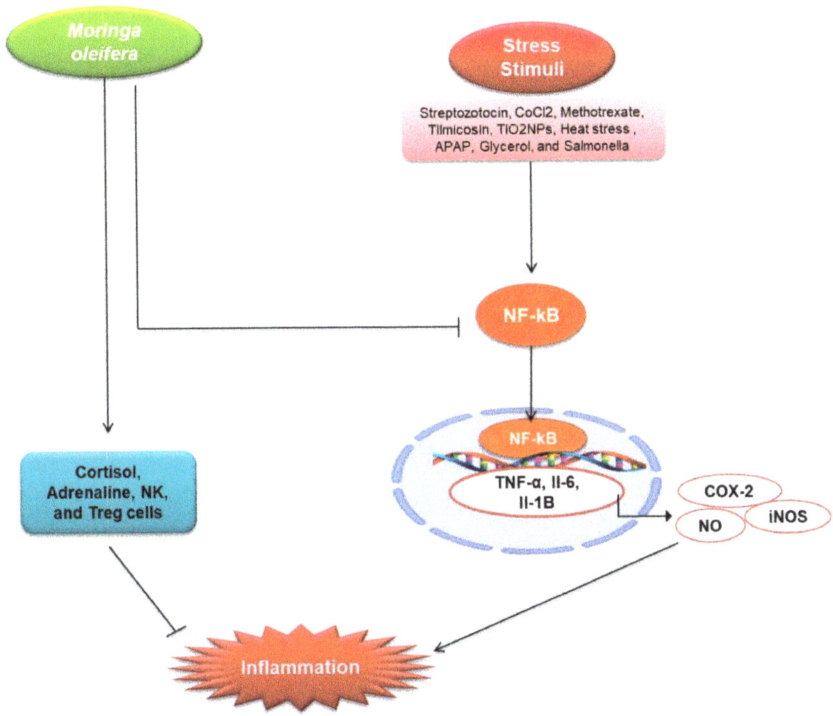

Figure 3. Renoprotective effects of *M. oleifera* against inflammation. The expression of C-reactive protein (CRP), which activates NF-kB in the cytosol, is linked to stress factors. TNF-, Il-6, Il-1B, iNOS, and COX-2 are all activated when NF-kB enters the nucleus and binds to DNA. All of these elements have been linked to the development of inflammation. NO is activated even more by iNOS. NO is thought to be a pro-inflammatory mediator that causes inflammation. In the cytosol, *M. oleifera* suppressed the expression of CRP and NF-kB. It also boosted cortisol, adrenaline, NK, and Treg cells, which helped reduce inflammation. Anti-inflammatory hormones Cortisol and Adrenaline Both NK cells and Treg cells are anti-inflammatory regulators.

3.1. Oxidative Stress

Oxidative stress is caused by an imbalance between the excessive free radical generation and insufficient antioxidant defense [29,30]. It is frequently observed in CKD [31–33], and has become a diagnostic factor [34]. A number of studies documented that *M. oleifera* has antioxidative properties to protect and/or alleviate cellular damage (Table 1 and Figure 2). *M. oleifera* extracts and compounds, particularly quercetin, kaempferol, isothiocyanates, rutin, myricetin, ascorbic acid, and β-carotene, showed antioxidant potentials either via direct scavenging of free radicals [35].

Table 1. Summary on the protective effects of *M. oleifera* against kidney diseases.

Sl. No.	Experimental Model	Treatment Dose of Moringa Extract	Major Research Outcomes	Molecular Markers	Ref.
1	STZ-induced nephrotoxic male Wister rats	250 mg/kg b wt for 6 weeks	Amelioration of oxidative stress and inflammation	↓MDA and ROS ↑CAT, SOD, GSH, and GPx ↓TNF-α and IL-6	[36]
2	*db/db* mice	150 mg/kg/day for 5 weeks	Oxidative stress and inflammation	↓LDL ↓TNF-a, ↓IL-1b, ↓IL-6, ↓COX-2, and ↓iNOS	[20]

Table 1. *Cont.*

Sl. No.	Experimental Model	Treatment Dose of Moringa Extract	Major Research Outcomes	Molecular Markers	Ref.
3	Ischemia-reperfusion induced Wistar rats	200 mg/kg for 7 days; 400 mg/kg, 7 days by flank incision	Oxidative stress	↓MDA, ↑PC, ↓AOPP, ↓NO, ↓H_2O_2, ↓GPx and GST, ↑GSH	[24]
4	$CoCl_2$-induced rats	Orally received 400 mg/kg bw/day for 6 weeks	Oxidative stress Inflammation and Apoptosis	↓MDA, ↓H_2O_2, ↓8-OHdG, ↓CRP, ↓MPO, ↓TNF-α, and ↓NO	[37]
5	Gentamicin (GENT) induced Wistar rats	Orally treated with 100, 200 and 400 mg/kg/day for 28 days	Oxidative stress	↓TNF-α, and NO↓ ↓K^+ level, ↓plasma creatinine, ↑Creatinine clearance, ↓MDA, ↑SOD	[23]
6	Nickel-induced Wistar rats	5% *M. oleifera* 10% *M. oleifera* 15% *M. oleifera*	Oxidative stress	↓plasma creatinine, ↓urea, and ↑potassium, ↑plasma level of sodium	[38]
7	Methotrexate (MTX)-induced Mice	300 mg/kg body weight, orally for 7 days	Oxidative stress Inflammation Apoptosis	↓urea and ↓creatinine, ↓total protein, ↓MDA, ↑SOD and ↑GSH, ↑HO-1, ↑Nrf-2 ↓NF-kB, ↓Caspase-9	[25]
8	Tilmicosin (Til) induced Sprague Dawley rats	400 or 800 mg/kg bw, by oral gavage for 7 days	Oxidative stress, inflammation	↓H_2O_2, ↓MDA, ↑SOD, ↑GPx, mRNA expression ↓TNF-α, ↓IL-1β	[39]
9	Hg-induced Male Wistar rats	1.798 mg/kg p.o three times per week for 21 days	Oxidative stress	↓MDA level, ↑SOD, and ↑CAT	[40]
10	TiO_2NPs induce male albino rats	Daily oral dose of 400 mg/kg b w for 60 days	Oxidative stress Inflammation	↓MDA, ↑SOD, ↑GSH, ↑GST,↑GPx, ↑Total thiol and ↑HO-1, ↑Nrf2 ↓KIM-1, ↓NF-B, ↓TNF-α, and ↓HSP-70	[41]
11	NaF induced *Oreochromis niloticus*	6.1 mg/L for 8 weeks	Oxidative stress	↓MDA, ↑SOD, ↑CAT, ↑GSH, ↑GPx, ↑TAC	[42]
12	Gentamicin-induced (80 mg/kg) Rabbit	150 mg/kg body for 10 days, 300 mg/kg wt. for 10 days	Oxidative stress	↓Serum urea and creatinine levels, ↓LPO	[43]
13	Lead treated Male Wistar rats	500 mg/kg for 7 days	Oxidative stress	↓ROS, ↓LPP, ↓TPCC, ↓metal content,	[44]
15	Beryllium-induced rats	150 mg/kg daily for 5 weeks	Oxidative stress	↓LPO, ↑GSH, ↑antioxidant enzymes activities, ↑G-6-Pase activity	[45]
16	Arsenic-induced toxicity in rats	500 mg/kg, orally, once daily	Oxidative stress	↑ALAD, ↑GSH,↓ROS, ↑SOD, ↑Catalase, ↓GSSG	[46]
17	Heat stress (HS)-induced rabbits	100, 200, and 300 mg, 6 weeks	Inflammation	↑cortisol, ↑adrenaline, ↑leptin, ↓IFN-γ, ↓TNF-α, ↓urea, and ↓creatinine, ↓IL-10, ↑NK, and ↑Treg	[47]
18	ML-induced male Sprague Dawley rats	Orally 800 mg/kg bw 800 mg/kg bw	Oxidative stress, Inflammation Apoptosis	↓Total bilirubin, ↓direct bilirubin, ↓indirect bilirubin, ↓urea, and ↓creatinine ↑serum levels of protein, ↑albumin, ↑globulin, ↑GPx, and ↑CAT ↓KIM-1, and ↓TNF-α and ↑Bcl-2, ↓TIMP-1	[48]

Table 1. *Cont.*

Sl. No.	Experimental Model	Treatment Dose of Moringa Extract	Major Research Outcomes	Molecular Markers	Ref.
20	Seabream (*Sparus aurata*)	10% *M. oleifera* 4 weeks	Inflammation	↓TGF-β and ↓TNF-α ↑ACH_{50} and ↑lysozyme activities and ↑IgM level ↑ (lyso and c3), ↑ (occludin and zo-1)	[49]
21	APAP-treated mice	100 mg/kg of bw, 200 mg/kg bw	Oxidative stress, inflammation	↑SOD, ↑CAT and ↑GPx, ↓MDA, ↓TNF-α, ↓IL-1β, ↓IL-6, ↓IL-10	[50]
22	Iodide injected Rabbit	50 mg/kg body weight, orally once daily for 27 sequential days	Oxidative stress	↓MDA, ↑GSH, ↓NO, ↓lipid peroxidation, ↓ROS	[51]
23	Glycerol induced rat	50 mg/kg and 100 mg/kg for 7 days	Oxidative stress Inflammation	↑SOD, ↑GST, ↑GP_X, ↑GSH ↓MPO, ↓Creatinine, ↓BUN, ↓NO ↓H_2O_2, ↓AOPP, ↓MDA, ↓PC,↑PT, ↑NPT,↓KIM-1 and ↓NF-B	[52]
24	Salmonella-induced mice	14, 42 and 84 mg/kg/day for 28 days	Oxidative stress inflammation	↑HO-1, ↑SOD-2 ↑Nrf-2	[53]
25	STZ-induced rats	250 mg/kg and SRC. 42 days	Oxidative stress inflammation	↓LDL, ↑HDL, ↓CHOL, ↑ORAC ↓IL-6, ↓TNF-α, and ↓MCP-1	[54]
26	TGF-β-treated rat kidney fibroblast cells	10, 50, and 100 μg/mL	Fibrosis	↓Type I collagen, fibronectin, and PAI-1 ↓TβRII and Smad4, and phospho-ERK	[55]
27	Gentamicin-induced Wistar rats	28 days at graded doses of 100, 200 and 400 mg/kg	Nephrotoxicity	↓Creatinine and MDA ↑SOD	[23]

MDA, Malondialdehyde; TNF-α, tumor necrosis factor-alpha; IL-6, interleukin-6; STZ, streptozotocin (C8H15N3O7); GSH: glutathione; CAT, catalase; SOD, superoxide dismutase; GPx, Glutathione peroxidase; IL-1β, Interleukin 1 beta; COX-2, cyclooxygenase-2; iNOS, inducible nitric oxide synthase; AOPP, advanced oxidation protein products; PC, protein carbonyls; NO, nitricoxide; H_2O_2, hydrogen peroxide; 8-OHdG, 8-hydroxy-2-deoxyguanosine; MPO, myeloperoxidase; CRP, C-reactive protein; MTX, methotrexate; HO-1, heme oxygenase-1; Nrf2, nuclear factor erythroid 2-related factor 2; TAC, total antioxidant capacity; LPP, lipid perioxidation products; TPCC, total protein carbonyl content; ALAD, delta-amino levulinic acid dehydratase; BUN, Blood urea nitrogen; KIM-1, transmembrane tubular protein; Bcl-2, B-cell lymphoma 2; TGF-β, transforming growth factor beta; CHOL, Cholesterol; ORAC, oxygen radical absorbance capacity; and APAP, acetaminophen. ↑, increased; ↓, decreased.

Methanol extract of *M. oleifera* reduced the oxidative stress in STZ induced male rats by lowering the production of MDA, ROS, LDL, and CHOL, which increase the risk of CKD [36,54]. Methanol extract also lowered the generation of MDA, AOPP, NO, H_2O_2, GPx, and GST, all of which induce oxidative stress in ischemia-induced Wistar rats [29]. Another study showed that metabolic extract reduced the levels of BUN and creatinine, and total protein is increased in CKD patients [42]. Ethanolic extract of *M. oleifera* inhibits oxidative stress and atherosclerosis in CKD by lowering LDL [20]. 8-OHdG causes oxidative stress to DNA and promotes cancer [56], ameliorated by the ethanolic extract of *M. oleifera* [56]. Ethanol extracts decrease the plasma creatinine level by enhancing the process of creatinine clearance [30]. Plasma sodium and potassium levels were raised after treating nickel-induced Wistar rats with ethanolic extract of *M. oleifera* [34]. Ethanolic extract detoxified plasma by reducing the bilirubin levels (indirect/direct), urea levels, etc., in ML-induced male Sprague Dawley rats [48]. HO-1 and Nrf2 expression were stimulated by leaf extract of *M. oleifera* at dosages of 300 and 400 mg/kg body weight, respectively [25,41]. Leaf extracts up-regulated the level of total thiol TiO_2NPs induced male albino rats, which play an important role in antioxidant protection [41]. Leaf extract of *M. oleifera* also downregulated the oxidative stress generating mediators in sodium fluoride (NaF)-induced *Oreochromis niloticus*, gentamicin-induced rabbit, and APAP-treated mice [23,42,57].

M. oleifera alcoholic extract reduced oxidative stress by lowering the lipid peroxidation, and ROS in iodide injected rabbits [51]. Furthermore, fermented leaf extract of *M. oleifera* boosts the antioxidant activity in bacteria-induced mice [53]. *M. oleifera* extract reduced the manifestation of MDA, indicating that the free radicle overproduction was reduced in both Tilmicosin and Hg induced rats. Abarikwu et al. showed that SOD level was increased after treatment with *M. oleifera* in tilmicosin induced rats [40]. Hydroalcoholic root extract raised blood sugar, antioxidant enzyme activities, and G-6-phase activities, which protect the kidney from nephropathy in Beryllium-induced rats [45]. Seed powder reduced free radical species, TPCC, metal content, and increased ALAD activity in lead-treated rats [57]. In arsenic-treated rats, seed powder of *M. oleifera* considerably increased antioxidant function including GSH, CAT, and ALAD [46].

3.2. Inflammation

The kidney is responsible for maintaining whole-body homeostasis. Kidney disease is characterized by inflammation as a major pathology [58–60]. Acute or chronic disease such as ischemia, toxins, or inflammation affects kidney tubules, causing kidney fibrosis that is associated with reduction of GFR in kidneys [61]. Kidney injury is linked to the production of cytokines levels, which prolongs the acute phase of kidney disease [62]. Moreover, chronic inflammation is regarded as a comorbid condition in CKD [63]. Many plants have an anti-inflammatory action through active substances such as hesperidin, diosmin, withaferin, fucoidan, thymoquinone, etc. [64–67]. Here, the anti-inflammatory effects of *M. oleifera* has been discussed. *M. oleifera* has been reported to exhibit strong inflammatory activity (Table 1 and Figure 3). Methanolic extract of *M. oleifera* reduced inflammation in STZ induced male Wister rats by down-regulating the tumor necrosis factor (TNF-α), IL-6, and MCP-1, an important chemokine [36,54]. Tang et al. investigated the effects of ethanolic extract of *M. oleifera* in metformin-induced mice and observed that the *M. oleifera* declines the production of inflammatory markers and the expression of cyclooxygenase-2 (COX-2) and nitric oxide synthase (iNOS) by reducing the phosphorylation of mitogen-activated protein kinase (MAPK) pathway [20]. Ethanolic extract of *M. oleifera* down-regulates the inflammatory cytokines in $CoCl_2$-induced rats, including NO, which is involved in the pathogenesis of inflammation [37]. Leaf extract of *M. oleifera* inhibits inflammatory cytokines production and regulates the inflammation by inhibiting NF-kB [25]. It was also observed that inflammation in Tilmicosin (Til) induced rats was reduced by *M. oleifera* extracts [39]. *M. oleifera* leaf extract protects against interstitial kidney inflammation with fibrosis by down-regulating KIM-1 in TiO_2NPs induced male albino rats [41]. *M. oleifera* extract increases the secretion of cortisol, adrenaline, Treg cells, NK, and leptin, promoting anti-inflammatory cytokines and regulating the immune system [47]. *M. oleifera* treatment reduced the expression of KIM-1, TIMP-1, and TNF-α in ML-induced male Sprague Dawley rats [48]. TNF-α, an inflammatory cytokine that stimulates IL-1; IL-6, downregulated by *M. oleifera* in Seabream (*Sparus aurata*); and activated TGF-β, elicits anti-inflammatory effects [49]. *M. oleifera* also reduced the inflammatory cytokines in APAP-treated mice, where APAP induces AKI [50]. Fermented extract of leaves also reduces the Nrf2 in *Salmonella*-induced mice [53].

Moringa seed's phytochemicals can reduce the production of nitric oxide (NO) and the gene expression of LPS-inducible iNOS and interleukins 1β and 6 (IL-1β and IL-6) compared to curcumin [68]. Flavonoids have been shown to be effective inhibitors of nitric oxide synthase type 2 (NOS-2) actions, and it also inhibits protein tyrosine kinase action that is involved in the NOS-2 expression at the molecular level [69–71]. Flower extract can cause the activation of pro-inflammatory proteins such as toll-like receptors. In the flowers, quercetin and kaempferol can inhibit the signal transducer and activator of transcription 1 (STAT-1) and the NF-κB pathways [72,73]. *M. oleifera* flowers contain 80% hydroethanolic, a potent agent of anti-inflammation in the NF-κB signaling pathway [74]. Scientists discovered that phenolic glycosides suppress inducible iNOS expression and NO production in mouse macrophage cells, as well as COX-2 and iNOS proteins [75,76]. *Moringa* extracts eventually down-regulate the inflammatory mediators because its seeds and flowers contain many bioactive compounds. Each of these compounds has its individual effects.

3.3. Fibrosis

Kidney fibrosis is defined as a radical harmful connective tissue deposition on the kidney parenchyma, which leads to renal dysfunction. Epithelial to mesenchymal transition (EMT) is the main mechanism of kidney fibrosis, and the TGFβ-1-SMAD pathway and hypoxia are known as the main modulator of EMT [32,77]. TGF-β-induced expression of fibronectin, type I collagen, and PAI-1 rat kidney fibroblast cells is reduced by *M. oleifera* extract [55]. Furthermore, moringa root extract selectively inhibited TGF-β-induced phosphorylation of SMAD4 and ERK expression. These results suggest that moringa root extract may reduce renal fibrosis by a mechanism related to its antifibrotic activity in rat kidney fibroblast cells. Oral administration of *M. oleifera* seed extract reduced CCl4-induced liver fibrosis in rats [78].

3.4. Other Pathologies Those Are Associated with Kidney Diseases

Autophagy has a critical role in kidney physiology and homeostasis [79], and, thus, its regulation is an important determinant of kidney diseases [61]. AKI or CKD causes mitochondrial damage, but damaged mitochondria begin to accumulate in response to these types of stimuli. Autophagy protects the kidney through the removal of ROS-producing mitochondria [80–82]. Apoptosis is a type of programmed cell death in which cells are killed by a controlled system. It is an energy-dependent complex process [83]. It contributes to develop AKI, even organ failure [84]. Ischemia/reperfusion (I/R) induces apoptosis or necrosis in the kidney and loss of tubular cells, leading to decreased GFR [85,86]. Renal tubular cells express cell surface 'death receptors' of TNF-α which is responsible for inducing apoptosis [87]. Also, ROS production in kidney disease is responsible for promoting apoptosis [86].

TNF-α inducer of apoptosis, also increased the expression of apoptosis-related molecules which was down-regulated by ethanol extract of *M. oleifera* in $CoCl_2$-treated rats [37,88]. Leaf extract at a dose of 300 mg/kg body weight reduced the expression of caspase-9, the precursor of caspase-3, leading to apoptosis [25,89]. Bcl-2 inhibited apoptosis by blocking cytochrome c release and preventing caspase activation [90] while it was up-regulated by ethanol extract of *M. oleifera* in ML-induced rats. *M. oleifera* also reduced the expression of TIMP-1, which is involved in renal fibrosis and apoptosis [48].

4. Prospects for *M. oleifera* in Drug Development

Researchers are targeting the development of drugs from natural sources instead of the synthetic drug because natural sources have fewer side effects than synthetic sources. Nigerian scientists proved that *M. oleifera* is a beneficial herb and causes no harm to the body and kidneys [91]. Another study reported that higher doses of *M. oleifera* created toxicity in rats, but a moderate level dose of *M. oleifera* is safe [92]. *M. oleifera* has been shown to alleviate diabetic nephropathy in alloxan-induced rats [93]. Acetaminophen causes hepato-renal toxicity, which can be cured by *M. oleifera* treatment at the dosage of 500 mg/kg [94]. *M. oleifera* reduced necrosis, dilatation of renal tubules in Cd-induced rats, where Saleh et al. suggested that *M. oleifera* could be used as an herbal drug [95]. *M. oleifera* leaf extracts reduced oxidative stress, kidney, and liver damage [96]. A randomized placebo-controlled study suggested that *M. oleifera* leaf capsules can be used to control blood sugar level and blood pressure level [97]. Moreover, aqueous extracts of *M. oleifera* can reduce metal (As (III), Cd, Ni and Pb) toxicity and showed the protective effects in *Saccharomyces cerevisiae* [98].

The rich phytochemical profile and advances in biotechnological techniques have made this tree indispensable for opening a new era in medical science. An in vitro propagation technique provides new insights into developing more effective, eco-friendly, and biodegradable products using mass multiplication and production techniques. Though efficiency in in vitro propagation techniques for *M. oleifera* has been established, there are still gaps in the production of metabolites and those specific metabolites in the human body. The use of biotechnological approaches will help in the commercialization of important plant products. There is no doubt that biotechnological protocols will allow great research to make *M. oleifera* one of the essential solutions for various health issues including kidney diseases.

5. Conclusions

Kidney function declines with age, and aging-related kidney complications proportionately increase. Their side effects limit the effectiveness of existing drugs for treating kidney diseases and, therefore, natural compounds with fewer side effects are being evaluated. The literature discussed in this review suggests that *M. oleifera* alleviates several pathological factors associated with kidney diseases, including inflammation and oxidative stress. However, a mechanism associated with protective potential of *M. oleifera* against kidney diseases has been provided in this study (Figure 4).

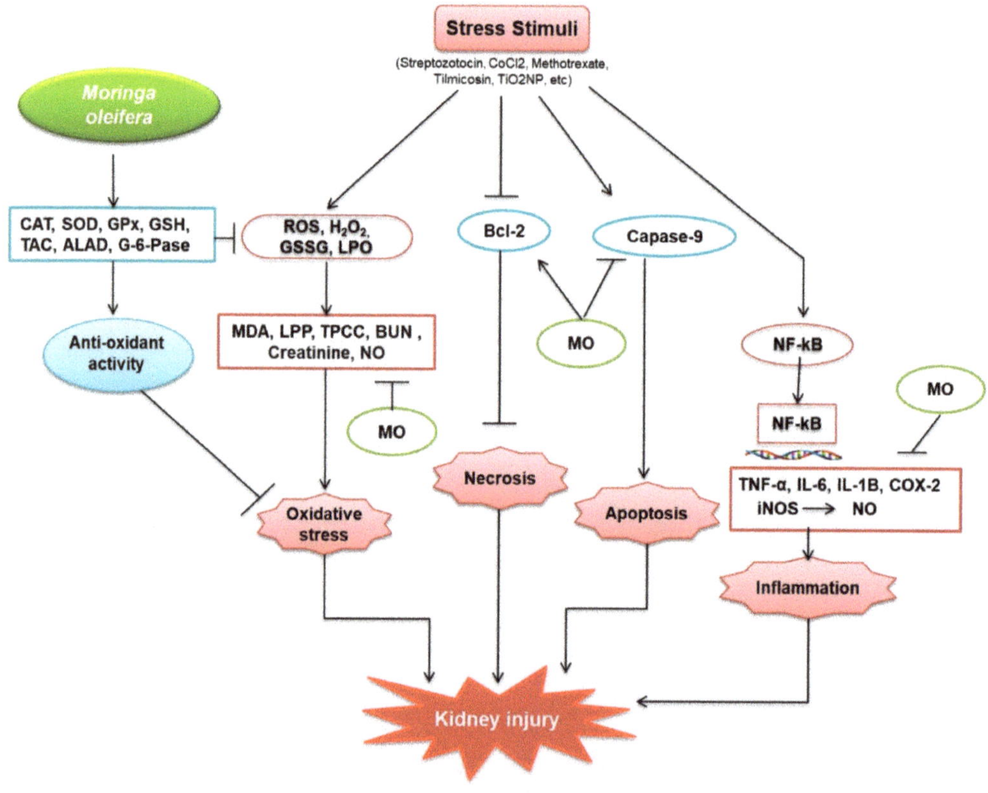

Figure 4. Protective mechanisms of *M. oleifera* against kidney injury. *M. oleifera* increased the production of catalase (CAT); superoxide dismutase (SOD); glutathione peroxidase (GPx); glutathione (GSH); total antioxidant capacity (TAC); delta-amino levulinic acid dehydratase (ALAD); and G-6-Pase, which facilitated oxidative stress reduction by activating glutathione (GSH), a non-protein thiol that suppresses free radicals. GSH suppresses the oxidative stress situation. *M. oleifera* also suppressed oxidative stressors caused by ROS, H_2O_2, GSSG, and LPO by inhibiting MDA, LPP, TPCC, BUN, Creatinine, and NO. Bcl-2 was similarly produced by stress stimuli and was linked to the suppression of necrosis, induced by *M. oleifera*. *M. oleifera* inhibited the expression of Caspase-9, a protein involved in the formation of caspases. Following NF-kB, stress stimuli also increased CRP expression. NF-kB then moved from the cytosol to the nucleus, bound to DNA, and activated inflammation-related proteins. *M. oleifera* inhibited the mechanism by which inflammation factors were produced, hence, reducing inflammation. *M. oleifera* has been linked to a reduction in the progression of kidney disease.

This study discusses the insights of *M. oleifera* against kidney diseases including AKI and CKD, which have not been reported previously. In addition, further studies are needed to confirm the effects of the bioactive phytochemicals (vitamins, alkaloids, polyphenols, isothiocyonates, glucosinolates, tannins, and saponins) of *M. oleifera* against kidney diseases. We anticipate that the points raised in this review will provide a future research direction for understanding how pharmacological interventions based on natural products could modulate kidney disease. In contrast, it would shed light on how *M. oleifera*-based drugs would potentially be a kidney protective agent in treating aging-associated kidney abnormalities. Considering the harmful effects of synthetic resources and their non-renewable nature, the use of natural resources as a source of medicine has received a lot of attention in recent years. *M. oleifera* based medicine would be an excellent protective agent against several risk factors associated with kidney diseases.

Author Contributions: Conceptualization, M.J.U., A.M. and M.A.R.; data curation, T.A. and M.A.I.A.; funding acquisition, M.J.U.; methodology, T.A., M.A.I.A. and A.F.; initial draft, T.A. and M.A.R.; supervision, M.J.U.; writing—review and editing, T.A., M.A.R., M.A.H. and M.J.U. All authors have read and agreed to the published version of the manuscript.

Funding: This study received no external funding.

Institutional Review Board Statement: Not applicable.

Informed Consent Statement: Not applicable.

Data Availability Statement: Not applicable.

Acknowledgments: This work acknowledges National Research Foundation (No. 2020R1I1A1A01072879 and 2015H1D3A1062189), and Brain Pool program funded by the Ministry of Science, and ICT through the National Research Foundation (No. 2020H1D3A2A02110924), Korea.

Conflicts of Interest: There are no conflicts of interest regarding this work.

References

1. Romagnani, P.; Remuzzi, G.; Glassock, R.; Levin, A.; Jager, K.J.; Tonelli, M.; Massy, Z.; Wanner, C.; Anders, H.J. Chronic kidney disease. *Nat. Rev. Dis. Primers* **2017**, *3*, 17088. [CrossRef]
2. Webster, A.C.; Nagler, E.V.; Morton, R.L.; Masson, P. Chronic kidney disease. *Lancet* **2017**, *389*, 1238–1252. [CrossRef]
3. Sugahara, M.; Tanaka, T.; Nangaku, M. Prolyl hydroxylase domain inhibitors as a novel therapeutic approach against anemia in chronic kidney disease. *Kidney Int.* **2017**, *92*, 306–312. [CrossRef] [PubMed]
4. Uddin, M.J.; Dorotea, D.; Pak, E.S.; Ha, H. Fyn kinase: A potential therapeutic target in acute kidney injury. *Biomol. Ther.* **2020**, *28*. [CrossRef]
5. Ahmed, S.R.; Romi, I.J.; Ahmed, J.; Hasan, M.; Roy, R.; Khan, M.M.H. Phytochemical profiling and antioxidant potentiality of medicinal plants along with their antibacterial efficacy. *JABET* **2019**, *2*, 140–145. [CrossRef]
6. Aktar, S.; Das, P.K.; Asha, S.Y.; Siddika, M.A.; Islam, F.; Khanam, J.A.; Rakib, M.A. *Moringa oleifera* leaves methanolic extract inhibits angiotensin converting enzyme activity in vitro which ameliorates hypertension. *JABET* **2019**, *2*, 73–77. [CrossRef]
7. Das, P.K.; Asha, S.Y.; Siddika, M.A.; Siddika, A.; M Tareq, A.R.; Islam, F.; Khanam, J.A.; Rakib, M.A. Methanolic extract of *Moringa oleifera* leaves mediates anticancer activities through inhibiting NF-kB and enhancing ROS in ehrlich ascites carcinoma cells in mice. *JABET* **2021**, *4*, 161–170.
8. Islam, A.; Mandal, C.; Habib, A. Antibacterial potential of synthesized silver nanoparticles from leaf extract of *Moringa oleifera*. *JABET* **2021**, *4*, 67–73. [CrossRef]
9. Gopalakrishnan, L.; Doriya, K.; Kumar, D.S. Moringa oleifera: A review on nutritive importance and its medicinal application. *Food Sci. Hum. Wellness.* **2016**, *5*, 49–56. [CrossRef]
10. Kou, X.; Li, B.; Olayanju, J.B.; Drake, J.M.; Chen, N. Nutraceutical or pharmacological potential of *Moringa oleifera* Lam. *Nutrients* **2018**, *10*, 343. [CrossRef] [PubMed]
11. Ma, Z.F.; Ahmad, J.; Zhang, H.; Khan, I.; Muhammad, S. Evaluation of phytochemical and medicinal properties of moringa (*Moringa oleifera*) as a potential functional food. *S. Afr. J. Bot.* **2020**, *129*, 40–46. [CrossRef]
12. Bhattacharya, A.; Tiwari, P.; Sahu, P.K.; Kumar, S. A review of the phytochemical and pharmacological characteristics of *Moringa oleifera*. *J. Pharm. Bioallied. Sci.* **2018**, *10*, 181–191. [PubMed]
13. Valdez-Solana, M.A.; Mejía-García, V.Y.; Téllez-Valencia, A.; García-Arenas, G.; Salas-Pacheco, J.; Alba-Romero, J.J.; Sierra-Campos, E. Nutritional content and elemental and phytochemical analyses of *Moringa oleifera* grown in Mexico. *J.Chem.* **2015**, *2015*, 860381. [CrossRef]

14. Glevitzky, I.; Dumitrel, G.-A.; Mirel, G.; Pasca, M.B.; Otřísal, P.; Bungau, S.; Cioca, G.; Carmen, P.; Maria, P. Statistical analysis of the relationship between antioxidant activity and the structure of flavonoid compounds. *Rev. Chim.* **2019**, *70*. [CrossRef]
15. Kashyap, D.; Sharma, A.; Tuli, H.S.; Sak, K.; Punia, S.; Mukherjee, T.K. Kaempferol—A dietary anticancer molecule with multiple mechanisms of action: Recent trends and advancements. *J. Funct. Foods* **2017**, *30*, 203–219. [CrossRef] [PubMed]
16. Lin, M.; Zhang, J.; Chen, X. Bioactive flavonoids in *Moringa oleifera* and their health-promoting properties. *J. Funct. Foods* **2018**, *47*, 469–479. [CrossRef]
17. Vargas-Sánchez, K.; Garay-Jaramillo, E.; González-Reyes, R.E. Effects of *Moringa oleifera* on glycaemia and insulin levels: A review of animal and human studies. *Nutrients* **2019**, *11*, 2907. [CrossRef]
18. Rahman, M.M.; Sheikh, M.M.I.; Sharmin, S.A.; Islam, M.S.; Rahman, M.A.; Rahman, M.M.; Alam, M.J.C.J.N.S. Antibacterial activity of leaf juice and extracts of *Moringa oleifera* Lam. against some human pathogenic bacteria. *CMU J. Nat. Sci.* **2009**, *8*, 219–227.
19. Khan, W.; Parveen, R.; Chester, K.; Parveen, S.; Ahmad, S. Hypoglycemic potential of aqueous extract of *Moringa oleifera* leaf and In Vivo GC-MS metabolomics. *Front. Pharmacol.* **2017**, *8*, 577. [CrossRef]
20. Tang, Y.; Choi, E.J.; Han, W.C.; Oh, M.; Kim, J.; Hwang, J.Y.; Park, P.J.; Moon, S.H.; Kim, Y.S.; Kim, E.K. *Moringa oleifera* from cambodia ameliorates oxidative stress, hyperglycemia, and kidney dysfunction in type 2 diabetic mice. *J. Med. Food* **2017**, *20*, 502–510. [CrossRef]
21. Pedraza-Chaverri, J.; Sánchez-Lozada, L.G.; Osorio-Alonso, H.; Tapia, E.; Scholze, A. New pathogenic concepts and therapeutic approaches to oxidative stress in chronic kidney disease. *Oxid. Med. Cell Longev.* **2016**, *2016*, 6043601. [CrossRef]
22. Meireles, D.; Gomes, J.; Lopes, L.; Hinzmann, M.; Machado, J. A review of properties, nutritional and pharmaceutical applications of *Moringa oleifera*: Integrative approach on conventional and traditional Asian medicine. *ADTM* **2020**, 1–21. [CrossRef]
23. Nafiu, A.O.; Akomolafe, R.O.; Alabi, Q.K.; Idowu, C.O.; Odujoko, O.O. Effect of fatty acids from ethanol extract of *Moringa oleifera* seeds on kidney function impairment and oxidative stress induced by gentamicin in rats. *Biomed. Pharmacother.* **2019**, *117*, 109154. [CrossRef]
24. Akinrinde, A.S.; Oduwole, O.; Akinrinmade, F.J.; Bolaji-Alabi, F.B. Nephroprotective effect of methanol extract of *Moringa oleifera* leaves on acute kidney injury induced by ischemia-reperfusion in rats. *Afr. Health Sci.* **2020**, *20*, 1382–1396. [CrossRef]
25. Soliman, M.M.; Aldhahrani, A.; Alkhedaide, A.; Nassan, M.A.; Althobaiti, F.; Mohamed, W.A. The ameliorative impacts of *Moringa oleifera* leaf extract against oxidative stress and methotrexate-induced hepato-renal dysfunction. *Biomed. Pharmacother.* **2020**, *128*, 110259. [CrossRef]
26. Abu-Zeid, E.H.; Abdel Fattah, D.M.; Arisha, A.H.; Ismail, T.A.; Alsadek, D.M.; Metwally, M.M.M.; El-Sayed, A.A.; Khalil, A.T. Protective prospects of eco-friendly synthesized selenium nanoparticles using *Moringa oleifera* or *Moringa oleifera* leaf extract against melamine induced nephrotoxicity in male rats. *Ecotoxicol. Environ. Saf.* **2021**, *221*, 112424. [CrossRef]
27. Page, M.J.; McKenzie, J.E.; Bossuyt, P.M.; Boutron, I.; Hoffmann, T.C.; Mulrow, C.D.; Shamseer, L.; Tetzlaff, J.M.; Akl, E.A.; Brennan, S.E.; et al. The PRISMA 2020 statement: An updated guideline for reporting systematic reviews. *BMJ.* **2021**, *372*, n71. [CrossRef] [PubMed]
28. Vergara-Jimenez, M.; Almatrafi, M.M.; Fernandez, M.L. Bioactive components in *Moringa oleifera* leaves protect against chronic disease. *Antioxidants* **2017**, *6*, 91. [CrossRef] [PubMed]
29. Daenen, K.; Andries, A.; Mekahli, D.; Van Schepdael, A.; Jouret, F.; Bammens, B. Oxidative stress in chronic kidney disease. *Pediatr. Nephrol.* **2019**, *34*, 975–991. [CrossRef] [PubMed]
30. Ling, X.C.; Kuo, K.-L. Oxidative stress in chronic kidney disease. *Renal. Replace. Ther.* **2018**, *4*, 53. [CrossRef]
31. Uddin, M.J.; Kim, E.H.; Hannan, M.A.; Ha, H. Pharmacotherapy against oxidative stress in chronic kidney disease: Promising small molecule natural products targeting nrf2-ho-1 signaling. *Antioxidants* **2021**, *10*, 258. [CrossRef]
32. Sohn, M.; Kim, K.; Uddin, M.J.; Lee, G.; Hwang, I.; Kang, H.; Kim, H.; Lee, J.H.; Ha, H. Delayed treatment with fenofibrate protects against high-fat diet-induced kidney injury in mice: The possible role of ampk autophagy. *Am. J. Physiol. Renal. Physiol.* **2017**, *312*, F323–F334. [CrossRef] [PubMed]
33. Hwang, I.; Uddin, M.J.; Lee, G.; Jiang, S.; Pak, E.S.; Ha, H. Peroxiredoxin 3 deficiency accelerates chronic kidney injury in mice through interactions between macrophages and tubular epithelial cells. *Free Radic. Biol. Med.* **2019**, *131*, 162–172. [CrossRef]
34. Rapa, S.F.; Di Iorio, B.R.; Campiglia, P.; Heidland, A.; Marzocco, S. Inflammation and oxidative stress in chronic kidney disease-potential therapeutic role of minerals, vitamins and plant-derived metabolites. *Int. J. Mol. Sci.* **2019**, *21*, 263. [CrossRef]
35. Pakade, V.; Cukrowska, E.; Chimuka, L. Comparison of antioxidant activity of *Moringa oleifera* and selected vegetables in South Africa. *S. Af. J. Sci.* **2013**, *109*. [CrossRef]
36. Omodanisi, E.I.; Aboua, Y.G.; Oguntibeju, O.O. Assessment of the anti-hyperglycaemic, anti-inflammatory and antioxidant activities of the methanol extract of Moringa oleifera in diabetes-induced nephrotoxic male Wistar rats. *Molecules* **2017**, *22*, 439. [CrossRef] [PubMed]
37. Abdel-Daim, M.M.; Khalil, S.R.; Awad, A.; Abu Zeid, E.H.; El-Aziz, R.A.; El-Serehy, H.A. Ethanolic extract of *Moringa oleifera* leaves influences NF-κB signaling pathway to restore kidney tissue from cobalt-mediated oxidative injury and inflammation in rats. *Nutrients* **2020**, *12*, 1031. [CrossRef] [PubMed]
38. Adeyemi, O.S.; Elebiyo, T.C. *Moringa oleifera* supplemented diets prevented nickel-induced nephrotoxicity in Wistar Rats. *J. Nutr. Metab.* **2014**, *2014*, 958621. [CrossRef]

39. Abou-Zeid, S.M.; Ahmed, A.I.; Awad, A.; Mohammed, W.A.; Metwally, M.M.M.; Almeer, R.; Abdel-Daim, M.M.; Khalil, S.R. *Moringa oleifera* ethanolic extract attenuates tilmicosin-induced renal damage in male rats via suppression of oxidative stress, inflammatory injury, and intermediate filament proteins mRNA expression. *Biomed. Pharmacother.* **2021**, *133*, 110997. [CrossRef]
40. Abarikwu, S.O.; Benjamin, S.; Ebah, S.G.; Obilor, G.; Agbam, G. Protective effect of *Moringa oleifera* oil against $HgCl_2$-induced hepato- and nephro-toxicity in rats. *J. Basic Clin. Physiol. Pharmacol.* **2017**, *28*, 337–345. [CrossRef]
41. Abdou, K.H.; Moselhy, W.A.; Mohamed, H.M.; El-Nahass, E.S.; Khalifa, A.G. *Moringa oleifera* leaves extract protects titanium dioxide nanoparticles-induced nephrotoxicity via Nrf2/HO-1 signaling and amelioration of oxidative stress. *Biol. Trace. Elem. Res.* **2019**, *187*, 181–191. [CrossRef]
42. Ahmed, N.F.; Sadek, K.M.; Soliman, M.K.; Khalil, R.H.; Khafaga, A.F.; Ajarem, J.S.; Maodaa, S.N.; Allam, A.A. *Moringa oleifera* leaf extract repairs the oxidative misbalance following Sub-chronic exposure to sodium fluoride in Nile tilapia *Oreochromis niloticus*. *Animals* **2020**, *10*, 626. [CrossRef]
43. Ouédraogo, M.; Lamien-Sanou, A.; Ramdé, N.; Ouédraogo, A.S.; Ouédraogo, M.; Zongo, S.P.; Goumbri, O.; Duez, P.; Guissou, P.I. Protective effect of *Moringa oleifera* leaves against gentamicin-induced nephrotoxicity in rabbits. *Exp. Toxicol. Pathol.* **2013**, *65*, 335–339. [CrossRef] [PubMed]
44. Velaga, M.K.; Daughtry, L.K.; Jones, A.C.; Yallapragada, P.R.; Rajanna, S.; Rajanna, B. Attenuation of lead-induced oxidative stress in rat brain, liver, kidney and blood of male Wistar rats by *Moringa oleifera* seed powder. *J. Environ. Pathol. Toxicol. Oncol.* **2014**, *33*, 323–337. [CrossRef]
45. Agrawal, N.D.; Nirala, S.K.; Shukla, S.; Mathur, R. Co-administration of adjuvants along with *Moringa oleifera* attenuates beryllium-induced oxidative stress and histopathological alterations in rats. *Pharm. Biol.* **2015**, *53*, 1465–1473. [CrossRef] [PubMed]
46. Gupta, R.; Kannan, G.M.; Sharma, M.; SJ, S.F. Therapeutic effects of *Moringa oleifera* on arsenic-induced toxicity in rats. *Environ. Toxicol. Pharmacol.* **2005**, *20*, 456–464. [CrossRef]
47. Abdel-Latif, M.; Sakran, T.; Badawi, Y.K.; Abdel-Hady, D.S. Influence of Moringa oleifera extract, vitamin C, and sodium bicarbonate on heat stress-induced HSP70 expression and cellular immune response in rabbits. *Cell Stress Chaperones* **2018**, *23*, 975–984. [CrossRef] [PubMed]
48. Abd-Elhakim, Y.M.; Mohamed, W.A.M.; El Bohi, K.M.; Ali, H.A.; Mahmoud, F.A.; Saber, T.M. Prevention of melamine-induced hepatorenal impairment by an ethanolic extract of *Moringa oleifera:* changes in KIM-1, TIMP-1, oxidative stress, apoptosis, and inflammation-related genes. *Gene* **2021**, *764*, 145083. [CrossRef]
49. Mansour, A.T.; Miao, L.; Espinosa, C.; García-Beltrán, J.M.; Ceballos Francisco, D.C.; Esteban, M. Effects of dietary inclusion of *Moringa oleifera* leaves on growth and some systemic and mucosal immune parameters of seabream. *Fish. Physiol. Biochem.* **2018**, *44*, 1223–1240. [CrossRef]
50. Karthivashan, G.; Kura, A.U.; Arulselvan, P.; Md Isa, N.; Fakurazi, S. The modulatory effect of *Moringa oleifera* leaf extract on endogenous antioxidant systems and inflammatory markers in an acetaminophen-induced nephrotoxic mice model. *Peer J.* **2016**, *4*, e2127. [CrossRef]
51. Altaee, R.A.; Fadheel, Q.J. The nephroprotective effects of moringa oleifera extract against contrast induced nephrotoxicity. *J. Pharm. Res. Int.* **2021**, *33*, 63–70. [CrossRef]
52. Adedapo, A.; Ue, E.; Falayi, O.; Ogunpolu, B.; Omobowale, T.; Oyagbemi, A.; Oguntibeju, O. Methanol stem extract of *Moringa oleifera* mitigates glycerol-induced acute kidney damage in rats through modulation of KIM-1 and NF-kB signaling pathways. *Sci. Afr.* **2020**, *9*, e00493. [CrossRef]
53. Widodo, N.; Widjajanto, E.; Jatmiko, Y.; Rifa'i, M. Red *Moringa oleifera* leaf fermentation extract protecting Hepatotoxicity in Balb/C mice injected with Salmonella typhi through Nrf-2, HO-1, and SOD-2 signaling pathways. *R. J. Pharm. Technol.* **2020**, *13*, 5947–5952.
54. Omodanisi, E.I.; Aboua, Y.G.; Chegou, N.N.; Oguntibeju, O.O. Hepatoprotective, antihyperlipidemic, and anti-inflammatory Activity of *Moringa oleifera* in diabetic-induced damage in male Wistar Rats. *Pharmacogn. Res.* **2017**, *9*, 182–187.
55. Park, S.-H.; Chang, Y.-C. Anti-fibrotic effects by *Moringa* root extract in rat kidney fibroblast. *J. Life Sci.* **2012**, *22*, 1371–1377. [CrossRef]
56. Valavanidis, A.; Vlachogianni, T.; Fiotakis, C. 8-hydroxy-2′-deoxyguanosine (8-OHdG): A critical biomarker of oxidative stress and carcinogenesis. *J. Environ. Sci. Health C. Environ. Carcinog. Ecotoxicol. Rev.* **2009**, *27*, 120–139. [CrossRef] [PubMed]
57. Oyagbemi, A.A.; Omobowale, T.O.; Azeez, I.O.; Abiola, J.O.; Adedokun, R.A.; Nottidge, H.O. Toxicological evaluations of methanolic extract of *Moringa oleifera* leaves in liver and kidney of male Wistar rats. *J. Basic Clin. Physiol. Pharmacol.* **2013**, *24*, 307–312. [CrossRef]
58. Anders, H.J.; Schaefer, L. Beyond tissue injury-damage-associated molecular patterns, toll-like receptors, and inflammasomes also drive regeneration and fibrosis. *J. Am. Soc. Nephrol.* **2014**, *25*, 1387–1400. [CrossRef]
59. Uddin, M.J.; Pak, E.S.; Ha, H. Carbon monoxide releasing molecule-2 protects mice against acute kidney injury through inhibition of ER stress. *Korean J. Physiol. Pharmacol.* **2018**, *22*, 567–575. [CrossRef]
60. Uddin, M.J.; Jeong, J.; Pak, E.S.; Ha, H. Co-releasing molecule-2 prevents acute kidney Injury through suppression of ROS-Fyn-ER stress signaling in mouse model. *Oxid. Med. Cell. Longev.* **2021**, *2021*, 9947772. [CrossRef]
61. Mackensen-Haen, S.; Bader, R.; Grund, K.E.; Bohle, A. Correlations between renal cortical interstitial fibrosis, atrophy of the proximal tubules and impairment of the glomerular filtration rate. *Clin. Nephrol.* **1981**, *15*, 167–171. [PubMed]

62. Akcay, A.; Nguyen, Q.; Edelstein, C.L. Mediators of inflammation in acute kidney injury. *Mediators Inflamm.* **2009**, *2009*, 137072. [CrossRef]
63. Akchurin, O.M.; Kaskel, F. Update on inflammation in chronic kidney disease. *Blood Purif.* **2015**, *39*, 84–92. [CrossRef]
64. Elhelaly, A.E.; AlBasher, G.; Alfarraj, S.; Almeer, R.; Bahbah, E.I.; Fouda, M.M.A.; Bungău, S.G.; Aleya, L.; Abdel-Daim, M.M. Protective effects of hesperidin and diosmin against acrylamide-induced liver, kidney, and brain oxidative damage in rats. *Environ. Sci. Pollut. Res. Int.* **2019**, *26*, 35151–35162. [CrossRef]
65. Behl, T.; Sharma, A.; Sharma, L.; Sehgal, A.; Zengin, G.; Brata, R.; Fratila, O.; Bungau, S. Exploring the multifaceted therapeutic potential of withaferin a and its derivatives. *Biomedicines* **2020**, *8*, 571. [CrossRef]
66. Abdel-Daim, M.M.; Abushouk, A.I.; Bahbah, E.I.; Bungău, S.G.; Alyousif, M.S.; Aleya, L.; Alkahtani, S. Fucoidan protects against subacute diazinon-induced oxidative damage in cardiac, hepatic, and renal tissues. *Environ. Sci. Pollut. Res.* **2020**, *27*, 11554–11564. [CrossRef]
67. Abdel-Daim, M.M.; Abo El-Ela, F.I.; Alshahrani, F.K.; Bin-Jumah, M.; Al-Zharani, M.; Almutairi, B.; Alyousif, M.S.; Bungau, S.; Aleya, L.; Alkahtani, S. Protective effects of thymoquinone against acrylamide-induced liver, kidney and brain oxidative damage in rats. *Environ. Sci. Pollut. Res.* **2020**, *27*, 37709–37717. [CrossRef] [PubMed]
68. Jaja-Chimedza, A.; Graf, B.L.; Simmler, C.; Kim, Y.; Kuhn, P.; Pauli, G.F.; Raskin, I. Biochemical characterization and anti-inflammatory properties of an isothiocyanate-enriched moringa (*Moringa oleifera*) seed extract. *PLoS ONE* **2017**, *12*, e0182658. [CrossRef]
69. Oblak, M.; Randic, M.; Solmajer, T. Quantitative structure-activity relationship of flavonoid analogues. 3. Inhibition of p56lck protein tyrosine kinase. *J. Chem. Inf. Comput. Sci.* **2000**, *40*, 994–1001. [CrossRef] [PubMed]
70. Olszanecki, R.; Gebska, A.; Kozlovski, V.I.; Gryglewski, R.J. Flavonoids and nitric oxide synthase. *J. Physiol. Pharmacol.* **2002**, *53*, 571–584.
71. Sulaiman, M.R.; Zakaria, Z.A.; Bujarimin, A.S.; Somchit, M.N.; Israf, D.A.; Moin, S. Evaluation of *Moringa oleifera* aqueous extract for antinociceptive and anti-inflammatory activities in animal models. *Pharm. Biol.* **2008**, *46*, 838–845. [CrossRef]
72. García-Mediavilla, V.; Crespo, I.; Collado, P.S.; Esteller, A.; Sánchez-Campos, S.; Tuñón, M.J.; González-Gallego, J. The anti-inflammatory flavones quercetin and kaempferol cause inhibition of inducible nitric oxide synthase, cyclooxygenase-2 and reactive C-protein, and down-regulation of the nuclear factor kappaB pathway in Chang Liver cells. *Eur. J. Pharmacol.* **2007**, *557*, 221–229. [CrossRef] [PubMed]
73. Hämäläinen, M.; Nieminen, R.; Vuorela, P.; Heinonen, M.; Moilanen, E. Anti-inflammatory effects of flavonoids: Genistein, kaempferol, quercetin, and daidzein inhibit STAT-1 and NF-kappaB activations, whereas flavone, isorhamnetin, naringenin, and pelargonidin inhibit only NF-kappaB activation along with their inhibitory effect on iNOS expression and NO production in activated macrophages. *Mediators. Inflamm.* **2007**, *2007*, 45673. [PubMed]
74. Tan, W.S.; Arulselvan, P.; Karthivashan, G.; Fakurazi, S. Moringa oleifera flower extract suppresses the activation of inflammatory mediators in lipopolysaccharide-stimulated raw 264.7 macrophages via NF-κB Pathway. *Mediators. Inflamm.* **2015**, *2015*, 720171. [CrossRef] [PubMed]
75. Park, E.J.; Cheenpracha, S.; Chang, L.C.; Kondratyuk, T.P.; Pezzuto, J.M. Inhibition of lipopolysaccharide-induced cyclooxygenase-2 and inducible nitric oxide synthase expression by 4-[(2′-O-acetyl-α-L-rhamnosyloxy)benzyl] isothiocyanate from *Moringa oleifera. Nutr. Cancer* **2011**, *63*, 971–982. [CrossRef]
76. Fard, M.T.; Arulselvan, P.; Karthivashan, G.; Adam, S.K.; Fakurazi, S. Bioactive extract from *Moringa oleifera* inhibits the pro-inflammatory mediators in lipopolysaccharide stimulated macrophages. *Pharmacogn. Mag.* **2015**, *11*, S556. [CrossRef] [PubMed]
77. Efstratiadis, G.; Divani, M.; Katsioulis, E.; Vergoulas, G. Renal fibrosis. *Hippokratia* **2009**, *13*, 224–229.
78. Hamza, A.A. Ameliorative effects of *Moringa oleifera* Lam seed extract on liver fibrosis in rats. *Food Chem. Toxicol.* **2010**, *48*, 345–355. [CrossRef]
79. Lin, T.A.; Wu, V.C.; Wang, C.Y. Autophagy in chronic kidney diseases. *Cells* **2019**, *8*, 61. [CrossRef]
80. Kimura, T.; Takabatake, Y.; Takahashi, A.; Kaimori, J.Y.; Matsui, I.; Namba, T.; Kitamura, H.; Niimura, F.; Matsusaka, T.; Soga, T.; et al. Autophagy protects the proximal tubule from degeneration and acute ischemic injury. *J. Am. Soc. Nephrol.* **2011**, *22*, 902–913. [CrossRef]
81. Takahashi, A.; Kimura, T.; Takabatake, Y.; Namba, T.; Kaimori, J.; Kitamura, H.; Matsui, I.; Niimura, F.; Matsusaka, T.; Fujita, N.; et al. Autophagy guards against cisplatin-induced acute kidney injury. *Am. J. Pathol.* **2012**, *180*, 517–525. [CrossRef]
82. Liu, S.; Hartleben, B.; Kretz, O.; Wiech, T.; Igarashi, P.; Mizushima, N.; Walz, G.; Huber, T.B. Autophagy plays a critical role in kidney tubule maintenance, aging and ischemia-reperfusion injury. *Autophagy* **2012**, *8*, 826–837. [CrossRef]
83. Elmore, S. Apoptosis: A review of programmed cell death. *Toxicol. Pathol.* **2007**, *35*, 495–516. [CrossRef]
84. Bonegio, R.; Lieberthal, W. Role of apoptosis in the pathogenesis of acute renal failure. *Curr. Opin. Nephrol. Hypertens.* **2002**, *11*, 301–308. [CrossRef] [PubMed]
85. Molitoris, B.A. Acute renal failure. *Drugs Today (Barc)* **1999**, *35*, 659–666. [CrossRef]
86. Havasi, A.; Borkan, S.C. Apoptosis and acute kidney injury. *Kidney Int.* **2011**, *80*, 29–40. [CrossRef] [PubMed]
87. Feldenberg, L.R.; Thevananther, S.; del Rio, M.; de Leon, M.; Devarajan, P. Partial ATP depletion induces Fas- and caspase-mediated apoptosis in MDCK cells. *Am. J. Physiol.* **1999**, *276*, F837–F846. [CrossRef] [PubMed]

88. Levey, A.S.; de Jong, P.E.; Coresh, J.; El Nahas, M.; Astor, B.C.; Matsushita, K.; Gansevoort, R.T.; Kasiske, B.L.; Eckardt, K.U. The definition, classification, and prognosis of chronic kidney disease: A KDIGO controversies conference report. *Kidney Int.* **2011**, *80*, 17–28. [CrossRef]
89. Mace, P.D.; Riedl, S.J. Molecular cell death platforms and assemblies. *Curr. Opin. Cell Biol.* **2010**, *22*, 828–836. [CrossRef]
90. Tzifi, F.; Economopoulou, C.; Gourgiotis, D.; Ardavanis, A.; Papageorgiou, S.; Scorilas, A. The role of BCL2 family of apoptosis regulator proteins in acute and chronic leukemias. *Adv. Hematol.* **2012**, *2012*, 524308. [CrossRef]
91. Adebola, A.O.; Oluwatoyin, O.; Toyin, A.I.; Nnena Linda, N. Histological variances of *Moringa olifer*a on the kidney of adult Wistar rats. *Innov.J. Med. Sci.* **2021**, *9*, 36–38. [CrossRef]
92. Hassan, I.M.; Saidu, B.; Afaru, J.; Ishaq, A.; Dahiru, A.; Abdulazeez, N.; Yusuf, H.; Karofi, D.; Pilau, N.; Abubakar, A.A.; et al. Effects of *Moringa oleifer*a biochemical constituents on kidney, liver and brain of Wister rats. *SJAC* **2020**, *8*, 128–132. [CrossRef]
93. Akpan, H.; Akande, A.; Ojewale, A.; Oladipupo, F.; Akinpelu, O.F.; Jimoh, S. *Moringa oleifera* ameliorates nephropathic changes in alloxaninduced diabetic adult Wistar rats. *J. Afr. Assoc. Physiol. Sci.* **2018**, *6*, 110–118.
94. Abaekwume, C.O.; Kagbo, H.D. Hepato-renal-curative effect of the herbal supplement of *Aloe vera* Linn Gel versus *Moringa oleifera* on acetaminophen-induced damage on the liver and Kidney of Wistar rats (*Rattus novergicus*). *JAMPS* **2021**, *23*, 12–23. [CrossRef]
95. Saleh, A. Evaluation of hepatorenal protective activity of *Moringa oleifera* on histological and biochemical parameters in cadmium intoxicated rats. *Toxin Rev.* **2018**, *38*, 1–8. [CrossRef]
96. Arafat, N.; Awadin, W.; ElShafei, R.; El-Metwalley, V.; Saleh, R. Protective Role of Moringa oleifera Leaves Extract Against Gentamicin-induced Nephro- and Hepato- Toxicity in Chickens. *Alex. J. Vet. Sci.* **2018**, *58*, 173. [CrossRef]
97. Taweerutchana, R.; Lumlerdkij, N.; Vannasaeng, S.; Akarasereenont, P.; Sriwijitkamol, A. Effect of *Moringa oleifera* leaf capsules on glycemic control in therapy-naïve type 2 diabetes patients: A randomized placebo controlled study. *Evid. Based. Complement. Alternat. Med.* **2017**, *2017*, 6581390. [CrossRef] [PubMed]
98. Kerdsomboon, K.; Chumsawat, W.; Auesukaree, C. Effects of *Moringa oleifera* leaf extracts and its bioactive compound gallic acid on reducing toxicities of heavy metals and metalloid in *Saccharomyces cerevisiae*. *Chemosphere* **2021**, *270*, 128659. [CrossRef] [PubMed]

plants

MDPI

Review

Countering Triple Negative Breast Cancer via Impeding Wnt/β-Catenin Signaling, a Phytotherapeutic Approach

Laleh Arzi [1,*], Homa Mollaei [2] and Reyhane Hoshyar [3]

1 Department of Microbiology, Shahr-e-Qods Branch, Islamic Azad University, Tehran 37515-374, Iran
2 Department of Biology, Faculty of Sciences, University of Birjand, Birjand 97175-615, Iran
3 Cell biology Department, GenEdit, South San Francisco, CA 94080, USA
* Correspondence: laleh.arzi@ut.ac.ir; Tel.: +98-9122574325

Abstract: Triple negative breast cancer (TNBC) is characterized as a heterogeneous disease with severe malignancy and high mortality. Aberrant Wnt/β-catenin signaling is responsible for self-renewal and mammosphere generation, metastasis and resistance to apoptosis and chemotherapy in TNBC. Nonetheless, in the absence of a targeted therapy, chemotherapy is regarded as the exclusive treatment strategy for the treatment of TNBC. This review aims to provide an unprecedented overview of the plants and herbal derivatives which repress the progression of TNBC through prohibiting the Wnt/β-catenin pathway. Herbal medicine extracts and bioactive compounds (alkaloids, retinoids. flavonoids, terpenes, carotenoids and lignans) alone, in combination with each other and/or with chemotherapy agents could interrupt the various steps of Wnt/β-catenin signaling, i.e., WNT, FZD, LRP, GSK3β, Dsh, APC, β-catenin and TCF/LEF. These phytotherapy agents diminish proliferation, metastasis, breast cancer stem cell self-renewal and induce apoptosis in cell and animal models of TNBC through the down-expression of the downstream target genes of Wnt signaling. Some of the herbal derivatives simultaneously impede Wnt/β-catenin signaling and other overactive pathways in triple negative breast cancer, including: mTORC1; ER stress and SATB1 signaling. The herbal remedies and their bioactive ingredients perform essential roles in the treatment of the very fatal TNBC via repression of Wnt/β-catenin signaling.

Keywords: herbal medicine; bioactive derivative; triple negative breast cancer; Wnt/β-catenin; anti-cancer

Citation: Arzi, L.; Mollaei, H.; Hoshyar, R. Countering Triple Negative Breast Cancer via Impeding Wnt/β-Catenin Signaling, a Phytotherapeutic Approach. *Plants* **2022**, *11*, 2191. https://doi.org/10.3390/plants11172191

Academic Editors: Juei-Tang Cheng, I-Min Liu and Szu-Chuan Shen

Received: 19 July 2022
Accepted: 21 August 2022
Published: 24 August 2022

1. Introduction

Breast cancer was reported as the most diagnosed cancer and the leading cause of cancer mortality worldwide among women in 2020 [1]. Advances in molecular techniques have prompted a conventional classification of breast cancer. The St. Gallen International Breast Cancer Conference 2011 allocated four molecular subtypes to breast cancers: Luminal A (ER+/PR+/HER2-/lowKi-67); Luminal B (ER+/PR+/HER2-/+/high Ki-67); HER2-overexpression (ER-/PR-/HER2+) and triple negative breast cancers (TNBCs) (ER-/PR-/HER2-) [2]. The TNBC is characterized as the most aggressive and fatal subtype, particularly in young women. It is famed as "the kiss of death", being attributed with a poor chance of survival, early relapse, high proliferation and metastatic potential, heterogeneity and lack of efficient, approved targeted therapies [3–5]. Molecular analysis has evidenced that various signaling pathways are activated in the TNBC cells, such as Wnt/β-catenin, Hedgehog, Notch, TNF-α, Hippo and JAK-STAT [6]. However, since aberrant the Wnt/β-catenin pathway was reported as a predisposing factor of TNBC, prohibition of this pathway could be a worthwhile target for conquering TNBC progress [7–9]. The Wnt-signaling pathway is associated with cell proliferation, survival, invasion, metastasis and chemotherapy resistance in TNBC [10].

The Wnt signaling commences with the binding of Wnt ligands to the N-terminal extra-cellular cysteine-rich domain of the Frizzled (FZD) seven-pass transmembrane receptors and the LDL receptor-related proteins (LRP coreceptor). The trimetric complex (Wnt, FZD and LPR) recruits phosphorylated Disheveled (Dsh) and Axin, to prevent β-catenin phosphorylation. Therefore the APC/CK1/GSK-3β/Axin/β-catenin degradation complex is inactivated and the phosphorylation of β-catenin by GSK-3β is repressed. The accumulated β-catenin in the cytoplasm translocates into the nucleus, where it modulates the target gene expression with the T cell factor/lymphoid enhancer factor (Tcf/Lef) family of transcription factor [11]. Besides the classical canonical β-catenin-dependent mechanism of Wnt signaling, this pathway might progress through alternative signaling called "β-catenin independent mechanisms or non-canonical pathways" [12].

Although chemotherapeutic agents have been employed as exclusive treatments of TNBC, the resistance to chemotherapy agents (anthracyclines, taxanes, capecitabine, gemcitabine, eribulin), biomarker-based treatments and checkpoint-based immunotherapy has led to inconsequential responses to the medications. An extensive spectrum of clinical investigations has claimed the favorable influence of herbal remedies on the survival, immune enhancement and quality of life of cancer sufferers [13]. Moreover, due to their pharmacological safety, these compounds can be applied alone or complementary to chemotherapy in order to boost the therapeutic efficacy and diminish the chemotherapy-induced toxicity [14,15]. A myriad of investigations have verified the effectiveness of medicinal plant extracts and their bioactive components in targeting the Wnt pathway as novel therapeutic agents [16–18]. Accordingly, this paper is dedicated to the review and appraisal of plants and natural derivatives which suppress the progression of TNBC through prohibiting the Wnt/β-catenin pathway.

2. Natural Derivatives Targeting TNBC via Impeding the Wnt/β-Catenin Pathway

2.1. Saikosaponin D

Saikosaponin D, one of the main bioactive triterpenes, is extracted from the root of *Radix bupleuri* L. (Chaihu)—a known Chinese traditional medicine with broad applications. The Chinese Food and Drug Administration (CFDA) has approved fifteen clinical preparations of *Radix bupleuri* [19]. Saikosaponin D has been reported to suppress cancer cells through the downregulation of TNF-α-mediated NF-κB, activation of autophagy or the blocking of the Wnt/β-catenin signaling pathway [20–22]. Wang et al. evinced the theory that Saikosaponin D reduced proliferation and activated apoptosis in various types of TNBC cell lines. The dynamic mass redistribution assay, TopFlash, and Western blot were applied to clarify the corresponding molecular mechanisms. They detailed that Saikosaponin D remarkably suppressed β-catenin and its downstream target genes (c-Myc and CyclinD1), leading to caspase-dependent cell death. The molecular docking of Saikosaponin D to the crystal structure of β-catenin suggests that it attaches to β-catenin via hydrogen bonds and hydrophobic interaction [23].

2.2. Echinacoside

Echinacoside is a phenylethanoid glycoside, isolated from the species of genus *Cistanches* (Orobanchaceae) and *Echinacea* (Asteraceae) [24]. Recent investigations have indicated the anti-proliferation, anti-migration and anti-invasion potencies of Echinacoside on TNBC cells (MDA-MB-231 and MDA-MB-468 cells), and illustrated that it exerted its effects via the inhibition of the Wnt/β-catenin signaling pathway; as the protein expression of crucial Wnt/β-catenin signaling factors (phosho-LRP6, total LRP6, phosho-Dvl2, active β-catenin, and total β-catenin) and the mRNA and protein expression levels of Wnt target genes (CD44, LEF1 and Cyclin D1) were reduced in the TNBC cell lines and the MDA-MB-231 xenograft mice model. The assessment of the murine tumor sizes and weights revealed the tumor-growth inhibitory activity of Echinacoside in the treated animals [25].

2.3. *Sulforaphane*

Sulforaphane, 4-methylsulfinylbutyl isothiocyanate, naturally originates from specific species of the *Brassica* vegetable family, most noteworthy, broccoli. The young sprouted broccoli seeds contain an immense amount of glucosinolate, which is metabolized by the myrosinase enzyme to Sulforaphane [26]. The potential of Sulforaphane to target breast cancer stem cells (BCSCs) in the cell and xenograft models of TNBC was evaluated. The sulforaphane treatment led to decreasing cell viability and induction of apoptosis in the SUM159 cells, via activation of caspase-3. The sulforaphane prevented mammosphere formation in the TNBC cells and remarkably reduced the Aldehyde Dehydrogenase (ALDH)-positive population cells in the cell and tumor models of TNBC. In addition, the tumor sizes in Sulforaphane-treated mice were lessened by 50% compared to the control mice. It should be mentioned that the Sulforaphane therapy had approximately no toxicity, as determined by measuring the mice body weight. Furthermore, the suppressive role of Sulforaphane on Wnt/β-catenin pathway was demonstrated, since β-catenin and Cyclin D1, the Wnt/β-catenin target genes, were down-expressed in SUM159 cells [27].

2.4. *Gigantol*

Gigantol, 4-[2-(3-hydroxy-5-methoxyphenyl) ethyl]-2-methoxyphenol, is a bibenzyl-type phenolic biomolecule extracted from various medicinal orchids, with numerous pharmaceutical benefits [28]. An investigation into the TNBC cell lines illustrated that Gigantol reduced viability and migration of the MDA-MB-231 and MDA-MB-468 cells, through impeding the Wnt/β-catenin signaling. The SuperTOPFlash assay reported that Gigantol prohibited the Wnt/β-catenin signaling via decreasing the level of phosphorylated LRP6, total LRP6 and cytosolic β-catenin, leading to a reduction in the expression of the Wnt target genes, Axin2 and Survivin [29].

2.5. *Naringin*

Naringin, a flavanone glycoside composed of the flavanone naringenin and neohesperidose, is a crucial bioactive ingredient of *Drynaria fortunei* (Kunze) J. Sm., *Citrus aurantium* L. and *Citrus medica* L. It occurs in citrus fruit and endows bitterness to the citrus juices [30]. The anti-tumor potential of Naringin on TNBC was assessed in both the cell (MDA-MB-231, MDA-MB-468 and BT-549 cells) and xenograft mice models. Western blot and immunohistochemistry assays demonstrated that Naringin elevated the p21 expression and diminished the Survivin and β-catenin levels. It was also found that the p21 and Survivin levels were regulated by β-catenin. The over-expression of β-catenin in the MDA-MB-231 and BT-549 cells remarkably reduced the repressive effect of Naringin on cell proliferation, whilst the knock-out of β-catenin caused the inhibition of TNBC cell growth. Shrinkage of the tumor was detected in the Naringin-treated mice. Therefore, it was supposed that Naringin could promote apoptosis (augmenting the activity of caspase 3) and G1 phase-arrest via hindering the Wnt/β-catenin pathway [31]. The nuclear β-catenin binds with and activates TCF4/LEF, which may switch on the transcription of p53 and c-Myc. C-Myc can elevate the expression of p14ARF, fas, Trail, fasR and DR4/5. FasR and DR4/5 promote the apoptotic extrinsic pathway, which is initiated by the binding of their respective ligands. This leads to the autoactivation of caspases-8 and -10, which sequentially boosts the catalytic activation of the effector caspase-3 [32].

2.6. *Oxymatrine*

Oxymatrine, a quinolizidine alkaloid compound isolated from the roots of *Sophoraflavescens Ait*, possesses diverse medicinal qualities [33]. Oxymatrine reinforced the anti-tumor activity of Bevacizumab, by restraining invasion and metastasis induced by Bevacizumab in TNBC cells, via inactivating the Wnt/β-Catenin pathway [34]. The monoclonal antibody, Bevacizumab, is applied, in combination with neoadjuvant chemotherapy, as the first-line treatment for metastatic breast cancer. It demonstrated effective anti-angiogenesis potential, whilst also augmenting the metastatic tendency of the TNBC cells [35,36]. Xie et al. showed that Oxymatrine inhibited the migration and invasion via reverting the EMT phenotype. In the treated group, the expression level of the mesenchymal-associated genes, N-Cad and Vim, and the EMT-related transcription factors including ZEB, Snail and Slug, were elevated, whereas the expression of the epithelial-associated gene E-Cad declined in comparison to the control. Oxymatrine decreased the subpopulation of the TNBC stem-like cells through repressing the Wnt/β-Catenin pathway, since the expressions of β-Catenin and its downstream oncoproteins were attenuated. In addition, Oxymatrine reduced the tumor growth in tumor-bearing mice and reduced the risk of relapse and metastasis by abating the self-renewal capacity of the cancer stem cells (CSCs). It is worth noting that Oxymatrine synergistically amplified the anti-angiogenic potency of Bevacizumab. This alkaloid suppressed the cell viability of HUVEC cells dose-dependently and reduced the formation of tumor neovascularization in the combination-treated mice [34].

2.7. *Silibinin*

Silibinin, an essential bioactive flavonolignan extracted from milk thistle (*SilybumMarianum* L. Gaertn), elucidated chemopreventive and chemo-sensitizing properties against various cancers [37]. It was demonstrated that Silibinin abated the proliferation of TNBC cells and suppressed the phosphorylation and expression of endogenous LRP6 in the MDA-MB-231 cells [38]. Growing evidence indicated that the LRP6 is a pivotal Wnt co-receptor, valued in therapeutic strategies for TNBC [39]. Therefore, it was speculated that Silibinin, as an inhibitor of Wnt/β-catenin signaling, could be considered as a natural propitious anti-tumor remedy.

2.8. *Rottlerin*

Rottlerin, a polyphenol extracted from the Asian Kamala plant *Mallotus philippinensis,* is regarded as a multifaceted anti-cancer natural agent in various cancers [40]. Lu's research group surveyed the molecular mechanism of the anti-proliferative potency of this polyphenol on various prostate and breast cancers, including the TNBC cell line (MDA-MB-231). Rottlerin impaired the Wnt/β-catenin signaling, since it diminished the expression of the cytosolic-free human β-catenin, total cellular human β-catenin, human LRP6, phospho-LRP6 and axin2. Subsequently, it was illustrated that Rottlerin-mediated LRP6 downregulation was unassociated with AMP-activated protein kinase (AMPK). Furthermore, Rottlerin disrupted the mTORC1 signaling, as it remarkably repressed P70-S6K, phospho-P70-S6K, S6 and phospho-S6in TNBC cells. The Rottlerin lowered the expression of the common target oncogenes of the Wnt/β-catenin- and mTORC1-signaling pathways, Cyclin D1 and Survivin [41]. Mounting evidence supported the crosstalk between the Wnt/β-catenin and the mTORC1 signaling, the activation of the Wnt/β-catenin-signaling upregulates the mTORC1 signaling in the cancer cells [42]. Hence, this investigation assumed that the anti-cancer potential of Rottlerin was associated with the dual suppression of the Wnt/β-catenin and mTORC1 signaling.

2.9. *Baicalin*

Baicalein is the flavonoid glucoside of *Scutellaria baicalensis* Georgi, Lamiaceae, which has anticancer potential. This medicinal plant is native to China, Korea, Mongolia and in the Russian Far East and Siberia [43]. The anti-metastatic potency of Baicalin and its underlying mechanisms have been evaluated in the most aggressive type of breast cancer. Baicalin attenuated the survival, migration and invasion of TNBC cells (MDA-MB-231 and 4T1), whereas it illustrated no noticeable effect on the proliferation of the cancer cells. Moreover, Baicalin-treated tumor-bearing mice possessed less liver and lung metastatic lesions compared to the control mice. In addition, it was elucidated that this flavonoid functioned as an anti-metastatic agent via inverting EMT and downregulating the expression of β-catenin in the cell and mice models of TNBC. These findings recommend Baicalin in conjunction with conventional chemotherapy agents in the treatment of TNBC patients [44].

2.10. Baicalein

Baicalein is a crucial bioactive flavonoid found in the dried root of *Scutellaria baicalensis* Georgi, a traditional Chinese medicine belonging to the *Lamiaceae* family. Extensive research displayed that Baicalein activated apoptosis, induced cell cycle arrest and suppressed metastasis in different cancers [45]. A survey assessed the effect of Baicalein on the cell and animal models of TNBC. They confirmed that Baicalein decreased the viability, migration and invasion of the MDA-MB-231 cells dose- and time-dependently. In accordance with the cell assays' findings, it was revealed that Baicalein gavage in mice reduced the metastasis rate in the liver and lung tissues of the prevention and therapy group compared to the control mice. In addition, in the Baicalein-treated MDA-MB-231 cells and mice, the EMT process was inhibited, since the expression of E-Cad (molecular marker of epithelial cells) was elevated, while Vim (molecular marker of mesenchymal cells) and Snail (transcription factor) expressions diminished. In addition, Baicalein exposure promoted the downregulation of Wnt1, β-catenin and AT-rich sequence-binding protein-1 (SATB1) in the MDA-MB-231 cells and tumor-bearing mice. It should be mentioned that Baicalein reduced the expression of Cyclin D1 and Axin2 in TNBC cells. It can be deduced that Baicalein restricted metastasis in the cell and animal models by reverting EMT, which may be associated with impeding SATB1 and the Wnt pathway [46].

2.11. Epigallocatechin Gallate

Epigallocatechin gallate, a biopolyphenol, is the most abundant and influential antioxidant extracted from green tea [47]. Hong et al.'s investigation into breast cancer patients (females suffering from invasive ductal carcinoma, having undergone a curative operation) illustrated that β-catenin is upregulated in breast cancer tissue compared with its expression in normal adjacent tissues, and that the β-catenin overexpression is correlated with lymph node involvement, high tumor stage and ER-negative status. They also showed that Epigallocatechin gallate treatment reduced the cell viability and expressions of β-catenin, P-AKT and Cyclin D1 in the MDA-MB-231 cells, and that pretreatment with phosphatidylinositol-3 (PI3) kinase inhibitors (25 μm LY294002 or 5 μM Wortmannin) elevated the suppressive effect of Epigallocatechin gallate on the β-catenin expression. These results imply that Epigallocatechin gallate's cytotoxic potential on TNBC is related to the inactivation of Wnt/β-catenin [48]. This finding is in accord with a previous survey illustrating that β-catenin is able to attach to PI3 kinase, which may play a role in the stabilization of β-catenin. Therefore, the repression of PI3 kinase may lead to the declined expression of β-catenin [49].

Supporting findings were obtained by Kim's research group. They displayed that Epigallocatechin gallate obstructed growth, migration and invasion. This polyphenolic compound blocked Wnt signaling and its target gene, c-Myc in MDA-MB-231. They previously cited the HMG-Box Transcription Factor 1 (HBP1) as a negative transcriptional regulator of Wnt signaling [50]. HBP1 is a high mobility group (HMG) box transcription factor, similar to LEF and TCF in the Wnt pathway [51]. Therefore, they evaluated the capability of Epigallocatechin gallate in targeting the Wnt signaling through HBP1. They proposed that the anti-proliferative and anti-metastatic properties of Epigallocatechin gallate emerged from the suppression of the Wnt signaling via targeting the HBP1 transcriptional repressor [50].

2.12. Cardamonin

Cardamonin, (2,4-dihydroxy-6-methoxychalcone) is a naturally occurring chalcone, commonly derived from different plants of the *Zingiberaceae* family. There is mounting evidence supporting the multi-potency anti-cancer activities of Cardamonin via modulating various signaling pathways, transcriptional factors, cytokines and enzymes, such as mTOR, NF-κB, Akt, STAT3, Wnt/β-catenin and COX-2 [52]. The breast tumor-suppressive role of Cardamonin and its underlying molecular mechanism was assessed in cell and mice models of TNBC. The MTT assay revealed the selective cytotoxicity of Cardamonin on TNBC cells, without influencing the normal breast epithelial cells, MCF-10A. The Cardamonin activated the mitochondrial pathway of apoptosis through modulation of Bax, Bcl-2, Cyt-C, caspase-3 and PARP in MDA-MB-231 and BT-549 cells. Cardamonin impaired the migration and invasion of TNBC cells via impeding EMT. Furthermore, the assays have illuminated the role of Cardamonin reduction in the stability and nuclear translocation of β-catenin via activating GSK3b. Measuring the tumor volumes demonstrated that the tumor sizes were reduced in the murine model of TNBC [53].

2.13. *Inotodiol*

Inotodiol is a natural lanostane-type triterpenoid isolated from *Inonotus obliquus* (*Chaga mushroom*), an edible fungus which inhabits birch trees [54]. A previous piece of research on the cell and mice models of colorectal cancer claimed that the aqueous extract of *Inonotus obliquus* exhibited potent anti-inflammatory and anti-proliferative properties through the downregulation of Wnt/β-catenin [55]. The effect of Inotodiol was assessed on the progress of breast cancer in 7, 12-dimethylbenz(a)anthracene (DMBA)-administered diabetic rats [56]. It was demonstrated that the DMBA induced breast cancer in mice through the activation of Wnt/β-catenin signaling [57]. The results indicated that Inotodiol lowered blood glucose (fasting blood glucose levels and oral glucose tolerance test) and plasma levels of cholesterol, triglyceride and HDL, and elevated the antioxidant enzymes activities in Sprague-Dawley (SD) rats. It was observed that in the Inotodiol-treated rats, the expression of the proliferating cell nuclear antigen (PCNA), a tumor proliferation marker, decreased. Subsequently, the deregulation of β-catenin and its downstream target genes (c-Myc and Cyclin D1) and the induction of apoptosis (Caspase-3 and PARP gene expressions were strikingly upregulated) eventuated. It was concluded that Inotodiol regulated the blood glucose in diabetic rats and subsequently repressed tumor progression in breast cancer by activating apoptosis via the downregulation of the Wnt/β-catenin pathways [56]. This finding is in accord with research that established Wnt/β-catenin as a linkage between high glucose and cancer [58].

2.14. *Schisandrin A*

Schisandrin A is the most dominant lignan (polyphenolic compounds) present in the fruit of *Schisandra chinensis* (Turcz.) Baill. [59]. The anti-tumor efficacy of Schisandrin on TNBC cells has been studied. It has been demonstrated that Schisandrin (25, 50 and 100 μM) reduces viability, inhibits colony formation, induces nuclear fragmentation, condenses chromatin, arrests the cell cycle in G1 phase and promotes apoptosis by elevating Bax and P53 and reducing the Bcl2 proteins in TNBC cell lines. The oral gavages of Schisandrin (25 mg/kg) induced no apparent toxicity in the organs of female BALB/c nude mice and the animals showed no appreciable changes in body weight and possessed smaller tumors. Following Schisandrin administration, an increase in the endoplasmic reticulum stress-specific protein, ATF4, CHOP and phosphorylated elf2α, and decrease in the Wnt signaling-associated proteins, β-catenin and phosphorylated GSK3β, were detected in the MDA-MB-231 and BT-549 cells and tumor tissues [60]. The accumulation of extra ER stress enhanced cancer cell death and the ER stress activators were assumed as the potential candidates for the remedy of TNBC [61]. Therefore, it can be concluded that Schisandrin exerts its anti-tumorigenic effects through the regulation of the Wnt/ER stress signaling pathway.

2.15. *Resveratrol*

Resveratrol (3,4′, 5-trihydroxy-trans-stilbene) is a plant polyphenolic derivative, widely found in grapes, berries and peanuts [62]. The clinical finding of CSCs in breast cancers have displayed a relation between the fraction of CSCs and poor prognosis [63]. Accordingly, Fu et al. assessed the repressive potential of Resveratrol on BCSCs. They found that Resveratrol inhibited the proliferation of SUM159. It diminished the augmented ALDH-positive breast cancer cells and reduced the number and size of mammospheres. The injection of Resveratrol (100 mg/kg) to female non-obese diabetic/severe combined immunodeficiency (NOD/SCID) tumor-burden mice reduced the tumor volume and ALDH population in tumor cells. The induction of autophagy by resveratrol was displayed in the BCSCs from the SUM159 cells, since the number of autophagosomes and the expression of Lc3-II, Beclin1 and Atg7 genes required for autophagosome formation were elevated. The Western blot results illustrated that Resveratrol attenuated the expression of β-catenin and Cyclin D1 in BCSCs and xenograft mice. The overexpression of β-catenin by transected plasmid of pcDNA3-S33Y β-catenin or repression of autophagy by chloroquine eliminated the suppressive effect of Resveratrol on the Wnt/β-catenin signaling [64]. The Wnt/β-catenin-signaling pathway is pivotal in the management of BCSCs self-renewal and autophagy [65], therefore it was inferred that Resveratrol suppressed the BCSCs and activated autophagy via impeding the Wnt/β-catenin-signaling pathway, while exhibiting no cytotoxicity on the noncancerous cells and the treated mice [64].

2.16. Deguelin

Deguelin is one of the major naturally occurring rotenoids, extracted from *Mundulea sericea* L., belonging to the *Leguminosae* family [66,67]. It was observed that Deguelin suppressed the growth of the MDA-MB-231 cells dose-dependently, while no significant growth restriction was reported for up to one week in the MCF-12F cells (normal immortalized human mammary epithelial cells). The Deguelin treatment prompted cell cycle arrest at the S phase and induced apoptosis in the TNBC cells. The microarray analysis indicated that Deguelin upregulated the Wnt/β-catenin inhibitors (WIF-1, DDK4) and some of the cadherin family members (CDH3, CDH7, CDH9), while downregulating the Wnt/β-catenin activators (Wnt14, Wnt2B, Wnt3) and Snail. The Western blot results demonstrated that following Deguelin exposure, no appreciable changes were observed in GSK-3β(repressor of β-catenin), in spite of the decreased expression of p-GSK-3β (inactive form of GSK-3β), β-catenin and Cyclin D1.Therefore, it was concluded that Deguelin exerted its therapeutic benefits on TNBC via inhibition of Wnt/β-catenin by activation of GSK-3β, and subsequently β-catenin destruction [67].

2.17. Hydroxytyrosol

Hydroxytyrosol (3,4-dihydroxyphenylethanol) is a phenolic alcohol extracted from olive oil [68]. Cruz-Lozano et al.'s investigation into triple negative breast cancer cell lines displayed that hydroxytyrosol diminished the mammosphere-formation efficiency (MSFE) and decreased the volume of the second generation of mammospheres (assembled from the detached primary mammospheres), thus reducing BCSC self-renewal. This herbal phenol decreased $ALDH^+$ and mesenchymal–like $CD44^+/CD24^{-/low}$ BSCS subpopulations. In addition, it suppressed the migration and invasion of various TNBC cell lines (SUM159PT, BT549, MDA-MB-231, and Hs578T) [69]. These results are in line with the role of cells with high CD44 and low CD24 expressions and high ALDH activity in vast incidences of metastasis, therapy resistance and tumor relapse in breast cancers [70]. Hydroxytyrosol reduced the expression of the EMT-related transcription factors (Zeb, Slug and Snail) and the mesenchymal marker (Vim), and increased the epithelial marker (ZO-1), as well as downregulating the Wnt/β-catenin proteins (p-LRP6, LRP6, β-catenin, and Cyclin D1). Hence, Hydroxytyrosol, as a chemopreventive agent, reduced BCSCs and the metastatic potential of TNBC via disturbing EMT and the Wnt/β-catenin pathway [69].

2.18. Fucoidan

Fucoidan, a polysaccharide obtained from brown seaweed and some marine invertebrates, consists of considerable amounts of L-fucose and sulfate ester groups. It may also contain other monosaccharides (mannose, galactose, glucose, xylose) and uronic acids, acetyl, sulfate and protein [71]. It was reported that Fucoidan (50, 100, and 200 μg/mL) inhibited 4T1 cell growth, induced apoptosis in the 4T1 cells and the tumor tissue of BALB/c mice (G1 arrest in cell cycle was detected) and inhibited the TCF/LEF reporter activity dose-dependently. The Western blot analysis showed that fucoidan treatment downregulated the expression of β-catenin and its downstream target genes, c-Myc, Cyclin D1 and Survivin, in 4T1 and tumor-bearing mice. The immunohistochemical staining confirmed the Western blot results, as Fucoidan attenuated the β-catenin-positive cells [72]. These results are consistent with the evidence that the downregulation of the β-catenin proteins prohibits the TCF/LEF reporter activity [73]. Therefore, it can be inferred that the anti-proliferative and apoptotic potency of Fucoidan is associated with restraining the Wnt/β-catenin signaling pathway.

2.19. *Jatrophone*

Jatrophone is a macrocyclic diterpene, consisting of an oxaspiro core and several electrophilic centers. Jatrophone is isolated from the *Euphorbiaceae* family, such as *Jatropha isabelli* and *Jatropha gossypiifolia* [74]. Fatima's research group determined the cytotoxicity of Jatrophone on different subtypes of TNBC: mesenchymal stem-like (MDA-MB-231 and MDA-MB-157), basal-like-1(HCC-38 and MDA-MB-468) and patient-derived xenograft. They observed that it arrested the cell cycle in the S phase even more efficiently than the classic Wnt inhibitor, ICG-001, and induced late apoptosis in MDA-MB-231. The Topflash reporter and immunofluorescence outcomes demonstrated that Jatrophone impeded the Wnt/β-catenin signaling between the receptor complex and β-catenin (early level of Wnt/β-catenin pathway). The qPCR and immunoblot assays revealed that Jatrophone reduced the expression of Wnt/β-catenin direct target genes (reducing the mRNA level of Birc5, Axin2, Hmga2, Myc, PCNA and Ccnd1 and lowering the protein level of Axin2, Hmga2, Myc, PCNA, CyclinD1 and activated β-catenin). Likewise it decreased the migration of MDA-MB-231cells by downregulation of EMT markers, such as Slug, fibronectin and Vim. Hence, the anti-proliferative and anti-migration potency of Jatrophoneis applied through inhibiting the Wnt/β-catenin pathway by the elimination of nuclear-activated β-catenin and the downregulation of the downstream Wnt target genes [75].

2.20. *Luteolin*

Luteolin, 3,4,5,7-tetrahydroxy flavone, is a natural flavonoid, extensively present in many fruits and vegetables, such as celery, sweet bell peppers, carrots, broccoli, onion leaves and parsley [76]. The anti-metastatic potential of Luteolin on MDA-MB-231 and BT5-49 were evinced by wound healing and Transwell chamber assays. Luteolin caused an alteration in the morphology of TNBC cells from mesenchymal to oval epithelial. Immunofluorescent staining, Western blot and qPCR results confirmed the reversion of the EMT phenotype, i.e., the mesenchymal markers (N-Cad and Vim), and the EMT related transcription factors (Snail and Slug) were downregulated, while the epithelial markers (E-Cad and claudin) were upregulated, and the down-expression of β-catenin in the TNBC cells and mice tumor tissues was observed. The mice model assays exhibited that the Luteolin-treated tumor-burden mice possessed less metastatic colonies in the lungs. Thus, the anti-metastatic behavior of Luteolin on TNBC stemmed from its ability to reverse EMT through the downregulation of β-catenin [77].

2.21. *Triptolide*

Triptolide, diterpene triepoxide, a natural component, is extracted from the traditional Chinese medicinal plant *Tripterygium wilfordii* Hook F. [78]. It was observed that triptolide (10, 25 and 50 nM) lowers cell proliferation and increases the apoptosis rate in the MDA-MB-231 cells dose-dependently. Triptolide treatment downregulated the expression level of β-catenin. The therapeutic efficacy of Triptolide (anti-proliferative and apoptotic potency) on TNBC cells was applied via interfering with the Wnt/β-catenin signaling [79].

2.22. *Astragalus Polysaccharide*

Astragalus polysaccharide, a water-soluble hetero polysaccharide, consisting of α-1,4-(1,6)-glucan, rhamnus-galacturonic acid polysaccharide I, arabinogalactan polysaccharide and arabinogalactan protein polysaccharide. It is an essential bioingredient isolated from the roots of *Astragalus membranaceus* L., a traditional Chinese remedy [80,81]. The anti-proliferative property of *Astragalus* polysaccharide on human triple negative breast cancer cells (MDA-MB-231) was evaluated and evidenced by MTT and Ki67 immunofluorescence staining assays. Furthermore, it was found that Astragalus polysaccharide therapy prohibited the migration and invasion of these cells. These findings were consistent with RT-qPCR, Western blot and immunofluorescence staining analysis, as they confirmed the impediment of EMT via the downregulation of the mesenchymal marker Vim and the EMT-related transcription factor Snail, and the upregulation of E-Cad. Following Astragalus polysaccharide exposure, the expression of β-catenin and its downstream oncoproteins c-Myc and Cyclin D1 were reduced. The repressive potential of this hetero-polysaccharide on metastasis was exerted via altering the expression of the Wnt/β-catenin- and EMT-related genes. Moreover, it was demonstrated that lithium chloride (LiCl), an agonist of the Wnt/β-catenin pathway, inverted the suppressive effect of Astragalus polysaccharide on the Wnt/β-catenin pathway [81].

2.23. *Quercetin*

Quercetin (3,3′,4′,5,7-pentahydroxyflavone) is a yellow pigment bioflavonoid contained in numerous fruits, vegetables, seeds, nuts, green tea and red wine [82,83]. Srinivasan's group investigated the effect of this flavonoid on human triple negative breast cancer cells and showed that quercetin reduced the survival rate, proliferation, migration and invasion of the MDA-MB-231 and MDA-MB-468 cells. Co-treatment of Doxorubicin amplified the anti-migration potency of Quercetin. It induced a phenotype transformation in human TNBC cells from a fibroblast spindle-like form to a cobblestone epithelial shape. Accordingly, Vim protein expression was reduced and E-Cad protein and mRNA expressions were enhanced. The Quercetin exposure of MDA-MB-231 localized the β-catenin in the cytoplasm and reduced the expressions of Cyclin D1, C-Myc and P-AKT [83]. It is reported that the abnormal aggregation and nuclear localization of β-catenin is related to tumorigenic activation [84]. It could be assumed that the anti-proliferative and anti-metastatic properties of Quercetin results from inhibiting Wnt/β-catenin signaling via impeding the presence of β-catenin in the nucleus and downregulation of its downstream target genes.

2.24. *Quinacrine*

Quinacrine (Atabrine, Mepacrine, 4-*N*-(6-chloro-2-methoxyacridin-9-yl)-1-*N*,1-*N*-diethylpentane-1, 4-diamine), is a quinine derivative extracted from the bark of the cinchona tree [85]. Preet et al.'s survey on human TNBC cells demonstrated that Quinacrine attenuated the activity of the Wnt transcription factor TCF/LEF, diminished the expressions of β-catenin and Cyclin D1 and elevated the expression of adenomatous polyposis coli (APC). It reduced survival and proliferation, and induced DNA damage and apoptosis in the MDA-MB-231 cells. Moreover, it was found that the knock-down of APC by siRNA might block the effects of Quinacrine, suggesting a correlation between Quinacrine function and APC. It was proved that Quinacrine inhibited the topoisomerase activity. The research team demonstrated that Quinacrine elevated the APC expression via suppression of topoisomerase activity, eventually repressing WNT/TCF singling in TNBC [86,87].

Lycopene, the main carotenoid in tomatoes, synergistically reinforced the anti-proliferative capability of Quinacrine. It is worth noting that lycopene, quinacrine or their combination possessed considerably less toxicity on MCF-10A (normal epithelial cells). The treatment of cells with Quinacrine followed by Lycopene, decreased the TCF/LEF activity compared to only Quinacrine therapy. The co-treatment with Quinacrine and Lycopene upregulated the APC expression while it downregulated β-catenin and Cyclin D1. Lycopene improved the anti-proliferation potential of Quinacrine by prohibiting the WNT/TCF signaling in the MDA-MB-231 cells [87].

2.25. Curcumin

Curcumin (1,7-bis (4-hydroxy-3-methoxyphenyl)-1,6-heptadiene-3,5-dione) is a yellow-colored natural hydrophobic polyphenol occurring in the rhizome of the *Curcuma longa* L. [88]. Prasad's research team examined the growth repressive activity of curcumin on the MDA-MB-231 cells. They exhibited that Curcumin inhibited the proliferation and promoted apoptosis by arresting cell cycle at the G2/M phase in human TNBC cells. In addition, the Curcumin therapy attenuated the protein expression of the Wnt/β-catenin components Disheveled, β-catenin, Cyclin D1 and Slug, with negligible modification in the GSK3β and E-Cad levels after 12 h. The immunofluorescence analysis supported the Western blot results, since following the Curcumin treatment of the MDA-MB-231 cells, a decrease in the cytoplasmic and nuclear expressions of Disheveled and a reduction in the nuclear expression of β-catenin, Cyclin D1 and Slug were reported [89].

In another piece of research on SUM159 cells, it was displayed that the Curcumin treatment reduced the tumor sphere formation and downregulated the BCSCS markers (CD44, ALDH1A1, Nanog and Oct4), cell-proliferation proteins (PCNA and Cyclin D1) and the anti-apoptotic protein (Bcl2), and upregulated the apoptotic proteins (Bax, Caspase8, Caspase9 and cleaved Caspase3) [90]. It is worth noting that β-catenin activates TCF4/LEF, which may promote the transcription of p53 and c-Myc. C-Myc overexpressed p14ARF, fas, Trail, fasR and DR4/5. FasR and DR4/5 initiate the apoptotic extrinsic pathway by the attaching of their respective ligands. This leads to the autoactivation of caspases-8 and -10, which in turn stimulate the catalytic activation of the effector caspase-3. Another goal of caspase-8 is the pro-apoptotic protein Bid, which is hydrolyzed to tBid, inducing Bax oligomerization and mitochondrial depolarization with a release of cyt c. In addition to the activation of caspase-9, these events amplify the apoptotic pathway [32]. Curcumin repressed Wnt/β-catenin through the downregulation of p-GSK3β, β-catenin and c-Myc. The effect of Curcumin on Wnt/β-catenin was assessed by applying LiCl, which activated the Wnt/β-catenin pathway by blocking GSK3β. It reduced the restrictive effects of Curcumin on tumor sphere formation and the expression of the BCSCs markers. Furthermore, the upregulation of β-catenin in the SUM159 cells by transfection with the control vector and β-catenin plasmids abrogated the Curcumin-induced CD44 and Nanog down-expression [90]. Consistent with the above findings, a survey on normal breast tissue reported that Curcumin could serve as a potent inhibitor of breast cancer. It prohibited stem cell self-renewal through the suppression of Wnt signaling. The deregulation and acquisition of self-renewal potency in stem cells may interfere with carcinogenesis [91].

2.26. Crocin

Crocin, 8, 8-diapocarotene-8, 8-dioic acid, is the essential water-soluble carotenoid extracted from the dried stigmas of *Crocus sativus* L. (saffron) [92,93]. Arzi et al. surveyed the anti-metastatic effect of Crocin on 4T1 cells, establishing the anti-proliferative, anti-migration and anti-invasion properties of Crocin on TNBC cells. In another study, our team demonstrated that the Crocin-treated tumor-burden mice had more body weight, higher survival chances, smaller tumor sizes and less metastatic colonies in their livers and lungs compared to the control animals. It should be noted that Crocin generated no toxicity on BALB/c mice, since no apparent alteration in biochemical markers and body weight was observed. In addition, this carotenoid was reported to repress Wnt signaling via the down-expression of Wnt target genes, Fzd7, Nedd9 and Vim and the VEGF-α genes in 4T1 cells and mice tumor and lung tissues [94,95].

In a complementary piece of research, our group investigated the anti-metastatic efficiency of the combination of Crocin and Crocetin, another major carotenoid of saffron, on the cell and mice models of TNBC. It was shown that the combination therapy of 4T1 cells reduced proliferation, migration, invasion and adhesion to ECM, even more effectively than Crocin and/or Crocetin. The xenograft mice assays confirmed these outcomes, as the combination-treated mice possessed less metastatic deposits in the liver and lung tissues. The Western blot and real-time PCR findings exhibited that the combination therapy reduced the expressions of Fzd7, Nedd9, VEGF-α, Vim and Mmp9 and upregulated E-Cad in 4T1 cells and liver and tumor tissues, thus preventing Wnt/β-catenin signaling and averting metastasis in TNBC [96].

The chemical structures of the natural derivatives combating TNBC through interfering with the Wnt/β-catenin pathway are demonstrated in Figure 1.

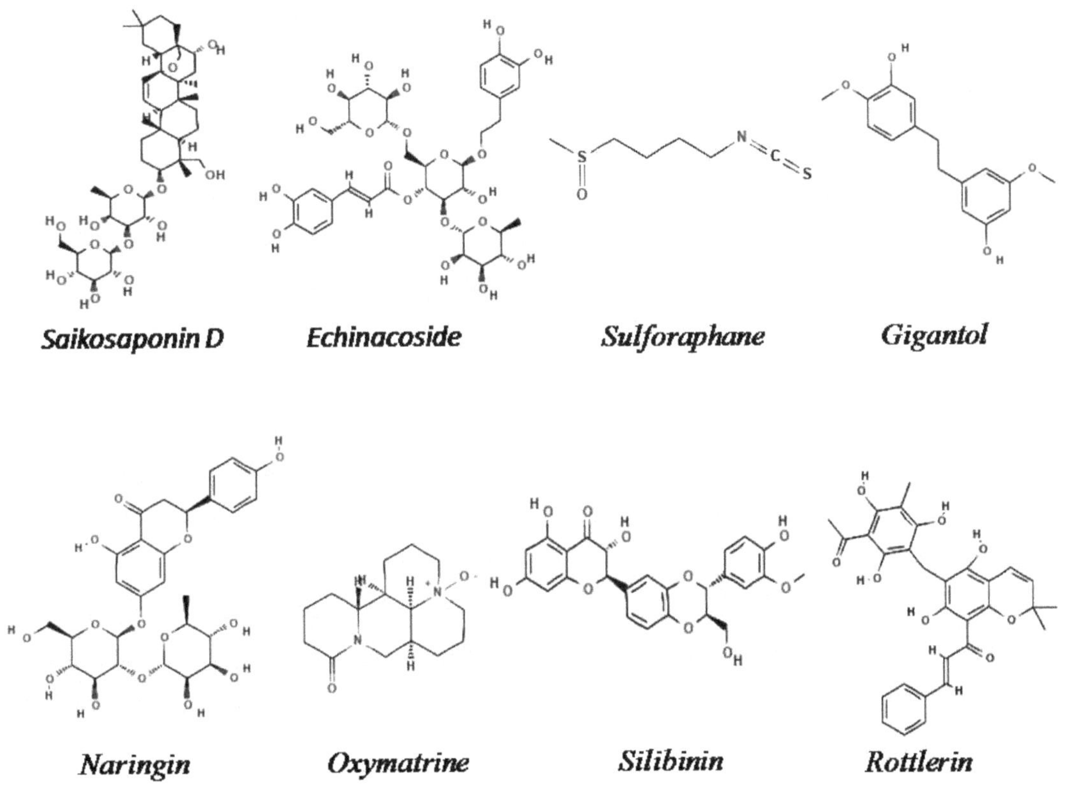

Figure 1. *Cont.*

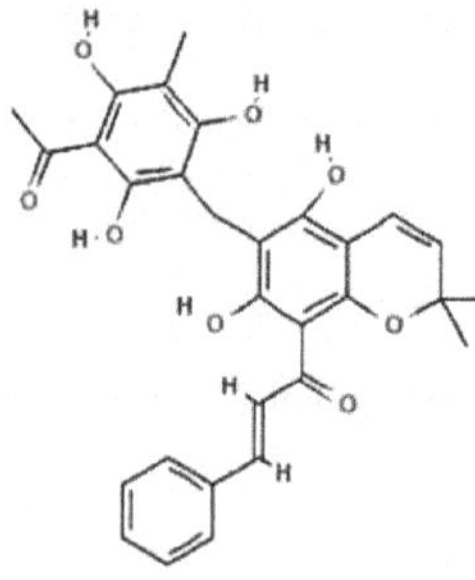

Baicalin

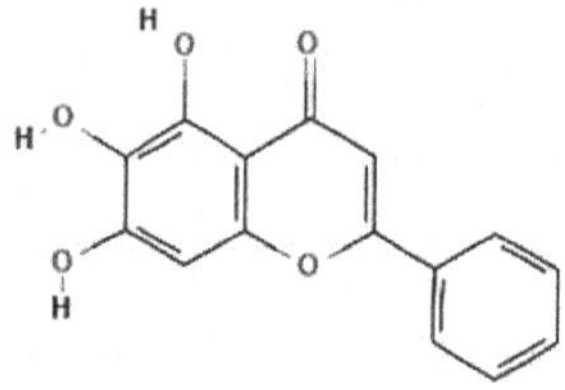

Baicalein

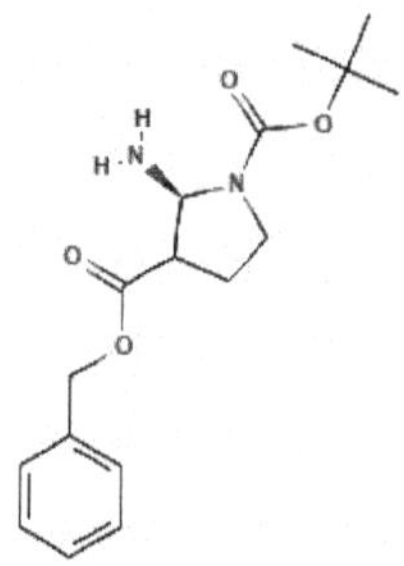

Epigallocatechin Gallate

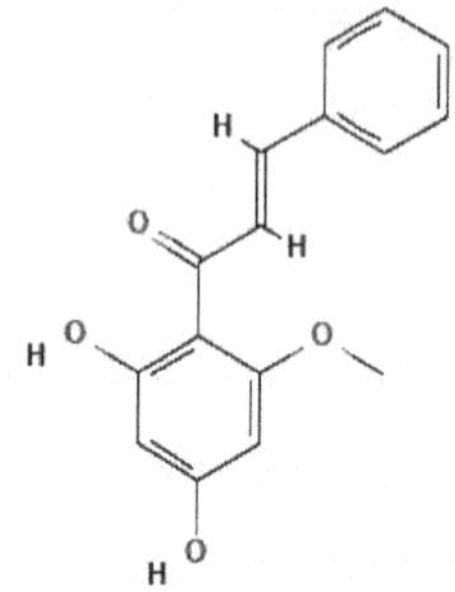

Cardamonin

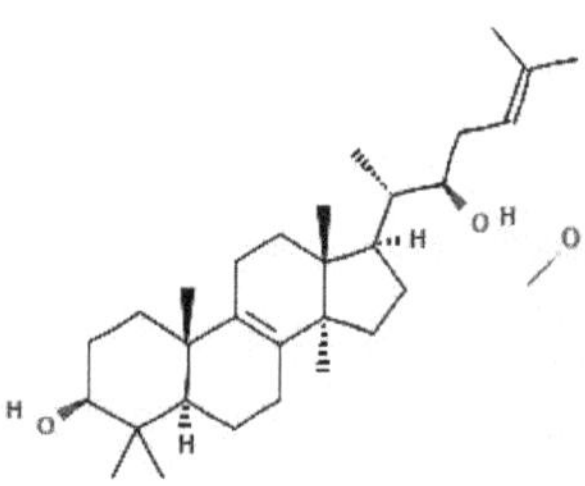

Inotodiol

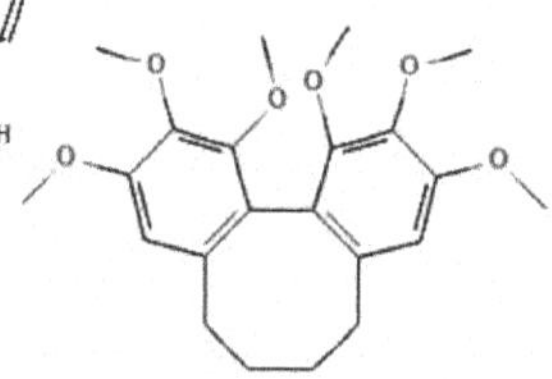

Schisandrin A

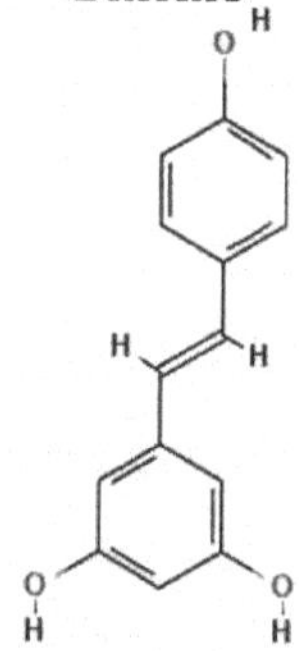

Resveratrol

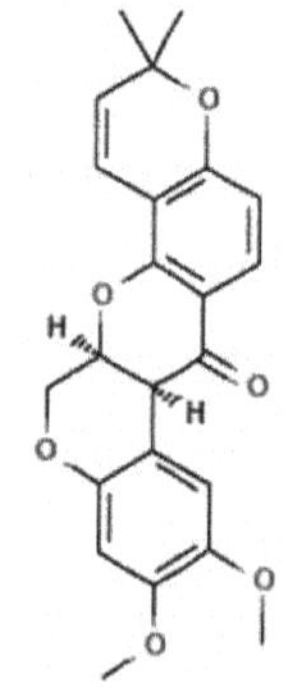

Deguelin

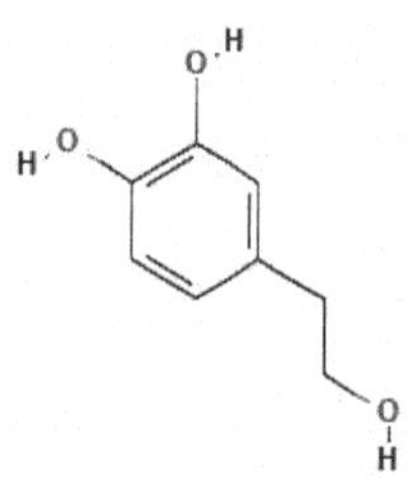

Hydroxytyrosol

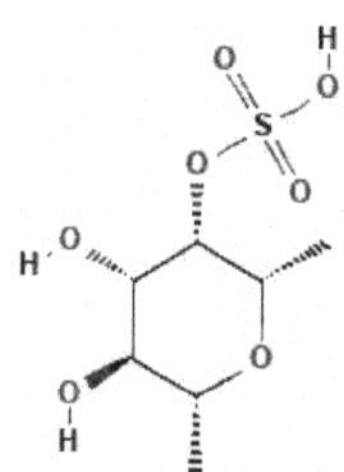

Fucoidan

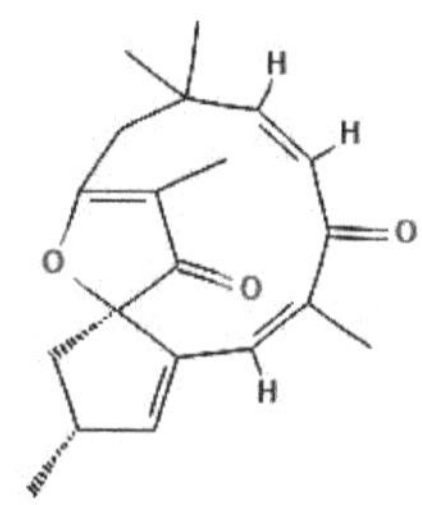

Jatrophone

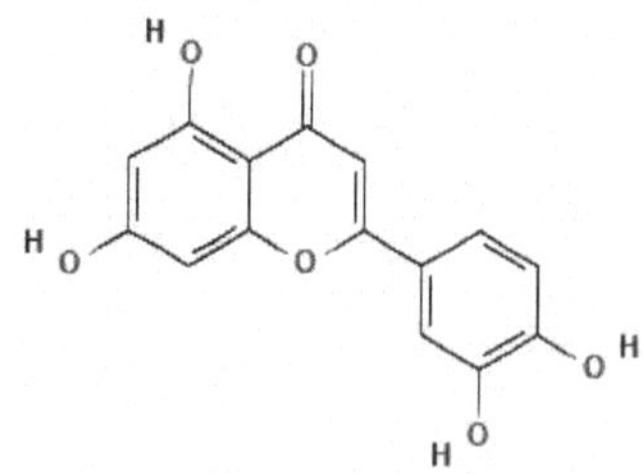

Luteolin

Figure 1. *Cont.*

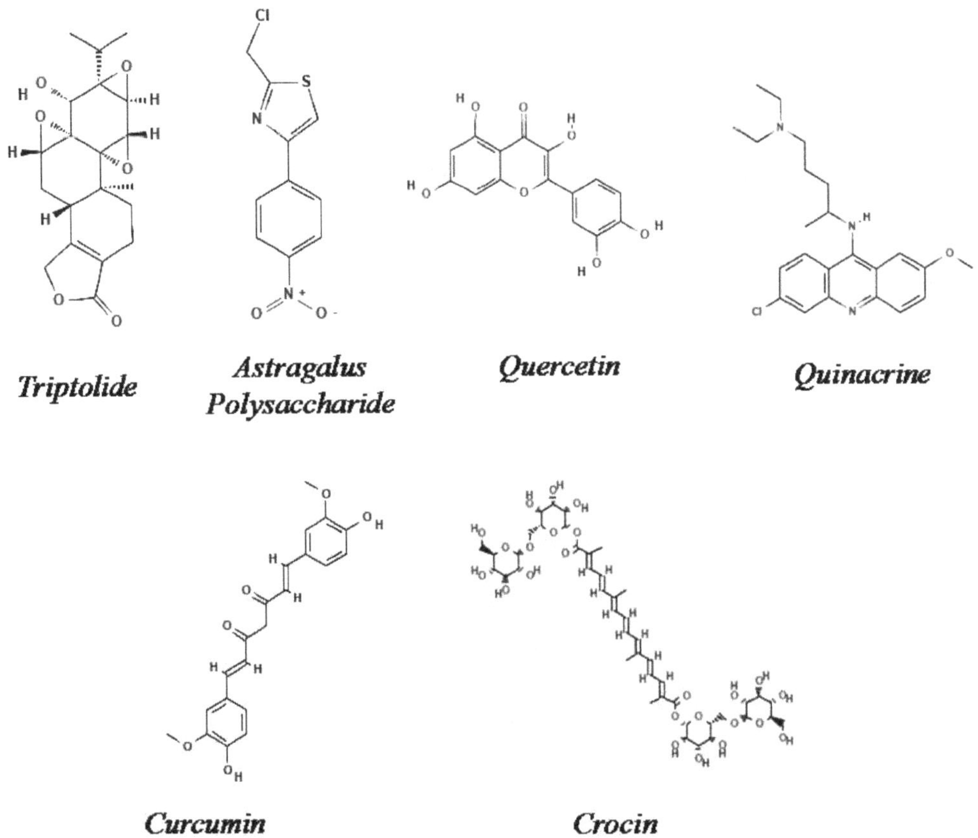

Figure 1. Chemical structure of natural derivatives combating TNBC through interfering with the Wnt/β-catenin pathway (Pubchem).

3. Natural Pill or Extracts Targeting TNBC via Impeding the Wnt/β-Catenin Pathway

3.1. Liuwei Dihuang Pill

The Liuwei Dihuang pill (LWDHP) is a classic Chinese herbal medicine containing six herbs: *Rehmannia glutinosa* (Gaetn.) Libosch. ex Fisch. et Mey (Shudi Huang); *Dioscorea opposita* Thunb (Shan Yao); *Cornus officinalis Sieb.* et Zucc (Shanzhu Yu); *Poriacocos* (Schw.) Wolf (Fu Ling); *Alisma orientalis* (Sam) Juzep (ZeXie) and *Paeonia suffruticosa* Andr (Mudan Pi) [97]. It has been prescribed for the prevention and treatment of breast cancer, particularly TNBC [98,99]. Lixiang et al.'s research showed that LWDHP therapy restricted the tumor volume and elevated the survival rate of spontaneous breast carcinoma mice [100]. Moreover, in 2019, they confirmed that the LWDHP reduced the cancer tissue size and weight, increased the cancer inhibitory rate, prolonged the survival time and prohibited lung or liver metastasis in TNBC-bearing mice compared to the control mice. Their results suggested that the LWDHP modulates the deregulation of the Wnt signaling pathway since, β-catenin, CyclinD1, TCF-1 and VEGF expressions decreased. However, high doses of LWDHP augmented the expression of Axin-2. In addition, this study confirmed that LWDHP disrupted the β-catenin/TCF-1 interaction in the nuclei. It should be noted that the activation of the β-catenin/TCF-1 complexes in Wnt signaling promoted the transcription of the downstream genes e.g., VEGF and CyclinD1, and elevated cell proliferation [101].

3.2. *Syzygium guineense*

Syzygium guineense Wall., a member of the *Myrtaceae* family, is a widespread plant native to the regions of Australia, Asia and Africa. It has been prescribed as a medicinal plant due to its anti-tumor activity against TNBC cells [102,103]. Koval et al. reported that Wnt3a is strikingly overexpressed in TNBC. They demonstrated that the *S. guineense* extract therapy of malignant cells lead to an inhibition of cell proliferation through blocking the Wnt3a-induced β-catenin stabilization and suppression of the Wnt-dependent transcription. It is also noteworthy that phyto-analytical studies have introduced tannins, a group of polyphenols, as an effective inhibitor of Wnt3a-stimulated signaling and TNBC cell proliferation [103].

3.3. *Ganoderma lucidum*

Ganoderma lucidum (Reishi or Lingzhi), the most common species in the genus *Ganoderma*, is a famous Chinese traditional medicinal plant, which has found novel applications as a cancer adjunct-therapy agent. The phytochemical research elucidated that the antitumor property of Ganoderma extract is due to its constituent triterpenes [104]. The studies on SUM-149 and mice models have provided evidence of Reishi's therapeutic potential for breast cancer treatment [105]. Zhang et al., in 2017, elucidated its anti-cancer molecular mechanism, indicating that Reishi diminished cell viability and migration via suppressing the phosphorylation of the Wnt co-receptor LRP6, leading to blocking the Wnt/β-catenin signaling pathway in human and mouse breast cancer cell lines [106].

4. Conclusions

Natural compounds are proposed to compensate for the undesirable effects of chemotherapeutic drugs. Frequently, the bio-accessibility and bioavailability of natural agents, and their nontoxicity on normal cells and anti-tumor/anti-metastatic potentials nominates natural agents as propitious, multipotent chemopreventive and chemotherapeutic remedies. Various surveys on cells and animal models of TNBC revealed the suppressive efficacy of herbal medicine extracts and bioactive compounds including: flavonoids; terpenes; carotenoids; alkaloids; retinoids and lignans on various molecules of the Wnt/β-catenin signaling pathway, WNT, FZD, LRP, GSK3β, APC, Dsh, β-catenin and TCF/LEF. These phytotherapy agents exerted anti-proliferation, proapoptotic, anti-migration and anti-invasion effects and the inhibition of BCSC self-renewal via modulation of the Wnt downstream genes (c-Myc, Cyclin D1, Survivin, Axin2, and EMT-related genes). The administration of a number of these herbal medications simultaneously repressed Wnt and mTORC1, ER stress and SATB1 signaling (Table 1). The co-administration of natural compounds and common chemotherapy agents ameliorates the function of chemotherapeutics. However, it should be considered that the combination therapy may have adverse consequences through pharmacodynamic and pharmacokinetic herb–chemotherapeutic interactions. Therefore, an outlook and prospect for the future would be the evaluation of natural bioactive compounds in clinical trials. In the absence of efficient approved targeted therapies, these findings open a new avenue on the role of herbal extracts and derivatives as an auspicious medication for the treatment of TNBC sufferers.

Table 1. Anticancer potentials of active phyto-therapeutic ingredients against triple negative breast cancer.

Active Pharmaceutical Ingredient	Plant	Model of Study	Dose	Targeted Genes	Impact on TNBC	Reference
Saikosaponin D	*Radix Bupleuri*	HCC1937 MDA-MB-468 MDA-MB-231	10, 15 and 20 μM	*β-Catenin*↓ *Cyclin D1*↓ *c-Myc*↓	Proliferation↓ Apoptosis↑	[23]
Echinacoside	*Cistanche* and *Echinacea*	MDA-MB-468 MDA-MB-231	25–100μM	*p-LRP6*↓ *total LRP6*↓ *p-Dvl2*↓ *active β-Catenin*↓ *total β-Catenin*↓ *CD44*↓ *LEF1*↓ *Cyclin D*↓	Proliferation↓ Migration↓ Invasion↓ Tumor sizes↓ Tumor weights↓	[25]
		Nude mice	10 mg/kg			
Sulforaphane	*Brassica*	SUM159	0.5, 1 and 5 μM	*β-Catenin*↓ *Cyclin D1*↓	Proliferation↓ Apoptosis↑ Mammosphere formation↓ BCSC activity↓ Tumor sizes↓	[27]
		NOD/SCID mice	50 mg/kg			
Gigantol	*Orchidaceae*	MDA-MB-231 MDA-MB-468	0–100 μM	*p-LRP6,*↓ *total LRP6*↓ *Cytosolic β-catenin*↓ *Axin2*↓ *Survivin*↓	Proliferation↓ Migration↓	[29]
Naringin	*Drynaria fortunei Citrus aurantium Citrus medica and* citrus fruit	MDA-MB-231 MDA-MB-468 BT-549	0–200 μM	*Survivin*↓ *P21*↓ *β-Catenin* ↓ *Cyclin E*↓ *Rb* ↓ *p-Rb*↓	Proliferation↓ Apoptosis↑ Arrest cell cycle in G1 Tumor sizes↓ Tumor weights↓	[31]
		SCID hairless mice	100 mg/kg			
Oxymatrine	*Sophora Flavescens* Ait	MDA-MB-231 MDA-MB-468	0, 1, 2 and 4 mM Bevacizumab 200 nM	*N-Cad*↓ *Vim*↓ *ZEB*↓ *Snail*↓ *Slug*↓ *E-Cad*↑ *β-Catenin*↓ *c-Myc*↓ *Cyclin D1*↓ *CD44*↓ *VegfA*↓	Migration↓ Invasion↓ EMT↓ BCSC self-renewal↓ Tumor size↓ Angiogenesis↓	[34]
		BALB/c nude mice	25 mg/kg + 5 mg/kg Bevacizumab			
Silibinin	*Silybum marianum.* Gaertn	MDA-MB-231	0–200 μM	*LRP6*↓ *p-LRP6*↓ *Axin2*↓	Proliferation↓	[38]

Table 1. *Cont.*

Active Pharmaceutical Ingredient	Plant	Model of Study	Dose	Targeted Genes	Impact on TNBC	Reference
Rottlerin	*Mallotus philippinensis*	MDA-MB-231	0.1–31.6 μM	*cytosolicβ-catenin*↓ *totalβ-Catenin*↓ *LRP6*↓ *p–LRP6*↓ *Axin2*↓ *P70-S6K*↓ *p-P70-S6K*↓ *S6*↓ *p-S6*↓ *CyclinD1*↓ *Survivin*↓	Proliferation↓	[41]
Baicalin	*Scutellaria baicalensis* Georgi	MDA-MB-231 4T1	10, 30 and 100 μM	*β-Catenin*↓ *E-Cad*↑ *Claudin*↑ *N-Cad*↓ *VIM*↓ *Snail*↓ *Slug*↓	Proliferation↓ Migration↓ Invasion↓ EMT↓ Metastatic colonies in liver and lung↓	[44]
		BALB/c mice	100 mg/kg			
Baicalein	*Scutellaria baicalensis* Georgi	MDA-MB-231	10, 20, and 40 μM/L	*E-Cad*↑ *Vim*↓ *Snail*↓ *Wnt1*↓ *β-Catenin*↓ *SATB1*↓ *Cyclin D1*↓ *Axin2*↓	Proliferation↓ Migration↓ Invasion↓ EMT↓ Metastatic colonies in liver and lung↓	[46]
		BALB/c nude mice	50 or 100 mg/kg			
Epigallocatechin gallate	Green tea	MDA-MB-231	25, 50, 75, 100 and 200 μM	*β-Catenin*↓ *p-AKT*↓ *Cyclin D1*↓	Proliferation↓	[48]
		MDA-MB-231	25–100 μM	*HBP1*↓ *β-Catenin*↓ *c-Myc*↓	Proliferation↓ Migration↓ Invasion↓	[50]
Cardamonin	*Zingiberaceae*	MDA-MB-231 BT-549	0–100 μM	*Bax*↑ *Bcl-2*↓ *Cyt-C*↑ *Caspase-3*↑ *PARP*↑ *β-Catenin*↓ *E Cad*↑ *N-Cad*↓ *Vim*↓ *Slug*↓ *Snail*↓ *Cyclin D1*↓ *c-Myc*↓ *Vegf*↓ *CDK4*↓	Proliferation↓ Migration↓ Invasion↓ EMT↓ Apoptosis↑ Arrest cell cycle in G2/M Tumor size↓	[53]
		BALB/c mice	2.5 and 5 mg/kg			

Table 1. *Cont.*

Active Pharmaceutical Ingredient	Plant	Model of Study	Dose	Targeted Genes	Impact on TNBC	Reference
Inotodiol	*Inonotus obliquus*	Sprague-Dawley rats	10 mg/kg	*PCNA*↓ *β-Catenin*↓ *c-Myc*↓ *Cyclin D1*↓ *Caspase-3*↑ *PARP*↑	Proliferation↓ Apoptosis↑ Body weight↑ Antioxidant enzyme activities↑ Abnormal histological signs of pancreas↓ Glucose↓ Cholesterol↓ Triglyceride↓ HDL↓	[56]
Schisandrin	*Schisandra chinensis* Baill	MDA-MB-231 BT-549	25,50 and 100 μM	*ATF4*↑ *CHOP* ↑ *p-Elf2α*↑ *β-Catenin*↓ *p-GSK3β*↓ *Bax*↑ *P53*↑ *Bcl2*↓	Proliferation↓ Arrest cell cycle in G1 Apoptosis↑ ER stress↑ Tumor size↓	[60]
		BALB/c mice	25 mg/kg			
Resveratrol	Grapes, Berries and Peanuts	SUM159	10, 20 and 40μM	*LC3-II*↑ *Beclin1*↑ *Atg 7*↑ *β-Catenin*↓ *Cyclin D1*↓	Proliferation↓ BCSC activity↓ BCSC self-renewal↓ Autophagy↑ Tumor size↓	[64]
		NOD/SCID mice	100 mg/kg			
Deguelin	*Mundulea sericea*	MDA-MB-231	0.1–10 μM/L	*WIF-1*↑ *DDK4*↑ *CDH3*↑ *CDH7*↑ *CDH9*↑ *Wnt14*↓ *Wnt 2B*↓ *Wnt 3*↓ *Snail*↓ *p-GSK-3β*↓ *β-Catenin*↓ *Cyclin D1*↓	Proliferation↓ Apoptosis↑ Arrest cell cycle in S	[67]
Hydroxytyrosol	Olive oil	SUM159PT BT549 MDA-MB-231 Hs578T	0–100 μM	*Zeb*↓ *Slug*↓ *Vim*↓ *Zo-1*↑ *p-LRP6*↓ *LRP6*↓ *β-Catenin*↓ *Cyclin D1*↓	BCSC self-renewal↓ Migration↓ Invasion↓ EMT↓ BCSC activity↓	[69]
Fucoidan	Brown seaweed and Marine invertebrates	4T1	50, 100 and 200 μg/mL	*β-Catenin*↓ *c-Myc*↓ *Cyclin D1*↓ *Survivin*↓ *TCF/LEF*↓	Proliferation↓ Apoptosis↑ Arrest cell cycle in G1 Tumor size↓ Tumor weight↓	[72]
		BALB/c mice	5, 10 mg/kg			

Table 1. *Cont.*

Active Pharmaceutical Ingredient	Plant	Model of Study	Dose	Targeted Genes	Impact on TNBC	Reference
Jatrophone	Euphorbiaceae	MDA-MB-231 MDA-MB-157 HCC38 MDA-MB-468 Patient-derived xenograft	100 nM-30 μM	*BIRC5*↓ *Axin2*↓ *HMGA2*↓ *Myc*↓ *PCNA*↓ *CCND1*↓ *Cyclin D1*↓ *β-Catenin*↓ *Slug*↓ *Fibronectin*↓ *Vim*↓	Proliferation↓ Apoptosis↑ Arrest cell cycle in S-phase EMT↓ Migration↓	[75]
Luteolin	Celery Sweet bell peppers, Carrots Broccoli Onion leaves Parsley	MDA-MB-231 BT5–49	10, 30 and 100 μM	*N-Cad*↓ *Vim*↓ *Snail*↓ *Slug*↓ *E-Cad*↑ *Claudin*↑ *β-Catenin*↓	Migration↓ Invasion↓ EMT↓ Metastatic colonies↓	[77]
		Nude mice	100 mg/kg			
Triptolide	*Tripterygium wilfordii Hook F*	MDA-MB-231	10, 25 and 50 nM	*β-Catenin*↓	Proliferation↓ Apoptosis↑	[79]
Astragalus polysaccharide	*Astragalus membranaceus*	MDA-MB-231	25, 50, 100, 200, 400, 800 and 1600 μg/mL	*Vim*↓ *Snail*↓ *E-Cad*↑ *β-Catenin*↓ *c-Myc*↓ *Cycline D1*↓	Proliferation↓ Migration↓ Invasion↓ EMT↓	[81]
Quercetin	Fruits, Vegetables, Seeds, Nuts, Green tea, and Red wine	MDA-MB-231 MDA-MB-468	10 and 50 μM	*Vim*↓ *E-Cad*↑ *Cyclin D1*↓ *c-Myc*↓ *p-AKT*↓	Proliferation↓ Survival rate↓ Migration↓ Invasion↓ Reshape mesenchymal to epithelial shape	[83]
Quinacrine	Cinchona tree	MDA-MB-231	5, 10, 15 and 20 μM	*β-Catenin*↓ *Cyclin D1*↓ *APC*↑	Proliferation↓ Survival rate↓ Apoptosis↑ DNA damage↑ Topoisomerase activity↓	[86,87]
Quinacrine + Lycopen	Cinchona tree Tomato		5 μM + 2, 4, 8, 10, 12, 14 and 16 μM	*β-Catenin*↓ *Cyclin D1*↓ *APC*↑	Proliferation↓ Survival ↓	

Table 1. *Cont.*

Active Pharmaceutical Ingredient	Plant	Model of Study	Dose	Targeted Genes	Impact on TNBC	Reference
Curcumin	*Curcuma longa*	MDA-MB-231	20 µM	*Dsh*↓ *β-Catenin*↓, *Cyclin D1*↓ *Slug*↓	Proliferation↓ Apoptosis↑ Arrest cell cycle in G2/M	[89]
		SUM159	10, 20 and 40 µM	*CD44*↓ *ALDH1A1*↓ *Nanog*↓ *OCT4*↓ *PCNA*↓ *Cyclin D1*↓ *Bcl2*↓ *Bax*↑. *Caspase 8*↑ *Caspase9*↑ *cleaved Caspase 3*↑ *p-GSK3β*↓ *β-Catenin*↓ *c-Myc*↓	Proliferation↓ Apoptosis↑ Mammosphere formation↓ BCSC activity↓	[90]
Crocin	*Crocus sativus*	4T1	2.5 and 3 mM	*FzD7*↓ *Nedd9*↓ *Vegf-α*↓ *Mmp9*↓ *Vim*↓ *E-Cad*↑	Proliferation↓ Invasion↓ Migration↓ Cell–ECM adhesion↓ Tumor size↓ Metastatic colonies↓	[94]
		BALB/c mice	200 mg/kg			[95]
Crocin + Crocetin	*Crocus sativus*	4T1	Crocin 2.5 mM + Crocetin 0.05 mM & Crocetin 0.1 mM + Crocin 2 mM	*Fzd7*↓ *Nedd 9*↓ *Vegf-α*↓ *Mmp9*↓ *Vim*↓	Proliferation↓ Invasion↓ Migration↓ Tumor size↓ Cell–ECM adhesion↓ Metastatic colonies↓	[96]
		BALB/c mice	Crocin 200 mg/kg + Crocetin 5 mg/kg			
Liuwei Dihuang pill	*Rehmannia glutinosa* *Dioscorea opposita* *Cornus officinalis* *Poria cocos* *Alisma orientalis* *Paeonia suffruticosa*	Kunming mice	2.3, 4.6 and 9.2 g/kg	*β-Catenin*↓ *Cyclin D1*↓ *TCF-1*↓ *Vegf*↓	Tumor sizes↓ Tumor weights↓ Survival time↑ Metastatic colonies in liver and lung↓	[101]

Table 1. *Cont.*

Active Pharmaceutical Ingredient	Plant	Model of Study	Dose	Targeted Genes	Impact on TNBC	Reference
Tannins	*Syzygium guineense* Wall	BT-20 HCC38 MDA-MB-231, HCC1806 HCC1395 MDA-MB-468	0–100 µg/mL	*Wnt3a*↓ *β-Catenin*↓ *LRP6*↓	Proliferation↓	[103]
Ganoderma Lucidum	-	MDA-MB-231 4T1	0–200 µg/mL	*β-Catenin*↓ *p-LRP6*↓ *p-Dvl2*↓ *Axin2*↓	Proliferation ↓ Migration↓	[106]

Author Contributions: L.A. conceived of the idea, performed the literature search, wrote and edited the manuscript, and had the responsibility of supervision; H.M. and R.H. performed the literature search and drafted for the manuscript. All authors have read and agreed to the published version of the manuscript.

Funding: This research did not receive any specific grant from funding agencies in the public, commercial, or not-for-profit sectors.

Institutional Review Board Statement: Not applicable.

Informed Consent Statement: Not applicable.

Data Availability Statement: Not applicable.

Acknowledgments: The author would like to thank Reza Rajaie Khorasani for editing the manuscript.

Conflicts of Interest: The authors declare that the research was conducted in the absence of any commercial or financial relationships that could be construed as a potential conflict of interest.

References

1. Sung, H.; Ferlay, J.; Siegel, R.L.; Laversanne, M.; Soerjomataram, I.; Jemal, A.; Bray, F. Global cancer statistics 2020: GLOBOCAN estimates of incidence and mortality worldwide for 36 cancers in 185 countries. *CA Cancer J. Clin.* **2021**, *71*, 209–249. [CrossRef] [PubMed]
2. Al-Thoubaity, F.K. Molecular classification of breast cancer: A retrospective cohort study. *Ann. Med. Surg.* **2020**, *49*, 44–48. [CrossRef]
3. Yin, L.; Duan, J.J.; Bian, X.W.; Yu, S.C. Triple-negative breast cancer molecular subtyping and treatment progress. *Breast Cancer Res.* **2020**, *22*, 1–13. [CrossRef] [PubMed]
4. Foulkes, W.D.; Smith, I.E.; Reis-Filho, J.S. Triple-negative breast cancer. *N. Engl. J. Med.* **2010**, *363*, 1938–1948. [CrossRef] [PubMed]
5. Jitariu, A.A.; Cîmpean, A.M.; Ribatti, D.; Raica, M. Triple negative breast cancer: The kiss of death. *Oncotarget* **2017**, *8*, 46652. [CrossRef]
6. Fultang, N.; Chakraborty, M.; Peethambaran, B. Regulation of cancer stem cells in triple negative breast cancer. *Cancer Drug Resist.* **2021**, *4*, 321–342. [CrossRef]
7. Pohl, S.G.; Brook, N.; Agostino, M.; Arfuso, F.; Kumar, A.P.; Dharmarajan, A. Wnt signaling in triple-negative breast cancer. *Oncogenesis* **2017**, *6*, e310. [CrossRef]
8. Ghosh, N.; Hossain, U.; Mandal, A.; Sil, P.C. The Wnt signaling pathway: A potential therapeutic target against cancer. *Ann. N. Y. Acad. Sci.* **2019**, *1443*, 54–74. [CrossRef]
9. Harb, J.; Lin, P.J.; Hao, J. Recent development of Wnt signaling pathway inhibitors for cancer therapeutics. *Curr. Oncol. Rep.* **2019**, *21*, 12. [CrossRef]
10. Merikhian, P.; Eisavand, M.R.; Farahmand, L. Triple-negative breast cancer: Understanding Wnt signaling in drug resistance. *Cancer Cell Int.* **2021**, *21*, 1–8. [CrossRef]
11. Komiya, Y.; Habas, R. Wnt signal transduction pathways. *Organogenesis* **2008**, *4*, 68–75. [CrossRef] [PubMed]
12. Pai, S.G.; Carneiro, B.A.; Mota, J.M.; Costa, R.; Leite, C.A.; Barroso-Sousa, R.; Giles, F.J. Wnt/beta-catenin pathway: Modulating anticancer immune response. *J. Hematol. Oncol.* **2017**, *10*, 1–12. [CrossRef]
13. Yin, S.Y.; Wei, W.C.; Jian, F.Y.; Yang, N.S. Therapeutic applications of herbal medicines for cancer patients. *Evid. Based Complementary Altern. Med.* **2013**, *2013*, 302426. [CrossRef] [PubMed]

14. Tavakoli, J.; Miar, S.; Zadehzare, M.M.; Akbari, H. Evaluation of effectiveness of herbal medication in cancer care: A review study. *Iran. J. Cancer Prev.* **2012**, *5*, 144. [PubMed]
15. Ghorbani-Abdi-Saedabad, A.; Hanafi-Bojd, M.Y.; Parsamanesh, N.; Tayarani-Najaran, Z.; Mollaei, H.; Hoshyar, R. Anticancer and apoptotic activities of parthenolide in combination with epirubicin in mda-mb-468 breast cancer cells. *Mol. Biol. Rep.* **2020**, *47*, 5807–5815. [CrossRef]
16. Blagodatski, A.; Sova, V.; Gorovoy, P.; Usov, A.; Katanaev, V. Novel activators and inhibitors of the Wnt signaling pathway from medicinal plants. *Planta Med.* **2014**, *80*, SL29. [CrossRef]
17. Richard, T.S.; Kamdje, A.H.N.; Mukhtar, F. Medicinal plants in breast cancer therapy. *J. Dis. Med. Plants* **2015**, *1*, 19–23.
18. Yang, Z.; Zhang, Q.; Yu, L.; Zhu, J.; Cao, Y.; Gao, X. The signaling pathways and targets of traditional Chinese medicine and natural medicine in triple-negative breast cancer. *J. Ethnopharmacol.* **2021**, *264*, 113249. [CrossRef]
19. Yuan, B.; Yang, R.; Ma, Y.; Zhou, S.; Zhang, X.; Liu, Y. A systematic review of the active saikosaponins and extracts isolated from Radix Bupleuri and their applications. *Pharm. Biol.* **2017**, *55*, 620–635. [CrossRef]
20. Wong, V.K.W.; Zhang, M.M.; Zhou, H.; Lam, K.Y.C.; Chan, P.L.; Law, C.K.M.; Liu, L. Saikosaponin-d enhances the anticancer potency of TNF-via overcoming its undesirable response of activating NF-Kappa B signalling in cancer cells. *Evid. -Based Complementary Altern. Med.* **2013**, *2013*, 745295.
21. Wong, V.K.; Li, T.; Law, B.Y.; Ma, E.D.; Yip, N.C.; Michelangeli, F.; Liu, L. Saikosaponin-d, a novel SERCA inhibitor, induces autophagic cell death in apoptosis-defective cells. *Cell Death Dis.* **2013**, *4*, e720. [CrossRef] [PubMed]
22. Zhong, D.; Zhang, H.J.; Jiang, Y.D.; Wu, P.; Qi, H.; Cai, C.; Dang, Q. Saikosaponin-d: A potential chemotherapeutics in castration resistant prostate cancer by suppressing cancer metastases and cancer stem cell phenotypes. *Biochem. Biophys. Res. Commun.* **2016**, *474*, 722–729. [CrossRef] [PubMed]
23. Wang, J.; Qi, H.; Zhang, X.; Si, W.; Xu, F.; Hou, T.; Liang, X. Saikosaponin D from Radix Bupleuri suppresses triple-negative breast cancer cell growth by targeting β-catenin signaling. *Biomed. Pharmacother.* **2018**, *108*, 724–733. [CrossRef] [PubMed]
24. Liu, J.; Yang, L.; Dong, Y.; Zhang, B.; Ma, X. Echinacoside, an inestimable natural product in treatment of neurological and other disorders. *Molecules* **2018**, *23*, 1213. [CrossRef]
25. Tang, C.; Gong, L.; Qiu, K.; Zhang, Z.; Wan, L. Echinacoside inhibits breast cancer cells by suppressing the Wnt/β-catenin signaling pathway. *Biochem. Biophys. Res. Commun.* **2020**, *526*, 170–175. [CrossRef] [PubMed]
26. Houghton, C.A. Sulforaphane: Its "coming of age" as a clinically relevant nutraceutical in the prevention and treatment of chronic disease. *Oxidative Med. Cell. Longev.* **2019**, *2019*, 2716870. [CrossRef]
27. Li, Y.; Zhang, T.; Korkaya, H.; Liu, S.; Lee, H.F.; Newman, B.; Sun, D. Sulforaphane, a dietary component of broccoli/broccoli sprouts, inhibits breast cancer stem cells. *Clin. Cancer Res.* **2010**, *16*, 2580–2590. [CrossRef]
28. Wu, J.; Lu, C.; Li, X.; Fang, H.; Wan, W.; Yang, Q.; Wei, X. Synthesis and biological evaluation of novel gigantol derivatives as potential agents in prevention of diabetic cataract. *PLoS ONE* **2015**, *10*, e0141092. [CrossRef]
29. Yu, S.; Wang, Z.; Su, Z.; Song, J.; Zhou, L.; Sun, Q.; Lu, D. Gigantol inhibits Wnt/β-catenin signaling and exhibits anticancer activity in breast cancer cells. *BMC Complementary Altern. Med.* **2018**, *18*, 1–8. [CrossRef]
30. Chen, R.; Qi, Q.L.; Wang, M.T.; Li, Q.Y. Therapeutic potential of naringin: An overview. *Pharm. Biol.* **2016**, *54*, 3203–3210. [CrossRef]
31. Li, H.; Yang, B.; Huang, J.; Xiang, T.; Yin, X.; Wan, J.; Ren, G. Naringin inhibits growth potential of human triple-negative breast cancer cells by targeting β-catenin signaling pathway. *Toxicol. Lett.* **2013**, *220*, 219–228. [CrossRef] [PubMed]
32. Trejo-Solis, C.; Escamilla-Ramirez, A.; Jimenez-Farfan, D.; Castillo-Rodriguez, R.A.; Flores-Najera, A.; Cruz-Salgado, A. Crosstalk of the Wnt/β-catenin signaling pathway in the induction of apoptosis on cancer cells. *Pharmaceuticals* **2021**, *14*, 871. [CrossRef] [PubMed]
33. Lu, M.L.; Xiang, X.H.; Xia, S.H. Potential signaling pathways involved in the clinical application of oxymatrine. *Phytother. Res.* **2016**, *30*, 1104–1112. [CrossRef] [PubMed]
34. Xie, W.; Zhang, Y.; Zhang, S.; Wang, F.; Zhang, K.; Huang, Y.; Wang, J. Oxymatrine enhanced anti-tumor effects of Bevacizumab against triple-negative breast cancer via abating Wnt/β-Catenin signaling pathway. *Am. J. Cancer Res.* **2019**, *9*, 1796.
35. Bear, H.D.; Tang, G.; Rastogi, P.; Geyer Jr, C.E.; Robidoux, A.; Atkins, J.N.; Wolmark, N. Bevacizumab added to neoadjuvant chemotherapy for breast cancer. *N. Engl. J. Med.* **2012**, *366*, 310–320. [CrossRef] [PubMed]
36. Sun, Z.; Lan, X.; Xu, S.; Li, S.; Xi, Y. Efficacy of bevacizumab combined with chemotherapy in the treatment of HER2-negative metastatic breast cancer: A network meta-analysis. *BMC Cancer* **2020**, *20*, 1–17. [CrossRef]
37. Delmas, D.; Xiao, J.; Vejux, A.; Aires, V. Silymarin and cancer: A dual strategy in both in chemoprevention and chemosensitivity. *Molecules* **2020**, *25*, 2009. [CrossRef]
38. Lu, W.; Lin, C.; King, T.D.; Chen, H.; Reynolds, R.C.; Li, Y. Silibinin inhibits Wnt/β-catenin signaling by suppressing Wnt co-receptor LRP6 expression in human prostate and breast cancer cells. *Cell. Signal.* **2012**, *24*, 2291–2296. [CrossRef]
39. King, T.D.; Suto, M.J.; Li, Y. The wnt/β-catenin signaling pathway: A potential therapeutic target in the treatment of triple negative breast cancer. *J. Cell. Biochem.* **2012**, *113*, 13–18. [CrossRef]
40. Ma, J.; Hou, Y.; Xia, J.; Zhu, X.; Wang, Z.P. Tumor suppressive role of rottlerin in cancer therapy. *Am. J. Transl. Res.* **2018**, *10*, 3345.
41. Lu, W.; Lin, C.; Li, Y. Rottlerin induces Wnt co-receptor LRP6 degradation and suppresses both Wnt/β-catenin and mTORC1 signaling in prostate and breast cancer cells. *Cell. Signal.* **2014**, *26*, 1303–1309. [CrossRef] [PubMed]

42. Morris, S.A.L.; Huang, S. Crosstalk of the Wnt/b-catenin pathway with other pathways in cancer cells. *Genes Dis.* **2015**, *3*, 41–47. [CrossRef] [PubMed]
43. Li, K.; Liang, Y.; Cheng, A.; Wang, Q.; Li, Y.; Wei, H.; Wan, X. Antiviral properties of baicalin: A concise review. *Rev. Bras. De Farmacogn.* **2021**, *31*, 408–419. [CrossRef] [PubMed]
44. Zhou, T.; Zhang, A.; Kuang, G.; Gong, X.; Jiang, R.; Lin, D.; Li, H. Baicalin inhibits the metastasis of highly aggressive breast cancer cells by reversing epithelial-to-mesenchymal transition by targeting β-catenin signaling. *Oncol. Rep.* **2017**, *38*, 3599–3607. [CrossRef] [PubMed]
45. Gao, Y.; Snyder, S.A.; Smith, J.N.; Chen, Y.C. Anticancer properties of baicalein: A review. *Med. Chem. Res.* **2016**, *25*, 1515–1523. [CrossRef]
46. Ma, X.; Yan, W.; Dai, Z.; Gao, X.; Ma, Y.; Xu, Q.; Zhang, S. Baicalein suppresses metastasis of breast cancer cells by inhibiting EMT via downregulation of SATB1 and Wnt/β-catenin pathway. *Drug Des. Dev. Ther.* **2016**, *10*, 1419. [CrossRef]
47. Gan, R.Y.; Li, H.B.; Sui, Z.Q.; Corke, H. Absorption, metabolism, anti-cancer effect and molecular targets of epigallocatechin gallate (EGCG): An updated review. *Crit. Rev. Food Sci. Nutr.* **2018**, *58*, 924–941. [CrossRef]
48. Hong, O.Y.; Noh, E.M.; Jang, H.Y.; Lee, Y.R.; Lee, B.K.; Jung, S.H.; Youn, H.J. Epigallocatechin gallate inhibits the growth of MDA-MB-231 breast cancer cells via inactivation of the β-catenin signaling pathway. *Oncol. Lett.* **2017**, *14*, 441–446. [CrossRef]
49. Nelson, C.M.; Chen, C.S. Cell-cell signaling by direct contact increases cell proliferation via a PI3K-dependent signal. *FEBS Lett.* **2002**, *514*, 238–242. [CrossRef]
50. Kim, J.; Zhang, X.; Rieger-Christ, K.M.; Summerhayes, I.C.; Wazer, D.E.; Paulson, K.E.; Yee, A.S. Suppression of Wnt signaling by the green tea compound EGCG in breast cancer and the requirement for the HBP1 transcriptional repressor. *Cancer Res.* **2006**, *66* (Suppl. 8), 30.
51. Sampson, E.M.; Haque, Z.K.; Ku, M.C.; Tevosian, S.G.; Albanese, C.; Pestell, R.G.; Yee, A.S. Negative regulation of the Wnt–β-catenin pathway by the transcriptional repressor HBP1. *EMBO J.* **2001**, *20*, 4500–4511. [CrossRef] [PubMed]
52. Nawaz, J.; Rasul, A.; Shah, M.A.; Hussain, G.; Riaz, A.; Sarfraz, I.; Selamoglu, Z. Cardamonin: A new player to fight cancer via multiple cancer signaling pathways. *Life Sci.* **2020**, *250*, 117591. [CrossRef] [PubMed]
53. Shrivastava, S.; Jeengar, M.K.; Thummuri, D.; Koval, A.; Katanaev, V.L.; Marepally, S.; Naidu, V.G.M. Cardamonin, a chalcone, inhibits human triple negative breast cancer cell invasiveness by downregulation of Wnt/β-catenin signaling cascades and reversal of epithelial–mesenchymal transition. *Biofactors* **2017**, *43*, 152–169. [CrossRef]
54. Zhang, S.D.; Yu, L.; Wang, P.; Kou, P.; Li, J.; Wang, L.T.; Fu, Y.J. Inotodiol inhibits cells migration and invasion and induces apoptosis via p53-dependent pathway in HeLa cells. *Phytomedicine* **2019**, *60*, 152957. [CrossRef] [PubMed]
55. Mishra, S.K.; Kang, J.H.; Song, K.H.; Park, M.S.; Kim, D.K.; Park, Y.J.; Oh, S.H. Inonotus obliquus suppresses proliferation of colorectal cancer cells and tumor growth in mice models by downregulation of β-catenin/NF-κB-signaling pathways. *Eur. J. Inflamm.* **2013**, *11*, 615–629. [CrossRef]
56. Zhang, X.; Bao, C.; Zhang, J. Inotodiol suppresses proliferation of breast cancer in rat model of type 2 diabetes mellitus via downregulation of β-catenin signaling. *Biomed. Pharmacother.* **2018**, *99*, 142–150. [CrossRef]
57. Kwon, Y.J.; Ye, D.J.; Baek, H.S.; Chun, Y.J. 7,12-Dimethylbenz [α] anthracene increases cell proliferation and invasion through induction of Wnt/β-catenin signaling and EMT process. *Environ. Toxicol.* **2018**, *33*, 729–742. [CrossRef]
58. Garcia-Jimenez, C.; García-Martínez, J.M.; Chocarro-Calvo, A.; De la Vieja, A. A new link between diabetes and cancer: Enhanced WNT/β-catenin signaling by high glucose. *J. Mol. Endocrinol.* **2014**, *52*, R51–R66. [CrossRef]
59. Kopustinskiene, D.M.; Bernatoniene, J. Antioxidant Effects of Schisandra chinensis Fruits and Their Active Constituents. *Antioxidants* **2021**, *10*, 620. [CrossRef]
60. Xu, X.; Rajamanicham, V.; Xu, S.; Liu, Z.; Yan, T.; Liang, G.; Wang, Y. Schisandrin A inhibits triple negative breast cancer cells by regulating Wnt/ER stress signaling pathway. *Biomed. Pharmacother.* **2019**, *115*, 108922. [CrossRef]
61. McGrath, E.P.; Logue, S.E.; Mnich, K.; Deegan, S.; Jäger, R.; Gorman, A.M.; Samali, A. The unfolded protein response in breast cancer. *Cancers* **2018**, *10*, 344. [CrossRef] [PubMed]
62. Berman, A.Y.; Motechin, R.A.; Wiesenfeld, M.Y.; Holz, M.K. The therapeutic potential of resveratrol: A review of clinical trials. *NPJ Precis. Oncol.* **2017**, *1*, 1–9. [CrossRef] [PubMed]
63. Chuthapisith, S.; Eremin, J.; El-Sheemey, M.; Eremin, O. Breast cancer chemoresistance: Emerging importance of cancer stem cells. *Surg. Oncol.* **2010**, *19*, 27–32. [CrossRef]
64. Fu, Y.; Chang, H.; Peng, X.; Bai, Q.; Yi, L.; Zhou, Y.; Mi, M. Resveratrol inhibits breast cancer stem-like cells and induces autophagy via suppressing Wnt/β-catenin signaling pathway. *PLoS ONE* **2014**, *9*, e102535. [CrossRef] [PubMed]
65. Lindeman, G.J.; Visvader, J.E. Insights into the cell of origin in breast cancer and breast cancer stem cells. *Asia-Pac. J. Clin. Oncol.* **2010**, *6*, 89–97. [CrossRef] [PubMed]
66. Boyd, J.; Han, A. Deguelin and its role in chronic diseases. *Drug Discov. Mother Nat.* **2016**, *929*, 363–375.
67. Murillo, G.; Peng, X.; Torres, K.E.; Mehta, R.G. Deguelin inhibits growth of breast cancer cells by modulating the expression of key members of the Wnt signaling pathway. *Cancer Prev. Res.* **2009**, *2*, 942–950. [CrossRef]
68. Sánchez, R.; Bahamonde, C.; Sanz, C.; Pérez, A.G. Identification and Functional Characterization of Genes Encoding Phenylacetaldehyde Reductases That Catalyze the Last Step in the Biosynthesis of Hydroxytyrosol in Olive. *Plants* **2021**, *10*, 1268. [CrossRef]

69. Cruz-Lozano, M.; González-González, A.; Marchal, J.A.; Muñoz-Muela, E.; Molina, M.P.; Cara, F.E.; Granados-Principal, S. Hydroxytyrosol inhibits cancer stem cells and the metastatic capacity of triple-negative breast cancer cell lines by the simultaneous targeting of epithelial-to-mesenchymal transition, Wnt/β-catenin and TGFβ signaling pathways. *Eur. J. Nutr.* **2019**, *58*, 3207–3219. [CrossRef]
70. Vikram, R.; Chou, W.C.; Hung, S.C.; Shen, C.Y. Tumorigenic and metastatic role of CD44−/low/CD24−/low cells in luminal breast cancer. *Cancers* **2020**, *12*, 1239. [CrossRef]
71. Li, B.; Lu, F.; Wei, X.; Zhao, R. Fucoidan: Structure and bioactivity. *Molecules* **2008**, *13*, 1671–1695. [CrossRef]
72. Xue, M.; Ge, Y.; Zhang, J.; Liu, Y.; Wang, Q.; Hou, L.; Zheng, Z. Fucoidan inhibited 4T1 mouse breast cancer cell growth in vivo and in vitro via downregulation of Wnt/β-catenin signaling. *Nutr. Cancer* **2013**, *65*, 460–468. [CrossRef]
73. Zhurinsky, J.; Shtutman, M.; Ben-Ze'ev, A. Plakoglobin and beta-catenin: Protein interactions, regulation and biological roles. *J. Cell Sci.* **2000**, *113*, 3127–3139. [CrossRef]
74. Devappa, R.K.; Makkar, H.P.; Becker, K. Nutritional, biochemical, and pharmaceutical potential of proteins and peptides from Jatropha. *J. Agric. Food Chem.* **2010**, *58*, 6543–6555. [CrossRef]
75. Fatima, I.; El-Ayachi, I.; Taotao, L.; Lillo, M.A.; Krutilina, R.; Seagroves, T.N.; Miranda-Carboni, G.A. The natural compound Jatrophone interferes with Wnt/β-catenin signaling and inhibits proliferation and EMT in human triple-negative breast cancer. *PLoS ONE* **2017**, *12*, e0189864. [CrossRef]
76. Imran, M.; Rauf, A.; Abu-Izneid, T.; Nadeem, M.; Shariati, M.A.; Khan, I.A.; Mubarak, M.S. Luteolin, a flavonoid, as an anticancer agent: A review. *Biomed. Pharmacother.* **2019**, *112*, 108612. [CrossRef]
77. Lin, D.; Kuang, G.; Wan, J.; Zhang, X.; Li, H.; Gong, X.; Li, H. Luteolin suppresses the metastasis of triple-negative breast cancer by reversing epithelial-to-mesenchymal transition via downregulation of β-catenin expression. *Oncol. Rep.* **2017**, *37*, 895–902. [CrossRef]
78. Tong, L.; Zhao, Q.; Datan, E.; Lin, G.Q.; Minn, I.; Pomper, M.G.; Liu, J.O. Triptolide: Reflections on two decades of research and prospects for the future. *Nat. Prod. Rep.* **2021**, *38*, 843–860. [CrossRef]
79. Shao, H.; Ma, J.; Guo, T.; Hu, R. Triptolide induces apoptosis of breast cancer cells via a mechanism associated with the Wnt/β catenin signaling pathway. *Exp. Ther. Med.* **2014**, *8*, 505–508. [CrossRef]
80. Zheng, Y.; Ren, W.; Zhang, L.; Zhang, Y.; Liu, D.; Liu, Y. A review of the pharmacological action of Astragalus polysaccharide. *Front. Pharmacol.* **2020**, *11*, 349. [CrossRef]
81. Yang, S.; Sun, S.; Xu, W.; Yu, B.; Wang, G.; Wang, H. Astragalus polysaccharide inhibits breast cancer cell migration and invasion by regulating epithelial-mesenchymal transition via the Wnt/β-catenin signaling pathway. *Mol. Med. Rep.* **2020**, *21*, 1819–1832. [CrossRef]
82. David, A.V.A.; Arulmoli, R.; Parasuraman, S. Overviews of biological importance of quercetin: A bioactive flavonoid. *Pharmacogn. Rev.* **2016**, *10*, 84.
83. Srinivasan, A.; Thangavel, C.; Liu, Y.; Shoyele, S.; Den, R.B.; Selvakumar, P.; Lakshmikuttyamma, A. Quercetin regulates β-catenin signaling and reduces the migration of triple negative breast cancer. *Mol. Carcinog.* **2016**, *55*, 743–756. [CrossRef]
84. Curtin, J.C.; Lorenzi, M.V. Drug discovery approaches to target Wnt signaling in cancer stem cells. *Oncotarget* **2010**, *1*, 563. [CrossRef]
85. Oien, D.B.; Pathoulas, C.L.; Ray, U.; Thirusangu, P.; Kalogera, E.; Shridhar, V. Repurposing quinacrine for treatment-refractory cancer. *Semin. Cancer Biol.* **2021**, *68*, 21–30. [CrossRef]
86. Preet, R.; Mohapatra, P.; Mohanty, S.; Sahu, S.K.; Choudhuri, T.; Wyatt, M.D.; Kundu, C.N. Quinacrine has anticancer activity in breast cancer cells through inhibition of topoisomerase activity. *Int. J. Cancer* **2012**, *130*, 1660–1670. [CrossRef]
87. Preet, R.; Mohapatra, P.; Das, D.; Satapathy, S.R.; Choudhuri, T.; Wyatt, M.D.; Kundu, C.N. Lycopene synergistically enhances quinacrine action to inhibit Wnt-TCF signaling in breast cancer cells through APC. *Carcinogenesis* **2013**, *34*, 277–286. [CrossRef]
88. Vallée, A.; Lecarpentier, Y.; Vallée, J.N. Curcumin: A therapeutic strategy in cancers by inhibiting the canonical WNT/β-catenin pathway. *J. Exp. Clin. Cancer Res.* **2019**, *38*, 1–16. [CrossRef]
89. Prasad, C.P.; Rath, G.; Mathur, S.; Bhatnagar, D.; Ralhan, R. Potent growth suppressive activity of curcumin in human breast cancer cells: Modulation of Wnt/β-catenin signaling. *Chem. -Biol. Interact.* **2009**, *181*, 263–271. [CrossRef]
90. Li, X.; Wang, X.; Xie, C.; Zhu, J.; Meng, Y.; Chen, Y.; Zhao, Y. Sonic hedgehog and Wnt/β-catenin pathways mediate curcumin inhibition of breast cancer stem cells. *Anti-Cancer Drugs* **2018**, *29*, 208–215. [CrossRef]
91. Kakarala, M.; Brenner, D.E.; Korkaya, H.; Cheng, C.; Tazi, K.; Ginestier, C.; Wicha, M.S. Targeting breast stem cells with the cancer preventive compounds curcumin and piperine. *Breast Cancer Res. Treat.* **2010**, *122*, 777–785. [CrossRef]
92. Hoshyar, R.; Mollaei, H. A comprehensive review on anticancer mechanisms of the main carotenoid of saffron, crocin. *J. Pharm. Pharmacol.* **2017**, *69*, 1419–1427. [CrossRef]
93. Arzi, L.; Hoshyar, R. Saffron anti-metastatic properties, ancient spice novel application. *Crit. Rev. Food Sci. Nutr.* **2021**, *62*, 1–12. [CrossRef]
94. Arzi, L.; Riazi, G.; Sadeghizadeh, M.; Hoshyar, R.; Jafarzadeh, N. A comparative study on anti-invasion, antimigration, and antiadhesion effects of the bioactive carotenoids of saffron on 4T1 breast cancer cells through their effects on Wnt/β-catenin pathway genes. *DNA Cell Biol.* **2018**, *37*, 697–707. [CrossRef]

95. Arzi, L.; Farahi, A.; Jafarzadeh, N.; Riazi, G.; Sadeghizadeh, M.; Hoshyar, R. Inhibitory effect of crocin on metastasis of triple-negative breast cancer by interfering with Wnt/β-catenin pathway in murine model. *DNA Cell Biol.* **2018**, *37*, 1068–1075. [CrossRef]
96. Arzi, L.; Hoshyar, R.; Jafarzadeh, N.; Riazi, G.; Sadeghizadeh, M. Anti-metastatic properties of a potent herbal combination in cell and mice models of triple negative breast cancer. *Life Sci.* **2020**, *243*, 117245. [CrossRef]
97. Feng, P.; Che, Y.; Chen, D.Q. Molecular mechanism of action of LiuweiDihuang pill for the treatment of osteoporosis based on network pharmacology and molecular docking. *Eur. J. Integr. Med.* **2020**, *33*, 101009. [CrossRef]
98. Limopasmanee, W.; Chansakaow, S.; Rojanasthien, N.; Manorot, M.; Sangdee, C.; Teekachunhatean, S. Effects of the Chinese herbal formulation (Liu Wei Di Huang Wan) on the pharmacokinetics of isoflavones in postmenopausal women. *BioMed Res. Int.* **2015**, *2015*, 902702. [CrossRef]
99. Wang, J.; Yao, K.; Yang, X.; Liu, W.; Feng, B.; Ma, J.; Xiong, X. Chinese patent medicine liuwei di huang wan combined with antihypertensive drugs, a new integrative medicine therapy, for the treatment of essential hypertension: A systematic review of randomized controlled trials. *Evid. -Based Complementary Altern. Med.* **2012**, *2012*, 714805.
100. Lixiang, Z.; Hongning, L.; Yan, G.; Xianming, M.; Runde, J.; Xiaomin, W.; Yue, W. Effect of Liuweidihuang pill and Jinkuishenqi pill on inhibition of spontaneous breast carcinoma growth in mice. *J. Tradit. Chin. Med.* **2015**, *35*, 453–459. [CrossRef]
101. Lixiang, Z.; Qing, Z.; Zhipeng, Y.; Jian, W.; Xiaoying, R.; Yan, G.; Haizhou, L. LiuweiDihuang pill suppresses metastasis by regulating the wnt pathway and disrupting β-catenin/T cell factor interactions in a murine model of triple-negative breast cancer. *J. Tradit. Chin. Med.* **2019**, *39*, 826–832.
102. Tadesse, S.A.; Wubneh, Z.B. Antimalarial activity of Syzygiumguineense during early and established Plasmodium infection in rodent models. *BMC Complementary Altern. Med.* **2017**, *17*, 1–7. [CrossRef] [PubMed]
103. Koval, A.; Pieme, C.A.; Queiroz, E.F.; Ragusa, S.; Ahmed, K.; Blagodatski, A.; Katanaev, V.L. Tannins from *Syzygium guineense* suppress Wnt signaling and proliferation of Wnt-dependent tumors through a direct effect on secreted Wnts. *Cancer Lett.* **2018**, *435*, 110–120. [CrossRef] [PubMed]
104. Kladar, N.V.; Gavarić, N.S.; Božin, B.N. Ganoderma: Insights into anticancer effects. *Eur. J. Cancer Prev.* **2016**, *25*, 462–471. [CrossRef] [PubMed]
105. Suarez-Arroyo, I.J.; Rosario-Acevedo, R.; Aguilar-Perez, A.; Clemente, P.L.; Cubano, L.A.; Serrano, J.; Martínez-Montemayor, M.M. Anti-tumor effects of Ganoderma lucidum (reishi) in inflammatory breast cancer in in vivo and in vitro models. *PLoS ONE* **2013**, *8*, e57431. [CrossRef]
106. Zhang, Y. Ganoderma lucidum (Reishi) suppresses proliferation and migration of breast cancer cells via inhibiting Wnt/β-catenin signaling. *Biochem. Biophys. Res. Commun.* **2017**, *488*, 679–684. [CrossRef]

MDPI
St. Alban-Anlage 66
4052 Basel
Switzerland
Tel. +41 61 683 77 34
Fax +41 61 302 89 18
www.mdpi.com

Plants Editorial Office
E-mail: plants@mdpi.com
www.mdpi.com/journal/plants

www.ingramcontent.com/pod-product-compliance
Lightning Source LLC
Chambersburg PA
CBHW061828210926
51850CB00006B/835